Principles and Practices of Physical Chemistry

Principles and Practices of Physical Chemistry

Edited by **Bernard Wilde**

NY RESEARCH PRESS

New York

Published by NY Research Press,
23 West, 55th Street, Suite 816,
New York, NY 10019, USA
www.nyresearchpress.com

Principles and Practices of Physical Chemistry
Edited by Bernard Wilde

International Standard Book Number: 978-1-63238-483-6 (Hardback)

Printed in China.

Contents

Preface

Physical chemistry refers to that branch of chemistry which combines physical and chemical studies to deduce the physical properties of molecules and how these affect their chemical properties and bonding characteristics. It collaborates many branches like electrochemistry which studies the ways ions, electrons bond under electrical current, thermochemistry which is the study of why heat is produced in a reaction, spectroscopy which refers to the study of radiation emitting matter and chemical kinetics which is the study of the speed of a chemical reaction, etc. This book provides comprehensive insights into this field. It gives detailed explanations of all the various branches of this subject and their applications. It presents researches and studies performed by experts across the globe. In this book, using case studies and examples, constant effort has been made to make the understanding of the difficult concepts of physical chemistry as easy and informative as possible, for the readers. It will prove to be a beneficial reference guide for researchers and students alike.

After months of intensive research and writing, this book is the end result of all who devoted their time and efforts in the initiation and progress of this book. It will surely be a source of reference in enhancing the required knowledge of the new developments in the area. During the course of developing this book, certain measures such as accuracy, authenticity and research focused analytical studies were given preference in order to produce a comprehensive book in the area of study.

This book would not have been possible without the efforts of the authors and the publisher. I extend my sincere thanks to them. Secondly, I express my gratitude to my family and well-wishers. And most importantly, I thank my students for constantly expressing their willingness and curiosity in enhancing their knowledge in the field, which encourages me to take up further research projects for the advancement of the area.

Editor

Quasicrystals

Feridoun Samavat[*], Mohammad Hossein Tavakoli, Safdar Habibi, Babak Jaleh, Parisa Taravati Ahmad
Department of Physics, Bu-Ali Sina University, Hamedan, Iran
Email: [*]fsamavat@yahoo.com

ABSTRACT

Quasicrystals are material with perfect long-range order, but with no three-dimensional translation periodicity. They are typically binary and ternary metallic alloys. Quasicrystals have characteristic physical properties. Some resemble those of periodic crystals while others have similar properties to amorphous alloys. Many of their mechanical physical properties are quite unusual by the standards of common metals. The peculiar physical properties of quasicrystals certainly give rise to the hope that they may become of some practical importance in the future. Some of these properties are surface related. This provides the main motivation for surface scientists to study this material. In this paper; quasicrystals, types of them, physical properties, surface science and potential applications of them have been discussed.

Keywords: Quasicrstals; Icosahedral Quasicrystals; Polygonal Quasicrystals; Aperiodic Intermetallics

1. Introduction

A crystal is defined as a three dimensional periodic arrangement of atoms or molecules with transitional periodicity along its three principal axes [1]. Thus it is possible to obtain an infinitely extended crystal structure by aligning building blocks called unit-cells until the space is filled up. In 1984, Dan Shechtman, Blech, Gratias & Cahn discovered quasi-crystals (QCs) [2,3] which are well ordered but aperiodic intermetallics. These are typically binary and ternary metallic alloys often containing 60% - 70% aluminium which exhibit unusual physical properties, such as low friction, low adhesion, high hardness and high wear resistance [4,5].

QCs also present unique structural features. To a surface scientist, quasicrystals are quite unusual because of several factors, the first of which is their structure. They are neither periodically ordered, like crystals, nor disordered or amorphous solids. They have a well defined, discrete group symmetry, like crystals [6], but one that is explicitly incompatible with three-dimensional periodic translation order (e.g. exhibiting 5, 8, 10, or 12-fold symmetry axes). Crystal lattices can be carried or mapped into themselves by various symmetry operations. A typical symmetry operation is that of rotation about an axis that passes through a lattice point. Lattices can be found such that 1, 2, 3, 4, and 6-fold rotation axes carry the lattice into itself. But, we can not find a lattice that goes into itself under $2\pi/5$ radians. Therefore we should not expect the lattice to have 5-fold rotation axis because it is not possible to fill all space with a connected array of pentagons. However, quasicrystals do exhibit 5-fold symmetry.

The other element of strangeness about quasicrystals is that many of their mechanical-physical properties are quite unusual by the standards of common metals. Some of these properties are surface related. This intriguing element provides the main motivation for surface scientists to study these materials.

2. The Concept of Higher Dimensional Space

Quasicrystals are materials with perfect long-range order, but with no three-dimensional translational periodicity. Since quasicrystals lose periodicity in at least one dimension it is not possible to describe them in 3D-space as easily as normal crystal structures. Thus it becomes more difficult to find mathematical formalisms for the interpretation and analysis of diffraction data. For normal crystals we can assign three integer values (Miller indices) to label the observable reflections [1]. This is due to the three-dimensional translational periodicity of the structure. Quasicrystals are periodic if one goes to six (or higher) dimensional space [7]. In order to assign integer indices to the diffraction intensities of quasicrystals, however, at least 5 linearly independent vectors are necessary. So we need 5 indices for polygonal quasicrystals and 6 indices for icosahedral quasicrystals. Atomic order is best defined in terms of the Fourier transform of the mass density of the solid. In an ordinary crystal, this transform can be written as a Fourier series [1]:

[*]Corresponding author.

$$\rho(r) = 1/V \sum_{G} \rho(G) \exp(iG \cdot r) \quad (1)$$

The set of wavevectors G define a discrete reciprocal lattice in which each wavevector in the sum can be written as an integer linear combination of three "basis" vectors a_i^*

$$G = ha_1^* + ka_2^* + la_3^* \quad (2)$$

The a_i^* are linearly independent integer, which means that G = 0 if and only if all the h, k, l = 0. The a_i^* are said to span the reciprocal lattice. They are related to the "basis" vectors a_i that define the unit cell if crystal in physics space [1]:

$$a_i^* a_j = \begin{cases} 0 & \text{if } i \neq j \\ 2\pi & \text{if } i = j \end{cases}$$

In a quasi-crystal, the Fourier transform of the mass density is also a Fourier series and the wavevectors in the Fourier sum also form a discrete reciprocal lattice. However, the number of integer linearly independent basis vectors required to "span" the reciprocal lattice exceeds the spatial dimension, and the point symmetry of the reciprocal lattice is incompatible with periodic translational order. For example, six basis vectors are required to span the reciprocal lattice for three-dimensional quasi-crystals with icosahedral symmetry:

$$G = n_1a_1^* + n_2a_2^* + n_3a_3^* + n_4a_4^* + n_5a_5^* + n_6a_6^* \quad (3)$$

where the a_i^* can be selected to point along the fivefold axes of an icosahedron. They are called generalized Miller indices. The necessary n vectors span an nD-reciprocal space. Therefore there is also a nD-direct space in which a structure can be built that gives rise to a diffraction pattern as it is observed for quasicrystals. To put it simply it can be said that in the higher-dimensional space a quasi periodic structure can be described as a periodic one. Thus it is enough to define a single unit cell of the nD-structure. This enables us to describe the whole quasicrystal structure with a finite set of parameters. If we described it in 3D-space only, we needed thousands of atoms to obtain a representative volume segment of the whole structure.

3. Experimental Discovered

Quasicrystals were discovered in 1982, when Dan Shechtmen observed patterns of sharp spots with icosahedral symmetry in selected area electron diffraction patterns of a rapidly cooled Al-Mn alloy [2]. It took about two years before an account of this icosahedral phase appeared in print.

Independently, another type of quasicrystal with a twelvefold rotational axis was found in a Ni-Cr alloy [8]. This dodecagonal phase was also characterized by sharp electron diffraction spots, a clear evidence of an ordered structure. Subsequently, two more types of quasicrystals with ten-and eightfold rotational symmetry were found. Again by selected area election diffraction, a decagonal phase was observed in a rapidly solidified Al-Mn alloy [9] with a somewhat higher Mn content than those used by Shechtman *et al.* [2]. An octagonal phase was first found in rapidly solidified V-Ni-Si and Cr-Ni-Si alloy [10].

Thus, the peculiar property of quasicrystals that led to their experimental discovery is the occurrence of sharp spots in electron diffraction patterns with symmetries that are not found in "conventional" crystals. In other words, quasicrystals are well-ordered solids with unusual symmetry properties.

4. Non-Crystallographic Symmetry

The crystallographic restriction or crystallographic lemma states that, in two and three dimensions, a lattice can only have rotational symmetry axes of order one, two, three, four or six, corresponding to angles of 360°, 180°, 120°, 90° and 60°, respectively and integer multiphase of these elementary rotation angles. The proof of this lemma is elementary. If a rotation is a symmetry of the lattice; it has to map the vectors that span the lattice onto integer linear combinations of themselves. This immediately restricts the set of possible angles P to those where 2Cos (P) is integer, which gives precisely the list above with their multiples. Clearly, there is no three-dimensional lattice showing icosahedral symmetry, and none with a symmetry axis of order fire, eight, ten, or twelve. Nevertheless, the sharpness of the observed diffraction spots shows that the atomic positions in these solids must be long-range ordered; quasicrystals are clearly distinct from amorphous solids. The only solution that remains is to give up the idea that the atoms are arranged according to a periodic lattice, that is, the arrangement of atoms in a quasicrystals is ordered, but aperiodic. For the icosahedral phase, this is true in all directions of space; there is no periodicity whatsoever.

5. Types of Quasicrystals

The icosahedral quasicrystals form one group and the polygonal quasicrystals another (8, 10, 12-fold symmetry). The former is manifested in the occurrence of sharp diffraction spots and the latter in the presence of a non-crystallographic rotational symmetry. The polygonal or dihedral quasicrystals are periodic in one direction, so they consist of a periodic sequence of slices with twelve-, ten-, or eightfold symmetry. They are quasi periodic in two dimensions. There is one periodic direction perpendicular to the quasi-periodic layers. For the icosahedral phase, this is true in all directions of space; there is no periodicity whatsoever. They are quasiperiodic in three dimensions. There is no periodic direction. The symmetry that determines the type of the quasicrystals is first seen in its diffraction pattern. Simulations of some diffraction patterns are shown in **Figure 1**. These could represent either electron diffraction patterns or the Zeroth layers of precession photographs (X-ray). A simulation for a Laue pattern (X-ray) from an icosahedral quasicrystal is shown in **Figure 2**, where the X-ray beam is along one of the five-fold axes.

6. Systems with Quasicrystals

The first quasicrystalline samples were found in rapidly solidified binary alloys, an icosahedral and a decagonal phase in Al-Mn and the dodecagonal phase in Ni-Cr. These phases are metastable, as is the case for a sizable fraction of the known quasicrystals. The first thermodynamically stable icosahedra quasicrysrtals were found in a ternary Al-Cu-Li alloy, and today quite a collection of stable icosahedral and a decagonal phase is known.

Typically, stable quasicrystals occur in ternary or quaternary alloys. A large class consists of Al-based alloys, the most prominent being Al-Cu-Fe, Al-Pd-Mn, and Al-Pd-Re, which shows highly, ordered icosahedral phases. There is also a stable decagonal phase in the Al-Pd-Mn system.

Although Al-based quasicrystals play a major role, it is by no means true that these exhaust all quasicrystalline materials. A new class of stable icosahedral quasicrystals was discovered in Mg-Zn alloys containing rare earth elements. These are very interesting from the physical point of view because the rare earth elements, due to their large magnetic moments, are expected to lead to interesting magnetic properties of these materials. Other materials that may be important for applications are Ti-Zr-based quasicrystals.

octagonal QC:	**decagonal QC:**
V-Ni-Si	**Al-Cu-Mn**
Mn-Si	**Al-Cu-Fe**
Mn-Si-Al	**Al-Cu-Ni**
Mn-Fe-Si	**Al-Cu-Co**
	Al-Cu-Co-Si
	Al-Mn-Pd
	V-Ni-Si
	Cr-Ni

dodecagonal QC:	**icosahedral QC:**
Cr-Ni	**Al-Mn**
V-Ni	**Al-Mn-Si**
V-Ni-Si	**Al-Li-Cu**
	Al-Pd-Mn
	Al-Cu-Fe
	Al-Mg-Zn
	V-Ni-Si
	Pd-U-Si

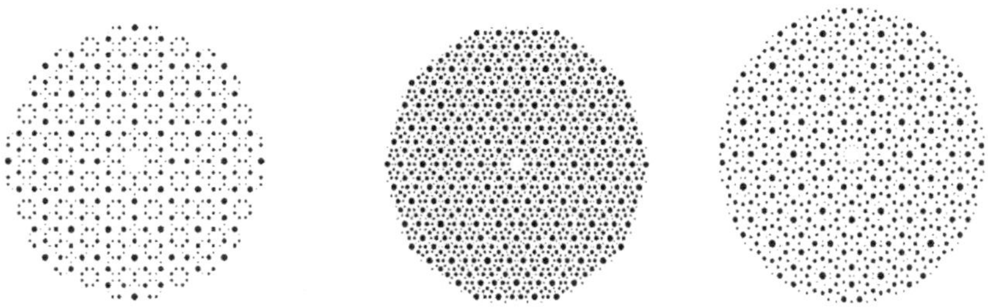

octagonal QC **decagonal QC** **dodecagonal QC**

Figure 1. Simulations of some diffraction patterns.

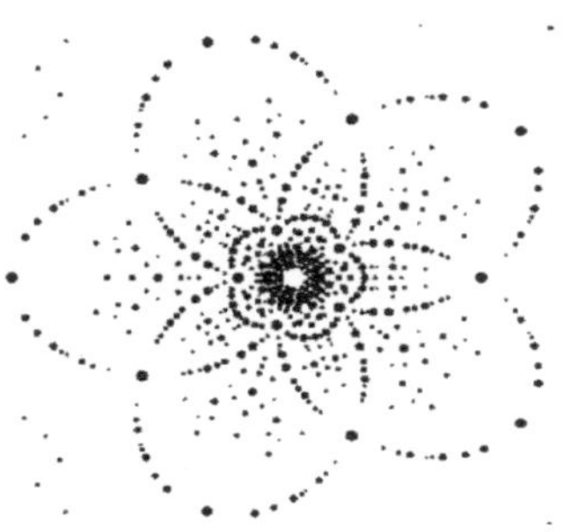

icosahedral QC

Figure 2. A simulation from an icosahedral quasicrystal.

7. Physical Properties

Quasicrystals have characteristic physical properties. Some resemble those of periodic crystals while others have similar properties to amorphous alloys. Many of their mechanical physical properties are quite unusual by the standards of common metals. Some of these properties are surface related. This provides the main motivation for surface scientists to study this material. Some of the physical properties of quasicrystals, relative to selected benchmark materials, are shown in **Table 1**. Two ternary quasicrystal line materials were chosen for **Table 1**. They are icosahedral (i-)-Al-Pd-Mn and i-Al-Cu-Fe.

The latter alloy is of some commercial interest because it is an equilibrium phase with inexpensive components. The values of Vickers' hardness, H_v [11,12] in **Table 1** demonstrate that this quasicrystal is harder than any of its individual components and comparable in hardness to silica. Quasicrystals also exhibit a low coefficient of friction and is reputed to be relatively unreactive (*i.e.* non-stick) [28,29]. The combined properties of high hardness, low friction and inertness lead naturally to speculation that quasicrystals maybe useful as low-friction or release coating in industrial settings where abrasion resistance is also important. However, to date, such an application has not reached the marketplace.

The results in **Table 1** also illustrate other properties, including brittleness, elasticity, conductivity, and surface energy. Quasicrystals are quite brittle, illustrated by the low value of fracture toughness, which is a drawback for many applications [12,15]. Its elasticity, judged by its Young's modulus, is comparable to that of typical metals [12,16]. Although metallic, quasicrystals are poor thermal and electrical conductors, which can be put to good use in thermal barrier applications. Quasicrystals exhibit a rather low surface energy which is much lower than of a typical clean metal and significantly lower than that of an oxidized metal (alumna) [14,19]. Finally, although not shown in **Table 1**, qualitative observations suggest that the quasicrystals are resistant to oxidation. For the first studies of surface oxidation [20,21] i-Al-Pd-Mn were chosen because, starting in the early 1990s, it could be grown in large, single-grain form [22,23]. Later studies have indicated, however, that oxidation phenomena on i-Al-Pd-Mn and i-Al-Cu-Fe are quite similar [24]. Hence, the i-Al-Pd-Mn, although not commercially attractive because of its precious metal content, may serve as a useful model for understanding oxidation of the other Al-rich quascrystalline alloys.

8. Surface Science of Quasicrystals

Early interest in surfaces of quasicrystals was sparked by the observation of the beautifully facetted growth structures. At that time, a variety of shapes had been observed. The current generation of work on surfaces of quasicrystals was ushered in by the use of ultrahigh vacuum (UHV), and associated spectroscopies and microscopies, to examine these materials. A group of experimentalists at AT & T Bell laboratories were the pioneers, performing

Table 1. Physical properties of icosahedral alloys, compared with relevant benchmark materials at room temperature.

Property	value	Material
Hardness (H_v)	6000 – 10,000	Diamond [17]
	800 - 1000	i-Al-Cu-Fe [11,18]
	700 - 800	i-Al-Pd-Mn [12]
	70 - 200	low-carbon steel [17]
	40 -105	Copper [17]
	25 - 45	Aluminum [17]
coefficient of friction (unlubricated with a diamond pin)	0.42	Copper [13]
	0.37	aluminum alloy [13]
	0.32	low-carbon steel [13]
	0.05 - 0.2	i-Al-Cu-Fe [13,14]
Young's modulus (10^6 Psi)	31	stainless steel [16]
	29	i-Al-Pd-Mn [12]
	19	Copper []
	10	Aluminum [16]
thermal conductivity ($W \cdot m^{-1} \cdot K^{-1}$)	390	Copper [16]
	170	Aluminum [18]
	50	low-carbon steel [16]
	2	i-Al-Cu-Fe [18]
surface energy (mJ/m^2)	2480	Iron [19]
	1830	Copper [19]
	24 -25	i-Al-Pd-Mn(air-oxidized) [14]

the first LEED and STM experiments on surfaces of decagonal AL-Co-Cu [25]. Subsequent work focused mainly upon icosahedral Al-Pd-Mn, primarily because of availability of good samples, and icosahedral Al-Cu-Fe. Thus, all work to date concerns Al-rich alloys. The research is driven by the fact that several attractive properties of quasicrystals (low friction, good oxidation resistance, low surface energy) involve surface phenomena. A main goal is to determine whether and how these properties may be related to the unusual atomic structure. Furthermore, in most proposed applications, quasicrystals would be in the form of low-dimensional solids-coatings, thin films, precipitates, or composites-where the ratio of surface area to volume is high, and where surface or interface properties could be crucial to the success of the application [5,14]. This section first discusses the preparation of clean surfaces, then the chemical composition of surfaces, their structure and their surface chemistry.

8.1. Preparation of Clean Surfaces

Typically, a clean surface can only be prepared and preserved for a reasonable length of time within an Ultra-high Vacuum (UHV) environment. Quasicrystals can undergo surface oxidation and contamination, so that a sample prepared in air by any means contains the same contaminants observed on crystalline metals-oxide(s), as well as carbon- and sulfur-containing compounds. Furthermore, heating a sample can cause surface segregation of trace impurities from the bulk. One of the approaches for cleaning the sample is, transferring the sample to UHV and bombarded with noble gas ions (sputtered) to remove surface material, then annealing to remove the structural and chemical damage left by sputtering. This can be done repeatedly until segregation of bulk contaminants is eliminated.

8.2. Surface Composition

Many studies of surface composition have been reported, using X-ray Photoelectron Spectroscopy (XPS) and Auger Electron Spectroscopy (AES), techniques which provide a depth-weighed average over approximately the top 10 nm of material. All such studies have been on sputter-annealed surfaces, one is represented in **Figure 3**, demonstrating the AES-based compositions as a function of temperature after Ar sputtering of the fivefold surface of i-Al-Pd-Mn [33].

8.3. Surface Structure

In studies of the surface structure of quasicrystals, the type of information obtained depends sensitively upon surface preparation and surface composition. Sputtering usually leaves a non-stoichiometric surface which can return to equilibrium by annealing to very high temperatures.

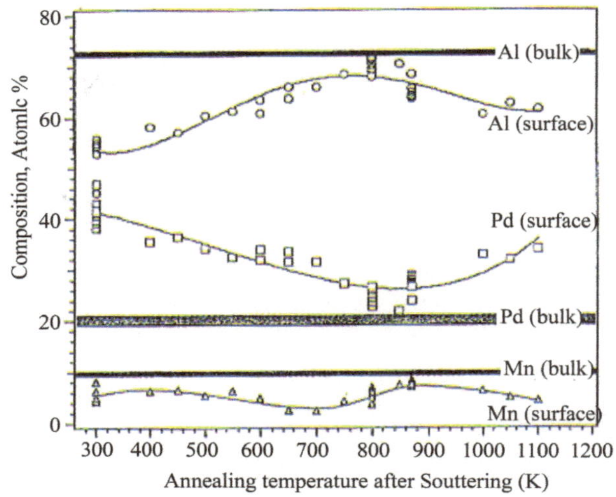

Figure 3. Surface atomic compositions, based upon Auger Electron Spectroscopy, as a function of annealing temperature in UHV [33].

8.4. Surface Chemistry

Another fundamental surface property is chemical reactivity. Studies under various environmental conditions suggest that quasicrystals are relatively unreactive and/or surface energy is very low, suggesting a non-stick character which maybe useful as a coating. The surface energy of a clean metal is typically in the range of a few thousand mJ/m^2, of aluminum oxide about 50 mJ/m^2 and of polytetrafluoro ethylene (PTFE) about 17 - 18 mJ/m. Air-oxidized i-Al-Cu-Fe has a surface energy close to that of PTFE, about 24 - 25 mJ/m^2, and significantly lower than that of alumina [14,16]. It has been proposed that the surface chemistry of quasicrystals should be dominated by the suppression of the density of states at the fermi level [27]. This hypothesis is supported by measurements which reveal that, under atmospheric condition, the surface of quasicrystal behaves more like a covalently bonded material (e.g. Teflon) than like a metal; the polar component of the surface energy is much lower even than that of crystalline phases which are similar in composition and local structure [28]. A low polar component suggests low surface polarizability, *i.e.* low free-electron density.

9. Potential Applications

Perhaps the fastest growing sub field of quasicrystal research in the 1990s has been the area of technological applications based upon the interesting combinations of physical properties that many of these systems offer. The desirable mechanical properties of quasicrystalline alloys, discussed in the previous section, include low surface energies and coefficients of friction, good oxidation and corrosion resistance, and high hardness. Furthermore, at least in the case of Al-Cu-Fe, the constituent elements are readily available, low-cost, and non-toxic. Together with their low thermal and electrical conductivity and some unusual optical behavior, the mechanical properties listed above show great potential for application in the realm of coatings.

Nowadays, the question of potential industrial applications of non-materials plays a major role. The peculiar physical properties of quasicrystals certainly give rise to the hope that they may become of some practical important in the future. The three main applications are outlined below.

9.1. Surface Coatings

The most important area of applications concerns the use of quasicrystals as materials for surface coatings. This is favoured against the use of bulk material because bulk quasicrystals tend to be rather brittle, while coatings benefit from the hardness of qusicrystals. The most prominent example is the application of quasicrystalline coatings to frying pans—an application that has been famous in the quasicrystal community for years though it has not been realised until just recently that a quasicrystal-coated frying pan appeared on the market. They are now sold by the French Company Sitram under the trademark cybernox [29].

9.2. Alloys Containing Quasicrystalline Nanoparticles

A different way to circumvent the brittleness of quasicrystalline bulk material while preserving some of its useful properties is the utilization of an Al-based alloy (typically Al-Mn-Ce) reinforced by precipitation of icosahedral particles in the nanometer range. Such materials, which are now commercially available in Japan, are of great technological interest because they can be very strong but much lighter than other material with comparable physical properties. Examples of applications that have already been realized are razor blades and surgeon's instruments, though this may have been more by chance than being an intentional application of quasicrystals. Nevertheless they will undoubtedly find their way to applications in many different areas such as the aviation industry [30].

9.3. Hydrogen Storage

A third, and maybe more speculative, application concerns the use of quasicrystals as a reversible storage medium for hydrogen. The most promising quasicrystal materials for hydrogen storage are Zr-based quasicrystals [31]. For such systems, storage capabilities of almost two hydrogen per metal atom have been reported, comparable to the storage capability of the Ti-Fe hybrids, which have already been applied in non-polluting internal combustion engines.

A less desirable feature of QCs, however, is that at temperatures lower than about 70% - 80% of the melting temperature, these alloys are very brittle, limiting the range of applications of bulk material. The hardness of icosahedral Al-Pd-Mn at room temperature, for example, exceeds that of most steels, but it has very low fracture toughness.

One way to avoid the difficulties inherent to the extreme brittleness of bulk quasicrystalline alloys is to deposit them as coating on more robust substrates. In this manner, the desirable properties can be exploited while maintaining the mechanical integrity of the substrate. Most of the interest in applications so far has focused on the use of thick or thin coatings of quasicrystals. Applications can make use of the high strengh of these alloys as wear resistant coatings for steel and aluminium parts in manufacturing and the automotive industries.

Moving beyond the chemical properties of quasicrystal surface, there are other physical properties such as low thermal conductivity and low diffusivity [32], which show promise for applications as thermal barrier coatings.

10. Diffusion in Quasicrystals

Diffusion properties are important for the production of quasicrystals and for application of these materials at elevated temperatures. Diffusion in solid materials, irespective of whether they are crystalline or quasicrystalline, is an important topic of materials science. Diffusion processes play a key role in the production and heat treatments of materials and in the kinetics of many microstructural changes and their associated properties. The atomic mechanisms of diffusion in crystalline materials are closely linked to defects. Point defects such as vacancies and self-interstitials are the simplest defects. A new kind of defect called a phason exists in quasiperiodic structures. Phasons are violations of the matching rules of a quasiperiodic structure. Kalugin and Katz have proposed a self-diffusion mechanism via phason flips without the presence of vacancies [34]. It has been suggested that the diffusion rate via the phason mechanism can be much faster than conventional diffusion processes.

11. Summary

A Quasicrystal is a material which shows diffraction patterns with rotational symmetries that are a "forbidden" by classical crystallography. The structure of quasicrystals is comprised of atoms that are arranged in nonperiodic fashion. Quasicrystal structures show long-range order, but no translational periodicity. Quasicrystals are typically binary and ternary metallic alloys, often containing 60% - 70% aluminium. They present unique structural features, coupled with unusual combinations of physical properties. Some of the interesting properties of quasicrystals, such as low friction and "non-stick" character, involve surface phenomena. Studies have shown that quasicrystalline materials exhibit low coefficients of friction, are wear resistant, and are hard. In most proposed applications, QCs would be in the form of low-dimensional solids-coatings, thin films, precipitates, or composites where the areas in quasicrystal research that have seen great advances in the past 10 years, each of these topics represent new horizons.

REFERENCES

[1] C. Kittel, "Introduction to Solid State Physics," John Wiley, New York, 1983.

[2] D. Shechtman, I. Blech and D. Gratias, "Metallic Phase with Long-Range Orientational Order and No Translational Symmetry," *Physical Review Letters*, Vol. 53b, No. 20, 1984, pp. 1951-1953.

[3] A. I. Goldman and M. Widom, "Quasicrystal Structure and Properties, "*Annual Reviews of Physical Chemistry*, Vol. 92, 1991, pp. 685-729. doi:10.1146/annurev.pc.42.100191.003345

[4] S. S. Kang, J. M. Dubois and J. Von Stebut, "Tribological Properties of Quasicrystalline Coatings," *Journal of Materials Research*, Vol. 8, No. 10, 1993, pp. 2471-2481. doi:10.1557/JMR.1993.2471

[5] J. M. Dubois, S. S. Kaing and A. Perrot, *Materials Science and Engineering*, Vol. A179/A180, 1994, pp. 122-126.

[6] A. I. Goldman and M. Widom, "Quasicrystal Structure Abd Properties," *Annual Review of Physical Chemistry*, Vol. 42, 1991, pp. 685-729. doi:10.1146/annurev.pc.42.100191.003345

[7] http://mcs.open.ac.uk/ugg2/quasi_intro2.html-ssi

[8] T. Ishimasa, H. Nissen and Y. Fukano, "New Ordered State between Crystalline and Amorphous in Ni-Cr Particles," *Physical Review Letters*, Vol. 55, No. 5, 1985, pp. 511-513. doi:10.1103/PhysRevLett.55.511

[9] L. Bendersky, "Quasicrystal with One-Dimensional Translational Symmetry and a Tenfold Rotation Axis," *Physical Review Letters*, Vol. 55, No. 14, 1985, pp. 1461-1463. doi:10.1103/PhysRevLett.55.1461

[10] N. Wang, H. Chen and K. H. Kuo, "Two-Dimensional Quasicrystal with Eightfold Rotational Symmetry," *Physical Review Letters*, Vol. 59, No. 9, 1987, pp. 1010-1013. doi:10.1103/PhysRevLett.59.1010

[11] U. Koster, W. Liu, H. Liebertz and M. Michel, "Mechanical Properties of Quasicrystalline and Crystalline Phases in Al_3Cu_2Fe Alloys," *Journal of Non-Crystalline Solids*, Vol. 153-154, 1993, pp. 446-452. doi:10.1016/0022-3093(93)90393-C

[12] Y. Yokoyama, A. Inoue and T. Masumoto, *Materials Transactions*, Vol. 34, 1993, pp. 135-145.

[13] S. S. Kang, J. M. Dubois and J. Von Stebut, "Tribological Properties of Quasicrystalline Coatings," *Materials Research*, Vol. 8, No. 10, 1993, pp. 2471-2481. doi:10.1557/JMR.1993.2471

[14] J. M. Dubois, In: A. I. Goldman, D. J. Sordelet and P. A. Thiel, Eds., *Proceedings of the Conference New Horizons in Quasicrystals*, World Scientific, Singapore City, Vol. 77, 1997, pp. 1928-1938.

[15] D. Kovar and M. J. Ready, "Role of Grain Size in Strength Variability of Alumina," *American Ceramic Society*, Vol. 77, No. 7, 1994, pp. 1928-1938. doi:10.1111/j.1151-2916.1994.tb07073.x

[16] H. E. Boyer, T. L. Gall, "Metals Handbook," Desk Edition, American Society for Metals, Metals Park, 1984.

[17] I. M. Hutchings, "Tribology: Frictions and Wear Engineering Materials," CRC, Boca Raton, 1992.

[18] J. M. Dubois and P. Weinland, "CNRS, Nancy, France, Coating Materials for Metal Alloys and Metals and Method," Eur. Patent Ep 0356287 Al and U.S. Patent 5, Vol. 204, 1993, p. 191.

[19] F. R. de Boer, R. Boom, W. C. M Mattens, A. R. Miedema and A. K. Niessen, "Cohesion in Metals," North-Holland, Amsterdam, 1988.

[20] S. L. Chang, W. B. Chin, C. M. Zhang, C. J. Jenks and P. A. Thiel, "Oxygen Adsorption on a Single-Grain, Quasicrystal Surface," *Surface Science*, Vol. 337, No. 1-2, 1995, pp. 135-146. doi:10.1016/0039-6028(95)00564-1

[21] S. L. Chang, C. M. Zhang, C. J. Jenks, J. W. Andereg and P. A. Thiel, In: C. Janot and R. Mosseri, Eds., *Procceding of the 5th International Conference on Quasicrystals*, World Scientific, Singapore City, 1995, pp. 786-789.

[22] Y. Yokoyama, A. P. Tsai, A. Inoue and T. Masumoto, *Materials Transactions*, Vol. 32, 1991, pp. 1089-1097.

[23] Y. Yokoyama, T. Miura, A. P. Tsai, A. Inoue, T. Masumotos, *Materials Transactions*, Vol. 33, 1992, pp. 97-101.

[24] C. J. Jenks, P. J. Pinhero, S. L. Chang, J. W. Anderegy, M. F. Besser, D. J. Sordelet and P. A. Thiel, In: A. I. Goldman, D. J. Sordelet, P. A. Thiel and J. M. Dubois, Eds., *Proceedings of the Conference New Horizons in Quasicrystals*, World Scientific, Singapore City, 1997, pp. 157-164.

[25] A. R. Kortan, R. S. Becker, P. A. Thiel and H. S. Chen, "Real-Space Atomic Structure of a Two-Dimensional Decagonal Quasicrystal," *Physical Review Letters*, Vol. 64, No. 2, 1990, pp. 200-203. doi:10.1103/PhysRevLett.64.200

[26] F. R. Deboer, R. Boom, W. C. M. Mattens, A. R. Miedema and A. K. Niessen, "Cohesion in Metals," North-Holland, Amsterdam, 1988.

[27] N. Rivier, *Proceeding of the Conference*, *New Horizons in Quasicrystals*, *Research and Applications*, World Scientific, Singapore City, 1997, p. 188.

[28] J. M. Dubois, P. Plaindoux, E. Belin-Ferre, N. Tamure and D. J. Sordelet, *Proceeding of the 6th International Conference on Quasicrystals*, World Scientific, Singapore City, 1998, p. 733.

[29] http://www.inductionsystems.com/cookware/cybernox.htm.

[30] S. Takeuchi and T. Fujiwara (Eds.), *Proceedings of the 6th International Conference on Quasicrystals*, World Scientific, Singapore City, 1998.

[31] G. Coddens, A. M. Viano, P. C. Gibbons, K. F. Kelton and M. J. Kramer, "Time-of-Flight Neutron Scattering Study of Hydrogen Dynamics in Icosahedral Ti45Zr38 Ni17H150 Quasicrystals," *Solid State Communications*, Vol. 104, 1997, pp. 179-182. doi:10.1016/S0038-1098(97)00110-5

[32] P. Archamboult and C. Janot, *MRS Bulletin*, Vol. 22, 1997, p. 48.

[33] A. I. Goldman and P. A. Thiel, *Horizons in Quasicrysral Research*, 1999, pp. 561-601.

[34] P. A. Kalugin and A. Katz, "A Mechanism for Self-Diffusion in Quasi-Crystals," *Europhysics Letters*, Vol. 21, No. 9, 1993, p. 921. doi:10.1209/0295-5075/21/9/008

2

Circular Dichroism and Derivative Spectra Study of the Excitonic Aggregation of Pinacyanol by Aerosol-OT

Sa'ib J. Khouri[1*], Volker Buss[2]
[1]Department of Chemistry, American University of Madaba, Madaba, Jordan
[2]Fachbereich Chemie, Universität Duisburg-Essen, Essen, Germany
Email: *sbkhouri@yahoo.com

ABSTRACT

Different structures are proposed for the complexes formed from the interaction between the cationic cyanine dye 1, 1'-diethyl-2, 2'-carbocyanine chloride (Pinacyanol chloride) and the anionic surfactant sodium bis (2-ethylhexyl) sulfosuccinate (Aerosol-OT) in ethanol-water solutions. UV/vis and electronic circular dichroism (CD) spectra provide tools to study the concentration and solvent dependence of this interaction. The aggregation spectra of the dye are dominated by a strongly blue-shifted, sharp and single visible band, which appears at concentrations much below the critical micelle concentration (cmc) of Aerosol-OT. Above the cmc, the spectra in pure aqueous solution indicate dissociation into monomer and dimer species, and the system becomes completely empty of chirality. Two different CD spectra with a distinct isosbestic point are observed for complexes with different surfactant to dye ratios. Both the addition of ethanol (more than 7.5%) and increasing the temperature retards the metachromatic process between the dye and Aerosol-OT. By using the peakFit program, the two overlapping excitonic absorption bands together with the optically inactive absorption band for one spectrum (the most optimal one) at a specific Aerosol-OT concentration were separated.

Keywords: Cyanine Dye; Pinacyanol; Surfactant; Aerosol-OT; Aggregation; Circular Dichroism; Derivative Spectra

1. Introduction

Aerosol-OT (AOT for short, **Figure 1(a)**) is an anionic surfactant with two branching hydrophobic tails. The surfactant has three asymmetric carbon atoms; one of them is next to the sulfonate group, which makes this group optically active that can induce chirality in a symmetric cationic molecule after they bond [1]. Symmetric cationic cyanine dyes play a major role in the aggregation process of anionic surfactants even below its critical micelles concentration (cmc), in which these cationic dyes act as mutual counter ion to reduce the electrostatic repulsion between the groups that carry the negative charge [2]. Many researchers have been reported on the micelles aggregation's type of AOT surfactant by using different physical methods [3-6]. AOT is soluble in any type of solvents and can form micro-emulsions with or without co-surfactants with oils [7,8].

Cyanine dyes have a natural tendency to aggregate spontaneously in pure aqueous solutions or in mixed solvents when their concentration is increased or the temperature is lowered. Cyanine dyes have two types of aggregates; H- and J-types that can be formed depending on the conformation of the aggregates. H-type is formed when the dye molecules arrange themselves face-to-face in near vertical stacks with blue shifts of the absorption maxima that depend quantitatively on the number of monomers in the aggregates [9]. J-aggregates are formed when the dye molecules arrange themselves into slanted stacks with red shifts of the absorbance [10].

Anionic surfactants (above or below their cmc) can also affect in contrast the aggregation of cationic cyanine dyes, in which ion-pairing is a prerequisite for this process [11,12]. The aggregated species that formed from a dye in solution lead to the development of new blue-shifted spectral bands relative the monomer band [13]. 1, 1'-diethyl-2, 2'-carbocyanine chloride (Pinacyanol chloride or PIN for short, **Figure 1(b)**) is the dye that was studied in this work. This dye is a symmetric cationic cyanine dye, where its absorption spectrum in solution shows three overlapping spectral bands; for monomer, dimer and higher aggregates [14]. The interaction of cyanine dyes with various types of surfactants was interested by many researchers [6,15,16]. Pal and Pal [15] have shown that AOT induces strong dichroism in the cationic pinacyanol dye aqueous solutions, and that was the first report of the induction of circular dichroism in a

*Corresponding author.

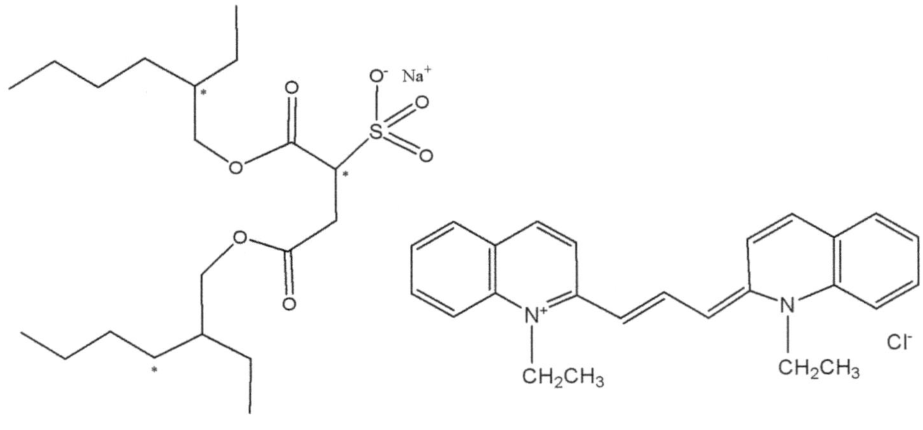

(a) (b)

Figure 1. (a) The structure of Aerosol-OT; (b) The structure of Pinacyanol chloride.

cationic dye by AOT or by other detergents in dilute pure aqueous solutions (as they reported).

In the present work, we intend to study the UV/vis and elecrtronic induced CD absorption spectra of the symmetrical cyanine dye, PIN, in the presence of different concentrations of AOT with different experimental conditions. The resulted spectra were treated using derivative spectroscopy [17]. Peak analysis was used to analyze and separate the various new absorption bands that resulted by such interaction for the most optimal spectrum. All measurements in this study are conducted mainly in 7.5% v/v ethanol/water solutions, at which the excitonic absorption band was the optimum one.

2. Materials and Methods

2.1. Materials

AOT was obtained from Merck and PIN was obtained from Sigma. Spectrophotometric grade ethanol was from Merck. Triple distilled water was used throughout the study.

2.2. Experimental Methods and Instruments Used

Standard dye and AOT solutions were prepared in 25 ml volumetric flasks. 4.00 ml of AOT-dye solutions were prepared in stoppered rolled rim glasses of 10 cm^3 capacity. To prevent the dye to precipitate at the glass walls, AOT solutions were added first into a rolled rim glass, followed by the required amount of the dye solution.

The UV/vis spectra were recorded with a Perkin-Elmer Lambda 5 spectrophotometer, and CD-spectra were measured with AVIV circular dichroism spectrophotometer, Model 62DS. Both instruments were connected to a personal computer for data collection in ASCII-file format.

3. Results and Discussions

3.1. UV/Vis and CD Spectra

Figure 2 presents the UV/vis of PIN (11 μM) in ethanol-aqueous solution (7.5% v/v) in the presence of five different concentrations of AOT (5 - 500 μM) corresponding to molar concentration ratios of surfactant to dye from 0.45:1 to 45:1 respectively. The effect of changing AOT concentration on the spectral shape of PIN was clear in **Figure 2**. The absorption maxima of the free dye are found at 600 and 546 nm in addition to a shoulder over 505 nm, which are attributed to the monomer, dimer, and higher aggregate forms of PIN respectively [9,18]. The intensity of the apparent bands at 600 and 552 nm decreases as the surfactant to dye ratio increases, and the formation of a new broad blue-shifted band is visible, with intensity and maximum wavelength deeding

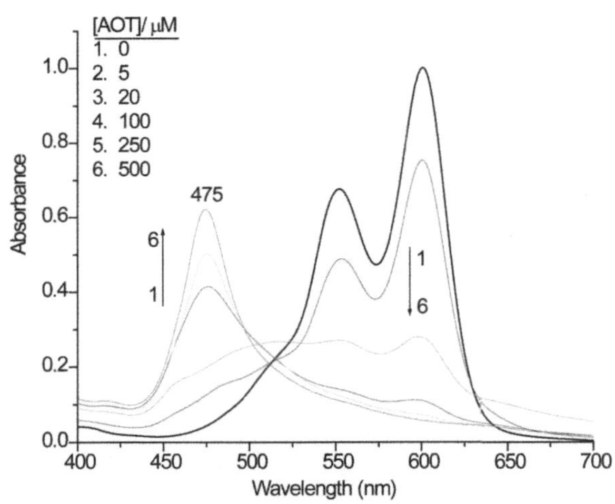

Figure 2. The visible absorption spectra of an ethanol-aqueous solution (7.5% v/v) of 11 μM PIN in the presence of five different concentrations of AOT at 25°C in a 1.00 cm cell.

on the AOT to dye ratio (**Figure 2**). At an AOT/PIN ratio of 9:1 (spectrum 4 in **Figure 2**), the blue-shifted band over about 475 nm is clearly shown. The absorbance over 600 nm reaches its lowest intensity at an AOT/PIN molar ratio of 45:1 (spectrum 6 in **Figure 2**). At even higher AOT/dye molar ratios, the intensity of the induced blue shifted band at 475 nm remains the same. There is a clear isosbestic point at about 497 nm for the spectra 4 - 6 in **Figure 2**.

Figure 3(a) shows the fourth derivative spectra of the UV/vis-spectra shown in **Figure 2**. The changes in the intensity of the derivative bands in **Figure 3(a)** become clearer and they were measured quantitatively [17,19]. **Figure 3(b)** shows the most obvious bands of PIN over 475, 550, and 601 nm as increasing or decreasing in their intensities against the change in AOT concentration. The increase of band over 475 nm indicates a favourable interaction between the higher aggregates of PIN and the surfactant. Due to irrelevance, there is nothing to say in **Figure 3(b)** about the intensity changes for the other components presented in the fourth derivative spectrum (**Figure 3(a)**).

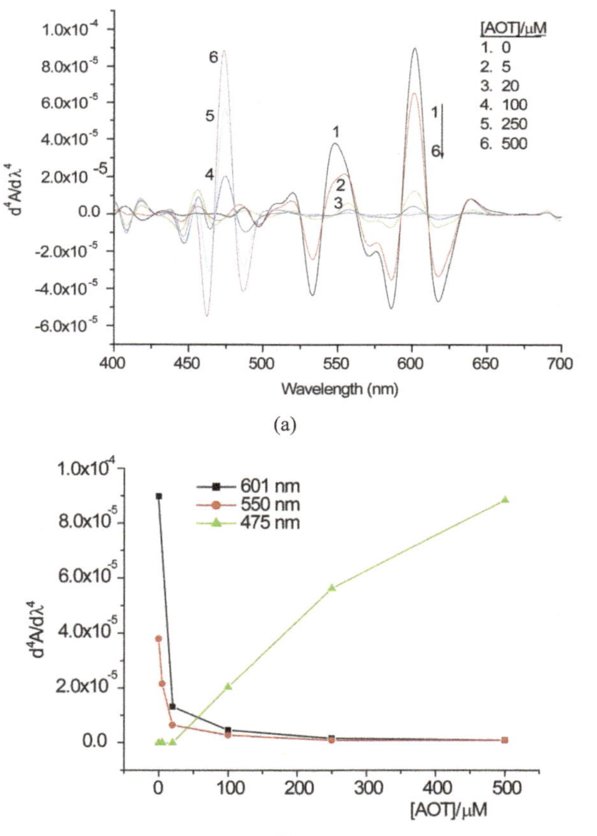

Figure 3. (a) The fourth derivative spectra that shown in Figure 2; (b) The peak heights of the fourth derivative spectra for the main absorption bands as a function of AOT concentration.

Figure 4 shows the CD spectra of PIN (11 μM) in ethanol-aqueous solution (7.5% v/v) in the presence of four different concentrations of AOT (5 - 750 μM) corresponding to molar concentration ratios of surfactant to dye from 0.45:1 to 68:1 respectively. No CD spectrum is observed in the absence of AOT. A negative band at 467 nm and a positive one at about 499 nm are clearly shown. This broad couplet reaches its maximum amplitude in spectrum 3, which corresponds to Spectrum 6 in **Figure 2**.

The reported value of the critical micelle concentration (cmc) for AOT in water at room temperature is 2400 μM [20]. All spectra in **Figures 2** and **4** regarding AOT concentrations are below the cmc value. Cmc values show rise in the case of adding solvents less polar than water like ethanol. The two overlapping excitonic absorption bands breaks down completely and loses its optical activity when AOT concentration exceeding its cmc, which we have observed experimentally at AOT/PIN molar ratio of 270:1.

Figure 5 shows the resolved absorption bands of spectrum 6 in **Figure 2** using PeakFit program, which is non-linear peak-fitting program from Systat software Inc. Voigt function was chosen for the pure analytical computation and the line shape arising as a result of the convolution of the Lorentzian and Gaussian components within a spectroscopic peak. The three λ_{max} values over 467, 475, and 499 nm that obtained from PeakFit program for the most excitonic obvious band (spectrum 6 in **Figure 2**) are in very good agreement with the values obtained from the fourth-derivative spectrum (**Figure 3(a)**) and from the excitonic CD band in **Figure 4**. There is nothing to say about the other λ_{max} values more than 500 nm due to irrelevance here. The graph of the residuals (not shown here), which depicts the difference between the experimental absorption spectrum and the spectrum calculated with PeakFit program on the basis of

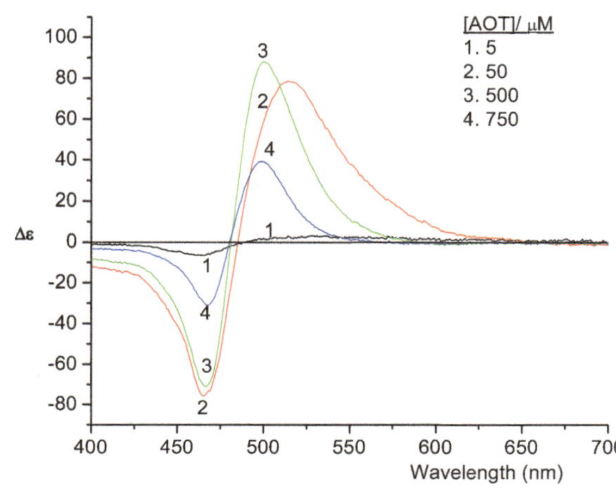

Figure 4. CD spectra of an ethanol-aqueous solution (7.5% v/v) of 11 μM PIN in the presence of four different concentrations of AOT at 25°C in a 1.00 cm cell.

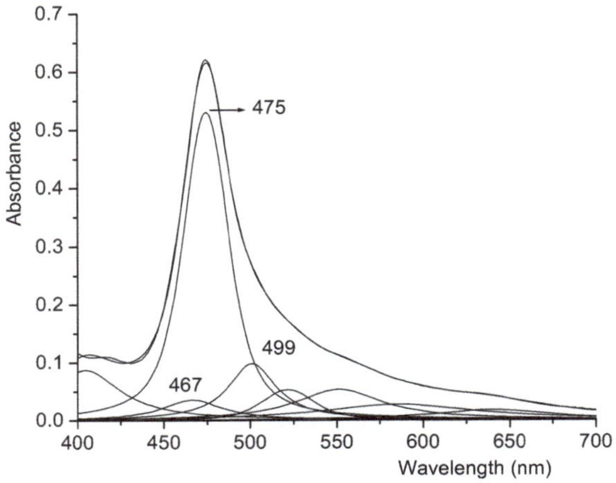

Figure 5. The resolved components of spectrum 6 in Figure 2 by using PeakFit program.

the different components, shows absolute values that are statistically highly acceptable, in which their percentage in their maximum values to the values of the absorption are tried during this work to be less than 1%.

3.2. Effect of Ethanol on the Spectra

Ethanol is known to affect the aggregation behavior of dyes, and it will affect the aggregation behavior of the surfactant, including its cmc. **Figure 6(a)** shows the effect of added ethanol (from 0% to 22.5%) to an aqueous solution of an AOT/PIN molar ratio of 45:1 on the UV/vis spectral shape. Except for an initial weak increase in the absorption intensity, the band at 468 nm is seen to decrease strongly in intensity as ethanol is added to the solution. The associated red-shift of the band maximum can be assumed to be due to the overlap with the developing bands at the low-energy side of the spectrum. We also note the strong increase of the monomer dye absorption at 600 nm and the dimer absorption at 555 nm. There is, even at the highest ethanol concentration, a significant residual absorption at around 480 nm, which indicates the presence of a higher dye aggregate.

In the corresponding CD spectra (**Figure 6(b)**) the couplet between 467 and 499 nm, which dominated the spectra shown in **Figure 4**, is present at ethanol concentrations below 10%. It reaches its largest amplitude at 7.5% ethanol content. The amplitude is about three-fold compared to the spectra without ethanol. Increasing the ethanol concentration, however, has the opposite effect, where the couplet breaks down, just as the blue-shifted band in the visible absorption. The residual band at 485 nm, which is seen under these conditions in the visible spectrum, displays a positive absorption in the CD.

It should be concluded that ethanol in low concentrations promotes formation of the strongly CD active aggregate resulting from the interaction between the AOT and PIN. At higher ethanol concentrations the dye becomes mostly dissolved in monomeric form, but a part of it remains aggregated, though in a different form.

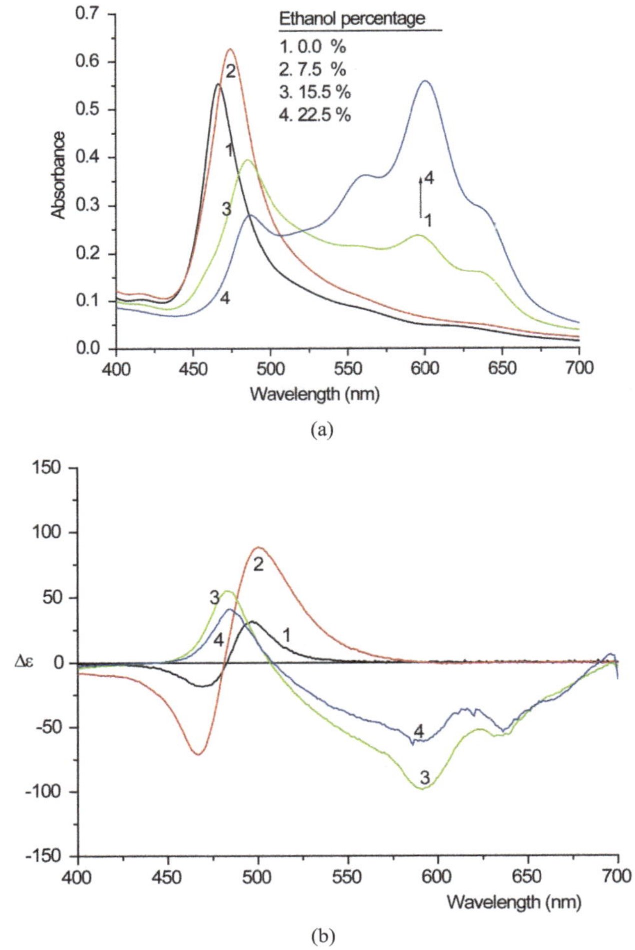

Figure 6. (a) The visible absorption an aqueous solution of PIN (11 μM) and AOT (500 μM) with different ethanol percentages (0% to 22.5%) at room temperature; (b) The CD spectra of Figure 6(a).

3.3. Effect of Temperature on the Spectra

Figure 7 shows how the visible absorption and the CD spectra of an AOT/PIN ethanol-aqueous solution (7.5%) of a 45:1 molar ratio respond to changes of the temperature. At lower temperature (5˚C), the intensity of the absorption band has its largest value; we also note a slight blue-shift of the band maximum to 470 nm. The absorbance of the monomeric dye at 600 nm is minimal. Increasing the temperature has two effects: the intensity of the blue-shifted band decreases, and the band is shifted to the red. At the same time most of the absorbance is shifted into a broad band reaching from 500 to 650 nm, in which peaks corresponding to monomeric dye and to other aggregated species are clearly visible.

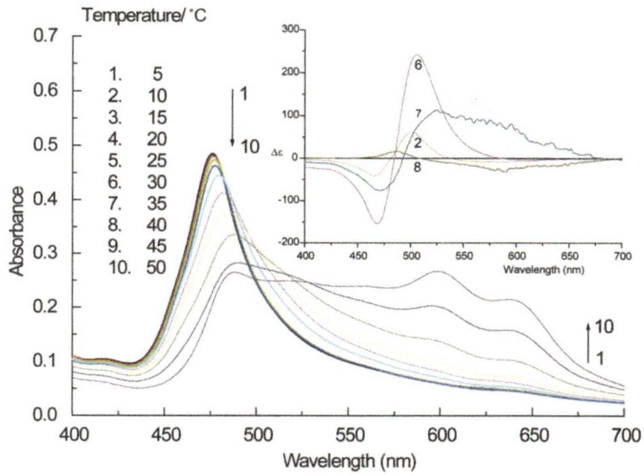

Figure 7. The visible absorption of an ethanol-aqueous solution (7.5% v/v) of PIN (11 μM) and AOT (500 μM) at different temperatures (5°C to 50°C) (insets: corresponding CD spectra for spectra 2, 6, 7, and 8).

The CD spectra (insets) reflect this somewhat unstructured behavior. The designate CD band at low temperature collapses into one positive band at 460 nm when a temperature of 45°C is reached, and the broad band shows weak negative CD absorption. Both the visible absorption and the CD spectra taken at 45°C are similar in shape to the spectrum 3 in **Figure 6**, indicating that excess ethanol and excess heat affect the aggregating behavior of PIN in the presence of AOT in a similar way.

4. Discussion

AOT induces very strong blue-shifted band in PIN solutions at concentrations below the cmc (**Figure 2**). The reported value for the cmc (2400 μM, [20], of AOT in pure water at room temperature would correspond to an AOT/PIN molar ratio of 220:1 when the PIN concentration is 11 μM. In the figure, there is a developed progressive blue shifted band, which its intensity increases until the AOT/PIN molar ratio is reached to 45:1 with a wavelength of 475 nm. It is obvious that AOT induces a sharp and single absorption band with PIN, indicating that the dye cations aggregate systematically with reasonable distances between each other, and they are not overcrowded because there are no multiple bands.

Conductometric titrations show a 1:1 stoichiometry interaction between the cationic dye and AOT [15]. The question arises whether PIN binds at alternating anionic groups of aggregated AOT, and whether these aggregates are already present in the solution before PIN is added. The cmc tells us that AOT is also present in monomeric form when the AOT/PIN molar ratio is less than 220:1, because the transition from premicellar to micellar solution at the cmc occurs dramatically over a narrow concentration range, and additional surfactant exists as aggregates or micelles [21]. As the surfactant concentration increases further, micelles can be deformed and can change into different shapes [22]. The presence of cationic dye molecules in the solution may induces the formation of AOT aggregates, which can easily interact with the anionic site in AOT molecules. Ionic coupling yields larger molecular structures, which become salt-like macromolecules and have the capability to aggregate in solution. The enhancement of this aggregation depends directly on the AOT/PIN molar ratio.

The intense and sharp positive and negative bands in the CD spectra (**Figure 4**) reflect a presence of exciton splitting [23], which indicates to a strong dye-dye interaction with an optimum status at the AOT/PIN molar ratio of 45:1. Regardless of the interaction between a part of AOT and PIN, the cmc of the free excess AOT in the solution should not change significantly; however, the new structure may behaves different from the free excess AOT itself. This behavior is noted from the spectral shape of the UV/vis band at an AOT/PIN molar ratio of 270:1 by using pure water solvent, in which the surfacetant is present in a concentration above the cmc, which has an absorbance band shape very similar to the one observed at the ratio of 45:1. This makes us conclude that the mode of aggregation of PIN molecules that induced by AOT molecules less than cmc is similar to that induced by the surfactant at or above its cmc.

The aggregated PIN molecules at the AOT/PIN molar ratio of 270:1 loses its optical activity because the micelles itself that formed at this ratio are optically inactive. In homogeneous solution, 270 molecules of AOT are distributed with one PIN molecule. The micelle size that expressed experimentally as the number of monomers making up the spherical micelles (aggregation number), is generally between 50 and 100 for ionic surfactant with a radius similar to that of the length of an extended hydrocarbon chain, which associated to form regions from which the solvent (water) is effectively excluded and has properties closely related to the liquid hydrocarbon [24]. According to this reported micelle size, the micelle/PIN molar ratio here is about 4:1, and PIN molecules almost present out of the spherical micelle cavity due to location of the sulfonate groups outside the sphere and to the steric consideration with respect to the size of the dye molecules relative to the cavity presents inside the sphere. At the 270:1 molar ratio, the mode of aggregation, in the presence of achiral micelle and probably deformed micelle, forms absence of real induced CD bands. Among the attempts to explaine why AOT/PIN molar ratio over cmc loses its optical activity is the PIN molecules may enter within micelles side by side to the sulfonate groups. Thus, on the basis of steric considerations and also the distortion of the symmetrical shape of the micelle from the presence of PIN molecules inside the micelle cavity, it is to be expected that this explanation in the absence of any optical activity should be excluded.

As a suggested additional work based on our practical results with respect to the shifts in maximum wavelengths of the excitonic absorption bands that caused by the addition of AOT, a study and investigation to this change may be come from extensive theoretical calculations to find out further interpretations and conclusions regarding the mode of AOT/PIN interactions at various conditions, which may draw attentions for promising further important information on this system.

5. Acknowledgements

This work was supported financially by KAAD in Bonn. The technical assistance of Mr. Robert Knierim from Duisburg-Essen University is highly acknowledged.

REFERENCES

[1] A. N. Maitra and H. F. Eicke, "Effect of Rotational Isomerism on the Water-Solubilizing Properties of Aerosol OT as Studied by Hydrogen-1 NMR Spectroscopy," *Journal of Physical Chemistry*, Vol. 85, No. 18, 1981, pp. 2687-2691. doi:10.1021/j150618a024

[2] E. Pramauro and E. Pelizzetti, "Surfactants in Analytical Chemistry: Applications of Organized Amphiphilic Media," Elsevier, Amsterdam, 1996.

[3] F. Candau, Y. S. Leong, G. Pouyet and S. Candau, "Inverse Microemulsion Polymerization of Acrylamide: Characterization of the Water-in-Oil Microemulsions and the Final Microlatexes," *Journal of Colloid and Interface Science*, Vol. 101, No. 1, 1984, pp. 167-183. doi:10.1016/0021-9797(84)90017-1

[4] J. Eastoe, B. H. Robinson, A. J. W. G. Visser and D. C. Steytler, "Rotational Dynamics of AOT Reversed Micelles in Near-Critical and Supercritical Alkanes," *Journal of the Chemical Society*, *Faraday Transactions*, Vol. 87, No. 12, 1991, pp. 1899-1903. doi:10.1039/ft9918701899

[5] G. Calvaruso, A. Minore and V. Turco Liveri, "FT-IR Investigation of the Urea State in AOT Reversed Micelles," *Journal of Colloid and Interface Science*, Vol. 243, No. 1, 2001, pp. 227-232. doi:10.1006/jcis.2001.7878

[6] L. M. Nikolenko, A. V. Ivanchihina, S. B. Brichkin and V. F. Razumov, "Ternary AOT/Water/Hexane Systems as 'Micellar Sieves' for Cyanine Dye J-Aggregates," *Journal of Colloid and Interface Science*, Vol. 332, No. 2, 2009, pp. 366-372. doi:10.1016/j.jcis.2008.12.022

[7] A. Jada, J. Lang and R. Zana, "Relation between Electrical Percolation and Rate Constant for Exchange of Material between Droplets in Water in Oil Microemulsions," *Journal of Physical Chemistry*, Vol. 93, No. 1, 1989, pp. 10-12. doi:10.1021/j100338a004

[8] S. Bisal, P. K. Bhattacharya and S. P. Moulik, "Conductivity Study of Microemulsions: Dependence of Structural Behavior of Water/Oil Systems on Surfactant, Cosurfactant, Oil, and Temperature," *Journal of Physical Chemistry*, Vol. 94, No. 1, 1990, pp. 350-355. doi:10.1021/j100364a060

[9] W. West and S. Pearce, "The Dimeric State of Cyanine Dyes," *Journal of Physical Chemistry*, Vol. 69, No. 6, 1965, pp. 1894-1903. doi:10.1021/j100890a019

[10] D. L. Akins and J. W. Macklin, "Dependence of Raman Scattering by Aggregated 2,2-Cyanine on pH and Excitation Wavelength," *Journal of Physical Chemistry*, Vol. 93, No. 16, 1989, pp. 5999-6007. doi:10.1021/j100353a014

[11] P. Bilski, R. N. Holt and C. F. Chignell, "Premicellar Aggregates of Rose Bengal with Cationic and Zwitterionic Surfactants," *Journal of Photochemistry and Photobiology A*, Vol. 110, No. 1, 1997, pp. 67-74. doi:10.1016/S1010-6030(97)00166-4

[12] M. Sarkar and S. Poddar, "Studies on the Interaction of Surfactants with Cationic Dye by Absorption Spectroscopy," *Journal of Colloid and Interface Science*, Vol. 221, No. 2, 2000, pp. 181-185. doi:10.1006/jcis.1999.6573

[13] H. Von Berlepsch and C. Böttcher, "Network Superstructure of Pseudoisocyanine J-Aggregates in Aqueous Sodium Chloride Solution Revealed by Cryo-Transmission Electron Microscopy," *Journal of physical chemistry B*, Vol. 106, No. 12, 2002, pp. 3146-3150. doi:10.1021/jp0143701

[14] P. Mukerjee and K. J. Mysels, "A Re-Evaluation of the Spectral Change Method of Determining Critical Micelle Concentration," *Journal of the American Chemical Society*, Vol. 77, No. 11, 1955, pp. 2937-2943. doi:10.1021/ja01616a003

[15] M. K. Pal and P. K. Pal, "Dichroic Probe of Different Modes of Aggregation of Pinacyanol by Bis(2-Ethylhexyl)sulfosuccinate below and above Its Critical Micelle Concentration," *Journal of Physical Chemistry*, Vol. 94, No. 6, 1990, pp. 2557-2559. doi:10.1021/j100369a062

[16] R. Sabaté and J. Estelrich, "Determination of Micellar Microenvironment of Pinacyanol by Visible Spectroscopy," *Journal of physical chemistry B*, Vol. 107, No. 17, 2003, pp. 4137-4142. doi:10.1021/jp027409q

[17] L. Antonov and D. Nedeltchera, "Resolution of Overlapping UV-Vis Absorption Bands and Quantitative Analysis," *Chemical Society Reviews*, Vol. 29, No. 3, 2000, pp. 217-227. doi:10.1039/a900007k

[18] H. Min, J. Park, J. Yu and D. Kim, "The Spectroscopic Studies on the Aggregation Behavior of Cyanine Dyes," *Bulletin of the Korean Chemical Society*, Vol. 19, No. 6, 1998, pp. 650-654.

[19] S. J. Khouri, and V. Buss, "UV/Vis Spectral Study of the Self-Aggregation of Pinacyanol Chloride in Ethanol-Water Solution," *Journal of Solution Chemistry*, Vol. 39, No. 1, 2010, pp. 121-130. doi:10.1007/s10953-009-9476-2

[20] E. F. Williams, N. T. Woodberry and J. K. Dixon, "Purification and Surface Tension Properties of Alkyl Sodium Sulfosuccinates," *Journal of Colloid and Interface Science*, Vol. 12, No. 4, 1957, pp. 452-459.

[21] W. C. Presto and W. Preston, "Some Correlating Principles of Detergent Action," *Journal of Physical and Colloid Chemistry*, Vol. 52, No. 1, 1948, pp. 84-97. doi:10.1021/j150457a010

[22] C. Burda, X. Chen, R. Narayanan and M. A. El-Sayed, "Chemistry and Properties of Nanocrystals of Different Shapes," *Chemical Reviews*, Vol. 105, No. 4, 2005, pp. 1025-1102. doi:10.1021/cr030063a

[23] N. Berova and K. Nakanishi, "Exciton Chirality Method: Principles and Applications," In: N. Berova, K. Nakanishi and R. Woody, Eds., *Circular Dichroism: Principles and Applications*, Wiley-VCH, New York, 2000, p. 337.

[24] J. L. Lynn and B. H. Bory, "Surfactants," In: R. Kirk, Ed., *Concise Encyclopaedia of Chemical Technology*, Wiley, New York, 1999, p. 1949.

3

Buffer Standards for the Biochemical pH of the Zwitterionic Buffer *N*-Tris-(Hydroxymethyl) Methyl-2-aminoethanesulfonic Acid (TES) from 5˚C to 55˚C

Rabindra N. Roy*, Lakshmi N. Roy, Blake M. Bodendorfer, Zachary M. Downs, Stephen D. Rocchio, Joshua T. Wollen, Jessica M. Stegner, Isaac B. Henson, Nicholas W. Grove, Lauren A. Dieterman

Hoffman Department of Chemistry, Drury University, Springfield, USA

Email: *rroy@drury.edu

ABSTRACT

The authors have undertaken the determination of pH values for one buffer solution of TES without NaCl and nine buffer solutions with NaCl yielding an ionic strength $I = 0.16$ mol·kg^{-1} similar to that of blood. These buffer solutions have been evaluated in the temperature range of 5˚C to 55˚C using an extended version of the Debye-Hückel equation. The pH values are reported using 1) the Debye-Hückel extension of the Bates-Guggenheim convention in the temperature range 5˚C to 55˚C and 2) with and without liquid junction correction at 25˚C and 37˚C. These TES buffer solutions are recommended as secondary standard references for pH measurements in the range of pH 7.2 to 7.5 for physiological application with an ionic strength of $I = 0.16$ mol·kg^{-1}.

Keywords: Ionic Strength; Liquid Junction Potentials; Second Dissociation Constant; Buffer Solutions; Activity Coefficients

1. Introduction

Good *et al.* [1,2] have proposed a select group of zwitterionic buffers for physiological use. The authors have previously reported the pH values of N-[tris(hydroxymethyl) methyl-3-amino]propansulfonic acid (TAPS) [3] from 5 to 55˚C, including 37˚C. Vega and Bates [4] have reported the second dissociation constant, pK_2, and related thermodynamic data of TES from 5˚C to 55˚C. Roy and associates [5] also determined pK_2 values of TES with thermodynamic quantities from 5˚C to 55˚C. The purpose of the current investigation is to provide pH values for the amino acid TES, the structure of which is given in **Figure 1**.

The universally used NIST certified physiological phosphate primary standard buffer solution has pH values of 7.415 and 7.395 at 25˚C and 37˚C, respectively [6]. This phosphate buffer contains 0.008695 mol·kg^{-1} KH_2PO_4 and 0.03043 mol·kg^{-1} Na_2HPO_4. Because of problems associated with the current phosphate buffer [7-9], we have been prompted to investigate some zwitterionic buffers including TES for use in physiological pH range 6.5 - 7.5.

The following compositions were used for the determination of pa_H: 1) (0.08 mol·kg^{-1}) TES + (0.08 mol·kg^{-1}) NaTES, $I = 0.08$ mol·kg^{-1};
2) (0.02 mol·kg^{-1}) TES + (0.02 mol·kg^{-1}) NaTES + (0.14 mol·kg^{-1}) NaCl, $I = 0.16$ mol·kg^{-1};
3) (0.04 mol·kg^{-1}) TES + (0.04 mol·kg^{-1}) NaTES + (0.12 mol·kg^{-1}) NaCl, $I = 0.16$ mol·kg^{-1};
4) (0.06 mol·kg^{-1}) TES + (0.06 mol·kg^{-1}) NaTES + (0.10 mol·kg^{-1}) NaCl, $I = 0.16$ mol·kg^{-1};
5) (0.08 mol·kg^{-1}) TES + (0.08 mol·kg^{-1}) NaTES + (0.08 mol·kg^{-1}) NaCl, $I = 0.16$ mol·kg^{-1};

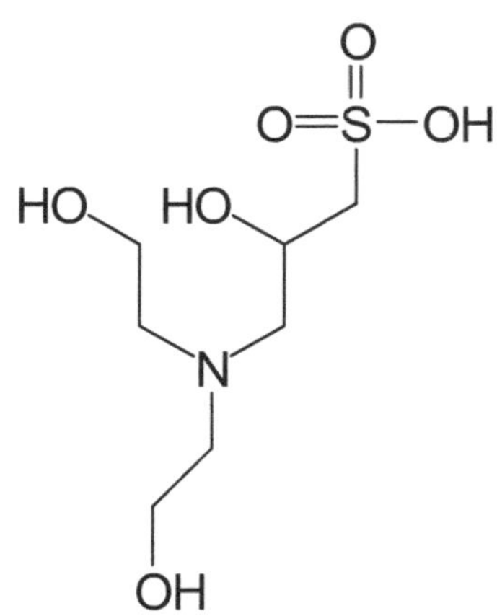

Figure 1. ***N*-tris-(hydroxymethyl)methyl-2-aminoethanesulfonic acid (TES).**

*Corresponding author.

6) (0.05 mol·kg^{-1}) TES + (0.05 mol·kg^{-1}) NaTES + (0.11 mol·kg^{-1}) NaCl, I = 0.16 mol·kg^{-1};
7) (0.02 mol·kg^{-1}) TES + (0.04 mol·kg^{-1}) NaTES + (0.12 mol·kg^{-1}) NaCl, I = 0.16 mol·kg^{-1};
8) (0.03 mol·kg^{-1}) TES + (0.06 mol·kg^{-1}) NaTES + (0.10 mol·kg^{-1}) NaCl, I = 0.16 mol·kg^{-1};
9) (0.04 mol·kg^{-1}) TES + (0.08 mol·kg^{-1}) NaTES + (0.08 mol·kg^{-1}) NaCl, I = 0.16 mol·kg^{-1};
10) (0.05 mol·kg^{-1}) TES + (0.10 mol·kg^{-1}) NaTES + (0.06 mol·kg^{-1}) NaCl, I = 0.16 mol·kg^{-1}.

2. Experimental

The TES was obtained from Research Organics in Cleveland, OH. The experimental method for further crystallization and the assay have been previously reported in detail [5]. Buffer solutions 1) to 10) were prepared by weighing recrystallized TES buffer, ACS reagent grade NaCl, a standard NaOH solution, and calculated amounts of double distilled CO_2-free water. Vacuum corrections were made for all masses.

The cell design, preparation of the chloroplatinic acid, hydrogen gas purification, silver-silver chloride, solution preparation, and other experimental details have been described previously [10-13].

3. Methods and Results

Cell (A) contains one solution without the Cl^- and nine solutions containing NaCl to yield an ionic strength of I = 0.16 mol·kg^{-1}. The emf values have been corrected to a hydrogen pressure of 1 atm. The values for the pa_H calculations are given in **Table 1**. The error in the emf, on an average of two readings, lies within 0.02 mV.

For accurate calculations of the pH values of the buffer solutions, the following Cell (A) was used:

$$\mathrm{Pt(s),H_2(g,1\ atm)|TES}(m_1)+\mathrm{NaTES}(m_2)+\mathrm{NaCl}(m_3)| \cdot\mathrm{AgCl(s),Ag(s)} \quad \text{(A)}$$

where m_1, m_2, and m_3 denote the molalities of the respective species at 1 atm. The Harned-type cell was employed.

Cell (B) involves liquid junction as shown with a double vertical line:

$$\mathrm{Pt(s),H_2(g,1\ atm)|TES}(m_1)+\mathrm{NaTES}(m_2)+\mathrm{NaCl}(m_3)\| \cdot\mathrm{KCl(satd),Hg_2Cl_2(s),Hg(l)} \quad \text{(B)}$$

where the abbreviations "s", "l", and "g" indicate the solid, liquid, and gaseous states, respectively.

For Cell (C), the phosphate salts were NIST standard reference materials. Its solutions are recommended for pH measurements in physiological media. The schematic diagram of the Cell (C) is shown below:

$$\mathrm{Pt(s),H_2(g,1\ atm)|KH_2PO_4+Na_2HPO_4\|KCl(satd), Hg_2Cl_2(s),Hg(l)} \quad \text{(C)}$$

The E°_{SCE} values of the saturated calomel electrode were taken as –0.2415 and –0.2335 at 25°C and 37°C, respectively [12,14]. The δE_j values of the buffer solution for Cell (B) were calculated using the following equation:

$$\delta E_j = E + E^{\circ}_{SCE} - k\mathrm{pH} \quad (1)$$

where k = 0.059156 and pH = 7.415 at 25°C; k = 0.061538 and pH = 7.395 at 37°C. These values are for phosphate primary standard and buffer solutions. The operational definition of pH is denoted as pH(x) which can be calculated by use of the equation:

Table 1. Electromotive force of Cell A: $\mathrm{Pt(s);H_2(g,1\ atm)|TES}(m_1),\mathrm{NaTES}(m_2),\mathrm{NaCl}\ (m_3)|\mathrm{AgCl(s),Ag(s)}$.

m_1[a]	m_2[a]	m_3[a]	5°C	10°C	15°C	20°C	25°C	30°C	35°C	37°C	40°C	45°C	50°C	55°C
0.08	0.08	0.005	0.79716	0.79847	0.79946	0.80037	0.80099	0.80206	0.80295	0.80296	0.80380	0.80417	0.80532	0.80577
0.08	0.08	0.010	0.78204	0.78302	0.78377	0.78449	0.78480	0.78585	0.78633	0.79628	0.78673	0.78697	0.78740	0.78765
0.08	0.08	0.015	0.77406	0.78302	0.77540	0.77587	0.77594	0.77698	0.77715	0.77712	0.77747	0.77740	0.77747	0.77738
0.08	0.08	0.020	0.76875	0.76939	0.76996	0.77034	0.77000	0.77120	0.77128	0.77105	0.77131	0.77110	0.77105	0.77078
0.02	0.02	0.14	0.72121	0.72111	0.72087	0.72050	0.71991	0.71933	0.71864	0.71839	0.71783	0.71692	0.71591	0.71466
0.04	0.04	0.12	0.72597	0.72594	0.72576	0.72547	0.72506	0.72450	0.72385	0.72358	0.72307	0.72224	0.72129	0.72019
0.06	0.06	0.10	0.73153	0.73162	0.73149	0.73122	0.73085	0.73034	0.72975	0.72949	0.72903	0.72823	0.72734	0.72634
0.08	0.08	0.08	0.73804	0.73830	0.73825	0.73808	0.73781	0.73741	0.73688	0.73666	0.73627	0.73553	0.73467	0.73377
0.05	0.05	0.11	0.72998	0.72980	0.72939	0.72890	0.72809	0.72735	0.72640	0.72624	0.72566	0.72492	0.72404	0.72288
0.02	0.04	0.12	0.73954	0.73974	0.73986	0.73989	0.73960	0.73935	0.73898	0.73883	0.73853	0.73798	0.73733	0.73674
0.03	0.06	0.10	0.74487	0.74523	0.74544	0.74547	0.74527	0.74515	0.74481	0.74468	0.74440	0.74395	0.74337	0.74263
0.04	0.08	0.08	0.75155	0.75189	0.75223	0.75236	0.75232	0.75224	0.75205	0.75193	0.75165	0.75140	0.75087	0.75036
0.05	0.10	0.06	0.75903	0.75958	0.76027	0.76050	0.76080	0.76088	0.76087	0.76078	0.76060	0.76038	0.76004	0.75969

[a]Units of m, mol·kg^{-1}.

$$\mathrm{pH}(\mathrm{x}) = \mathrm{pH}(\mathrm{s}) + \frac{E_{\mathrm{x}} - E_{\mathrm{s}} + \delta E_{\mathrm{j}}}{k} \quad (2)$$

where "x" refers to the unknown buffer solution (TES + NaTES in this case), "s" is the NIST standard phosphate buffer solution of known pH, and $\delta E_{\mathrm{j}} = E_{\mathrm{j\,(s)}} - E_{\mathrm{j(x)}}$.

To calculate the pH values for the buffer solutions, the determination of the acidity function, p($a_{\mathrm{H}}\gamma_{\mathrm{Cl}}$), in the temperature range 5°C to 55°C is required. These calculations were made using the emf (E) values listed in **Table 1**. The following equation was used:

$$p(a_{\mathrm{H}}\gamma_{\mathrm{Cl}}) = \frac{E - E^{\circ}}{k} + \log m_{\mathrm{Cl}} \quad (3)$$

where "k" is the Nernst slope.

The plot of $p(a_{\mathrm{H}}\gamma_{\mathrm{Cl}})$ as a function of $m_{\mathrm{Cl^-}}$, a linear plot was obtained. The y-intercept gives a $p(a_{\mathrm{H}}\gamma_{\mathrm{Cl}})^{\circ}$ value at $m_{\mathrm{Cl^-}} = 0$. The $p(a_{\mathrm{H}}\gamma_{\mathrm{Cl}})^{\circ}$ values for the chloride-free buffer solution are listed in **Table 2**. The $p(a_{\mathrm{H}}\gamma_{\mathrm{Cl}})$ values for the buffer solutions of high ionic strength containing $\mathrm{Cl^-}$ are entered in **Tables 2** and **3** from 5°C to 55°C.

Table 2. $p(a_{\mathrm{H}}\gamma_{\mathrm{Cl}})^{\circ}$ **of** $(\mathrm{TES} + \mathrm{NaTES})$ **buffer solutions from 5°C to 55°C, computed using Equation 5**[a] $p(a_{\mathrm{H}}\gamma_{\mathrm{Cl}})$ **of** $(\mathrm{TES} + \mathrm{NaTES})$ **buffer solutions from 5°C to 55°C, computed using Equation (4)**[a].

t (°C)	0.08 m TES + 0.08 m NaTES + 0.00 m NaCl $I = 0.08$ m	0.02 m TES + 0.02 m NaTES + 0.14 m NaCl $I = 0.16$ m	0.04 m TES + 0.04 m NaTES + 0.12 m NaCl $I = 0.16$ m	0.06 m TES + 0.06 m NaTES + 0.10 m NaCl $I = 0.16$ m	0.08 m TES + 0.08 m NaTES + 0.08 m $I = 0.16$ m
5	7.867	7.971	7.991	8.012	8.033
10	7.759	7.861	7.881	7.902	7.924
15	7.650	7.756	7.774	7.795	7.817
20	7.550	7.654	7.673	7.692	7.713
25	7.455	7.555	7.575	7.594	7.615
30	7.364	7.462	7.481	7.499	7.520
35	7.277	7.372	7.390	7.407	7.427
37	7.241	7.338	7.355	7.372	7.392
40	7.194	7.285	7.302	7.319	7.339
45	7.115	7.202	7.219	7.235	7.254
50	7.043	7.121	7.138	7.153	7.171
55	6.972	7.041	7.059	7.074	7.091

[a]Units of m, mol·kg^{-1}.

Table 3. $p(a_{\mathrm{H}}\gamma_{\mathrm{Cl}})$ **of** $(\mathrm{TES} + \mathrm{NaTES})$ **buffer solutions from 5°C to 55°C, computed using Equation (4)**[a].

t (°C)	0.05 m TES + 0.05 m NaTES + 0.11 m NaCl $I = 0.16$ m	0.02 m TES + 0.04 m NaTES + 0.12 m NaCl $I = 0.16$ m	0.03 m TES + 0.06 m NaTES + 0.10 m NaCl $I = 0.16$ m	0.04 m TES + 0.08 m NaTES + 0.08 m NaCl $I = 0.16$ m	0.05 m TES + 0.10 m NaTES + 0.06 m NaCl $I = 0.16$ m
5	8.026	8.237	8.254	8.278	8.289
10	7.912	8.126	8.145	8.166	8.178
15	7.800	8.021	8.039	8.061	8.077
20	7.694	7.921	7.937	7.959	7.974
25	7.589	7.821	7.838	7.860	7.879
30	7.491	7.728	7.745	7.766	7.785
35	7.394	7.637	7.654	7.675	7.694
37	7.361	7.603	7.619	7.640	7.659
40	7.306	7.551	7.566	7.586	7.605
45	7.224	7.468	7.484	7.505	7.522
50	7.143	7.389	7.404	7.424	7.442
55	7.062	7.313	7.324	7.346	7.364

[a]Units of m, mol·kg^{-1}.

The pH(s) values for solutions without liquid junction in the absence of NaCl were determined using the next Equation:

$$\mathrm{pH}(s) = p(a_{\mathrm{H}}\gamma_{\mathrm{Cl}})^{\circ} + \log \gamma_{\mathrm{Cl}}^{\circ} \quad (4)$$

where the single-ion activity coefficient, $\gamma_{\mathrm{Cl}}^{\circ}$, cannot be experimentally measured. A previous publication gives the method used for obtaining $\gamma_{\mathrm{Cl}}^{\circ}$ [10,15,16]. The pH values from the liquid junction cell are indicated by pH whereas the "conventional" pH calculated from Equayon. 5 is designated as pH(s). The Bates-Guggenheim convention [10,15] is commonly known as "pH convention" and is expressed by the Equation:

$$\log_{10} \gamma_{\mathrm{Cl}}^{\circ} = -\frac{A\sqrt{I}}{1 + Ba^{\circ}\sqrt{I}} \quad (5)$$

where Equation (5) is valid only up to 0.1 mol·kg^{-1}. Since blood and other physiological solutions have I = 0.16 mol·kg^{-1}, there is a need for a better choice to calculate $\log \gamma_{\mathrm{Cl}}^{\circ}$. Thus, a linear-dependent "CI" term is added in the Bates-Guggenheim convention as shown in Equation (6):

$$\log_{10} \gamma_{\mathrm{Cl}}^{\circ} = -\frac{A\sqrt{I}}{1 + Ba^{\circ}\sqrt{I}} + CI \quad (6)$$

where "I" is the ionic strength of the buffer solution, the constants "A", "B", and a° have the usual physical significances. The following empirical equation is used for the calculation of the adjustable parameter "C" [8,10].

$$C = C_{25} + (6.2\times10^{-4})(t-25) - (8.7\times10^{-6})(t-25)^2 \quad (7)$$

where C_{25}= 0.032 kg·mol^{-1} at 25°C and t is the temperature in degrees Celsius.

The pa_{H} values listed in **Tables 4** and **5** for the TES buffer solution with and without the presence of NaCl

Table 4. pa_{H} **of** $(\mathrm{TES} + \mathrm{NaTES})$ **buffer solutions from 5°C to 55°C, computed using Equations (4)-(6)[a].**

t (°C)	0.08 m TES + 0.08 m NaTES + 0.00 m NaCl I = 0.08 m	0.02 m TES + 0.02 m NaTES + 0.14 m NaCl I = 0.16 m	0.04 m TES + 0.04 m NaTES + 0.12 m NaCl I = 0.16 m	0.06 m TES + 0.06 m NaTES + 0.10 m NaCl I = 0.16 m	0.08 m TES + 0.08 m NaTES + 0.08 m NaCl I = 0.16 m
5	7.767	7.846	7.865	7.887	7.908
10	7.659	7.736	7.755	7.777	7.799
15	7.549	7.630	7.649	7.669	7.691
20	7.450	7.529	7.548	7.567	7.588
25	7.354	7.429	7.449	7.468	7.488
30	7.262	7.335	7.354	7.372	7.393
35	7.174	7.224	7.262	7.279	7.299
37	7.138	7.210	7.227	7.244	7.264
40	7.091	7.156	7.174	7.190	7.210
45	7.011	7.073	7.090	7.105	7.124
50	6.938	6.991	7.008	7.023	7.041
55	6.867	6.909	6.927	6.943	6.960

[a]Units of m, mol·kg^{-1}.

Table 5. pa_{H} **of** $(\mathrm{TES} + \mathrm{NaTES})$ **buffer solutions from 5°C to 55°C, computed using Equations (4)-(6)[a].**

t (°C)	0.05 m TES + 0.05 m NaTES + 0.11 m NaCl I = 0.16 m	0.02 m TES + 0.04 m NaTES + 0.12 m NaCl I = 0.16 m	0.03 m TES + 0.06 m NaTES + 0.10 m NaCl I = 0.16 m	0.04 m TES + 0.08 m NaTES + 0.08 m NaCl I = 0.16 m	0.05 m TES + 0.10 m NaTES + 0.06 m NaCl I = 0.16 m
5	7.900	8.111	8.128	8.153	8.163
10	7.786	8.001	8.019	8.041	8.053
15	7.674	7.895	7.914	7.937	7.951
20	7.569	7.796	7.812	7.834	7.849
25	7.462	7.695	7.711	7.734	7.752
30	7.364	7.601	7.618	7.639	7.658
35	7.266	7.510	7.526	7.547	7.567
37	7.233	7.475	7.491	7.512	7.531
40	7.178	7.423	7.438	7.458	7.477
45	7.094	7.339	7.354	7.376	7.393
50	7.013	7.258	7.273	7.293	7.311
55	6.931	7.182	7.193	7.215	7.233

[a]Units of m, mol·kg^{-1}.

were calculated using the following equation:

$$1)\ pa_{\mathrm{H}} = 7.35 - \left(1.90\times10^{-2}\right)(t-25) - \left(8.98\times10^{-5}\right)(t-25)^2 \quad (8)$$

in the temperature range of 5°C to 55°C. The standard deviation of regression for the pa_H of the chloride-free buffer solution was obtained using Equations (3), (4), (6), and (7). The standard deviation was 0.0012.

For the buffer solutions containing NaCl, with an isotonic saline media ionic strength of

$I = 0.16$ mol·kg^{-1}, the pa_H values were also calculated using Equations. 3 - 7. The acidity function data, $p(a_H\gamma_{Cl})$ from **Tables 2** and **3** were used to generate the pa_H data entered in **Tables 4** and **5**. These pa_H values are expressed by use of the quadratic equations:

$$2)\ pa_{\mathrm{H}} = 7.431 - \left(1.93\times10^{-2}\right)(t-25) - \left(6.80\times10^{-5}\right)(t-25)^2 \quad (9)$$

$$3)\ pa_{\mathrm{H}} = 7.449 - \left(1.94\times10^{-2}\right)(t-25) - \left(1.22\times10^{-5}\right)(t-25)^2 \quad (10)$$

$$4)\ pa_{\mathrm{H}} = 7.468 - \left(1.95\times10^{-2}\right)(t-25) - \left(6.86\times10^{-5}\right)(t-25)^2 \quad (11)$$

$$5)\ pa_{\mathrm{H}} = 7.489 - \left(1.96\times10^{-2}\right)(t-25) - \left(6.66\times10^{-5}\right)(t-25)^2 \quad (12)$$

$$6)\ pa_{\mathrm{H}} = 7.463 - \left(2.02\times10^{-2}\right)(t-25) - \left(8.46\times10^{-5}\right)(t-25)^2 \quad (13)$$

$$7)\ pa_{\mathrm{H}} = 7.696 - \left(1.93\times10^{-2}\right)(t-25) - \left(7.20\times10^{-5}\right)(t-25)^2 \quad (14)$$

$$8)\ pa_{\mathrm{H}} = 7.713 - \left(1.94\times10^{-2}\right)(t-25) - \left(6.91\cdot10^{-5}\right)(t-25)^2 \quad (15)$$

$$9)\ pa_{\mathrm{H}} = 7.735 - \left(1.94\times10^{-2}\right)(t-25) - \left(7.12\times10^{-5}\right)(t-25)^2 \quad (16)$$

10)

$$pa_{\mathrm{H}} = 7.752 - \left(1.92\times10^{-2}\right)(t-25) - \left(6.42\times10^{-5}\right)(t-25)^2 \quad (17)$$

The observed standard deviations of regression from Equations. 9 - 17 are 0.0014, 0.0009, 0.0006, 0.0006, 0.0009, 0.0008, 0.0010, 0.0009, and 0.0009, respectively.

The emf values of Cells (B) and (C), pH values with and without liquid junction, an the values of δE_j at 25 and 37°C are given in **Table 6**. These are various sources of error: 1) extrapolation of the $p(a_H\gamma_{Cl})^\circ$ plot for Cl^- free solutions; 2) the calculation of the single ion activity coefficient $\log\gamma_{Cl}^{o}$ using Equation (6); 3) the error in the experimental emf measurement; and 4) in the estimation of the liquid junction potential. For the four buffer solutions in **Table 6** the pH values obtained from the extended Debye-Hückel equations and liquid junction correction are in very good agreement (±0.001 pH unit). Thus these TES buffer solutions are recommended as useful secondary pH standard for blood serum and physiological application in the pH range 7.2 - 7.5. The authors have undertaken the calculation of the single ion

Table 6. Emf of Cell B and pH values with δE_j correction at 25°C and 37°C for TES buffer.

m_1	m_2	m_3	I	E/V		δE_j[b]/mV		Without[c] δE_j corr	With[d] δE_j corr	Extended[e] D-H eqn.	Without[c] δE_j corr	With[d] δE_j corr	Extended[e] D-H eqn.
				25°C	37°C	25°C	37°C		25°C			37°C	
0.08	0.08	0.00	0.08	0.67873	0.67515	2.2	2.4	7.347	7.354	7.354	7.131	7.138	7.138
0.04	0.04	0.12	0.16	0.68263	0.67889	0.5	0.7	7.413	7.448	7.449	7.191	7.226	7.227
0.06	0.06	0.10	0.16	0.68375	0.67991	0.5	0.7	7.432	7.467	7.468	7.208	7.243	7.244
0.08	0.08	0.08	0.16	0.68491	0.68117	0.5	0.7	7.452	7.487	7.488	7.227	7.263	7.264
Emf of Cell C[a]													
0.008695 m KH_2PO_4 + 0.03043 m Na_2HPO_4				0.68275	0.69147	2.6	2.9						

[a]Published data [4,6,8] for physiological phosphate buffer solutions; units of m, mol·kg^{-1}. [b] $E_j = E + -k\,\mathrm{pH(s)}$ from Equation (1) and the residual liquid junction potential, $\delta E_j = E_j(\mathrm{s}) - E_j(\mathrm{x})$, which can be calculated from the values of **Tables 4** and **5**. The pH of the primary standard phosphate buffer is 7.415 and 7.395 at 25°C and 37°C, respectively. [c]Values obtained from Equation (3) and **Table 6**. [d]Obtained from Equation (2) and emf values from **Table 6**. [e]Obtained from extended Debye-Hückel (DH) equation of the Bates-Guggenheim convention.

activity coefficient need for accurate pH calculation of several buffer solutions using Pitzer formalism [17,18].

4. Acknowledgements

The authors are grateful for the funding from the National Institutes of Health (AREA), under the grant 2-R15 GM 066866-03 and the diversity supplemental grant 3-R15 GM 066866-03 S1. The authors would also like to thank Ethan Graham and Chad Lunn for their dedication and hard work. The content of this paper is the sole responsibility of the authors and does not necessarily represent the official views of the National Institutes of Health or the National Institutes of General Medical Sciences.

REFERENCES

[1] N. E. Good, G. D. Winget, W. Winter, T. N. Connolly, S. Izawa and R. M. M. Singh, "Hydrogen Ion Buffers for Biological Research," *Biochemistry*, Vol. 5, No. 2, 1966, pp. 467 -477. doi:10.1021/bi00866a011

[2] W. J. Ferguson, K. L. Braunschweiger, W. R. Braunschweiger, J. R. Smith, J. J. McCormick, C. C. Wasmann, N. P. Jarvis, D. H. Bell and N. E. Good, "Hydrogen Ion Buffers for Biological Research," *Analytical Biochemistry*, Vol. 104, No. 2, 1980, pp. 300-310. doi:10.1016/0003-2697(80)90079-2

[3] L. N. Roy, R. N. Roy, C. E. Denton, S. R. Le Noue, C. A. Himes, S. J. Richards, A. N. Simon and V. S. Somal, "Buffer Standards for the Physiological pH of the Zwitterionic Compound, TAPS, from 5˚C to 55˚C," *Journal of Solution Chemistry*, Vol. 35, 2006, pp. 551-556. doi:10.1007/s10953-005-9010-0

[4] C. A. Vega and R. G. Bates, "Buffers for the Physiological pH Range: Thermodynamic Constants of Four Substituted Aminoethanesulfonic Acids from 5˚C to 50˚C," *Analytical Chemistry*, Vol. 48, No. 9, 1976, pp. 1293-1296.

[5] R. N. Roy, C. P. Moore, J. A. Carlsten, W. S. Good, P. Harris, J. M. Rook, L. N. Roy and K. M. Kuhler, "Second Dissociation Constants of Two Substituted Aminoethanesulfonic Acids (MES) and (TES) in Water from 5˚C to 55˚C," *Journal of Solution Chemistry*, Vol. 26, No. 12, 1997, pp. 1209-1216. doi:10.1023/A:1022937324983

[6] V. E. Bower, M. Paabo and R. G. Bates, "A Standard for the Measurement of the pH of Blood and Other Physiological Media," *Journal of Research of the National Bureau of Standards*, Vol. 65A, No. 3, 1961, pp. 267-270.

[7] R. A. Durst and B. R. Staples, "Tris/Tris HCl: Standard Buffer for Use in the Physiological pH Range," *Clinical Chemistry*, Vol. 18, No. 3, 1972, pp. 206-208.

[8] D. Feng, W. F. Koch and Y. C. Wu, "Second Dissociation Constant and pH of N-(2-Hydroxyethyl)-piperazine-N'-2-ethanesulfonic Acid from 0˚C to 50˚C," *Analytical Chemistry*, Vol. 61, 1989, pp. 1400-1405. doi:10.1021/ac00188a019

[9] Y. C. Wu, P. A. Berezansky, D. Feng and W. F. Koch, "Second Dissociation Constant of 3-(N-Morpholino)-2-hydroxypropanesulfonic Acid and pH of Its Buffer Solutions," *Analytical Chemistry*, Vol. 65, 1993, pp. 1084-1987. doi:10.1021/ac00056a023

[10] R. N. Roy, D. R. Mrad, P. A. Lord, J. A. Carlsten, W. S. Good, P. Allsup, L. N. Roy, K. M. Kuhler, W. F. Koch and Y. C. Wu, "Thermodynamics of the Second Dissociation Constant and Standards for pH of 3-(N-Morpholino)propanesulfonic Acid (MOPS) from 5˚C to 55˚C," *Journal of Solution Chemistry*, Vol. 27, No. 1, 1998, pp. 73-87. doi:10.1023/A:1022692629289

[11] R. G. Bates, C. A. Vega and D. R. White Jr., "Standards for pH Measurements in Isotonic Saline Media of Ionic Strength I = 0.16," *Analytical Chemistry*, Vol. 50, No. 9, 1978, pp. 1295-1300. doi:10.1021/ac50031a026

[12] R. G. Bates, "Determination of pH," 2nd Edition, Wiley, New York, 1973.

[13] R. N. Roy, L. N. Roy, C. E. Denton, S. R. LeNoue, S. Ashkenazi, M. S. Fuge, C. D. Dunseth, J. L. Durden, C. N. Roy, A. Bwashi, J. T. Wollen and S. J. DeArmon, "Buffer Standards for the Physiological pH of 3-[(1,1-Dimethyl-2-hydroxymethyl)-amino]-2-hydroxypropanesulfonic Acid from (278.15 to 328.15) K," *Journal of Chemical & Engineering Data*, Vol. 54, No. 2, 2009, pp. 428-435.

[14] W. M. Latimer, "Oxidation Potentials," 2nd Edition, Prentice-Hall, New York, 1952.

[15] R. G. Bates and E. A. Guggenheim, "Report on the Standardization of pH and Related Terminology," *Pure and Applied Chemistry*, Vol. 1, No. 1, 1960, pp. 163-168. doi:10.1351/pac196001010163

[16] R. N. Roy, L. N. Roy, S. Ashkenazi, J. T. Wollen, C. D. Dunseth, M. S. Fuge, C. N. Roy, H. M. Hughes, B. T. Morris and K. L. Cline, "Buffer Standards for pH Measurement of N-(2-Hydroxyethyl)piperazine-N'-2-ethanesulfonic Acid (HEPES) for I = 0.16 $mol \cdot kg^{-1}$ from 5˚C to 55˚C," *Journal of Solution Chemistry*, Vol. 38, 2009, pp. 449-458. doi:10.1007/s10953-009-9378-3

[17] K. S. Pitzer and G. J. Mayorga, "Thermodynamics of Electrolytes. II. Activity and Osmotic Coefficients for Strong Electrolytes with One or Both Ions Univalent," *Journal of Physical Chemistry*, Vol. 77, No. 19, 1973, pp. 2300-2307. doi:10.1021/j100638a009

[18] R. P. Buck, S. Rondinini, A. K. Covington, F. G. K. Baucke, C. M. A. Brett, M. F. Camões, M. J. T. Milton, T. Mussini, R. Naumann, K. W. Pratt, P. Spitzer and G. S. Wilson, "Measurement of pH: Definition, Standards, and Procedures," *Pure and Applied Chemistry*, Vol. 74, No. 11, 2002, pp. 2169-2200. doi:10.1351/pac200274112169

4

Utilization of Special Potential Scan Programs for Cyclic Voltammetric Development of Different Nickel Oxide-Hydroxide Species on Ni Based Electrodes

Diego E. Pissinis, Leónides E. Sereno, Juan M. Marioli*
Departamento de Química, Universidad Nacional de Río Cuarto, Río Cuarto, Argentina
Email: *juanmmarioli@gmail.com

ABSTRACT

Special potential scan programs were designed for cyclic voltammetric development of β-NiOOH or γ-NiOOH/β-NiOOH mixtures on the surface of nickel or nickel-chromium (80:20) alloy electrodes in NaOH 0.10 M. The programs consisted on changing the anodic or cathodic switching limit to facilitate the chemical reactions taking place either between Ni(II) hydroxides or between Ni(III) oxides-hydroxides. The electrochemical charge density under the oxidative wave, observed at Ni or Ni-Cr electrode surfaces at approximately 0.48 V (vs SCE), remained almost constant with the number of *cv* cycles after approximately 600 *cv* cycles at 0.050 V/s. Thus, it can be suggested that a stable proportion of Ni(II)/Ni(III) oxides-hydroxides was obtained on the electrode surfaces. The relative amounts of β-NiOOH or γ-NiOOH species were calculated from the electrochemical charges under their reduction waves in the voltammetric experiments. Higher charge densities were always obtained with Ni-Cr alloy electrodes as compared to pure Ni electrodes. Linear relationships were obtained in our study on the dependence of the oxidative peak current with the square root of the scan rate at a scan rate range between 0.01 V/s and 0.16 V/s.

Keywords: Nickel; Nickel-Chromium Alloy; Oxides; Cyclic Voltammetry

1. Introduction

Since the 1970s, the oxidation of organic compounds on nickel electrodes in alkaline solutions has been extensively investigated [1,2]. Recently, nickel hydroxide electrodes have shown good electrocatalytic properties for the oxidation of aspirin [3], methane [4], and acetylcholine [5]. The nickel electrode exhibit catalytic activity in the oxidation of a wide variety of organic compounds in alkaline solutions. The Ni(III) species is generated on the electrode surface at potentials in the region 0f 0.4 V to 0.5 V vs Ag/AgCl. The reactions are [6-11] (and references therein):

$$\mathrm{Ni} + 2\mathrm{OH}^- \rightarrow \mathrm{Ni(OH)_2} + 2\mathrm{e} \quad (1)$$

$$\mathrm{Ni(OH)_2} + \mathrm{OH}^- \rightleftarrows \mathrm{NiOOH} + \mathrm{H_2O} + \mathrm{e} \quad (2)$$

$$\mathrm{NiOOH} + \mathrm{Analyte} \rightarrow \mathrm{Ni(OH)_2} + \mathrm{Radical} \quad (3)$$

$$\mathrm{Radical} \rightarrow \mathrm{Product} + \mathrm{e} \quad (4)$$

Reaction (1) represents the oxidation of Ni(0) to Ni(II) in alkaline solutions, which occurs at approximately −0.8 V (vs Saturated Calomel Electrode, SCE). Reaction (2) represents the reversible oxidation of Ni(II) to Ni(III), which is observed at nearly 0.4 V to 0.6 V vs SCE. Reaction (3) stands for the electrocatalytic oxidation of an analyte to regenerate Ni(II) species and to produce a radical, which is further oxidized to products (Reaction 4). Taking into account that electrocatalytic processes depend on the ability of the electroactive molecules to interact with the electrode [12], it is then important to understand as far as possible the electrochemistry of the electrode surface itself.

An advance in the understanding of the reactions of the nickel hydroxide electrode was made by Bode [13], who established the existence of two forms of $Ni(OH)_2$ and two forms of NiOOH. According to Bode *et al.*, electrochemical reactions of the $Ni(OH)_2$/NiOOH electrode can be summarized as in **Scheme 1**.

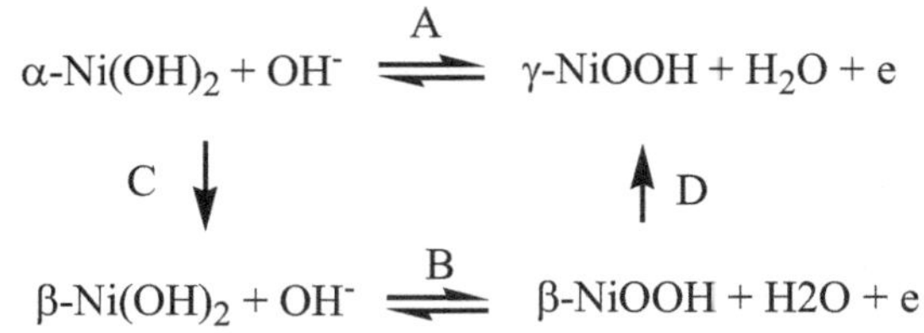

Scheme 1. Pathways A and B represent one electron electrochemical reactions, and pathways C and D represent chemical reactions; the arrows represent the most probable reaction path.

*Corresponding author.

It is well known that the alloying of some metals can markedly lower their corrosion rate. As regards nickel, it has been suggested that its corrosion resistance is considerably improved by alloying it with chromium [14-16]. A two-layer structure, with an inner oxide part formed directly at the metal surface and an outer hydroxide part in contact with the electrolyte, was proposed for Ni-Cr alloys in 1 M NaOH solutions as deduced from X-Ray photoelectron spectroscopic examinations of the electrochemically formed passive layers. The oxide layer consisted mainly of Cr_2O_3 and NiO entered only in the transpassive potential range. Moreover, it was suggested that Ni is enriched at the metal surface and it forms the outer hydroxide layer in contact with 1 M NaOH, especially at positive potentials and for long passivation times. Some soluble Cr(VI) was found within the hydroxide layer. However, due to the low solubility of $Ni(OH)_2$ in alkaline solutions, it has been suggested that Cr(VI) may be trapped within the hydroxide layer [16]. Above the passivation potential, Ni(III) species also accumulated within the passive layer, behaving as dopants and causing a shift of the Fermi level to lower energy.

We reported earlier that nickel-chromium alloys were good electrode materials for the electrochemical detection of molecules after HPLC separations using alkaline mobile phases [6,7,17]. It was observed that in alkaline solutions the Ni-Cr (80:20) alloy had a cyclic voltammetric (*cv*) response in the potential range from 0.00 V to 0.60 V (vs SCE) similar to that of a pure Ni electrode [18]. It was also reported there that the α, β and γ crystallographic forms of Ni oxides-hydroxides found at pure Ni electrodes were also present in the Ni-Cr (80:20) alloy electrode. However, the electrochemical charge density and peak current of the oxidative wave obtained at approximately 0.48 V (vs SCE) at the Ni-Cr alloy electrode were always higher than those obtained at pure Ni electrodes under similar experimental conditions. Thus, we considered it important to find the perturbation conditions under which the $Ni(OH)_2$/NiOOH species present at the Ni-Cr working electrodes may invariably be observed in a stable proportion. These perturbation conditions will allow us to begin surface studies to compare the oxides-hydroxides obtained at Ni or Ni-Cr alloy electrodes under similar potentiodynamic perturbations.

In this study, we developed a cyclic voltammetric procedure to obtain β-NiOOH or γ-NiOOH/β-NiOOH mixtures on Ni or Ni-Cr (80:20) alloy electrodes surfaces. The quantification of each NiOOH species was performed by fitting the *cv* response of the reverse scan with Lorentzian functions. This procedure allowed us to evaluate the effect of the working electrode perturbation waveform on the production of different nickel oxide-hydroxide species. The number of cv cycles needed to obtain a steady voltammogram (indicating that the oxide layer on the electrode did not grow noticeably) was also assessed. Finally, we studied the effect of the potential scan rate on the electrode response. Linear relation between the anodic peak current and the square root of the scan rate were studied and it was stated that the global electrode process is controlled by diffusion. The cyclic voltammetric electrode conditioning reported here allows obtaining a stable proportion of oxides-hydroxides on the surface of nickel or nickel-chromium alloys. This condition is important to begin with surface analysis experiments of nickel and nickel-chromium alloy electrodes.

2. Experimental

2.1. Electrode Preparation and Software

Working electrodes were made of Ni (Goodfellow, England, 99.99+%), or Ni-Cr (80:20) (Goodfellow, England, catalog number NI055150) wires of 0.10 cm diameter embedded in Teflon® shrinkable tubes (See **Scheme 2**). The metallic wire (approximately 10 cm long) was introduced into the Teflon® shrinkable tube (9 cm long), so nearly 1 cm was used for electric contact. This assembly was transferred to an electrical oven at approximately 300°C until a complete shrinkage of the Teflon® tube was observed. The electrodes were polished with sand paper and then with fine alumina down to 0.05 μm, rinsed with triply distilled water, sonicated during 5 minutes, rinsed again with triply distilled water and immersed in the electrochemical cell. The exposed geometrical area obtained after polishing was ca. 7.8×10^{-3} cm^2. Before cv experiments, −1.00 V was applied to the working electrodes for about 60 seconds in order to minimize oxides formation. A platinum foil of a large area (ca. 0.8 cm^2) was used as the counter electrode. All potentials were measured with reference to the Saturated Calomel Electrode (SCE). The SCE was held in a separate compartment of the electrochemical cell finishing in a Luggin capillary.

The electrochemical cell was a conventional two-compartment glass cell. Autolab models PGSTAT-30 or PGSTSAT-12 potentiostats were used to control the potential between the working and reference electrodes. The potentiostats were controlled by PC compatible computers using the PGSTAT software version 4.9. Graphics were obtained with a conventional printer. The temperature of the electrochemical cell was kept at 25.5°C ± 0.2°C.

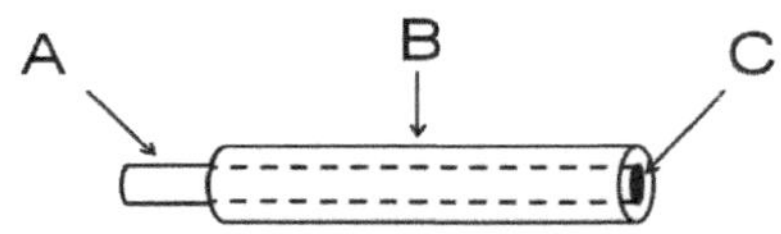

Scheme 2. Working electrode details. A: metal wire (either Ni or Ni-Cr); B: Teflon® shrinkable tube; C: circular exposed area of the electrode.

To facilitate the growth of each crystallographic form of NiOOH, two working electrode perturbation waveforms (**Figure 1**) were developed. The working electrode potential was swept over the potential range from 0.00 V to 0.60 V during 5 *cv* cycles, then a smaller potential range (0.50 V) was swept during 10 *cv* cycles, and finally an even smaller potential range (0.40 V) was swept during the remaining *cv* cycles. Thus, by shifting the smallest potential sweep range either in the anodic or in the cathodic direction (see **Figure 1**, panels A and B, from now on treatment A and treatment B, respectively), we were able to grow the crystallographic forms of NiOOH. The underlying principle of these waveforms was that the oxidized species remained in their oxidized form for a longer time by shifting the smallest potential sweep range toward the anodic direction, thus favoring the chemical reaction β-NiOOH → γ-NiOOH (reaction D in Scheme 1). On the other hand, reaction C could be favored by shifting the smallest potential sweep range toward the cathodic direction, where reduced species remained longer times. Thus, treatment A allowed us to obtain an electrode surface with almost even quantities of β-NiOOH and γ-NiOOH, while treatment B allowed us to obtain preferentially β-NiOOH. These results were obtained either with Ni or Ni-Cr alloy electrodes (See **Figure 1**).

The fitting of the voltammetric traces was performed by using Origin 6.0 built-in Lorentzian function:

$$y = y_0 + \frac{2A}{\pi} \frac{w}{4(x - x_c)^2 + w^2} \qquad (5)$$

where y_0 is the offset, x_c is the function center (peak potential in our case), w the width of the wave, and A represents the area under the curve.

2.2. Preparation of Chemicals

Sodium hydroxide solutions were prepared with triply distilled degassed water and analytical grade reagents (Merck, Argentina). They were standardized with potassium hydrogen phthalate (Merck, Argentina). The solutions were degassed by bubbling high purity N_2 prior to electrochemical experiments. Contact with atmospheric CO_2 was minimized by maintaining N_2 over the solutions during the electrochemical experiments.

3. Results and Discussion

In a previous study using Ni-Cr (80:20) alloy electrodes, we showed that the voltammetric behavior of the alloy in the potential range from –1.10 V to 0.00 V (vs SCE) in 0.10 M NaOH solution was similar to that of pure Ni electrodes in the same electrolyte [18]. The presence of chromium in the electrode was only noticeable on the i/E voltammetric response during the first few *cv* scans

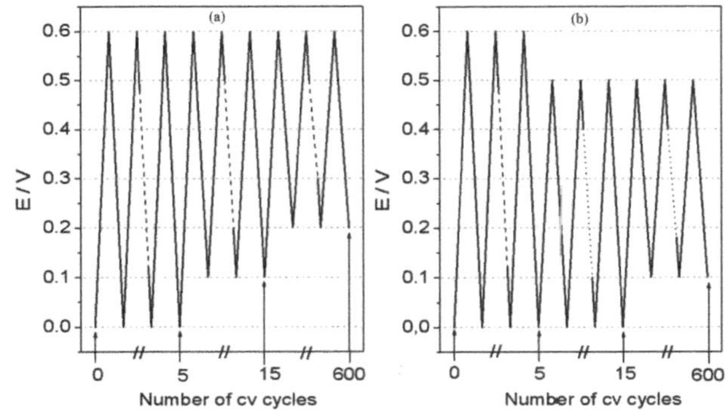

Figure 1. Potential-time programs used as working electrode perturbation waveforms for oxide growth at Ni and Ni-Cr alloy electrode in 0.10 M NaOH. (a) Waveforms used to grow γ-NiOOH and β-NiOOH; (b) Waveforms used to grow preferentially β-NiOOH.

starting at –1.0 V (vs ECS) [16]. Similar results were found by other authors in X-rays photoelectron spectroscopy (XPS) examinations of the electrochemically formed passive layers on Ni-Cr alloys [16]. They found that the XPS analysis of the passive layer formed at a potential range from –1.0 V to 0.0 V showed that Ni(II) and Cr(III) oxides and hydroxides were developed. An increase of the hydroxide and the oxide part of the XPS signal and a related decrease of the metal contribution was found when the passivation potential was higher than –0.5 V. We have found that no significant currents were observed on the voltammetric i/E traces after continuous cycling of the Ni-Cr electrode at the potential range from –1.0 V to 0.0 V in 0.1 M NaOH [18].

Ni(III) species are only formed at potentials above 0.0 V (vs SCE) [19]. The presence of NiOOH on the electrode surface is important for the electrocatalysis of organic molecules. Thus, the characterization of the electrochemical behavior of Ni and Ni-Cr alloy electrodes in alkaline solution at the potential range between 0.0 V and 0.6 V is of crucial importance for understanding the mechanism of electro-oxidation of organic molecules. In the following section, we show the results of a systematic study of the voltammetric behavior of Ni and Ni-Cr alloys in 0.10 M NaOH.

3.1. Voltammetric Response of Nickel and Nickel-Chromium Electrodes at the Potential Range from 0.00 V to 0.60 V

Figure 2 shows the i/E trace obtained with Ni and Ni-Cr alloy electrodes after different numbers of cv scans at a potential range between 0.00 V and 0.60 V in 0.10 M NaOH. An oxidative wave (peak I) was observed at the potential range between 0.40 V and 0.50 V at either Ni or Ni-Cr electrodes (**Figure 2**) during the positive going potential scan. This oxidative wave was ascribed [18,20] to the oxidation of $Ni(OH)_2$ to NiOOH, as represented in electrochemical steps A and B in Scheme 1. A reductive wave (peak II) was observed at a potential range from 0.35 V to 0.45 V during the negative going potential scan. This wave may be complementary to the anodic one (Equation (1)). We identified this reductive wave as due to the reduction of β-NiOOH [18].

Jabs *et al.* [16] reported that the formation of Ni(III) species at electrode potentials above 0.4 V was noticeable on both the i/E voltammetric trace and the XPS data. A redox system was observed on the i/E trace as a well-defined pair of anodic and cathodic waves at a potential range from 0.3 V to 0.5 V. On the other hand, a shift of the binding energies of all oxidic and hydroxidic XPS lines was informed. This shift could be related to the doping of the passive layer with higher valent Ni(III) species. They suggested a two-layer model for the passive layers formed at Ni-Cr (80:20) alloy electrodes, an inner oxide layer in contact with the metal and an outer hydroxide layer in contact with the electrolyte. The XPS data showed that $Ni(OH)_2$ and $Cr(OH)_3$ are present in the outer hydroxide layer, Ni(III) and some soluble Cr(VI) are only observed in the transpassive potential range [16]. Only the redox couple Ni(II)/Ni(III) was observed on the i/E voltammetric trace at a potential range from 0.0 V to 0.6 V.

A review of our *cv* results with Ni and Ni-Cr electrodes at the potential range from 0.0 V to 0.6 V indicates that the complexity of both the oxidative and the reductive waves increased with the number of *cv* scans. As shown in **Figure 2**, panels A and B, the peak current of the oxidative wave increased with the number of *cv* scans. On the other hand, a new reductive wave developed (peak III, Ep_{III} at a potential range between 0.25 V and 0.35 V) on both Ni and Ni-Cr electrode surfaces during the negative going potential scan. This new wave was identified as due to the reduction of γ-NiOOH [18].

Different authors have reported on the multiplicity of anodic and cathodic voltammetric current peaks at Ni oxide electrodes [20,21] (and references therein). The peak current multiplicity was related to the presence of different $Ni(OH)_2$ species on the electrode surface. Arvía *et al.* [20,21] developed experimental methodologies that enabled them to separate and study the processes occurring at $Ni(OH)_2$ surfaces under potentiodynamic conditions in alkaline solutions. By applying complex perturbation conditions to the working electrode, or by working at different temperatures, Arvía *et al.* were able to separate the anodic voltammetric current peak into two current peaks, indicating the presence of at least two different species on the electrode surface. They found that the electro-oxidation of the more unstable species was characterized by the first anodic current peak which was located at approximately 0.61 V and that the one corresponding to the most stable species was located at approximately 0.65 V (referred to the standard hydrogen electrode). Therefore, they suggested that the energy difference for electro-oxidation was in the range of 0.5 - 1.0 $kcal \cdot mol^{-1}$, a value too small for structural rearrangements involving bond breaks of any type. The splitting of the cathodic current peak occurred similarly to that of the anodic current peak.

In our work, a further analysis of the results obtained at Ni (**Figure 2**, panel (a)) and Ni-Cr (**Figure 2**, panel B) showed that the potential of the oxidative current peak decreased during the first few *cv* scans, then increased during approximately 150 *cv* scans, and finally became constant. At nickel electrodes, the oxidative current peak potential was nearly 0.47 V after 5 *cv* cycles, while it

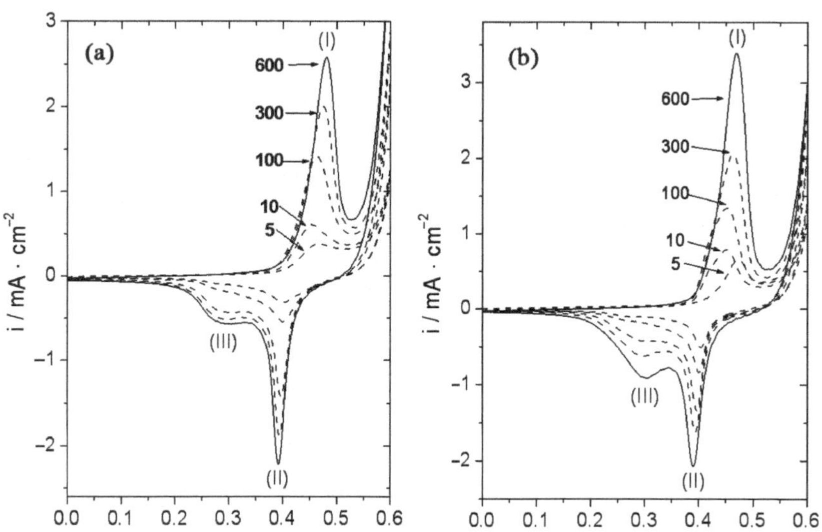

Figure 2. Evolution of the cyclic voltammetric response with the number of *cv* cycles at: a) Ni and b) Ni-Cr electrodes in 0.10 M NaOH. v = 0.050 V/s.

became 0.45 V after 10 *cv* cycles. Finally, it slowly increased and became constant at approximately 0.48 V after 600 *cv* cycles. At nickel-chromium alloys, the oxidative current peak potential observed after 5 *cv* cycles was nearly 0.47 V. It decreased to approximately 0.46 V after 10 *cv* cycles, and then slowly increased until it became constant at nearly 0.47 V after 600 *cv* cycles. Moreover, an analysis of the first derivative of the current with respect to potential ($\partial i/\partial E$, results not shown) clearly showed that a shoulder developed at potentials slightly smaller than Ep_I (*i.e.* 0.42 V) for the Ni-Cr electrode during the first 100 *cv* scans (see for example trace 10 in panel B, **Figure 2**), and then merged with peak I. This shoulder may be taken as indicative of an electrochemical process taking place at potentials very close to the main peak potential. On the contrary, this shoulder was not observed (or it was hardly distinguishable) at the Ni electrode. Finally, both anodic and cathodic peak charges and peak currents observed at the Ni-Cr electrode were higher (approximately 35% after 600 *cv* scans) than those observed with the Ni electrode.

The results obtained at Ni-Cr electrodes showed that at least two crystallographic forms of NiOOH were produced during the voltammetric cycles. Moreover, since the overall anodic and cathodic charges remained practically constant after the stabilization period, the relative proportion of oxidized Ni species (in principle β-NiOOH and γ-NiOOH) could be said to depend on the electrode perturbation waveform employed. Then, assuming that the charge under peak II and peak III represented each a single electrochemical reaction and that the dissolution of species was negligible, it could be suggested that the charge under each cathodic wave was indicative of the surface concentrations of species. Thus, the charges obtained under waves II and III were representative of the amount of β-NiOOH and γ-NiOOH produced during the positive going potential scan, respectively.

Considering that the cyclic voltammetric response of surface confined electron transfer processes is symmetrical [22], we tried various probability distribution equations to fit the experimental data. The best results were obtained with the Lorentz distribution (or Cauchy-Lorentz distribution). To obtain the electrochemical charge of the cathodic processes, each cathodic peak was fitted with a Lorentzian function with the following restraints: 1) the total electrochemical charge of the cathodic processes must be equal to that of the anodic process; 2) the addition of the individual electrochemical charges in the cathodic trace must be equal to that of the total cathodic process; and 3) the peak potentials obtained from the Lorentzian fitting process must be equal to those obtained experimentally. For the anodic trace, only the fitting obtained after 600 voltammetric *cv* scans is shown in **Figure 3** (panel (a), dotted line) as only one peak was observed during the positive going potential scan. In regards to the reductive waves, the cathodic i-E trace was deconvoluted into two Lorentzian components, and the area (charge) under each peak was then calculated. Thus, the individual contribution of each crystallographic form (β-NiOOH and γ-NiOOH) to the total cathodic current could be estimated at different numbers of *cv* scans. As an example, **Figure 3**, panel B, shows the reconstruction of the experimental cathodic trace observed after 400 *cv* scans. Dotted lines represent the individual reduction of β-NiOOH to β-$Ni(OH)_2$ and of γ-NiOOH to α-$Ni(OH)_2$ at approximately 0.37 V and 0.30 V, respectively. The dashed

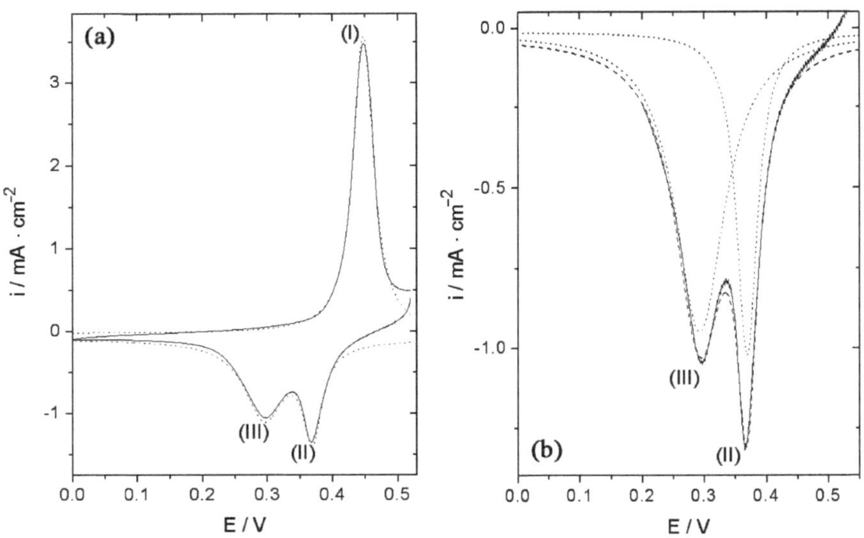

Figure 3. (a) Lorentzian fits of the anodic and cathodic traces obtained after 600 *cv* cycles. v = 0.050 V/s, 0.10 M NaOH. Experimental current (——), sum of individual Lorentzian fits (- - -); (b) Lorentzian fit of the cathodic trace obtained after 400 *cv* scans. Individual Lorentzian fits (····), sum of individual Loentzian fits (- - -), experimental cathodic current (——).

line represents the sum of the individual contributions. The experimental cathodic current is plotted as a solid line. A good match between the experimental cathodic current and the sum of the individual contributions was obtained. Good matches were obtained in all cases (different numbers of *cv* scans). Thus, the amount of each oxide-hydroxide crystallographic form could be estimated by assuming that one electron was transferred in each reductive step. This procedure was applied at different numbers of *cv* scans and allowed us to evaluate the differences in growth of the oxide-hydroxide species obtained by using each of the electrode potential sweep program developed.

3.2. Oxides-Hydroxides Formation at Ni Electrodes

Approximately equal amounts of β-NiOOH and γ-NiOOH were obtained at Ni electrodes (peaks II and III, **Figure 4**, panel A) by using treatment A. The first five cycles were performed by sweeping the electrode potential between 0.00 V and 0.60 V. The fifth cycle is shown in **Figure 4**, panel A (trace 5). Then, the cathodic inverting potential was changed to 0.10 V during the following 10 cycles. The fifteenth cycle is shown in **Figure 4**, panel A (trace 15). Finally, the cathodic inverting potential was changed to 0.20 V for the remaining part of the experiment (600 *cv* cycles). Traces obtained at 50, 100, 300, 400, and 600 *cv* cycles are also shown in **Figure 4**, panel A. The potential for peak current I shifted to slightly higher values with the number of *cv* cycles, and the charge under this oxidative peak grew until it remained almost constant at approximately 2.04 mC/cm^2 (**Figure 4**, panel B). On the other hand, the electrochemical response observed during the cathodic scan also evolved with the number of *cv* cycles. The charge under the reductive wave II increased until it remained almost constant. Besides, a new peak developed (peak III) at approximately 0.30 V. The charge under this wave also increased until it remained constant after 600 *cv* cycles. Thus, the total sum of charges under the cathodic waves remained constant after the 600 *cv* cycles at ca. 2.12 mC/cm^2, which was very close to the anodic charge, so the ratio QpI/(QpII + QpIII) = 0.96. This little difference may be attributed to the proximity of the waves to the oxygen evolution potential. Nevertheless, the closeness of this ratio to unity could be said to indicate that the electrochemical reactions were reversible, in the sense that all the species oxidized during the positive going potential scan were reduced during the reverse scan. It could also indicate that the dissolution of oxide species from the electrode surface was negligible. Moreover, approximately the same anodic and cathodic charges were obtained in the voltammetric experiments after the electrode was left in 0.1 M NaOH during a week, pointing to the stability of the film.

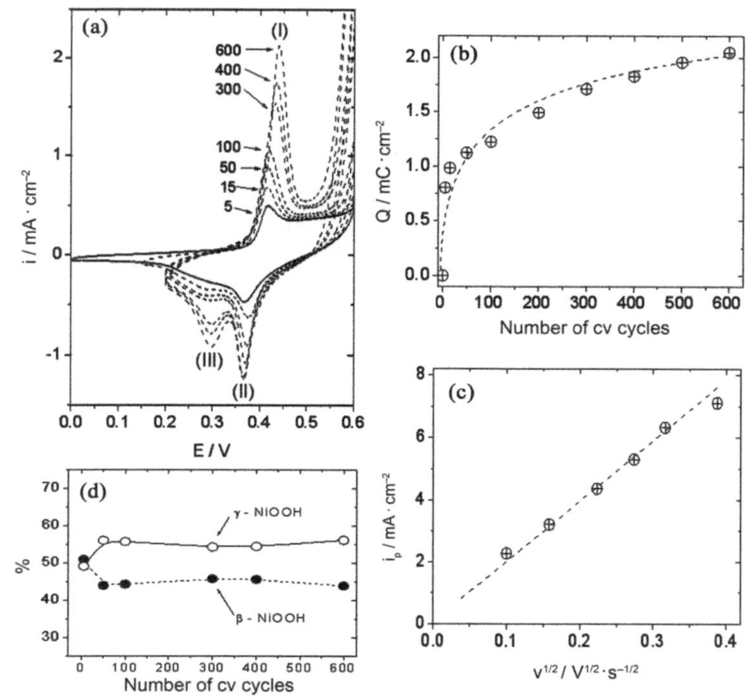

Figure 4. (a) Growth of γ-NiOOH and β-NiOOH at Ni electrodes in 0.10 M NaOH; (b) Evolution of peak I charge with the number of *cv* scans; (c) Electrode response (ip$_I$) as a function of scan rate. Oxide-hydroxide layer obtained after 600 *cv* scans; (d) Relative amounts of Ni oxide-hydroxide species obtained at different numbers of *cv* cycles.

We considered that after 600 *cv* cycles the species at the electrode surface had attained a stable concentration, and we studied the electrode response at different scan rates at a scan rate range between 0.01 and 0.16 V/s (**Figure 4**, panel (c)). The current at peak I increased linearly with the square root of the potential scan rate, which could be indicating that the global electrochemical process is diffusion controlled. It has been suggested in the literature that diffusion of species inside the oxide film is the dominating process at these scan rates [23]. Gosanlves *et al.* suggested that the redox processes in aqueous alkali metal hydroxides may involve incorporation/expulsion of H_3O^+ and/or OH^- and/or hydrated cations, as well as solvent transfer [24].

Figure 4, panel D, shows the variation of the relative amounts of β-NiOOH and γ-NiOOH calculated at different stages of the oxides-hydroxides layer growth. The calculations were performed from the Lorentzian fits of the cathodic traces, as explained before. The points (black dots and open circles) indicate the amounts of β-NiOOH and γ-NiOOH, respectively, calculated after 5, 15, 50, 100, 300, 400 and 600 *cv* cycles. It is clear from **Figure 4**, panel D, that γ-NiOOH grew preferentially during the first few *cv* cycles. After that, the relative amounts of γ-NiOOH and β-NiOOH remained almost constant with the number of voltammetric *cv* cycles, although the oxide-hydroxide layer grew, as can be deduced from the increase of the oxidative electrochemical charge with the number of *cv* cycles.

A clear growth of β-NiOOH was observed (peak II, **Figure 5**, panel (a)) when the working electrode was perturbed with treatment B. The first five *cv* cycles were performed by sweeping the electrode potential between 0.00 V and 0.60 V (trace 5 in **Figure 5**, panel (a)). Then, other ten *cv* cycles were performed at a potential range between 0.00 V and 0.50 V (**Figure 5**, trace 15). Finally, the working electrode potential was scanned between approximately 0.10 V and 0.50 V during the remaining part of the experiment. Traces obtained at 50, 100, 300, 400 and 600 *cv* cycles are also shown in **Figure 5**, panel

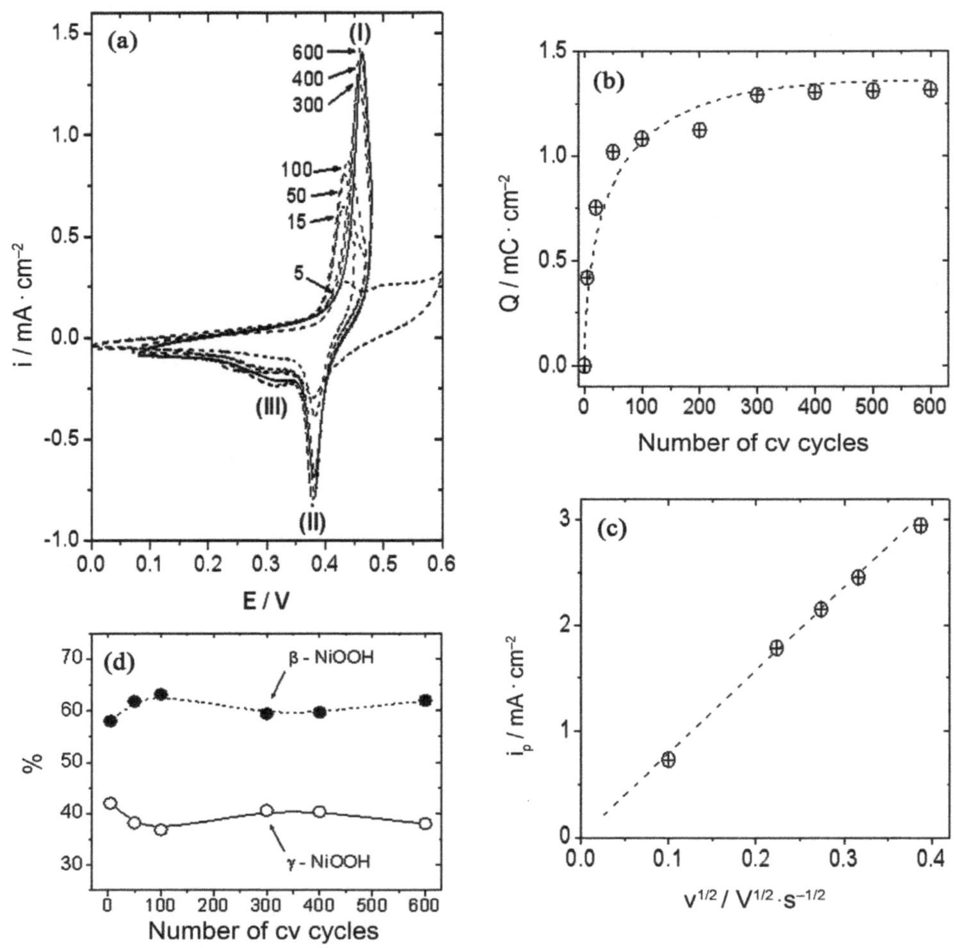

Figure 5. (a) β-NiOOH growth at Ni electrodes in 0.10 M NaOH; (b) Evolution of peak I charge with the number of *cv* cycles. (c) Electrode response (ipI) with scan rate. Oxide-hydroxide layer obtained after 600 *cv* scans; (d) Relative amounts of Ni oxide-hydroxide species obtained at different numbers of *cv* cycles.

A. The charge measured under the oxidative wave after 600 *cv* cycles was ca. 1.31 mC/cm^2. On the other hand, the electrochemical response observed during the negative going potential scan also changed with the number of *cv* cycles. The growth of the wave ascribed to the reduction of β-NiOOH was the main process observed on these scans. The cathodic trace also attained a stable electrochemical response after approximately 600 *cv* cycles, and the charge measured under the reductive wave II was **ca.** 1.37 mC/cm^2. Again, the ratio of anodic to cathodic charges was close to one (0.96), which may be interpreted as indicating the reversibility of the electrochemical system.

After approximately 300 *cv* cycles, the charge measured under the oxidative wave I remained almost constant, as observed in **Figure 5**, panel B. Then, it seemed reasonable to think that the species on the electrode surface had attained a stable concentration, and we studied the electrode response with the scan rate at a scan rate range between 0.01 and 0.16 V/s. Again, we observed a linear relationship between ip_I and the square root of the scan rate, which may be attributed to the diffusion of species inside the oxide film covering the electrode surface.

Although the formation of a little amount of γ-NiOOH may be deduced from the reductive wave observed at a potential range from 0.30 V to 0.35 V, it is clear that the development of β-NiOOH was the main process taking place by using treatment B. A preferential development of β-NiOOH against γ-NiOOH can be deduced from the analysis of the electrochemical charges under voltammetric peaks II and III, represented in **Figure 5**, panel D, as the relative amounts of oxide-hydroxide species. This development is evident during the first hundred *cv* cycles. After that, the relative amounts of γ-NiOOH and β-NiOOH remained almost constant until the end of the stabilization period (600 voltammetric *cv* cycles). The final relative amounts of β-NiOOH and γ-NiOOH obtained after the 600 cv cycles stabilization period by applying treatments A and B at Ni electrodes are listed in **Table 1**.

3.3. Oxides-Hydroxides Formation at Ni-Cr Electrodes

Figure 6, panel (a), shows the oxide-hydroxide development obtained with Ni-Cr (80:20) alloy electrodes by

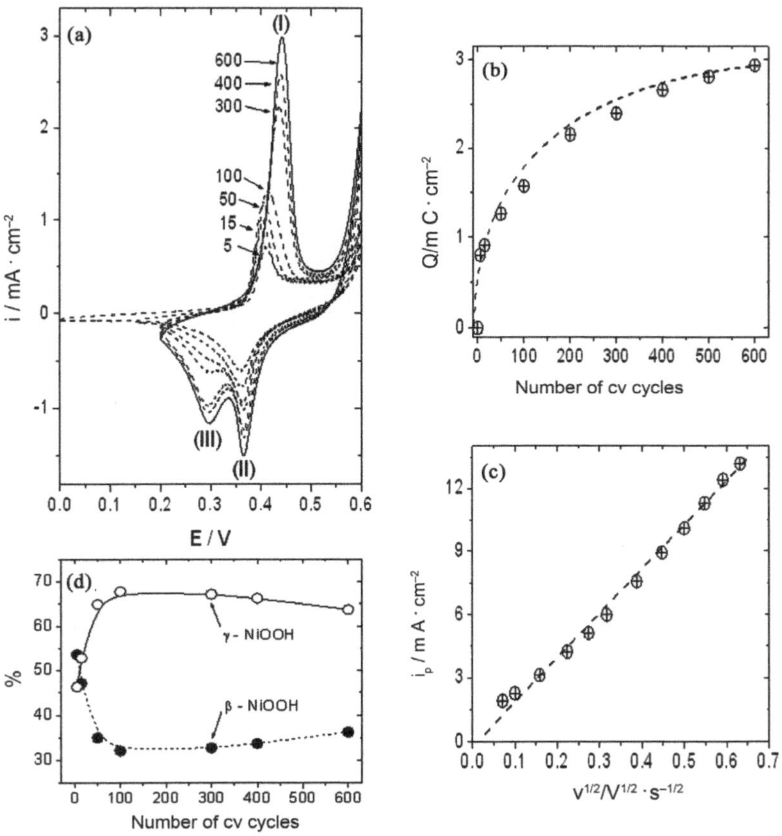

Figure 6. (a) Growth of γ-NiOOH and β-NiOOH at the Ni-Cr alloy electrode in 0.10 M NaOH; (b) Evolution of peak I charge with the number of *cv* scans; (c) Electrode response (ip_I) as a function of scan rate. Oxide-hydroxide layer obtained after 600 *cv* scans; (d) Relative amounts of Ni oxide-hydroxide species obtained at different numbers of *cv* cycles.

Table 1. Relative amounts[a] of oxide-hydroxides obtained with different working electrode perturbation programs.

	Ni		NiCr	
	1A[b]	1B[b]	1A[b]	1B[b]
β-NiOOH	44	62	37	72
γ-NiOOH	56	38	63	28

[a]Amounts (charge percent) of β-NiOOH and γ-NiOOH obtained at Ni and Ni-Cr electrodes using different working electrode perturbation programs. [b]Working electrode perturbation waveform (refer to **Figure 1**).

using treatment A. The formation of higher quantities of γ-NiOOH was observed at Ni-Cr alloy electrodes as compared to Ni electrodes (compare panels (a) in **Figures 4** and **6**). The electrochemical charge measured under peak I at Ni-Cr electrodes was higher than that related to the same process at pure Ni electrodes, *i.e.* approximately 2.95 mC/cm^2 vs 2.04 mC/cm^2, respectively (compare panels B in **Figures 4** and **6**). On the other hand, the total charge measured under the reductive waves during the negative going potential scan was also higher at Ni-Cr electrodes as compared to pure Ni electrodes, ca. 2.86 mC/cm^2 and 2.12 mC/cm^2, respectively. The ratio $Qp_I/(Qp_{II} + Qp_{III})$ observed at Ni-Cr alloy electrodes was also close to one, *i.e.* 1.03.

Finally, the peak current of the oxidative wave (ip_I), obtained after the stabilization period (approximately 600 cv cycles), had a linear relationship with the square root of the scan rate at a scan rate range between 0.01 and 0.16 V/s, pointing to a diffusion dependence of the process.

Figure 6, panel (d), shows the change in the relative amounts of γ-NiOOH and β-NiOOH oxides-hydroxides, as calculated from the Lorentzian fits of the cathodic traces, at different stages of the stabilization period. The relative amount of γ-NiOOH clearly increased during the first 100 voltammetric *cv* cycles, as compared to the relative amount of β-NiOOH. The relative amounts of γ-NiOOH and β-NiOOH remained almost constant after that number of *cv* scans, although the oxide-hydroxide layer thickened, as deduced from the increase in the anodic charges (**Figure 6**, panel (b)).

The growth of mainly β-NiOOH was observed when treatment B, was used (**Figure 7**, panel (a)). The potential

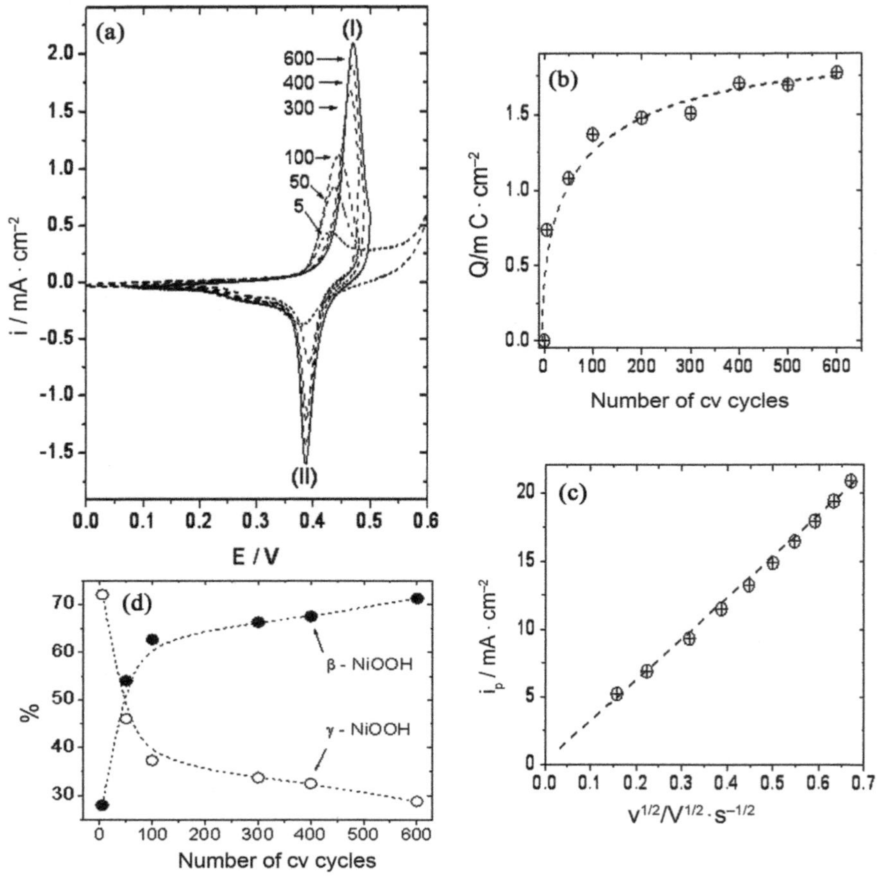

Figure 7. (a) β-NiOOH growth at Ni-Cr electrodes in 0.10 M NaOH; (b) Evolution of peak I charge with the number of *cv* scans; (c) Elecrode response (ip_I) with scan rate. Oxide-hydroxide layer obtained after 600 *cv* scans; (d) Relative amounts of Ni oxide-hydroxide species obtained at different numbers of *cv* cycles.

for peak current shifted to more positive values during the first few scans, and then remained constant up to the end of the experiment (approximately 600 *cv* cycles). The peak current and electrochemical charge under the oxidative wave observed during the positive going potential scan increased with the number of *cv* cycles up to approximately 400 *cv* cycles, and then remained almost constant until the end of the experiment (600 *cv* cycles). The electrochemical charge measured under this wave was higher than that obtained with pure Ni electrodes under the same perturbation program, ca. 1.74 mC/cm^2 vs 1.31 mC/cm^2. On the other hand, the reductive wave attributed to the reduction of β-NiOOH was the main process which could be observed during the negative going potential scan (**Figure 7**, panel (a)). The charge under this reductive wave was ca. 1.69 mC/cm^2, higher than that obtained when a pure Ni electrode was used (ca. 1.37 mC/cm^2). Again, the ratio Qp_I/Qp_{II} obtained at Ni-Cr alloy electrodes was close to one, *i.e.* 1.03.

The dependence of ip_I with the scan rate was studied at a scan rate range between 0.01 and 0.16 V/s once the proportion of Ni oxides-hydroxides on the electrode surface was stable, *i.e.* after approximately 600 *cv* cycles. A linear relationship was observed with the square root of the scan rate, indicating that the diffusion of species could be the main process observed at these scan rates (**Figure 7**, panel (c)).

Table 1 summarizes the results of the development of the different Ni oxide-hydroxide crystallographic species. Treatment A developed γ-NiOOH species slightly higher than β-NiOOH species at both Ni and Ni-Cr alloy electrodes. However, this difference was more evident at Ni-Cr alloy electrodes as compared to Ni electrodes. On the other hand, the preferential development of β-NiOOH species was the main process observed when treatment B was used in the *cv* experiments. Higher relative amounts of β-NiOOH species were obtained at Ni-Cr alloy electrodes as compared to Ni electrodes. These results indicate that the presence of Cr in the electrode material also has an influence on the proportion of Ni oxide-hydroxide species developed by the different working conditions proposed in this work. However, a stable proportion of Ni oxide-hydroxide species can be obtained when either treatment A or treatment B is applied in *cv* experiments using Ni or Ni-Cr alloy electrodes. This stable proportion of species will allow us to begin surface studies to further understand the electrocatalytic processes taking place when organic substances are oxidized by Ni or Ni-Cr alloy electrodes in alkaline solutions.

4. Conclusions

Special potential scan programs were developed for cyclic voltammetric growth of different nickel oxide-hydroxide species on Ni based electrodes. The proportion of oxide-hydroxide species on Ni, as well as on Ni-Cr alloy electrodes, can be easily controlled. β-NiOOH can preferentially be developed by allowing the Ni(II) hydroxides to remain longer times on the electrode surface, facilitating the chemical transformation of α-$Ni(OH)_2$ to β-$Ni(OH)_2$. On the contrary, a mixture of γ-NiOOH and β-NiOOH was obtained when Ni(III) oxides-hydroxides were allowed to stay longer times on the electrode surface. Deconvolution of the cathodic trace into Lorentzian components showed that the preferential development of a given crystallographic form of Ni oxide-hydroxides occurred during the first 100 - 200 voltammetric *cv* cycles. Then, the relative amounts of the crystallographic forms remained constant, although the oxide-hydroxide layer thickened, as deduced from the increases in charge of the electrochemical reactions.

A stable oxides-hydroxides proportion on the Ni or Ni-Cr electrode surfaces could be attained after approximately 400 to 500 *cv* cycles, as deduced from the shape of the voltammograms or the constancy of the oxidative charge with the number of *cv* cycles. Moreover, approximately the same anodic and cathodic charges were obtained in the voltammetric experiments after the electrode was left in 0.1 M NaOH during a week, pointing to the stability of the film. The electrochemical reactions occurring at the potential range studied were reversible, meaning that almost all of the Ni(III) formed during the positive going potential scan were reduced to Ni(II) during the reverse scan.

Finally, a linear dependence of the peak current with the square root of the scan rate was observed. This behavior was considered to indicate that the global electrochemical process is diffusion controlled.

5. Acknowledgements

The authors acknowledge Universidad Nacional de Río Cuarto for financial support. DEP acknowledges Consejo Nacional de Investigaciones Científicas y Técnicas (CONICET) for his doctoral fellowship. JMM is a permanent member of Carrera del Investigador Científico de CONICET (CIC-CONICET).

REFERENCES

[1] M. Fleischmann, K. Korinek and D. Pletcher, "The Kinetics and Mechanism of the Oxidation of Amines and Alcohols at Oxide-Covered Nickel, Silver, Copper, and Cobalt Electrodes," *Journal of the Chemical Society, Perkin Transactions*, Vol. 2, No. 10, 1972, pp. 1396-1403.

[2] B. E. Conway, "Reactions of Hydrogen and Organic Substances with and at Anodic Oxide Films at Electrodes," In: S. Trasatti, Ed., *Electrodes of Conducting Metallic Oxides*, Elsevier, Amsterdam, 1981, pp. 433-517.

[3] S. Majdi, A. Jabbari, and H. Heli, "A Study of the Elec-

trocatalytic Oxidation of Aspirin on a Nickel Hydroxide-Modified Nickel Electrode," *Journal of Solid State Electrochemistry*, Vol. 11, No. 5, 2007, pp. 601-607. doi:10.1007/s10008-006-0205-0

[4] M. Jafarian, M. G. Mahjani, H. Heli, F. Gobal and M. Heydarpoor, "Electrocatalytic Oxidation of Methane at Nickel Hydroxide Modified Nickel Electrode in Alkaline Solution," *Electrochemistry Communications*, Vol. 5, No. 2, 2003, pp. 184-188. doi:10.1016/S1388-2481(03)00017-1

[5] S. M. A. Shibli, K. S. Beenakumari and N. D. Suma, "Nano Nickel Oxide/Nickel Incorporated Nickel Composite Coating for Sensing and Estimation of Acetylcholine," *Biosenensor and Bioelectronics*, Vol. 22, No. 5, 2006, pp. 633-638. doi:10.1016/j.bios.2006.01.020

[6] J. M. Marioli and T. Kuwana, "Electrochemical Detection of Carbohydrates at Nickel-Copper and Nickel-Chromium-Iron Alloy Electrodes," *Electroanalysis*, Vol. 5, No. 1, 1993, pp. 11-15. doi:10.1002/elan.1140050104

[7] J. M. Marioli, "Electrochemical Detection of Carbohydrates in HPLC," *Current Topics in Electrochemistry*, Vol. 8, 2001, pp. 43-55.

[8] M. Hajjizadeh, A. Jabbari, H. Heli, A. A. Moosavi-Movahedi, A. Shafiee and K. Karimian, "Electrocatalytic Oxidation and Determination of Deferasirox and Deferiprone on a Nickel Oxyhydroxide-Modified Electrode," *Analalytical Biochemistry*, Vol. 373, No. 2, 2008, pp. 337-348. doi:10.1016/j.ab.2007.10.030

[9] Q. Yi, J. Zhang, W. Huang and X. Liu, "Electrocatalytic Oxidation of Cyclohexanol on a Nickel Oxyhydroxide Modified Nickel Electrode in Alkaline Solutions," *Catalysis Communications*, Vol. 8, No. 7, 2007, pp. 1017-1022. doi:10.1016/j.catcom.2006.10.009

[10] B. S. Hui and C. O. Huber, "Amperometric Detection of Amines and Amino Acids in Flow Injection Systems with a Nickel Oxide Electrode," *Analytica Chimica Acta*, Vol. 134, 1982, pp. 211-218. doi:10.1016/S0003-2670(01)84191-X

[11] I. L. de Mattos, D. Melo and E. A. G. Zagatto, "Nickel-Chromium Electrode as a Detector in Flow-Inyection Amperometry: Determination of Glycerol," *Analytical Sciences*, Vol. 15, No. 6, 1999, 537-541. doi:10.2116/analsci.15.537

[12] N. Alonso-Vante, "Generalidades Sobre Electrocatálisis," In: N. Alonso-Vante, Ed., *Electroquímica y Electrocatálisis*, Buenos Aires, 2003, pp. 18-90.

[13] H. Bode, K. Dehmelt and J. Witte, "Zur Kenntnis der Nickelhydroxidelektrode—I. Über das Nickel (II)-Hydroxidhydrat," *Electrochim Acta*, Vol. 11, No. 8, 1966, pp. 1079-1087. doi:10.1016/0013-4686(66)80045-2

[14] H. H. Huang, "Surface Characterization of Passive Film on NiCr-Based Dental Casting Alloys," *Biomaterials*, Vol. 24, No. 9, 2003, pp. 1575-1582. doi:10.1016/S0142-9612(02)00544-6

[15] M. Bojinov, G. Fabricius, P. Kinnunen, T. Laitinen, K. Mäkëla, T. Saario and G. Sundholm, "The Mechanism of Transpassive Dissolution of Ni-Cr Alloys in Sulphate Solutions," *Electrochimica Acta*, Vol. 45, No. 17, 2000, 2791-2802. doi:10.1016/S0013-4686(00)00387-X

[16] T. Jabs, P. Borthen and H. H. Strehblow, "X-Ray Photoelectron Spectroscopic Examinatios of Electrochemically Formed Passive Layers on Ni-Cr Alloys," *Journal of the Electrochemical Society*, Vol. 144, No. 4, 1997, pp. 1231-1243. doi:10.1149/1.1837577

[17] J. M. Marioli and L. E. Sereno, "Electrochemical Detection of Underivatized Amino Acids with a Ni-Cr Alloy Electrode," *Journal of Liquid Chromatography & Related Technologies*, Vol. 19, No. 15, 1996, pp. 2505-2515. doi:10.1080/10826079608014033

[18] J. M. Marioli and L. E. Sereno, "The Potentiodynamic Behavior of Nickel-Chromium (80:20) Alloy Electrodes in 0.10 N Sodium Hydroxide," *Electrochimica Acta*, Vol. 40, No. 8, 1995, pp. 983-989. doi:10.1080/10826079608014033

[19] S.-G. Sun, "Studying Electrocatalytic Oxidation of Small Organic Molecules with *in-Situ* Infra Spectroscopy," In: J. Lipkowski and P. N. Ross, Eds., *Electrocatalysis*, *Series of Frontiers in Electrochemistry*, Wiley-VCH, New York, 1998, pp. 243-290.

[20] R. S. Schrebler Guzmán, J. R. Vilche and A. J. Arvía, "Non-Equilibrium Effects in the Nickel Hydroxide Electrode," *Journal of Applied Electrochemistry*, Vol. 9, No. 2, 1979, pp. 183-189. doi:10.1007/BF00616088

[21] H. Gómez Meier, J. R. Vilche and A. J. Arvía, "The Influence of Temperature on the Current Peak Multiplicity Related to the Nickel Hydroxide Electrode," *Journal of Applied Electrochemistry*, Vol. 10, No. 5, 1980, pp. 611-621. doi:10.1007/BF00615484

[22] A. J. Bard and L. R. Faulkner, "Electrochemical Methods: Fundamentals and Applications," Chapter 13, 2nd Edition, John Wiley & Sons, New York, 2001.

[23] H. M. French, M. J. Henderson, A. R. Hillman and E. Vieil, "Ion and Solvent Transfer Discrimination at a Nickel Hydroxide Film Exposed to LiOH by Combined Electrochemical Quartz Crystal Microbalance (EQCM) and Probe Beam Deflection (PBD) Techniques," *Journal of Electroanalytical Chemistry*, Vol. 500, No. 1-2, 2001, pp. 192-207. doi:10.1016/S0022-0728(00)00373-9

[24] M. Gonsalves and A. R. Hillman, "Effect of Time Scale on Redox-Driven Ion and Solvent Transfers at Nickel Hydroxide Films in Aqueous Lithium Hydroxide Solutions," *Journal of Electroanalytcal Chemistry*, Vol. 454, No. 1-2, 1998, pp. 183-202. doi:10.1016/S0022-0728(98)00262-9

5

Temperature Dependent Phase Behavior of Pseudo-Ternary Thiourea X-100 Surfactant +1-Hexanol/Oil/Water Systems

Mohd Zul Helmi Rozaini[1,2]
[1]School of Environmental Sciences, University of East Anglia, Norwich, Norfolk, UK
[2]Department of Chemical Sciences, University Malaysia Terengganu, Kuala Terengganu, Terengganu, Malaysia
Email: zulhelmi@umt.edu.my

ABSTRACT

Temperature dependent phase behavior of Pseudo-ternary Thiourea X-100 +1-hexanol (1:5 molar ratios)/oil/water systems is reported. The influence of nature of hydrocarbon oil and type of electrolytes (weak as well as strong) has been investigated on the temperature induced phase behavior of the ternary system. At surfactant concentration, Φs = 40%, a "nose shaped" microemulsion region is observed. Below one-phase microemulsion region, $L\alpha$ phase appears. The presence of NaCl decreases the domain size of 1Φ micellar region whereas oxalic acid first decreases the domain below $\Phi w < 18$ and then increases above $\Phi w > 18$ in the lower boundry of the phase diagram. The critical weight fraction of water, Φwcri decreases in presence of both electrolytes. However, Φwmax increases in presence of oxalic acid and remains constant in presence of NaCl as compared to salt free system. Furthermore, when cyclohexane was replaced by a longer straight chain hydrocarbon, dodecane, the domain of the one-phase microemulsion region is tremendously increased.

Keywords: Surfactant; Amphiphile; Co-Surfactant; Microemulsion; Micelles; Reverse Micelles; Phase Behavior; Phase Diagram; Thiourea X-100; 1-Hexanol

1. Introduction

Microemulsions (MEs) are optically transparent, thermodynamically stable, nano-structured mixture of oil and water stabilized by surfactant and co-surfactant [1]. Microemulsions have attracted great interest because of their unique physiochemical characteristics such as large stabilization capacity, ultra low interfacial tension, a very large interfacial region, and because of their potential industrial applications such as enhanced oil recovery, biotechnology, nanotechnology, novel drug delivery, agriculture, beverages and chemical reaction [2].

Phase behavior and structural organization of microemulsions are known to play key roles in its industrial and technological applications. Phase behavior studies provide information on the phase boundaries of different phases as a function of composition and temperature and more important structural organization of surfactant aggregates can also be interred [3]. In addition, it allows comparison of efficiency of different surfactant for a given application. The boundaries of one-phase micellar region can be easily accessed by visual observation of the sample of known composition. However, long equilibration is required in multiphase regions especially if a liquid crystalline phase is involved and this makes phase determination tedious, time consuming, and difficult [1,2, 4].

The phase behavior and structural organization of surfactant aggregates are highly dependent on the elastic properties of surfactant monolayer separating oil and water domain [5]. The natural curvature of the surfactant monolayer depends upon the geometric properties of the hydrophilic and hydrophobic moiety of the surfactant molecules [4]. These properties in turn, depend on the temperature, pressure, ionic strength of water, and nature of oil etc. For a given surfactant, a change in the natural curvature from water (positive) to oil (negative) or vise-versa can be achieved by changing one or more formulation variables. The most pronounced effect is temperature for nonionic surfactant and salts for ionic surfactant [5,6].

These phase behavior of a ternary system is often studied as a function of temperature and composition and the position of different phases are determined within the phase prism by erecting a vertical section of the prism at constant surfactant concentration. The phase boundaries give rise to characteristic "channel" shown in **Figure**

1(a), frequently reported by Rozaini and Brimblecombe [7]. For a ternary system water (A)-Oil(B)-surfactant(C), at constant pressure the ternary system is specified by setting three independent variables. The three most common variables ate temperature T, weight percentage, Φ or volume fraction, α of the oil, either of which is given by B/(A + B) in the mixture of oil and water, and the weight fraction of the surfactant in the mixture, γ = C/(A+B+C). Each point in the three-dimensional phase space of temperature and composition is then defined by giving variables T, Φ or α and γ. It is easiest to discuss phase behavior if it is expressed in an upright triangle A-B-C as the base (**Figure 1(a)**), but now temperature is the field variable. This is done by mixing a known amount of surfactant in oil and making surfactant concentration constant and water is added gradually to increase α or Φ value from 0% to 100%. Several samples having γ constant are made. For each sample, temperature is raised and the phase sequence 2-1-2 is observed with increase in temperature. Plotting the result yields the phase diagram shown in **Figure 1(a)** and also shown in **Figure 1(b)** in the vertical section of the phase prism at constant γ. The shape of the phase diagram obtained, on such section looks like a "channel" The one-phase "channel" is sandwiched between two two-phase regions. The size and shape of the "channel" and the temperature at which it is located, provides key information about the system. At lower temperature, the nonionic amphiphile is better soluble in water than in oil, and o/w microemulsion is formed which equilibrates with excess oil phase (signified by 2). At higher temperature, the nonionic surfactant is more oil soluble than water, and w/o microemulsion is formed in equilibrium with excess aqueous phase (signified by 2). This is due to the fact that the solubilization of the nonionic amphiphile in water is driven by the formation of the hydrogen bonds [4]. These hydrogen bonds break with rising temperature, causing a better oil solubility of the amphiphile. Alternatively, nonionic surfactant becomes hydrophobic with rise in temperature due to conformational changes in the polyoxyethylene chain of the nonionic surfactant. As a result, solubility of the surfactant increases in hydrocarbon oil with rise in temperature [4,5].

Furthermore, nonionic microemulsions are of increasing importance in several industrial applications. They are highly used in liquid detergent formulations [6]. Nonionic surfactants are more biodegradable than their ionic counterparts and more suitable for applications in cosmetic, foods, and personal care products [9]. It is known that the control of phase state is very important in several applications [9] and this prompted to investigate the temperature dependent phase behavior of microemulsions of a well known nonionic surfactant, Thiourea which has been used for various chemical reaction [10]. In this context, in this paper temperature dependent phase behavior of pseudo-ternary Thiourea +1-Hexanol/oil/water systems is reported over a wide range of temperature and composition, The influence of nature of hydrocarbon oil and type of electrolytes (weak as well strong) were investigated on the temperature induced phase behavior of the ternary system. Crossed Polaroid microscope was used to distinguish between one-phase isotropic micellar region and anisotropic liquid crystalline or turbid phase.

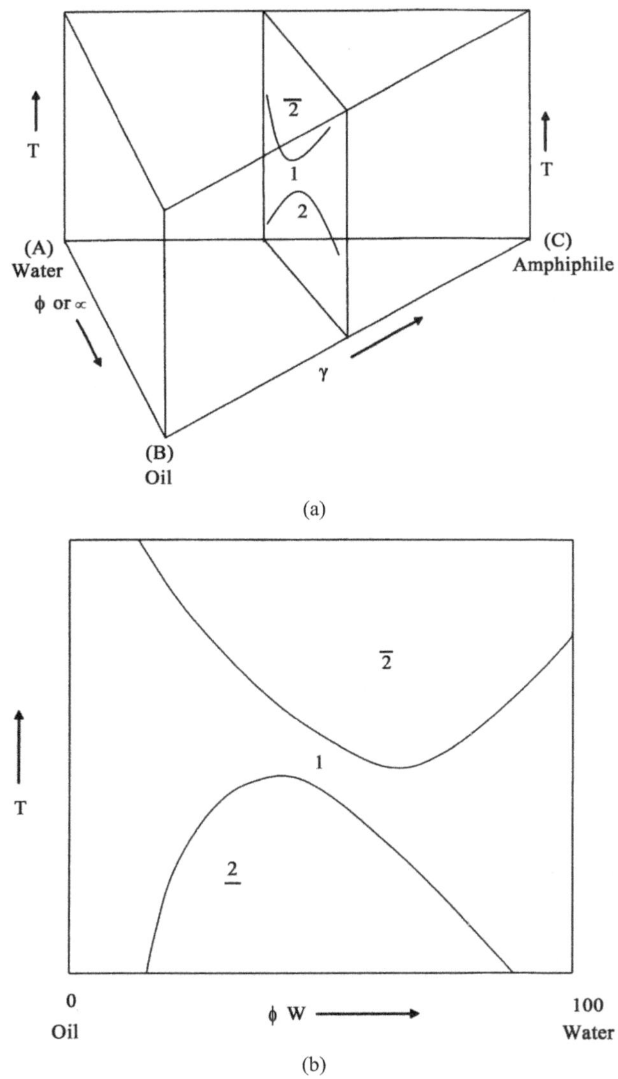

Figure 1. (a) Schematic "channel-shaped" phase prism of a ternary mixture of oil/water/nonionic surfactant in temperature-composition space at constant concentration, γ showing a single phase (1ϕ), two phase region; (b) Schematic "channel-shaped" phase diagram of a ternary mixture of oil/water/nonionic surfactant at vertical section of the phase prism at constant surfactant concentration, γ. The single phase (1ϕ) is surrounded by two two-phase region in which nonionic surfactant is dissolved either aqueous bottom phase (2) or organic top phase (2). For detail please see text.

2. Experimental

2.1. Materials

The nonionic surfactant, Thiourea X-100 (N-heksiloksifenil-N-(4-nitrobenzoil) thiourea) was obtained from Sigma Aldrich, Malaysia. Anhydrous dodecane and 1-hexanol were procured from Schuchardt, Germany. Sodium chloride (AR) was purchased from Fisher Scientific, Singapore. Oxalic acid (AR) was purchased from Sigma Aldrich Malaysia. Double distilled water was used throughout the experiment.

2.2. Methods

Constructing a Pseudo-Ternary Phase Diagram

Phase diagram of Thiourea X-100 +1-hexanol (1:5 molar ratio)/oil/water or aqueous electrolyte systems were constructed by titration method and the phase boundaries of one-phase micellar region of the ternary systems were determined at fixed surfactant concentration, $\gamma = 40\%$. A known amount of surfactant + co-surfactant and oil is taken and subsequently the mixture was titrated with water or aqueous electrolyte solution with the help of Hamilton microsyringe. The final mixture was gently stirred and then immersed in a temperature controlled water bath to equilibrate different phases at desired temperature. Characterization of the resulting isotropic one-phase micellar region or anisotropic phases were done visually and through crossed polaroid microscope after allowing sufficient time (typically 4 - 6 hours) for attainment of equilibrium at a particular temperature. The entire phase diagram was mapped in this manner. The detailed procedure for the determination of the phase diagram has been described elsewhere

3. Results and Discussions

One of the most common and convenient method 24 to study the phase behavior and related properties of a ternary system, surfactant/oil/water is to characterize the system at the vertical section of the prism at fixed amphiphile concentration. In this respect, phase behavior of pseudo-ternary Thiourea X-100 +1-hexanol (1:5 molar ratio)/cyclohexane/water system was determined as a function of temperature, T and weight fraction of water, Φw and shown in **Figure 2**. The weight fraction of surfactant, γ was kept constant and equal to 40%. It is evident from the figure that the domain of one-phase ME displays a "nosed-shaped" region. Similar "nose-shaped" micellar region were observed by Jeongt and co-workers [10], and Rozaini and Brimblecombe [11] for nonionic surfactant-oil-water ternary systems.

It is evident from the **Figure 2** that at $\Phi w < 5$, the temperature window for one-phase micellar region is from below 0°C to above 60°C. At $\Phi w > 5$, the upper boundary of micellar phase decreases with increase in water content. At $\Phi w \approx 0 - 15$, $L\alpha$ phase or two-phase region (2) does not exit even below 0°C. At $\Phi w > 15$, $L\alpha$ starts to appear in the lower portion of the phase diagram, and therefore, the weight fraction of water ($\Phi w \approx 0 - 15$) may be defined as the critical weight fraction of water, $\Phi wcri$ where phase transition occurs from 1Φ micellar region to $L\alpha$ phase. At $\Phi w \geq 15$, with rise in temperature, transition from $L\alpha$ to one-phase ME region takes place and the transition is very sharp (0.1°C) without formation of two-phase system. With further increase in temperature, one-phase ME region changes into two-phase region. Below one-phase micellar region, the domain of $L\alpha$ increases with increase in weight fraction of water and then finally changes into two-phase region (2). The exact boundary of the $L\alpha$ phase was not determined in this set of experiments. One-phase micellar region is present up to $\Phi w \approx 25$ and this may be defined as Φmax beyond which one-phase ME does not appear at any temperature and composition and only two-phase region was observed. The one-phase micellar region does not touch water axis. It means that $\Phi s \approx 40\%$, Thiourea-hexanol is unable to form normal micelle at any temperature in between 0°C - 60°C and also does not form oil swollen micelles (o/w microemulsions) near water axis. Otherwise one-phase micellar region must be extended up to water axis and one could expect a broad spectrum of micellar region in place of two-phase region beyond $\Phi w > 25$ in the ternary phase diagram. Further, at $\Phi w \approx 0 - 5$, the upper boundary of ME region is above 60°C. With increase in weight fraction of water (Φw), the upper boundry of temperature window of ME region decreases. As a result, a narrower temperature span of the ME region is observed and the domain of ME looks like "nosed-shaped".

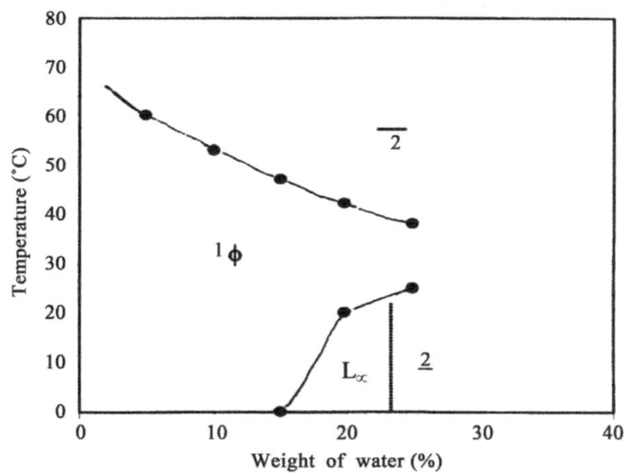

Figure 2. Temperature dependent Phase Diagram of Thiourea X-100 +1-hexanol/cyclohexane/water at surfactant concentration (γ = 40%) showing a single phase microemulsion region and L_α phase

Furthermore, it is known that the temperature dependent phase behavior of nonionic surfactant-oil-water system arises in the first place from the variation of local concentration of water in the ethylene glycol layer, which modulates the surfactant film curvature. 45 But it also originates from the solubility variation of the surfactant monolayer in the organic phase occurring as a function of temperature [12,13]. The phase diagram for the system, Thiourea X-100 +1-hexanol/cyclohexane/water plotted in **Figure 2** corresponds to the oil rich region of the prism diagram. The shape and size of the ME domain in **Figure 2** gives the key information about the ternary system. At Φw < 5, in the lower boundary, the one-phase micellar region is hydrated reversed micelles. 31 At Φw ≥ 15, Thiourea +1-hexanol forms lamellar structure resulting into Lα phase. With increase in temperature *Lα* phase turns into one-phase ME. With further increase in temperature, one-phase micellar system changes into two-phase region (2) in which water swollen reverse micelles equilibrates with excess aqueous phase.

Okochi and Brimblecombe (2009) had studied temperature dependent phase behavior of C12E4/dodecane/water and observed similar "nose-shaped" micellar region. However, critical weight fraction of water, Φwcri and maximum weight fraction of water, Φ wmax was 10 and 56 respectively [14,15]. *Lα* phase was also observed below one-phase microemulsion region in the lower boundary of the phase diagram. In the present studies, Φwcri is equal to ≈15, and Φwmax is equal to ≈25. By comparing this results, it is evident that the temperature dependent phase behavior of nonionic surfactant/oil/water is highly dependent on the amphiphilicity (hydrocarbon chain length) of the surfactant. If the amphiphilicity is decreased, one-phase ME region shrinks. That is what observed in the present investigations.

3.1. Influence of Electrolytes

The influence of 0.1 M NaCl and 0.3 M oxalic acid on the temperature induced phase behavior of the system, Thiourea X-100 +1-hexanol/cyclohexane/water was investigated and shown in **Figure 3**. In presence of both electrolytes, domain of 1Φ micellar phase display "nose-shaped" regions similar to the salt free system. However, the presence of electrolytes changes the phase boundaries of the one-phase micellar region. It is evident from the **Figure 3** that the critical weight fraction of water, Φwcri decreases in presence of both electrolytes, strong as well as weak. The critical weight fraction of water, Φwcri is equal to 10 in presence of both electrolytes whereas it is 15 for salt free system. However, Φwmax, increases in presence of oxalic acid and there is no change in Φwmax in presence of NaCl as compared to salt free system. At Φw ≈ 5, the temperature window for one-phase micellar region is from below 0°C - 54°C in presence of oxalic acid, from below 0°C - 50°C in presence of NaCl and from below 0°C - 60°C for salt free system.

That is, the presence of electrolytes contracts the temperature window of one-phase micellar region but the contraction is more in presence of strong electrolyte (NaCl) than the weak one (oxalic acid). In the lower boundary, at Φw > 10, the presence of NaCl and oxalic acid have opposite effect on the domain size of the micellar region. In the lower boundary, strong electrolyte, NaCl always decreases domain of the 1Φ microemulsion region whereas oxalic acid first decreases the domain of 1Φ micellar region below Φw < 18 and then increases the same above Φw > 18. In the upper boundary, the presence of oxalic acid expands temperature window of 1Φ micellar region above Φw >14 and the presence of NaCl increases the same at Φw > 19. Thus the presence of weak electrolyte, oxalic acid expands 1Φ micellar region more as compared to NaCl in the upper region of the phase diagram. As a result, a bigger domain 1Φ ME region is observed at higher water content.

The changes in the phase boundaries of one-phase micellar region (**Figure 3**) arises due to the presence of electrolytes in the aqueous domain of the ternary system. It is known that the phase behavior of a ternary system is highly influenced by the presence of electrolytes [16-18]. The presence of electrolytes reduces the solubility of the hydrophilic head group of the ionic surfactant amphiphile in water. In case of nonionic amphiphiles they compete for the hydrating water of the head groups. The presence of electrolytes (NaCl or oxalic acid) may decrease electrostatic interactions between the polar head groups of

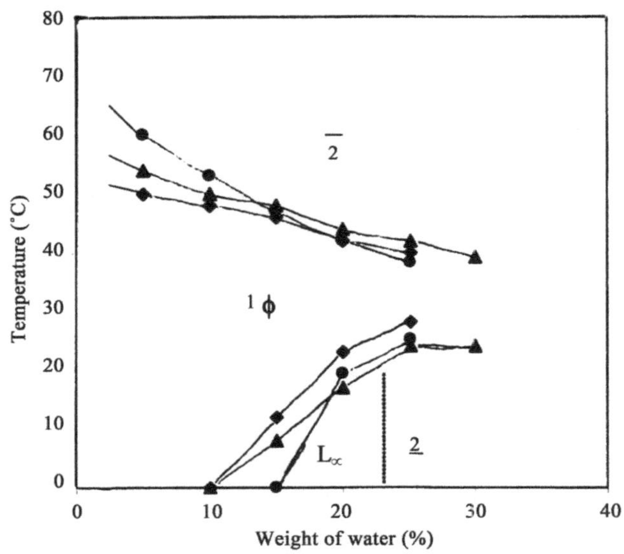

Figure 3. Temperature dependent Phase Diagram of Thiourea X-100 +1-hexanol/cyclohexane/ water at surfactant concentration (γ = 40%) in presence of 0.1 M NaCl (black diamond), 0.3 M oxalic acid (black triangle) and electrolyte free (black circle).

the surfactant and thereby coming closure to each other. This would lead more rigid interface, smaller inter-micellar interaction, decrease in size of the core of reverse micelles (RMs) and thus less amount of water is solubilized inside the core RMs in presence of electrolytes and this results in apparent decrease in Φwcrit from 15 (salt free syatem) to 10 (in presence of electrolytes). Further, it can be seen from the **Figure 3** that at Φw ≈ 5, 1Φ micellar region has upper critical temperature ~50°C in presence of 0.1 M NaCl, ~54°C in presence of 0.3 M oxalic acid and ~60°C for salt free system. That is, the presence ofelectrolytes decreases the upper critical transition temperature, Tuppercri and the decrease in Tuppercri is much more in presence of NaCl than oxalic acid. It is known 30 that on the oil-rich side, w/o MEs are present at low water content near upper critical temperature (near upper boundary of the phase diagram), which coagulate with decreasing temperature.

The presence of electrolyte decreases the size of RMs of Thiourea at any temperature 49 due to ion-dipole interaction or complexation between cations and polyoxyethylene chain of TX-100 that results in a closure packing of polar head groups of TX-100 and decreases the extent of solubilization of aqueous electrolyte solution inside the core of RMs. As a result, one-phase micellar region changes into two-phase region in presence of electrolytes relatively at a lower temperature as compared to salt free system. This causes an apparent decrease in upper critical temperature, Tuppercri in presence of electrolyte and the decrease in more in the presence of strong electrolyte (NaCl) than the weak one (oxalic acid). This may be due to a greater extent of ion-dipole interaction or complexation between the cations and polyoxyethylene chain of TX-100 in presence of the strong electrolyte (NaCl) as compared to the weak electrolyte (oxalic acid). As a result, the upper critical transition temperature, Tuppercri decreases much more in presence of strong electrolyte (NaCl) than the weak one (oxalic acid) and phase transition from one-phase micellar region to two-phase region occurs at relatively lower temperature (~50°C) in presence of NaCl than oxalic acid (~54°C). As a result, temperature window of one-phase micellar region becomes narrower in presence of both electrolytes as compared to salt free system.

Furthermore, it is known that the type of ME depends upon water-to-oil ratio and if Φw ≈ 25 - 70 wt%, a sponge like structure 31 is obtained which is known as bicontinuous ME in which surfactant monolayer separates into disordered, interpenetrating domains of oil and water. The size of the domains of water and oil in a bicontinuous ME is highly dependent on amphiphile concentration and water-to-oil ratio. 50 In the middle of phase diagram, at Φw > 18, the presence of electrolytes expands the temperature window of 1Φ micellar region in the upper region of the phase diagram. This may be due to the fact that the presence of electrolytes (weak or strong) makes the surfactant (TX-100) more hydrophobic. 45 That causes a better oil solubility of TX-100 in oil domain and therefore phase transition from 1Φ ME region to two-phase region takes place relatively at a higher temperature and higher water content as compared to salt free system. As a result, temperature window for one-phase micellar and Φmax increase in presence of both electrolytes. Further, at Φw > 18, the presence of oxalic acid expands the temperature window of 1Φ ME region more than NaCl and this may be due the fact that oxalic acid probably increases hydrophobicity of the TX-100 more than NaCl due to higher concentration of oxalic acid. This results an increase in the temperature window of one-phase micellar region in the middle of phase diagram.

3.2. Influence of Hydrocarbon Chain Length of Oil

The influence of the hydrocarbon chain length of oil on the temperature induced phase behavior of the ternary system. Thiourea X-100+1-hexanol (1:5 molar ratio)/ oil/0.1 M NaCl was investigated. Cyclohexane was replaced by a longer straight chain hydrocarbon, dodecane and domain of one-phase micellar region of Triton X-100+1-hexanol/dodecane/0.1 M NaCl was determined and shown in **Figure 4**. It is clear from the Figure that the general features of one-phase micellar regions are similar in presence of both oils, *i.e.* dodecane and cyclohexane. The critical weight fraction of water, Φwcri is 10 in presence of cyclohexane and 15 in presence of dodecane. Also, there is a tremendous increase in Φwmax

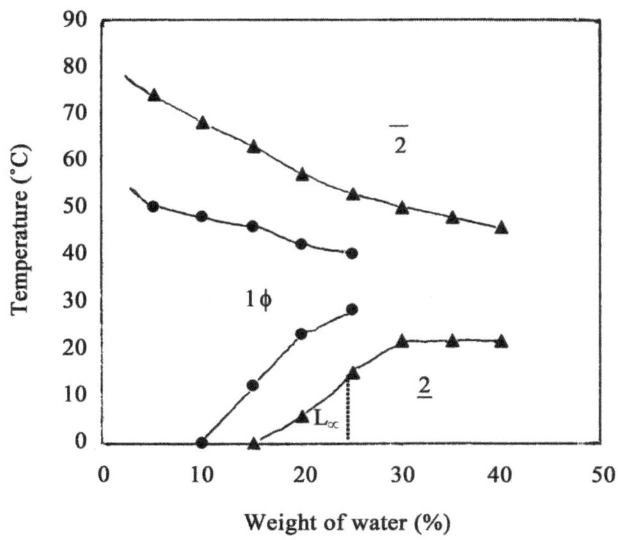

Figure 4. Temperature dependant phase diagram of Thiourea X surfactant+1-hexanol/oil/0.1 M NaCl at constant surfactant concentration (γ = 40%) in presence of cyclohexane (black circle) and dodecane (black triangle).

in presence of dodecane as compared to cyclohexane and is equal to 40% whereas it is 25% in presence of cyclohexane. At Φw = 5, the upper critical temperature, Tuppercri is increased and is equal to 70˚C in presence of dodecane. Thus the temperature window of one-phase micellar is very much larger in presence pf dodecane as compared to cyclohexane. In the lower boundary of the phase diagram, three is tremendous increase in domain of one-phase micellar in presence of dodecane. It is quite interesting to note that the one-phase micellar region does not touch the water axis in presence of dodecane also. It means TX-100 +1-hexanol is unable to form normal micelles or oil swollen ME near water axis in presence of oil, dodecane or cyclohexane.

The changes in phase boundary in one-phase micellar region occurs due to different nature of the oils used here in the presence studies. The presence of longer straight chain hydrocarbon, dodecane tremendously increases the domain of 1Φ micellar region as compared to cyclohexane. The critical weight fraction of water Φw cri and Φwmax are larger in presence of dodecane as compared to cyclohexane. Also Tuppercri has been inceased in presence of dodecane. As a result, the temperature window of one-phase micellar region increases. Thus the solubilization efficiency of the amphiphile, TX-100 is much more in presence of dodecane than cyclohexane and this may be attributed to higher hydrophobicity (longer hydrocarbon chain length) of dodecane. It is known that with rise in temperature, the nonionic amphiphile TX-100 becomes hydrophobic [18-20]. As a result, its solubility increases in more hydrophobic oil, dodecane than cyclohexane. Thus phase transition from one-phase micellar region to two-phase region takes place relatively at higher temperature. This causes an apparent increase in Tuppercri in presence of dodecane as compared to cyclohexane. As a result, the temperature window of one-phase micellar region becomes larger in presence of dodecane as compared to cyclohexane.

4. Conclusion

The temperature dependent phase behavior of Thiourea X-100+1-pentanol (1:5 molar ratio)/cyclohexane or dodecane/water or brine systems were investigated. At surfactant concentration, $\gamma s = 40\%$, a "nose-shaped" microemulsion region is observed. Below one-phase micellar region liquid crystalline, Lα appears. The one-phase micellar region does not touch water axis. At $\Phi w \approx 5$, the temperature window for one-phase micellar region is from below 0˚C - 50˚C in presence of NaCl, below 0˚C - 54˚C in presence of oxalic acid, and form below 0˚C - 60˚C for salt free system in presence of cyclohexane. That is, the presence of electrolytes contracts the domain of Φw micellar region but the contraction is more in presence of strong electrolyte (NaCl) than the weak one (oxalic acid). The presence of NaCl and oxalic acid has opposite effect on the one-phase micellar region. In the lower boundary of the phase diagram, presence of strong electrolyte, NaCl always decreases the domain of 1Φ micellar region whereas oxalic acid first decreases the domain $\Phi w < 18$ and then increases the temperature above $\Phi w > 18$. The critical weight fraction of water, Φwcri decreases in presence of both electrolytes. Further, when cyclohexane is replaced by dodecane, the temperature window of 1Φ ME region is tremendously increased.

5. Acknowledgements

The author thankfully acknowledges the financial support from Malaysian Ministry of Higher Education [MOHE] with Project No. FRGS59165.

REFERENCES

[1] R. Belcher and L. Gordon, "Kinetics of Precipitation," Pergamon Press, Oxford, 1964.

[2] A. Bonazza, P. Brimblecombe, C. M. Grossi and C. Sabbioni, "Carbon in Black Crusts from the Tower of London," *Environmental Science & Technology*, Vol. 41, No. 12, 2007, pp. 4199-4204. doi:10.1021/es062417w

[3] X. Liu, H. C. Monger and W. G. Whitford, "Calcium Carbonate in Termite Galleries-Biomineralization or Upward Transport," *Biogeochemistry*, Vol. 82, No. 3, 2007, pp. 241-250. doi:10.1007/s10533-006-9067-x

[4] L. Brecevic and D. Skrtic, "Transformation of Calcium Oxalate Hydrates," *Journal of Crystal Growth*, Vol. 74, 1986, pp. 399-408. doi:10.1016/0022-0248(86)90131-4

[5] H. H. M. Darweesh, "Building Materials from Siliceous Clay and Low Grade Dolomite Rocks," *Ceramics International*, Vol. 27, No. 1, 2001, pp. 45-50. doi:10.1016/S0272-8842(00)00040-7

[6] B. L. Davis and G. Jixiang, "Airborne Particulate Study in Five Cities of China," *Atmospheric Environment*, Vol. 34, No. 17, 2000, pp. 2703-2711. doi:10.1016/S1352-2310(99)00528-2

[7] M. Z. H. Rozaini and P. Brimblecombe, "The Solubility of Sodium Dicarboxylate in the Atmospheric Aerosols," *Journal of Chemical Thermodynamic*, Vol. 41, 2009, pp. 980-983. doi:10.1016/j.jct.2009.03.017

[8] B. Ford, I. Macleod and P. Haydock, "Rock Art Pigments from the Kimberly Region of Western Australia: Identification of the Minerals and Conversion Mechanisms," *Studies in Conservation*, Vol. 39, 1994, pp. 57-69. doi:10.2307/1506491

[9] J. H. Huang, Z. F. Mao and M. F. Luo, "Effect of Anionic Surfactant on Vaterite $CaCO_3$," *Materials Research Bulletin*, Vol. 42, No. 12, 2007, pp. 2184-2191. doi:10.1016/j.materresbull.2007.01.005

[10] T. S. Jeong, J. H. Kim, M. S. Han, K. Y. Lim and C. J. Youn, "X-Ray and Cathodoluminescence Study on the Effect of Intentional Long Time of Succinic Acid on Calcium Carbonate," *Journal of Crystal Growth*, Vol. 280,

No. 3-4, 2005, pp. 357-363. doi:10.1016/j.jcrysgro.2005.04.002

[11] M. Z. H. Rozaini and P. Brimblecombe, "The Odd-Even Behaviour of Dicarboxylic Acids Solubility in the Atmospheric Aerosols," *Water, Air & Soil Pollution (Springer)*, Vol. 198, 2009, pp. 65-75. doi:10.1007/s11270-008-9826-5

[12] S. Mann, J. M. Didymus, N. P. Sanderson and E. J. A. Samper, "Effects of Grazing and Topography on Dust Flux," *Journal of the Chemical Society, Faraday Transactions*, Vol. 86, 1990, pp. 1873-1880. doi:10.1039/ft9908601873

[13] M. G. S. Matteini, "The Protective Effect of Ammonium Oxalate Treatment on the Surface of Wall Paintings," Painted Facades Eurocare Project, Vienna, 1996, pp. 95-101.

[14] H. Okochi and P. Brimblecombe, "Potential Trace Metal-Organic Complexation in the Atmosphere," *Scientific World*, Vol. 2, 2002, pp. 767-786. doi:10.1100/tsw.2002.132

[15] C. Sabbioni, "Contribution of the Atmospheric Deposition to the Formation of Damage Layers," *Science of the Total Environment: The Deterioration of Monuments*, Vol. 167, No. 1-3, 1995, pp. 49-55. doi:10.1016/0048-9697(95)04568-L

[16] C. Sabbioni and G. Zappia, "Atmospheric-Derived Element Tracers on Damaged Stone," *The Science of the Total Environment*, Vol. 126, No. 1-2, 1992, pp. 35-48. doi:10.1016/0048-9697(92)90482-8

[17] S. Signorelli, C. Peroni, M. Camaiti and F. Fratini, "The Presence of Vaterite in Bonding Mortars of Marble Inlays from Florence Cathedral," *Mineralogical Magazine*, Vol. 60, No. 4, 1996, pp. 663-665. doi:10.1180/minmag.1996.060.401.13

[18] O. Söhnel and J. W. Mullin, "Long-Term and Thermally Stable Superhydrophobic Surfaces of Carbon Nanofibers," *Journal of Colloid Interface Science*, Vol. 123, 1988, pp. 43-51.

[19] H. Wei, Q. Shen, Y. Zhao, D.-J. Wang and D.-F. Xu, "Influence of Polyvinylpyrrolidone on the Precipitation of Calcium Carbonate and on the Transformation of Vaterite to Calcite," *Journal of Crystal Growth*, Vol. 250, No. 3-4, 2003, pp. 516-524. doi:10.1016/S0022-0248(02)02484-3

[20] A.-W. Xu, M. Antonietti, H. Colfen and Y.-P. Fang, "Uniform Hexagonal Plates of Vaterite $CaCO_3$ Mesocrystals Formed by Biomimetic Mineralization," *Advanced Functional Materials*, Vol. 16, No. 7, 2006, pp. 903-908. doi:10.1002/adfm.200500716

6

Stability Studies of Lysine Acetylsalicylate (Aspirin Derivative): Mechanisms of Hydrolysis

Oussama Kamal[1*], Abdelaziz Benlyamani[1], Farid Serdaoui[1], Mohammed Riri[1], Abdelmjid Cherif[1], Miloudi Hlaïbi[1,2]

[1]Laboratoire d'Interface Matériaux d'Electrochimie et Chimie de l'Environnement (LIME), Université Hassan II Faculté des Sciences Aïn Chock, Casablanca, Morocco

[2]Laboratoire des Polymères, Biopolymères, Surfaces, UMR 6270 du CNRS, Faculté des Sciences, Université de Rouen, Mont-Saint-Aignan, France

Email: [*]kamaloussama@hotmail.fr

ABSTRACT

To control the stability of the lysine acetylsalicylate compound (LAS) in aqueous solution, some studies of the hydronium ion-catalyzed, hydroxide ion-catalyzed, and spontaneous reactions of this active ingredient in water solutions have been carried out. The pH-rate profile (**log k_{obs} = f(pH)**), shows that the hydrolysis reaction of the LAS, is conducted by a catalysis acid-base mechanism, with multiple reaction pathways. The rate constants, k_H, k_{OH} and k_0 to the reaction pathways catalyzed by H_3O^+, HO^- ions and to the spontaneous reaction, for the hydrolysis reaction of the reagent LAS, were determined. The results show that the studied compound LAS is unstable in basic medium and the hydrolysis reaction catalyzed by HO^- ions is predominant.For a known acidity (**pH ≈ 10**), studies conducted for different temperatures of the medium, clearly indicate, that the experimental rate constant k_{obs} ,depends on the temperature according to the Arrhenius equation. The activation parameters: activation energy (E_a), enthalpy ($\Delta H^{\neq}$) and entropy ($\Delta S^{\neq}$), for the transition state were determined, The very negative value obtained for the activation parameter ΔS^*, first indicates that in the transition state there is gain in order, then this late state, resembles the products and that probably for the mechanism of the LAS hydrolysis reaction catalyzed by HO^- ions, the rate-determining step is a bimolecular reaction. Finally from all these results, the mechanism for the reaction pathway catalyzed by HO^- ions has been elucidated.

Keywords: Lysine Acetylsalicylate; Hydronium Ions Hydroxide Catalyzed; Catalysis Acid-Base Mechanism; Multiple Reaction Pathways; Activation Parameters

1. Introduction

The LAS compound is an active ingredient known to Aspegic. It is a molecule that is part of the family of salicylates including aspirin, which represents one of the most commonly, used drugs in the world. Aspegic is an acetyl derivative of salicylic acid that is a white, crystalline, weakly acidic substance. Acetylsalicylic acid decomposes rapidly in solutions of ammonium acetate or of the acetates, carbonates, citrates, phosphates or hydroxides of the alkali metals. Acetylsalicylic acid is stable in dry air, but gradually hydrolyses in contact with moisture to acetic and salicylic acids. In solution with alkalis, the hydrolysis proceeds rapidly and the clear solutions formed may consist entirely of acetate and salicylate. The principal cause of the hydrolysis of this family of the salicylic acid acyl esters (e.g. aspirin), in aqueous solution about pH 3, is the attack of these substances by water ions (H_3O^+ or HO^-) [1-6].

Since Edwards has shown that the rate of hydrolysis of aspirin is independent of pH between pH 4 and 8, the reaction has been the subject of a number of studies, particularly by Garrett [1,3-5]. Edwards suggested that in this pH range, the hydrolysis reaction resulting from the attack by a water molecule on the anion of aspirin (**Scheme 1**), but several authors have pointed out that this mechanism is not consistent with Edwards' own demonstration, showing that the hydrolysis reaction is not catalyzed by acetate ion, a considerably more powerful nucleophile than water.

They considered that the facts point rather to intramolecular nucleophilic catalysis by the ionized carboxyl group [7,8]. This mechanism has been set out in its most acceptable form by Bender (**Scheme 1**) [9]. This hypothesis was supported by the same authors [10], which showed that a small amount of labelled oxygen from solvent $H_2^{18}O$ appears in the salicylic acid produced. Only Garrett has

[*]Corresponding author.

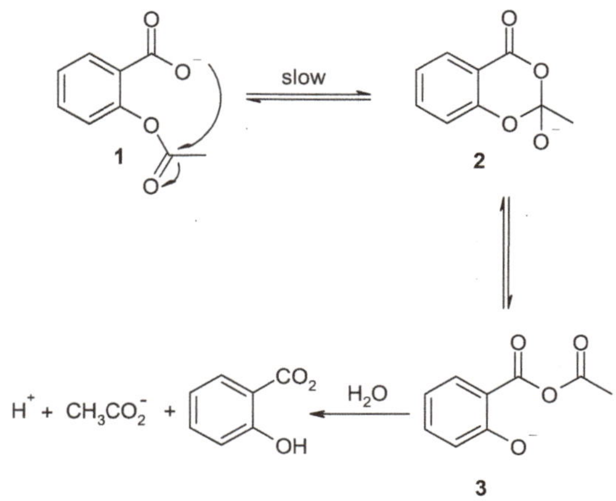

Scheme 1. Hydrolysis reaction of the LAS reagent.

seriously questioned this mechanism. He found that the addition of ethanol to the solvent increases the rate of solvolysis, and that ethyl acetate is then a product [3-6]. He ruled out the possibility that the rate increase is a generalized solvent effect by showing that the addition of dioxane has very little effect on the rate of hydrolysis of aspirin. Garrett tried to explain his results by proposing a mechanism involving nucleophilic attack by ethanol on the tetrahedral carbon atom of the intermediate (2) (**Scheme 1**), but this explanation has not been generally accepted [9]. Nevertheless, the demonstration that the addition of ethanol increases the rate of solvolysis does suggest strongly that the question of the involvement of a molecule of solvent in the transition state ought to be reopened. The work of Jencks and Gilchrist [11] shows clearly that ethanol and ethoxide ion are stronger nucleophiles than water and hydroxide ion in reactions at the carbonyl group, and a rate-determining step involving attack by either species would explain Garrett's results.

It has been recognized for many years that over a sufficiently wide pH range, three kinetically-distinguishable mechanisms for the hydrolysis of some compounds, such as epoxides and derivatives of acetylsalicylic acid [12-15]. The experimental rate expression for their hydrolysis is given by the following equation:

$$k_{obs} = k_{\mathrm{H}}\left[\mathrm{H}^{+}\right] + k_0 + k_{\mathrm{OH}}\left[\mathrm{OH}^{-}\right] \quad (1)$$

where k_H is the specific second-order rate constant for the hydronium ion-catalyzed reaction, k_0 is the specific first-order rate constant for the spontaneous reaction, and k_{OH} is the specific second order rate constant for the hydroxide-catalyzed reaction. This paper reports on the hydrolysis kinetics of the LAS compound (LAS), for pH values between 0.50 and 13. This detailed study for a wide range of pH, allows the one hand, to extract significant results and secondly, to have an overview on the various possible mechanisms for the hydrolysis of this important reagent.

2. Experimental Section

Inorganic salts were of analytical grade, and were used without further purification, as well used normadoses solutions of HCl and NaOH were analytical grade. Water was further glass distilled before use. The LAS was obtained commercially. Aqueous solutions of reagent LAS of known concentrations 5×10^{-4} M and 10^{-3} M were prepared, the desired pH values are determined by adding the quantities: HCl (0.5 M) or NaOH (1 M), respectively for acidic or basic mediums and these pH values are measured with a Hanna 210 microprocessor pH-meter, equipped with a combined glass electrode and calibrated with commercial buffers (pH 4.00 and 7.00). These solutions are maintained at fixed temperature 298 K, and small aliquots were withdrawn at known time intervals. These samples were analyzed by standard Helios γ UV-visible spectrometer, using quartz cells of optical path length l = 1 cm. The absorption measurements have been performed at wavelength λ_{max} = 300 nm. Indeed, the UV-visible spectra of **Figure 1**, clearly show that the salicylate ion formed by hydrolysis of the LAS, has a maximum band of absorption at λ_{max} = 300 nm, and at this maximum wavelength, the reagent LAS has a negligible absorption.

The initial concentration denoted "a" of the reagent has always been very low (5, 10^{-4} M), so that all the spectrophotometric measurements were carried out under conditions of pseudo-first-order, with higher concentrations of reacting species (H_3O^+, H_2O and HO^-), compared to this initial concentration a. All experimental values of optical densities were measured by reference to pure water.

3. Results and Discussion

3.1. Identification of the Catalysis Type for the Hydrolysis Reaction

The hydrolysis reaction of this reagent derivative of aspirin,

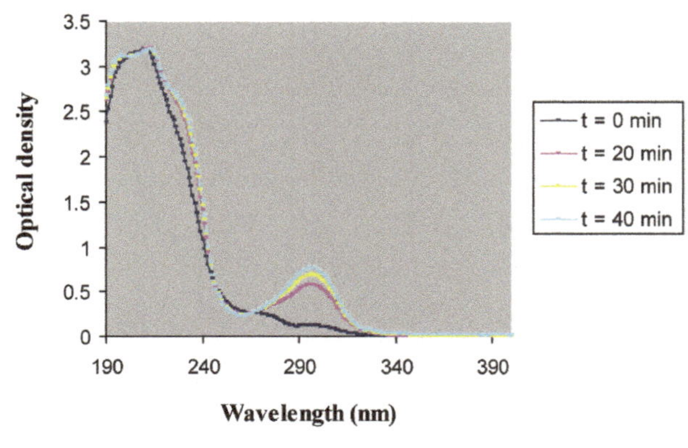

Figure 1. UV-Visible spectra for the hydrolysis of the reagent LAS, pH = 11.66, [LAS] = 5 × 10^{-4} M.

has been studied at 298 K in the pH range of 0.50 to 13 and the kinetics of this hydrolysis phenomenon have proved sensitive to the medium acidity, therefore, there is possibility of the existence of a catalysis phenomenon by ionic species of water (H_3O^+ and HO^-). Generally, the mechanism of hydrolysis may be conducted according to the following **Scheme 2**.

The rate constants of second-order k_H and k_{OH}, are associated with catalytic decomposition process, respectively, in acidic and basic environments. While the rate constant of first-order k_0, is involved in the process of hydrolysis under the effect of solvent molecules, however, it must be noted that this constant k_0 result of the operation k_{H_2O} * 55.55 (water concentration), with k_{H_2O} is the rate constant of second-order corresponding to the process of hydrolysis by an elementary molecule of the water.

If a is the initial concentration of the reagent LAS, and x the concentration of salicylate ions, formed at time t, and for a fixed acidity of the medium, the rate equation on the hydrolysis reaction is given by the following expression:

$$\frac{dx}{dt} = \left(k_H\left[H^+\right] + k_0 + k_{OH}\left[OH^-\right]\right)(a-x) \quad (1)$$

After separation of variables, we have the relation:

$$\frac{dx}{d(a-x)} = \left(k_H\left[H^+\right] + k_0 + k_{OH}\left[OH^-\right]\right)dt \quad (2)$$

The integration of this relationship allows obtaining the following kinetic equation:

$$\ln\left(\frac{a}{a-x}\right) = \left(k_H\left[H^+\right] + k_0 + k_{OH}\left[OH^-\right]\right)t = k_{obs}t \quad (3)$$

The apparent rate constant of pseudo first-order k_{obs}, on the reaction of hydrolysis reagent LAS, is given by the expression:

$$k_{obs} = \left(k_H\left[H^+\right] + k_0 + k_{OH}\left[OH^-\right]\right) \quad (4)$$

It is obvious to note that the kinetic law of pseudo first-order Equation (3) can be easily connected to changes in optical densities as a function of time, especially, only the hydrolyzed product absorbs at the measurement wavelength (λ_{max} = 300 nm). Indeed, if D_0, D_t and D_∞ are respectively the values of optical density at t = 0, at time t and at the end of the reaction and if ε_{sal} is the molar extinction coefficient of salicylate ion, we can write the following relations:

$$D_0 = 0;\ D_t = \varepsilon_{sal}lx;\ D_\infty = \varepsilon_{sal}la \quad (5)$$

where l is the optical path, a and x are the respective concentrations of LAS reagent and hydrolyzed product. From the kinetic law of Equation (3) and the preceding relations Equation (5), we can establish relationships that express the evolution of the optical density versus time t.

$$\ln\left(\frac{a}{a-x}\right) = \ln\left(\frac{D_\infty}{D_\infty - D_t}\right) = k_{obs}t \quad (6)$$

and we see immediately that the value of the apparent rate constant k_{obs} is obtained from the slope of the straight lines

$$\ln\left(\frac{D_\infty}{D_\infty - D_t}\right) = f(t) \quad (7)$$

The graphic in **Figure 2** illustrates the good linearity of the observed experimental lines.

Several experiments on the hydrolysis reaction of the reagent LAS, were performed for pH values from 0.50 to 13, with HCl and NaOH as agents, respectively to determine the pH of the studied solutions. For these experiments, the kinetic experimental lines Equation (6) were plotted and the apparent rate constants k_{obs} were calculated for all studied solutions, from the slopes of obtained

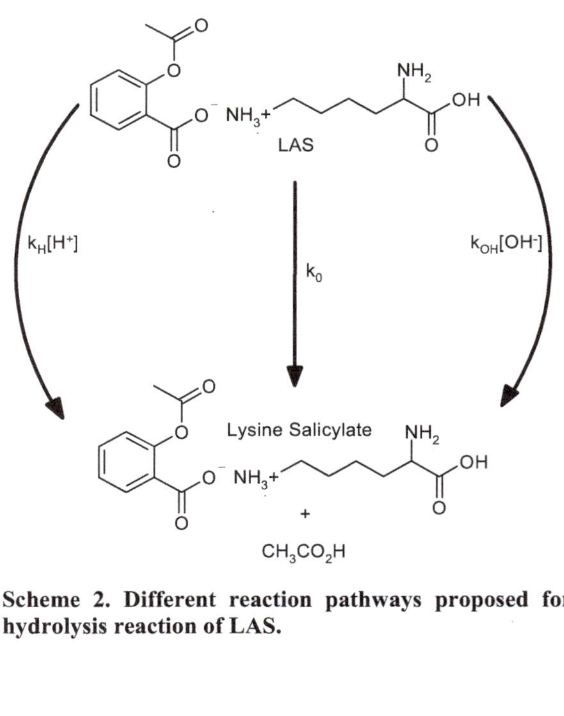

Scheme 2. Different reaction pathways proposed for the hydrolysis reaction of LAS.

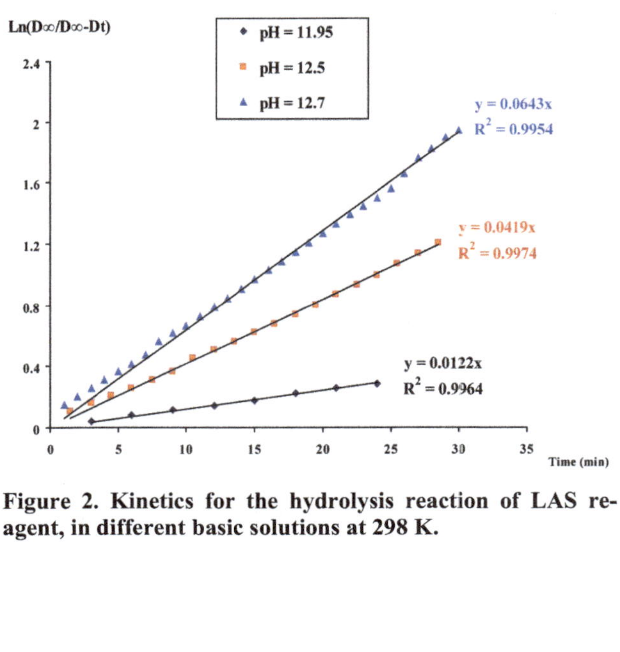

Figure 2. Kinetics for the hydrolysis reaction of LAS reagent, in different basic solutions at 298 K.

lines. **Table 1** includes all results related to the phenomenon of hydrolysis of the LAS compound.

These results show, firstly that the rate of hydrolysis reaction of this compound depends on the acidity of the medium and the reaction pathway catalyzed by the HO^- ions on the hydrolysis reaction is predominant, on the other hand, values of these apparent rate constants (k_{obs}) change in different ways, according to three pH ranges. From these values, the pH-rate profile graph ($\log k_{obs} = f(\text{pH})$), was established (**Figure 3**), and the obtained experimental curve shows three different slopes ($p = -1$, 0 and +1).

Table 1. Evolution of the apparent rate constants (k_{obs}) on the medium acidity.

pH	Log (k_{obs})
0.5	–3.222
0.7	–3.523
1.30	–3.886
3.00	–3.980
4.00	–3.990
5.30	–4.000
6.10	–4.097
7.04	–4.097
8.20	–4.000
11.33	–2.602
11.95	–1.921
12.18	–1.726
12.37	–1.404
12.50	–1.390
12.70	–1.211

pH values determined by adding concentrated solutions of HCl and NaOH, k_{obs} in mn^{-1}, [LAS] = 5×10^{-4} M.

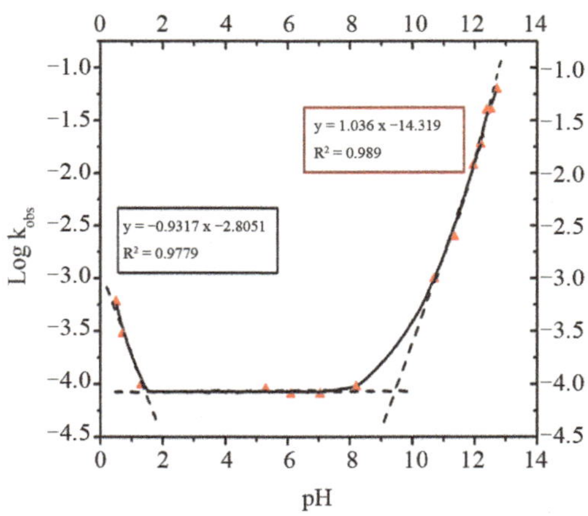

Figure 3. pH-rate profile for hydrolysis reaction of the LAS reagent at 298 K.

Therefore, the apparent rate constant k_{obs}, expression is composed of three terms and is identified with the Equation (4). The shape of this curve on the pH-rate profile clearly indicates that the hydrolysis reaction of the reagent LAS, is using three different reaction pathways, for three different pH ranges and therefore, this reaction presents a phenomenon of acid-base catalysis. For pH values below 2, the curve has a negative slope (–1) and in k_{obs} expression Equation (4), the term k_H is dominant, and this hydrolysis reaction proceeds according to the reaction pathway catalyzed by H_3O^+ ions. The zero slope of this curve, presented for pH values between 2 and 8, clearly indicates that in this pH range, the k_0 term is predominant Equation (4), the apparent rate constant k_{obs} depends on the concentration of HO^- ions, and the hydrolysis process of LAS, results from the intervention of these ions, according to the reaction pathway catalyzed by these ionic species HO^-. From the curve in **Figure 3**, we determined the values of three rate constants k_H, k_0, and k_{OH}, for different reaction pathways that constitute the global mechanism for the hydrolysis process of the LAS reagent. **Table 2** shows the values for these constants and the definitive expression of the apparent rate constant k_{obs} at 298 K, is given by the following equation:

$$k_{obs} = 1.73\times10^{-3}\left[H_3O^+\right] + 9.09\times10^{-5} + 1.58\left[OH^-\right] \quad (8)$$

These results indicate that the hydrolysis reaction of the LAS reagent, is achieved using an acid-base catalysis mechanism, the reaction pathway catalyzed by HO^- ions is dominant, and the degradation of this compound is important in basic medium. This value of the rate constant k_{OH} on the hydrolysis reaction of LAS in basic medium is the same order of magnitude as the values of the constants k_{OH} determined for the picryl chloride and trinitro-anisole hydrolysis reactions [16-18]. It is considerably less than the value calculated for the hydrolysis of 7-chloro-4,6-dinitrobenzofurazan [19], while this value is much higher than the values announced for the hydrolysis reactions of styrene oxides [20].

3.2. Determination of Activation Parameters and Reaction Mechanism

To elucidate the temperature effect on the evolution of the kinetic parameters relating to the reaction pathway catalyzed by HO^- ions, to determine the activation parameters, activation energy (E_a), enthalpy ($\Delta H^{\neq}$) and entropy ($\Delta S^{\neq}$) for the transition state of the rate-determining step and propose a mechanism for this predominant reac-

Table 2. The three rate constants for the hydrolysis phenomenon of LAS.

k_H ($M^{-1}\cdot mn^{-1}$)	k_0 (mn^{-1})	k_{OH} ($M^{-1}\cdot mn^{-1}$)
1.73×10^{-3}	9.09×10^{-5}	1.58

tion pathway in the hydrolysis process of studied reagent (LAS). We followed the kinetics of the studied hydrolysis reaction in basic medium of constant pH (10.12), by varying the temperature between 298 and 323 K. The experimental values of the apparent rate constant k_{obs} were calculated in the same way as the previous section, and **Table 3** summarizes all results.

The linear relationship between Ln k_{obs} and the inverse of absolute temperature (1/T) represented by the graph in **Figure 4**, indicated firstly, that a consistent solvolysis mechanism operated for this hydrolysis reaction pathway catalyzed by HO^- ions, on the other hand, the Arrhenius law is well verified for the studied temperatures and the activation parameters for this process were calculated from the slope and intercept on the experimental right.

Ln k_{obs} = f(1/T), plotted for the interval of studied temperatures, with the following values for the activation energy $E_a = (65.60 \pm 0.05)$ kJ·mol^{-1} and the preexponential factor in the Arrhenius relationship A = (1.803 ± 0.005) 10^6 s^{-1}.

From these activation parameters, we determined at temperature T = 298 K, the magnitude of activation enthalpy $\Delta H^{\neq}$ and entropy $\Delta S^{\neq}$, according to the following expressions derived from the theory of the transition state:

$$\Delta H^{\neq} = Ea - RT \quad \text{and} \quad \Delta S^{\neq} = R\left[\ln(A) - 30.46\right] \qquad (9)$$

After calculation we obtain the values:

$$\Delta H^{\neq} = (63.12 \pm 0.05)\ \text{kJ} \cdot \text{mol}^{-1}$$

$$\Delta S^{\neq} = -(133.42 \pm 0.05)\ \text{J} \cdot \text{K}^{-1} \cdot \text{mol}^{-1}$$

The $\Delta H^{\neq}$ and $\Delta S^{\neq}$ values associated with the solvolysis phenomenon of LAS reagent, can propose a structure of transition state on the studied reaction. In fact, a very negative value for the activation entropy $\Delta S^{\neq}$ reflecting a late transition state, very organized with a structure similar products and indicates that probably, for this reaction pathway catalyzed by HO^- ions, the intervention of the water molecule is in the rate-determining step, after the fast attack of the LAS reagent by very active HO^- ion, and that in the transition state there is a transfer of hydrogen atoms from one site to another. Indeed, some kinetic studies [21,22] on hydrolysis reactions of some organic compounds, have suggested mechanisms which involve a water molecule in the rate-determining step, and for this reaction pathway we propose the following mechanism:

Table 3. The apparent rate constant kobs evolution vs. the temperature.

$10^5 \cdot k_{obs}$ (s^{-1})	ln (k_{obs})	T (K)	$10^3 \cdot 1/T$ (K^{-1})
0.667	−11.917	298	3.36
1.160	−11.364	308	3.25
2.830	−10.472	318	3.15
5.330	−9.840	323	3.09

pH = 10.12, [LAS] = 5 × 10^{-4} M, k_{obs} in s^{-1}.

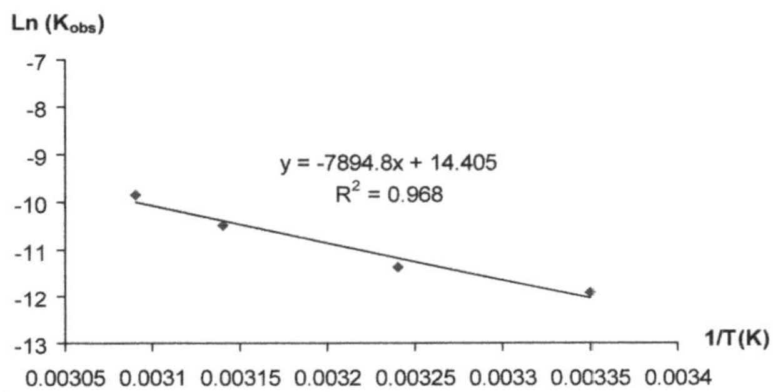

Figure 4. The Arrhenius representation Ln k_{obs} = f(1/T) for the hydrolysis reaction of the LAS reagent at pH = 10.12.

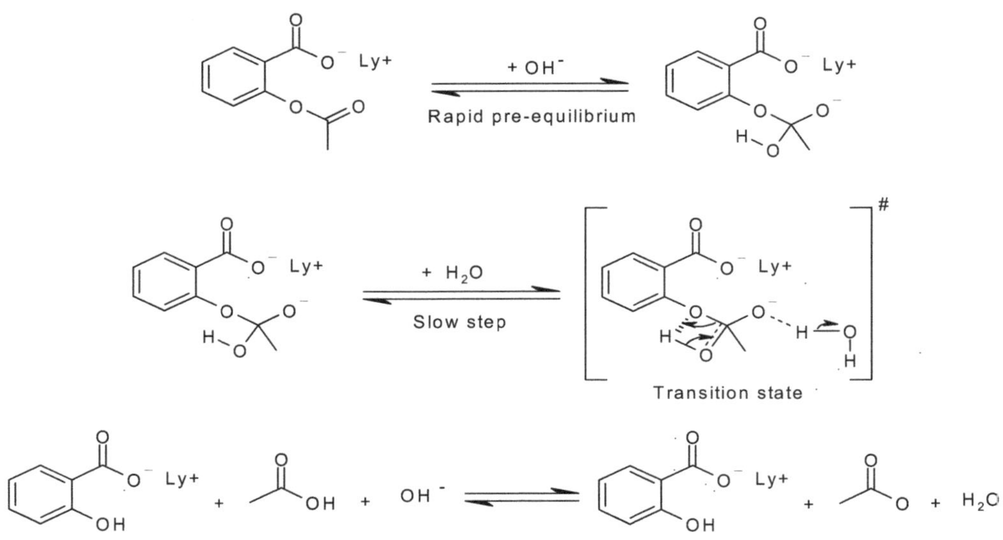

Figure 5. Mechanism for hydroxide ion catalyzed hydrolysis of the LAS reagent.

This mechanism for the reaction pathway catalyzed by HO^- ions, composed of three steps, indicating that Step 2 on the association of a water molecule with the carbanion produced by Step 1, is rate-determining step, with a highly ordered transition state whose structure resembles the products of the reaction. The passage through this highly ordered transition state with cyclic and internal transfer of hydrogen atoms from one site to another, well justifies the activation energy required for the formation of this transition state, and the very negative value of the activation entropy ($\Delta S^{\neq} = -133.42\ J \cdot K^{-1} \cdot mol^{-1}$). Moreover, this very negative value is in favor of a bimolecular process involving two species, to imagine that highly ordered transition state. Indeed, the value of this parameter ($\Delta S^{\neq}$) is very useful for identifying the rate-determining step molecularity. Recent studies on the degradation kinetics of some compounds [23,24], indicate clearly that for the unimolecular process, the values of the activation entropy $\Delta S^{\neq}$ are positive.

4. Conclusion

Studies in this work have identified the stability of the active reagent LAS as a function of the medium acidity, to determine the different rate constants for the three reaction pathways involved in the hydrolysis process of this compound. Finally a study of the temperature influence has been achieved for the hydrolysis reaction in basic medium, and the activation parameters for transition complex were determined. The analysis of these results has identified an acid-base catalysis type for the hydrolysis process of this studied compound, with a clear predominance for the reaction pathway catalyzed by HO^- ions in basic medium, and for this reaction pathway, the activation parameters on the transition complex was determined based on expressions derived from the transition state theory. On the one hand, the values of these parameters allowed us to propose a reaction mechanism consisting of three steps for the hydrolysis phenomenon of reagent LAS in basic medium, and secondly, to clearly specify the rate-determining step for this proposed mechanism.

REFERENCES

[1] L. J. Edwards, "The Hdrolysis of Aspirin. A Determination of the Thermodynamic Dissociation Constant and a Study of the Reaction Kinetics by Ultra-Violet Spectrophotometry," *Transactions of the Faraday Society*, Vol. 46, 1950, pp. 723-735. doi:10.1039/TF9504600723

[2] L. J. Edwards, "The Hydrolysis of Aspirin: Part 2," *Transactions of the Faraday Society*, Vol. 48, 1952, pp. 696-699. doi:10.1039/TF9524800696

[3] E. R. Garrett, "The Kinetics of Solvolysis of Acyl Esters of Salicyclic Acid," *Journal of the American Chemical Society*, Vol. 79, No. 13, 1957, pp. 3401-3408. doi:10.1021/ja01570a024

[4] E. R. Garrett, "Evidence for General Base Catalysis in an Ester Hydrolysis. I. Hydrolysis of an Alkyl Aminoacetylsalicylate," *Journal of the American Chemical Society*, Vol. 79, No. 19, 1957, pp. 5206-5209. doi:10.1021/ja01576a036

[5] E. R. Garrett, "The Solvolysis of Aspirin Anhydride," *J Journal of the American Chemical Society*, Vol. 82, No. 3, 1960, pp. 711-718. doi:10.1021/ja01488a052

[6] E. R. Garrett, "The Neutral Solvolysis of the Aspirin Anion in Aqueous and Mixed Solvents," *The Journal of Organic Chemistry*, Vol. 26, No. 10, 1961, pp. 3660-3663. doi:10.1021/jo01068a010

[7] J. D. Chanley, E. M. Gindler and H. Sobotka, "The Mechanism of the Hydrolysis of Salicyl Phosphate. I," *Journal of the American Chemical Society*, Vol. 74, No. 17, 1952, pp. 4347-4352. doi:10.1021/ja01137a030

[8] D. Davidson and L. Auerbach, "The Acid Anhydride Character of Aspirin," *Journal of the American Chemical Society*, Vol. 75, No. 23, 1953, pp. 5984-5986. doi:10.1021/ja01119a062

[9] M. L. Bender, "Mechanisms of Catalysis of Nucleophilic Reactions of Carboxylic Acid Derivatives," *Chemical Reviews*, Vol. 60, No. 1, 1960, pp. 53-113. doi:10.1021/cr60203a005

[10] M. L. Bender, F. Chloupek and M. C. Neveu, "Intramolecular Catalysis of Hydrolytic Reactions. III. Intramolecular Catalysis by Carboxylate Ion in the Hydrolysis of Methyl Hydrogen Phthalate," *Journal of the American Chemical Society*, Vol. 80, No. 20, 1958, pp. 5384-5387. doi:10.1021/ja01553a016

[11] W. P. Jencks and M. Gilchrist, "The Nucleophilic Reactivity of Alcoholate Anions toward p-Nitrophenyl Acetate," *Journal of the American Chemical Society*, Vol. 84, No. 15, 1962, pp. 2910-2913. doi:10.1021/ja00874a013

[12] J. N. Brönsted, M. Kilpatrick and M. Kilpatrick, "Kinetic Studies on Ethylene Oxides," *Journal of the American Chemical Society*, Vol. 51, No. 2, 1929, pp. 428-461. doi:10.1021/ja01377a012

[13] F. A. Long and J. G. Pritchard, "Hydrolysis of Substituted Ethylene Oxides in H_2O^{18} Solutions," *Journal of the American Chemical Society*, Vol. 78, No. 12, 1956, pp. 2663-2667. doi:10.1021/ja01593a001

[14] J. G. Pritchard and F. A. Long, "Hydrolysis of Ethylene Oxide Derivatives in Deuterium Oxide-Water Mixtures," *Journal of the American Chemical Society*, Vol. 78, No. 23, 1956, pp. 6008-6013. doi:10.1021/a01604a014

[15] A. R. Fersht and A. J. Kirby, "Hydrolysis of Aspirin. Intramolecular General Base Catalysis of Ester Hydrolysis," *Journal of the American Chemical Society*, Vol. 89, No. 19, 1967, pp. 4857-4863. doi:10.1021/ja00995a007

[16] R. H. De Rossi and E B. De Vargas, "Buffer Catalysis in the Hydrolysis of Picryl Chloride," *The Journal of Organic Chemistry*, Vol. 44, No. 23, 1979, pp. 4100-4104. doi:10.1021/jo01337a017

[17] J. Murto, A. Bovallius, G. Bergson, L. Ehrenberg, J. Brunvoll, E. Bunnenberg, C. Djerassi and R. Records,

"Nucleophilic Reactivity, Part 9: Kinetics of the Reactions of Hydroxide Ion and Water with Picrylic Compounds," *Acta Chemica Scandinavica*, Vol. 20, 1966, pp. 310-322. doi:10.3891/acta.chem.scand.20-0310

[18] B. Gibson and M. R. Crampton, "The Stabilities of Meisenheimer Complexes. Part 17. The Reactions of 1-X-2,4, 6-Trinitrobenzenes with Hydroxide Ions in Water," *Journal of the Chemical Society, Perkin Transactions* 2, No. 5, 1979, pp. 648-652. doi:10.1039/P29790000648

[19] F. Terrier, L. Xiao, M. Hlaibi and J. C. Halle, "Highly Eletro-Philic Heteroaromatics: The Hydrolysis of 7-Chloro-4, 6-Dinitrobenzofurane and 7-Chloro-4, 6-Dinitrobenzo-Furoxan in Aqueous Solution," *Journal of the Chemical Society, Perkin Transactions* 2, No. 3, 1993, pp. 337-341. doi:10.1039/P29930000337

[20] J. J. Blumenstein, V. C. Ukachukwu, R. S. Mohan and D. L. Whalen, "Effects of Para-Substituents on the Mechanisms of Solvolysis of Styrene Oxides," *The Journal of Organic Chemistry*, Vol. 58, No. 4, 1993, pp. 924-932. doi:10.1021/jo00056a027

[21] J.-P. Lee and T.-S. Uhm, "Kinetics and Mechanism of the Hydrolysis Reaction of N-Furoyl-2-Phenylimidazole," *Bull of the Korean Chemical Society*, Vol. 21, No. 1, 2000, pp. 29-30.

[22] J.-P. Lee and S.-S. Lee, "Kinetics and Mechanism for Hydrolysis Reaction of N-Benzoyl-2-Phenylimidazole," *Bull of the Korean Chemical Society*, Vol. 23, No. 1, 2002, pp. 151-153. doi:10.5012/bkcs.2002.23.1.151

[23] F. Boujelbane and N. B. Hamida, "Étude Cinetiaue de la Dégradation Hydrolytique d'un Insecticide: Le Bendiocarb," *Journal de la Société Chimique de Tunisie*, Vol. 9, No. 2, 2007, pp. 121-132.

[24] F. Boujelbane and N. B. Hamida "Étude par HPLC et Spectrophotométrie UV de la Dégradation Hydrolytique d'un Insecticide, le Zectran," *Journal de la Société Chimique de Tunisie*, Vol. 10, No. 1, 2008, pp. 39-51.

7

Hydroxyalkylation of Cyclic Imides with Oxiranes Part I. Kinetics of Reaction in Presence of Triethylamine as Catalyst

Jacek Lubczak
Faculty of Chemistry, Rzeszów University of Technology, Rzeszów, Poland
Email: jml@prz.edu.pl

ABSTRACT

Literature describes kinetics of reactions of alcohols, phenols, carboxylic acids, amines and amides with oxiranes such as ethylene oxide and propylene oxide. However, there is no information regarding kinetic of reaction of imides with oxiranes. In this article the kinetics of the reaction of cyclic monoimides: succinimide, phtalimide, and glutarimide, with ethylene and propylene oxides in presence of triethylamine in aprotic solvent was studied. The rate laws for those processes were established based upon on dilatometric measurements. I was said that cyclic monoimides react with oxiranes in presence of triethylamine to give N-(2-hydroxyalkyl)imides as major product. This product react further with oxiranes in consecutive reaction. The kinetics of the reaction of cyclic mono-imides with oxiranes obey the following rate law: $V = k_{1/2}\, c_{cat}^{1/2}\, c_{imide}^{3/2}\, c_{oxirane}^{1/2}$. Based upon kinetic data the following orders of reactivity of imides and oxiranes were obtained: phtalimide ≥ succinimide > glutarimide and ethylene oxide > propylene oxide. The solvent (DMF, DMSO and dioxane) effect was also studied. From temperature dependences the thermodynamic parameters: activation energy, enthalpy and entropy from linear Eyring plots were obtained.

Keywords: Imides; Oxiranes; Kinetic; Reactivity; Influence of Solvents

1. Introduction

Cyclic imides (I) reveal weak acidity of the O=C-NH-C=O group. Among those the succinimide (SI, R = $(-CH_2-)_2$), glutarimide (GI, R = $(-CH_2-)_3$), phtalimide (PI, R = Ph), and maleimide (MI, R = -CH=CH-) are common compounds [1]. The first three form N-(hydroxyalkyl) imides (II) in reactions with oxiranes [2] according to the following scheme:

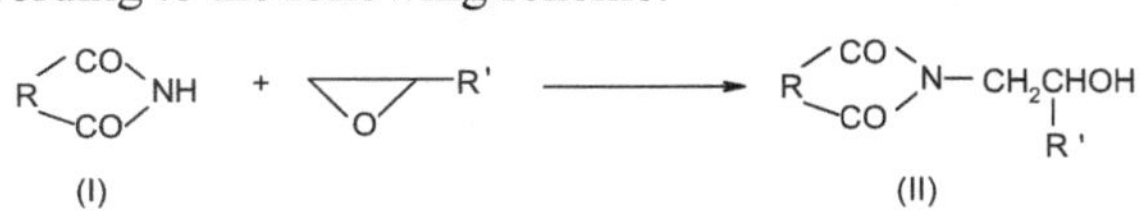

(1)

where: R' = -H, $-CH_3$, $-CH_2Cl$

The N-(hydroxyalkyl)imides can also be obtained by straightforward reaction of corresponding anhydride with aminoalcohol [3]. The products of the reaction of cyclic monoimides with ethylene carbonate and ethylene oxide were described [4,5]. These products have alcohol character which upon esterification with fatty acids appeared very early to be useful plastificators of polymers [6], while those obtained from multifunctional imides, like isocyanuric acid and ethylene oxide (EO) or propylene oxide (PO) were used for synthesis of polyurethanes, mainly to give polyurethane foams of remarkable thermal resistance [7]. The hydroxyalkylation of cyclic monoimides with oxiranes or formaldehyde was the matter of patents [2,8,9]. However the kinetics and mechanisms for those processes were not studied in detail. In contrary, the kinetics and mechanisms were described for the reactions of oxiranes with alcohols [10,11], phenols [12], thiols [13,14], carboxylic acids [15-17], and amines [18,19] depending on solvent, acidity of substrates and products and a catalyst used [18,20-25]. The reactions of amides with oxiranes were studied by Knaggs [26] and Grosmann [27]. Knaggs have shown that oxirane reacts with hydroxyl group of N-hydroxyalkyl substituent instead of amide group [26]. Grosmann has demonstrated the product distribution by preparative and analytical methods [27]. The mechanism of isomerization of imides substituted with oxirane and oxetane groups on nitrogen was also described [28].

In this paper the kinetics and products of the reaction of cyclic monoimides with oxiranes in presence of triethylamine (TEA) as catalyst were studied in order to establish the mechanism of the reaction. These processes can provide the kinetic models mimicking initial steps of the synthesis of polyetherols from oxiranes and azacyclic substrates containing few imide groups, for example isocyanuric acid [29], barbituric acid [30] or 6-amino-uracyl

[31]. These azacyclic substrates are sparingly soluble in organic solvents and therefore their reactivities cannot be studied in this class of solvents.

The method used for following the progress of the reactions studied in this work deserves a short comment. Our preliminary results indicated that determination of epoxide groups based on alkalimmetric titration of excess hydrochloric acid in dioxane in these systems failed due to the presence of amine catalyst. Therefore we have used the dilatometric method [32] by measuring the volume contraction that occurs with advancing reaction. Our preliminary studies indicated that the contraction of volume was proportional to the consumption of oxirane (c_{ox}). This was achieved by the kinetic studies of the reaction of imide with variable amount of EO or PO at the level of 0.2 to 1.0 equivalent and determination of the final volume of contraction (V_∞) to give a linear $V_\infty = f(c_{ox})$.

2. Experimental

2.1. Starting Materials

SI, PI, GI (pure, Sigma-Aldrich) were recrystallized from ethanol (reagent grade, POCH, Poland). Dimethylsulfoxide (DMSO, pure, Avocado, Germany) was dried over molecular sieves 4A (POCH, Poland), while EO and PO (racemate) (pure, Fluka, Switzerland), TEA (pure, Avocado, Germany) were used as received. Dimethyl-formamide (DMF, reagent grade, Merck) was purified by azeotropic distillation with water/benzene system and stored over molecular sieves 4A. Dioxane (reagent grade, Merck) was purified by 3 hours heating with 10 weight % concentrated hydrochloric acid, shaking with solid KOH and distillation from over the sodium.

2.2. Synthesis of N-hydroxyalkyl Derivatives of Cyclic Imides

N-(hydroxymethyl)succinimide (HMSI)

The HMSI was synthesized according to published protocol [9]. The crude product precipitated from reaction mixture, which was further purified by recrystallization from benzene.

HMSI: Yield—77%; $mp._{ref.}$ 64°C - 66°C [9], $mp._{found}$ 66°C; elemental analysis—% Calcd.: C 46.51; H 5.43; N 10.85; % Found: C 46.30; H 5.33; N 10.80; IR (in KBr pellet) [cm^{-1}], 3450 (OH), 1740 (C=O), 1440 - 1400 (CH_2), 1200 (-C-O-), 1100 (OH); ^{1}H-NMR (d_6-DMSO) [ppm], 4.6 (H, s, OH), 4.95 (2H, s, N-CH_2-) 2.70 (4H, s, CH_2CO).

N-(2-hydroxyethyl)succinimide (HESI) and N-(2-hydroxyethyl)phtalimide (HEPI)

HESI and HEPI were synthesized according to [4]. The products were purified by recrystallization from benzene or ethyl alcohol.

HESI: Yield—85%; $mp._{ref}$ 60°C [4], $mp._{found}$ 60°C; elemental analysis—% Calcd.: C 50.35; H 6.29; N 9.79; % Found: C 50.50; H 6.43; N 9.80; IR (in $CHCl_3$) [cm^{-1}], 3450 (OH), 1740 (C=O), 1440 - 1400 (CH_2), 1144 (C-O), 1100 (OH); ^{1}H-NMR (d_6-DMSO) [ppm], 4.2 (H, s, OH), 3.73 (2H, m, -CH_2-O-), 3.60 (2H, t, N-CH_2-), 2.78 (4H, s, CH_2CO).

HEPI: Yield—82%; $mp._{ref.}$ 128°C [4], $mp._{found}$ 127°C; elemental analysis—% Calcd.: C 62.83; H 4.71; N 7.33; % Found: C 62.63; H 4.53; N 7.51; IR (in $CHCl_3$) [cm^{-1}], 3350 (OH), 1720 (C=O), 1440 - 1400 (CH_2), 1225 (C-H in Ph), 1125 (C-O-), 1050 (OH); ^{1}H-NMR (d_6-DMSO) [ppm], 7.80 (4H, m, in Ph), 4. 25 (H, s, OH) 3.70 (2H, m, -CH_2-O-), 3.60 (2H, t, N-CH_2-).

2.3. Kinetic Measurements

Weighed sample of imide was placed in 50 cm^3 volumetric flask and dissolved in *ca* 30 cm^3 DMSO. After dissolving of imide and TEA as catalyst the temperature of the solution was raised to 25°C, 30°C, 35°C, 40°C, 45°C or 50°C (±0.05°C). Oxiranes (EO or PO) as liquids stored in refrigerator were transferred into the flask, which was immediately closed to avoid evaporation of boiling oxirane and weighed. The concentration of oxirane was calculated based on mass of substrates used. The solutions prepared in that way were then rapidly trans- ferred into thermostated dilatometer.

The influence of catalyst concentration on the rate of reaction was studied at constant concentration of imide and EO (1.0 mol/dm^3). The TEA concentration was varied within 0.1 - 0.5 mol/dm^3 region with 0.1 mol/dm^3 step. The influence of imide concentration on the rate of reaction was studied at fixed concentration of EO (0.25 mol/dm^3) and constant concentration of catalyst (0.5 mol/dm^3) with variable imide concentration within 1 - 3 mol/dm^3 region. The change of rate constant in the course of reaction and order for oxirane, calculated for the system with excess of imide is exemplified in **Table 1**.

The influence of oxirane concentration on the rate of reaction was studied varying its concentration within 0.25 - 1.00 mol/dm^3 with 0.25 mol/dm^3 step at constant concentration of imide (1.0 mol/dm^3) and TEA (0.5 mol/dm^3). Additionally, the kinetics was studied at variable concentration of imide (at 0.25 - 1.00 mol/dm^3) at fixed EO concentration (1.0 mol/dm^3). The results are given in **Tables 3** and **4**.

The temperature dependence was obtained for the system with 1.0 mol/dm^3 imide and oxirane and 0.5 mol/dm^3 TEA concentrations within 30°C - 45°C for the reaction with EO and 35°C - 50°C for the reaction with PO temperature range with 5°C step. The temperature range for kinetic measurements is limited from lower side by long reaction time (2 - 4 weeks for EO to 2 - 3 months for PO), which precludes the good accuracy of measurement, and from the higher side by boiling of oxiranes in reaction mixture (above 40° for EO and 50° for PO).

DMSO, DMF and dioxane were used as solvents. Due to low solubility of some imides and catalyst in dioxane, the imide and oxirane concentrations were 0.5 mol/dm^3 and that of TEA was 0.25 mol/dm^3.

The glass dilatometer of total volume of *ca.* 45 cm^3 was used equipped with 40 cm long capillary of inner diameter 1 mm. Before filled in it was kept at reaction temperature for at least 15 min. The reaction time was measured from the moment of mixing the reactants. The first reading of meniscus level was usually made after 10 min. needed to fill in the dilatometer and bringing the mixture to reaction temperature. The next readings were made after constantly increasing time intervals until the meniscus level stabilized. The initial level of meniscus was calculated by extrapolating the readouts to the time *t* = 0. Further reading were made in increasing time intervals until constant level of meniscus in capillary was attained.

Relative concentration of oxirane (b) was defined as:

$$b = \frac{c_B}{c_{0B}} = \frac{l_\infty - l}{l_\infty - l_o} \tag{2}$$

where: c_{0B}, c_B—corresponding to initial and instantaneous concentrations of oxirane, *mol/dm^3*. l - $l_{(t)}$—the meniscus level at time *t* and l_∞ and l_o are the levels at the end and beginning of experiment, respectively.

In order to establish the rate low, the substitution method was applied. The values of relative concentration of oxirane, *b* obtained for 0.2 - 0.8 region with 0.05 step and reaction time were put into the kinetic law equations of various orders (**Table 1**) to establish the order of reaction for oxirane. Standard deviation of k (determined 13 - 15 times in single kinetic run) and reproducibility (every determination of k was obtained in three independent kinetic runs) were within 6% - 7%, in some cases they were 10%.

Considering the system with EO as standard, the relative reactivity of PO (r_{PO}) was used as the ratio of the rate constants at the same temperature:

$$r_{PO} = \frac{k_{PO}}{k_{EO}} \tag{3}$$

where: k_{PO}, k_{EO}—rate constants for PO and EO, respecttively.

Relative reactivity of imides was analyzed in analogous manner using succinimide as standard.

The activation energy ($\Delta G^{\neq}$) was calculated from [33]:

$$\Delta G^{\neq} = RT \cdot \left(23.764 - \ln \frac{k}{T}\right) \tag{4}$$

where: R—gas constant [J/mol·K],

T—temperature of measurement, [K],

k—rate constant.

By plotting $\Delta G^{\neq}$ versus T a straight line was obtained with activation enthalpy and entropy of the reaction as the coefficients of the line [33]:

$$\Delta G^{\neq} = \Delta H^{\neq} - T \cdot \Delta S^{\neq} \tag{5}$$

2.4. Product Analysis

The products of reactions were isolated after full kinetic run. The DMSO was distilled off under reduced pressure. The oily N-(hydroxyalkyl)imides were extracted and purified by crystallization from ethanol or benzene (N-(hydroxyalkyl)imides of SI and PI) or by distillation under reduced pressure (N-(hydroxyalkyl)imides of GI). Pure products were characterized by elemental analysis (EA 1108, Carlo-Erba, Italy) and melting point, as well as the IR and ^{1}H-NMR spectra. The yield was estimated based on chromatographic analysis.

HESI: Yield—75.4%; $mp._{ref.}$ 60°C [4], $mp._{found}$ 59°C - 60°C; elemental analysis - % Calcd.: C 50.35; H 6.29; N 9.79; % Found: C 50.15; H 6.12; N 9.79; (the IR and ^{1}H-NMR spectra are analogous to those in *Synthesis of N-hy-droxyalkyl derivatives of cyclic imides* section).

HEPI: Yield—79.5%; $mp._{ref.}$ 128°C [4], $mp._{found}$ 127°C; elemental analysis - % Calcd.: C 62.83; H 4.71; N 7.33; % Found: C 62.94; H 4.58; N 7.22. % (the IR and ^{1}H-NMR spectra are analogous to those in *Synthesis of N-hydroxyalkyl derivatives of cyclic imides* section).

N-(2-hydroxypropyl)succinimide (HPSI): Yield—78.5%; $mp._{found}$ 82°C - 83°C; elemental analysis - % Calcd.: C 53.50; H 7.01; N 8.92; % Found: C 53.25; H 6.92; N 8.91. IR (in $CHCl_3$) [cm^{-1}], 3500 (OH), 1700 (C=O), 1450 (CH_2), 1345 (CH_3) 1180 (-C-O-), 1120 (OH); ^{1}H-NMR (d_6- DMSO) [ppm], 5.1 (H, s, OH), 4.10 (H, m CH-O), 3.75 (2H, d, N-CH_2-), 2.78 (4H, s, CH_2CO), 1.10 (3H, d, CH_3).

N-(2-hydroxypropyl)phtalimide (HPPI): Yield—79.9, $mp._{found}$ 81°C; elemental analysis—% Calcd.: C 64.39; H 5.37; N 6.83; % Found: C 64.28; H 5.27; N 6.94; IR (in $CHCl_3$) [cm^{-1}], 3280 (OH), 1720 (C=O), 1450 (CH_2), 1380 (CH_3), 1200 (-C-O-), 1050 (OH); ^{1}H-NMR (d_6-DMSO) [ppm], 7.85 (4H, m, in Ph), 4.25 (H, s, OH), 4.10 (H, m, -CH-O-), 3.80 (2H, d, N-CH_2-), 1.10 (3H, d, CH_3).

N-(2-hydroxyethyl)glutarimide HEGI: Yield—74.9%; elemental analysis—% Calcd.: C 53.50; H 7.01; N 8.92; % Found: C 53.35; H 6.83; N 8.80; IR (in $CHCl_3$) [cm^{-1}], 3490 (OH), 1720 (C=O), 1460 (CH_2), 1100 (-C-O-, OH); ^{1}H-NMR (d_6-DMSO) [ppm], 5.1 (H, s OH), 3.70 (2H, m CH_2-O), 3.60 (2H, t, N-CH_2-), 2.45 (4H, t, CH_2CO), 1.85 (2H, m, CH_2- CH_2- CH_2-).

N-(2-hydroxypropyl)glutarimide (HPGI): Yield—75.4%;

elemental analysis—% Calcd.: C 56.14; H 7.60; N 8.19; % Found: C 56.25; H 7.92; N 8.34; IR (in $CHCl_3$) [cm^{-1}], 3400 (OH), 1720 (C=O), 1460 (CH_2), 1370 (CH_3) 1175 (C-O), 1130 (OH); ^{1}H-NMR (d_6-DMSO) [ppm], 5.1 (H, s, OH), 4.10 (H, m CH-O), 3.75 (2H, d, N-CH_2-) 2.42 (4H, s, CH_2CO), 1.85 (2H, m, CH_2-CH_2-CH_2-), 1.12 (3H, d, CH_3).

In some cases the post-reaction mixtures were analyzed chromatographically (KB 5901 chromatograph, CORRABID, Poland) using: UV detector working at 253, 7 mm, 6 atm pressure, room temperature, steel columns of 100 mm length and 4 mm diameter, silicagel, 60%/40% v/v hexane/dioxane, 1 cm/min eluent flow. The system was calibrated using the imides and appropriate N-(hydroxyalkyl)imides, namely: HESI, HEPI, HPSI, HPPI, HEGI, HPGI as standards. The standards were synthesized from succinic or phthalic anhydrides and ethanoloamine (HESI, HEPI) [3] or SI, PI or GI and appropriate oxiranes or alkylene carbonates according to the procedures used before [4,5], and purified by recrystallization from benzene or ethanol and identified by elemental analysis and NMR spectroscopy.

Some reactions of imides with EO or PO in acetone in presence of TEA were also performed and the products were analyzed in order to isolate and determine quantitatively the products in solvent of different polarity.

IR spectra of N-(hydroxyalkyl)imides were recorded in KBr pellets or in $CHCl_3$, while the spectra of post-reaction mixtures were taken in films (spectrometer PARAGON 1000 FT IR, Perkin Elmer). ^{1}H-NMR spectra of (N-hydroxyalkyl)imides were recorded on a spectrometer BS586A, 80 MHz, TESLA, Czechoslovakia (in d_6-DMSO or in $CHCl_3$) with HMDS internal standard.

3. Results and Discussion

3.1. Kinetics

The rate law for the reaction of SI with EO was obtained by substituting the instantaneous values of relative oxirane concentration, *b* (Equation (2)) and reaction time into kinetic equations of various orders (**Table 1**). In presence of 12-fold excess of imide (as well as lower excess of imide) the following rate law was found:

$$V = -\frac{dc_B}{dt} = k_{1/2} c_B^{1/2} \quad (6)$$

where c_B is instantaneous concentration of oxirane, *mol/dm³*.

In columns 3 and the rate constants calculated for closest orders (zero and first) are shawn for comparison. In column 3 (zero order fit) the rate constant constantly decreased, while in column 5 (first order fit) it increases constantly, therefore the best fit was for 1/2 order. The rate constant depends on the catalyst concentration (c_k) according to the formula:

$$k_{1/2} = k'_{1/2} c_k^n \quad (7)$$

Plotting $\lg k_{1/2} = f(\lg c_K)$ the n ≈ 0.5 was obtained (**Figure 1**).

The reaction of monoimide with oxirane is composed of two steps: the reaction of imide groups AH with oxirane (reaction 1), and the reaction of hydroxyalkyl groups with remaining oxirane:

Table 1. Rate constants for the reaction of SI with EO at 313 K (40°C) calculated for various kinetic laws.

$c_{oSI} = 2.9997$ mol/dm³ $c_{oEO} = 0.2532$ mol/ dm³ $c_{TEA} = 0.4999$ mol/ dm³

Reaction time [min]	Relative conc. of EO b	$k_0 \cdot 10^6$ [mol/s·dm³]	$k_{1/2} \cdot 10^5$ [mol$^{1/2}$ s^{-1}. dm$^{-3/2}$]	$k_1 \cdot 10^5$ [s^{-1}]
98	0.8006	8.61	1.84	3.78
137	0.7240	8.50	1.82	3.93
156	0.6851	8.52	1.85	4.03
177	0.6450	8.46	1.86	4.13
198	0.6006	8.51	1.90	
226	0.5487	8.43	1.92	4.29
255	0.5097	8.11	1.87	
297	0.4610	7.66	1.81	4.43
340	0.4025	7.41	1.80	
368	0.3604	7.33	1.81	4.40
424	0.2987	6.98	1.79	4.35
446	0.2662	6.94	1.81	4.46
497	0.2012	6.78	1.85	4.62
				4.75
				4.95
				5.38
$k_{1/2}$ av·10^5 [mol$^{1/2}$ s^{-1}·dm$^{-3/2}$]			1.84 ± 0.04	

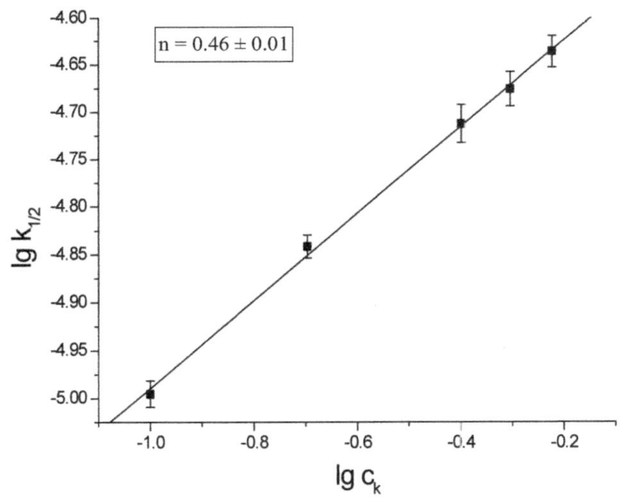

Figure 1. The plot of rate constant versus catalyst concentration for the reaction of SI with EO.

(8)

In order to avoid the second process, the reaction was performed at the excess of reactive groups of imide. This imposed $c_{AH} = const$. In such a condition the dependence:

$$k'_{1/2} = k''_{1/2} c^{p}_{AH} \tag{9}$$

operates. According to kinetic measurements the power of the imide group concentration was $p \approx 3/2$ (**Figure 2**). Thus the kinetic law describing the reaction of SI with EO is:

$$V = k''_{1/2} c_k^{1/2} c_{AH}^{3/2} c_B^{1/2} \tag{10}$$

Other imides (PI, GI) and also HMSI react with EO and PO in similar manner indicating 1/2 order on oxirane concentration (**Table 2**).

If the Equation (10) is valid, the $k_{1/2}$" should stay in agreement with the k calculated from the formula (11), in which the constant concentration of imide groups AH is not assumed, but the real momentary concentration is considered, according to the formula:

$$V = k c_k^{1/2} \left(c_{0AH} - x\right)^{3/2} \left(c_{0B} - x\right)^{1/2} \tag{11}$$

where: c_0—initial concentration of imide or oxirane, *mol/dm*^{3}, x—the decrease of oxirane concentration due to reaction with imide (consumption of oxirane due to side reactions was neglected), where:

$$x = c_0 \left(1 - b\right) \tag{12}$$

Integration of (11) led to:

$$k = \frac{2}{c_k^{1/2} \left(c_{0AH} - c_{0B}\right) t \sqrt{\dfrac{c_{0B}}{c_{0AH}}} - \sqrt{\dfrac{c_{0B} b}{c_{0AH} - c_{0B}\left(1 - b\right)}}} \tag{13}$$

This equation can be used only if oxirane reacts with imide groups, *i.e.* when excess imide is present. The results on rate constants obtained from (10) and (11) are collected in **Table 3**. Satisfactory agreement was obtained when at least 6-fold imide excess related to oxirane was applied.

The involvement of consecutive reactions was critical importance for presented kinetic studies of reaction of imide group with oxirane. The hydroxyl groups of N-(hydroxyalkyl)imide was involved in consecutive reaction with oxirane. The latter react as soon as the product of the first step appears. Thus, both reactions compete in further progress of reaction. Therefore the products were analyzed carefully.

1) It has been found the products of reaction of imide with oxiranes were oils, while N-hydroxyalkyl derivatives of studied imides (SI, PI) are solids of defined melting point. The N-hydroxyalkyl derivatives crystallized from post-reaction mixtures upon long time. The consecutive products are oily compounds. These were isolated and identified in the reaction of imides with oxiranes used in twice excess. From the comparison of the IR spectra of reaction mixtures with those of pure N-(hydroxyalkyl) imides one can see the difference in the intensity of the band centered at *ca* 1050 cm^{-1}, which is slightly higher for mixtures. This indicates the formation of ether bonds, *i.e.* the consecutive products. Moreover, the bands attributed to carbonyl groups (1700 cm^{-1}), and hydroxyl

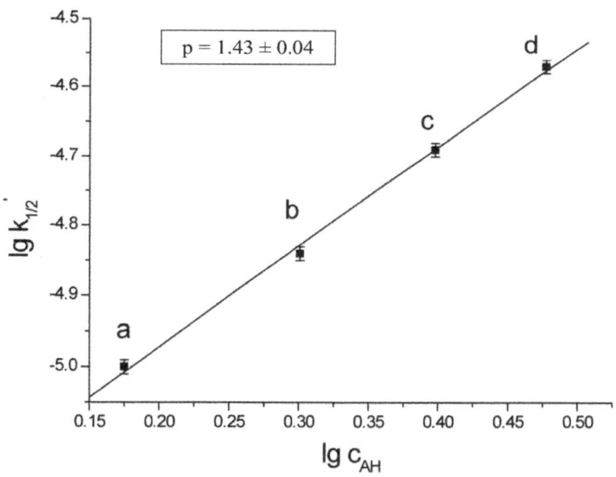

Figure 2. The plot of rate constant $k'_{1/2}$ versus concentration of imide groups AH(c_{0AH}/c_{0B} = 6 (a), 8 (b), 10 (c), 12 (d)).

Table 2. Rate constants $k_{1/2}$·10^6 of the reactions of oxiranes with imides in presence of TEA catalyst.

Imide	Oxirane	Rate constants $k_{1/2}$·10^6 [mol$^{1/2}$ s^{-1}·dm$^{-3/2}$] at temperature [K]					
		298	303	308	313	318	323
SI	EO	7.06 ± 0.28	10.0 ± 0.4	17.5 ±1.00	21.2 ± 0.8		
	PO			3.09 ± 0.15	4.80 ± 0.33	7.20 ± 0.49	10.4 ± 0.50
PI	EO	6.97 ± 0.68	10.2 ± 1.0	14.9 ± 1.2	22.9 ± 2.0		
	PO			3.97 ± 0.18	5.88 ± 0.11	7.05 ± 0.28	10.2 ± 0.80
GI	EO	5.45 ± 0.48	9.36 ± 0.48	12.3 ± 0.9	18.8 ± 1.8		
	PO			2.12 ± 0.20	2.66 ± 0.28	3.52 ± 0.38	4.87 ± 0.43
HMSI	EO	4.15 ± 0.12	6.21 ± 0.27	9.28 ± 0.45	14.3 ± 0.6		
	PO			1.98 ± 0.11	3.01 ± 0.2	4.28 ± 0.26	6.45 ± 0.39

Table 3. Comparison of rate constants for the SI + EO system at imide excess condition calculated from kinetic Equations (10) and (11) (c_B = 0.2530 ± 0.0006; c_{TEA} = 0.4997 ± 0.0003).

c_{0AH} [mol/dm^3]	Molar ratio SI to EO	10^6 $k''_{1/2}$ [dm^6/mol^2·s]	10^5 k [dm^6/mol^2·s]
2.9995	12:1	5.00 ± 0.03	5.23 ± 0.33
2.4999	10:1	4.58 ± 0.02	4.80 ± 0.29
1.9999	8:1	5.05 ± 0.02	5.36 ± 0.15
1.5001	6:1	5.44 ± 0.08	5.63 ± 0.12
0.9999	4:1	6.43 ± 0.24	7.24 ± 0.40

groups (3400, 1250 and 1100 cm^{-1}) were also observed.

2) Chromatographic analysis of oily product mixtures indicated that crude product consists of 75% - 80% N-(hydroxyalkyl)imides and 5% - 15% unreacted imide. The yields, elemental analysis and IR and ^{1}H-NMR spectra of products isolated after kinetic measurements are specified in *Product analysis* section.

3) For comparison the products were also readily isolated from the post-reaction mixtures for the systems with SI or PI with oxiranes in acetone. N-(hydroxyalkyl)imides were isolated by extraction with benzene or ethanol to give 75% - 80% pure products and 10% - 15% imides. The analysis of mixtures combined with yield indicated that consecutive products were formed with the yield not larger than 5% - 10% weight.

4) Control experiments of N-(hydroxyalkyl)imides reactivity towards oxiranes were performed using HESI or HEPI with EO at 1 mol/dm^3 imide derivative and oxirane and 0.5 mol/dm^3 TEA concentrations at 40°C. The following rate constants were obtained: (3.98 ± 0.12) 10^{-6} s^{-1} and (4.16 ± 0.23) 10^{-6}·s^{-1}, respectively. The order of reac- tion was first for oxirane. In contrary the order of reac- tion was 1/2 for oxirane and imide both at excess imide and equimolar conditions. If excess of oxirane was used (*i.e.* in conditions preferable for formation of consecutive products), the order of reaction changed to two for oxirane (**Table 4**), indicating clearly the change of a mechanism. In such condition the alcohol product prevails and the rate of oxirane consumption is determined by addition of alcohol to oxirane.

Due to slight involvement of consecutive reactions in equimolar imide and oxirane mixtures the negligible deviations of rate constants at late stages of reactions were found. Initially the rate constants do not depend on changes of oxirane concentration, while at the 60% - 70% of progress of reaction (b ≤ 0.4 - 0.3) the value of calculated constant decreases continuously (**Table 1**). Analogous dependence was observed for the reactions of PI, GI, and HMSI with oxiranes. Those deviations were less than 10%. However, in order to avoid the consecutive reactions which might influence the accuracy of obtained rate constants for N-hydroxyalkylation reaction, only the kinetic data for 20% - 60% progress of the process were taken for calculations (0.8 ≤ b ≤ 0.4, **Table 2**).

3.2. Reactivity of Monoimides and Oxiranes

Considering the electronic structure of oxiranes, the reactivity of these towards monoimides in DMSO should depend on the electropositive character of carbon 1:

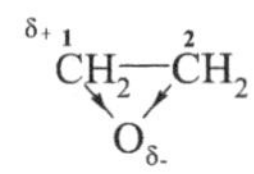
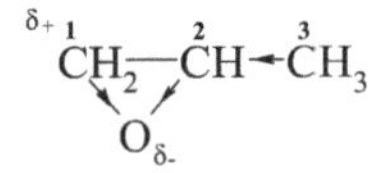

Higher reactivity of EO than PO was expected, because electrophilicity of C-1 in PO is weakened due to electro-donation from methyl group. Indeed, rate constants (**Table 2**) for EO with all studied imides were higher than those for PO.

Comparing relative reactivity of imides (**Table 2**) one can see that the rate constants for SI and PI were similar, while GI and HMSI showed lower reactivity than SI and PI. In case of HMSI, the hydroxymethyl group instead of imide group reacted with oxiranes. Its reactivity was therefore lower. In order to compare the relative reactivity of SI and PI the reactions with oxiranes with large excess of those substrates were run. It has been found out that PI is slightly more reactive than SI (**Table 5**) with EO, while the reactivity towards PO is the same. This will be discussed in details in Part II.

Based upon kinetic data the following orders of reactivity of imides and oxiranes were obtained:

PI ≥ SI > GI > HMSI and EO > PO

Table 4. Rate constants and reaction order for the SI + EO reaction at variable initial molar ratio at 313 K (40°C).

Molar ratio SI:EO	10^5 k	Order on reaction related to EO
1:0.25	0.61 ± 0.04	0.5
1:0.50	1.32 ± 0.04	0.5
1:0.75	1.74 ± 0.11	0.5
1:1.00	2.12 ± 0.08	0.5
0.75:1	2.63 ± 0.09	1
0.50:1	1.65 ± 0.14	2
0.25:1	1.00 ± 0.09	2

Table 5. Rate constants for the SI and PI + oxiranes at 8-fold imide excess in DMSO at 40°C. (c_{AH} = 2.0000 ± 0.0050; c_B = 0.2500 ± 0.0020; c_K = 0.5000 ± 0.0012).

Imide	Oxirane	10^5·$k_{1/2}$·[mol$^{1/2}$/dm$^{3/2}$s]
SI	EO	1.71 ± 0.10
	PO	0.26 ± 0.02
PI	EO	2.21 ± 0.01
	PO	0.29 ± 0.03

Activation energy ($\Delta G^{\neq}$) of the reactions were obtained from linear Eyring plots obtained from temperature dependence (**Figure 3**). Mean values of enthalpy and entropy of activation are collected in **Table 6**. From the values of entropy of activation one can deduce that the transition state in case of the reaction of PI and GI with PO is more ordered, probably because of rotational free-

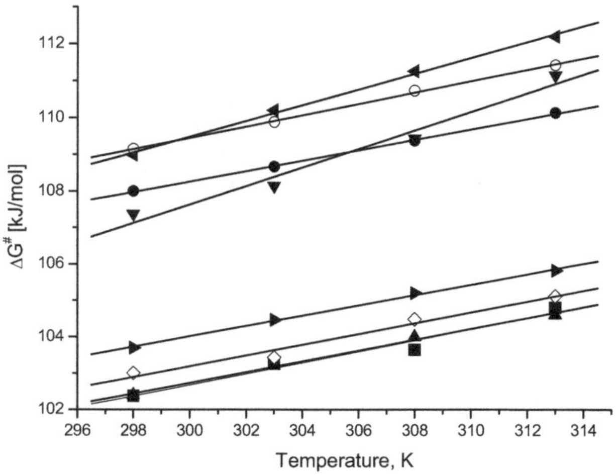

Figure 3. The changes of Gibbs energy of activation in function of temperature for the following reactions: SI + EO (■ r = 0.9839), SI + PO (● r = 0.9994), PI + EO (▲ r = 0.9977), PI + PO (▼r = 0.9961), GI + EO (◊ r = 0.9884), GI + PO (◀r = 0.9983), HMSI + EO (▶ r = 0.9989), HMSI + PO (○ r = 0.9992); r-correlation coefficient.

dom in case of the GI and PO system in comparison with other systems. This will be discussed later in Part II.

The discussed reactions were conducted in DMSO. The same reactions were also performed in DMF and dioxane and the results of kinetic measurements are collected in **Table 7**. The 1/2 order of concentration of EO and PO were also found in these solvents, which suggested the same mechanism operating in these solvents. The decreasing values of rate constants were found in the order: DMF > DMSO > dioxane. The highest reaction rate in DMF is presumably consistent with its basic character related to the presence of nitrogen atom, which might play the same catalytic role as TEA, *i.e.* enabling the proton transfer from imide into oxirane (cf. mechanism in Part II). The least polar dioxane cannot be involved in this step and therefore the rates of reactions were the least. Based upon rate constants the relative reactivity of substrates in all solvents was obtained (**Table 8**).

According to these data the relative reactivity of EO and PO depends only slightly on the solvent used. In polar solvents (DMF or DMSO) the subtle difference of reactivity of PI and SI with oxiranes was found; PI is about 1.1 - 1.2 times more reactive in DMF than is SI. In DMSO their reactivity depends on the kind of oxirane used, while in dioxane the PI is less reactive. The differences in reactivity can be explained probably in terms of solvation of transition state *PI-oxiranes* with polar solvent. The solvation of transition state is strong in DMF and DMSO causes the increase of the reactivity of PI.

4. Summary and Conclusions

1) Cyclic monoimides react with oxiranes in presence of triethylamine to give N-(2-hydroxyalkyl)imides as major product. This product react further with oxiranes in consecutive reaction.

2) The kinetics of the reaction of cyclic monoimides with oxiranes obey the following rate law:

$$V = k''_{1/2} c_k^{1/2} c_{AH}^{3/2} c_B^{1/2}$$

where $c_{K,} c_{AH}$ and c_B stay for concentrations of catalyst, imide, and oxirane, respectively.

3) In polar solvents the reactivity of imides and oxiranes decreases in the raw:

PI ≥ SI > GI > HMSI and EO > PO.

4) The decreasing values of rate constants were found in the order: DMF > DMSO > dioxane. The highest reaction rate in DMF is presumably consistent with its basic character related to the presence of nitrogen atom, which might play the same catalytic role as TEA, *i.e.* enabling the proton transfer from imide into oxirane The least po-

Table 6. Activation parameters for the reactions of imides with oxiranes.

Imide	Oxirane	Temperature range [K]	$\Delta G^{\#}$ [kJ/mol]	$\Delta H^{\#}$ [kJ/mol]	$\Delta S^{\#}$ [J/mol·s]	Correlation coefficient r
SI	EO	298 - 313	103.5 ± 8.8	56.8 ± 6.0	−153 ± 19	0.9839
	PO	308 - 323	110.5 ± 2.0	65.4 ± 1.1	−143 ± 3	0.9994
PI	EO	298 - 313	103.9 ± 4.3	58.4 ± 2.2	−149 ± 7	0.9977
	PO	308 - 323	111.6 ± 8.6	31.8 ± 9.2	−253 ± 30	0.9961
GI	EO	298 - 313	104.1 ± 9.7	58.9 ± 4.9	−148 ± 16	0.9884
	PO	308 - 323	112.7 ± 5.6	45.2 ± 2.8	−214 ± 9	0.9983
HMSI	EO	298 - 313	104.8 ± 2.9	61.4 ± 1.4	−142 ± 5	0.9989
	PO	308 - 323	111.8 ± 2.7	63.2 ± 1.4	−154 ± 4	0.9992

Table 7. The solvent dependence of rate constants for the reaction of imides with oxiranes at various solvents at 40˚C.

Imide	Oxirane	Solvent	$10^6 \cdot k_{1/2} \cdot [mol^{1/2}/dm^{3/2}s]$
SI	EO	DMSO	10.6 ± 0.55
		DMF	11.1 ± 0.60
		dioxane	3.80 ± 0.19
	PO	DMSO	2.10 ± 0.10
		DMF	2.60 ± 0.10
		dioxane	1.23 ± 0.10
PI	EO	DMSO	11.3 ± 0.15
		DMF	13.2 ± 0.20
		dioxane	2.76 ± 0.25
	PO	DMSO	2.73 ± 0.15
		DMF	3.20 ± 0.20
		dioxane	0.94 ± 0.02

Table 8. Relative reactivity of imides and oxiranes at 40˚C at various solvents.

Solvent	Substrate	System	Relative reactivity
DMSO	SI	EO:PO	1.00:0.20
	PI	EO:PO	1.00:0.24
	EO	SI:PI	1.00:1.07
	PO	SI:PI	1.00:1.30
DMF	SI	EO:PO	1.00:0.23
	PI	EO:PO	1.00:0.24
	EO	SI:PI	1.00:1.19
	PO	SI:PI	1.00:1.23
dioxane	SI	EO:PO	1.00:0.20
	PI	EO:PO	1.00:0.40
	EO	SI:PI	1.00:0.73
	PO	SI:PI	1.00:0.76

lar dioxane cannot be involved in this step and therefore the rates of reactions were the least.

5) The reaction of N-(hydroxyalkyl)imides with oxiranes was first order for oxirane. In contrary the order of reaction was 1/2 for oxirane and imide both at excess imide and equimolar conditions. If excess of oxirane was used, the order of reaction changed to two for oxirane indicating clearly the change of a mechanism. In such condition the alcohol product prevails and the rate of oxirane consumption is determined by addition of alcohol to oxirane.

REFERENCES

[1] A. Kirrmann, J. Cantacuzene and P. Duhamel, "Chimie Organique, Fonctions Simples," WNT, Warsaw, Vol. 2, 1980, pp. 225-226.

[2] R. Steele, A. Katzakian, J. Scigliano and E. Hamel, "Imide Oxirane Reactions," US Patent No. 3962182, 1976.

[3] J. Billman and E. Parker, "Amino Acids. I. Glycine," *Journal of the American Chemical Society*, Vol. 65, No. 5, 1943, pp. 761-762. doi:10.1021/ja01245a008

[4] K. Yanagi and S. Akiyoshi, "Notes: Hydroxyethylation of Imides," *The Journal of Organic Chemistry*, Vol. 24, No. 8, 1959, pp. 1122-1123. doi:10.1021/jo01090a601

[5] N. Schoenfeldt, "Grenzflaechenactive Aethylenoxid—Addukte," Wissenschaftliche Verlagsgesellschaft MBH, Stuttgart, 1976.

[6] R. Cornwell, "Composition Capable of Being Molded and Being Cast into Films," US Patent No. 2487105, 1949.

[7] M. Kucharski and J. Lubczak, "Polyurethane Foams Based on Reaction Products of Ethylene Oxide with 1,3,5-Tris(Hydroxymethyl) Isocyanurate," *Polimery*, Vol. 30, No. 9, 1985, pp. 354-361.

[8] P. Tawney, "N-methylol maleimide," US Patent No. 2526517, 1950.

[9] G.Weitzel, F. Schneider, A. Fretzdorf, K. Seynsche and H. Finger, "Futher Tumor Inhibiting Compounds. I. Cytostatic Effects of N- and S-Hydroxy-Methyl Compounds," *Zeitung Physiologische Chemie*, No. 334, 1963, pp. 1-25.

[10] S. Kucharski, "Nonionic Surfactants. I. Preparation of Poly(Oxyethylene) Alcohols," *Wiadomości Chemiczne*, Vol. 25, No. 8, 1971, pp. 579-594.

[11] N. Lebedev and I. Baranov, "Reaction Involving α-Oxides. X. Kinetics and Mechanism of the Reaction of Ethylene Oxide and Alcohols under the Conditions of Base Catalysis," *Kinetika i Kataliz*, Vol. 7, No. 4, 1966, pp. 619-626.

[12] N. Lebedev and V. Shvets, "Mechanism of Ethylene Oxide Reactions with Phenols," *Kinetika i Kataliz*, Vol. 9, No. 3, 1968, pp. 504-510.

[13] V. Shvets and I. Y. Lykov, "Kinetics and Mechanism of the Reaction of Ethylene Oxide with *n*-Decyl Mercaptan during Basic Catalysis," *Kinetika i Kataliz*, Vol. 12, No. 2, 1971, pp. 347-355.

[14] V. Shvets and I. Y. Lykov, "Kinetics and Mechanism of the Noncatalytic Reaction of Ethylene Oxide with *n*-Decyl Mercaptan," *Kinetika i Kataliz*, Vol. 12, No. 4, 1971, pp. 883-888.

[15] V. Shvets, A. Romashkin and V. Yudina, "Kinetics and Mechanism of the Reaction of Ethylene Oxide with Terephthalic Acid during Catalysis by Tetraal-Kylammonium Halides," *Kinetika i Kataliz*, Vol. 14, No. 4, 1973, pp. 928-932.

[16] N. Lebedev and K. Guskov, "Reaction of α-Oxides. II. Kinetics of the Reactions of Ethylene Oxides with Acetic and Monochloroacetic Acids," *Kinetika i Kataliz*, Vol. 5, No. 4, 1963, pp. 116-127.

[17] L. Shechter and J. Wynstra, "Glycidyl Ether Reactions with Alcohols, Phenols, Carboxylic Acids, and Acid Anhydrides," *Industrial and Engineering Chemistry*, Vol. 48, No. 1, 1956, pp. 86-93. doi:10.1021/ie50553a028

[18] N. Lebedev and M. Smirnova, "Mechanism of the Acid Catalysis of the Reaction of Ethylene Oxide with Amines," *Zhurnau Obshchei Khimii*, Vol. 39, No. 12, 1969, pp. 2732-2737.

[19] A. Weindenbacher, S. Serban and E. Weindenbacher, "Kinetics of Ethoxylation Reactions in Series of Primary Aliphatic-Amines," *Revista de Chimie*, Vol. 29, No. 3, 1978, pp. 204-209.

[20] N. Iranpoor, I. Mohammadpoor-Baltork and F. S. Zardaloo, "Ceric Ammonium Nitrate an Efficient Catalyst for

Mild and Selective Opening of Epoxides in the Presence of Water Thiols and Acetic Acid," *Tetrahedron*, Vol. 47, No. 47, 1991, pp. 9861-9866. doi:10.1016/S0040-4020(01)80723-7

[21] C. Moberg, L. Rakos, and L. Tottie, "Stereospecific Lewis Acid Catalyzed Methanolysis of Styrene Oxide," *Tetrahedron Letters*, Vol. 33, No. 16, 1992, pp. 2191-2194. doi:10.1016/0040-4039(92)88174-4

[22] C. C. Price and D. D. Carmelite, "Reactions of Epoxides in Dimethyl Sulfoxide Catalyzed by Potassium *t*-Butoxide," *Journal of the American Chemical Society*, Vol. 88, No. 17, 1966, pp. 4039-4044. doi:10.1021/ja00969a026

[23] A. Bukowska and W. Bukowski, "A Highly Regio-Selective Catalyst of Epichlorohydrin Acidolysis," *Journal of Chemical Technology & Biotechnology*, Vol. 73, No. 4, 1998, pp. 341-344. doi:10.1002/(SICI)1097-4660(199812)73:4<3C341::AID-JCTB960>3.0CO;2-6

[24] A. Guy, J. Doussot, C. Ferroud, R. Garreau and A. Godefroy-Falguieres, "Regioselective Ring Opening of Epoxides with Lithium Azide," *Synthesis*, Vol. 9, 1992, pp. 821-822. doi:10.1055/s-1992-26233

[25] M. Brunner, L. Mussmann and D. Vogt, "Kinetic Resolution of Oxiranes by Use of Chiral Lewis Acid Catalysts," *Synlett*, Vol. 12, 1993, pp. 893-894. doi:10.1055/s-1993-22641

[26] E. Knaggs, "Alkanoloamides in Soft Detergents," *Soap Chemical Specialties*, Vol. 40, No. 12, 1964, pp. 79-83.

[27] H. Grossmann, "Die Beeinflussung der Reaktion bei der Oxaethylierung von Fettsaeuremonoaethanolamiden," *Fette, Seifen, Anstrichmittel*, Vol. 74, No. 1, 1972, pp. 58-63. doi:10.1002/lipi.19720740111

[28] S. Kanoh, M. Naka, T. Nishimura and M. Motoi, "Isomerization of Cyclic Ethers Having a Carbonyl Functional Group: New Entries into Different Heterocyclic Compounds," *Tetrahedron*, Vol. 58, No. 35, 2002, pp. 7049-7064. doi:10.1016/S0040-4020(02)00701-9

[29] K. Frisch, D. Tummers and A. Nijenhuis, "Tris-(N-*β*-Hydroxypropyl) Isocyanurate," Europe Patent No. 3620, 1979.

[30] A. Ślączka and J. Lubczak, "Hydroxyalkylation of Barbituric Acid. II. Synthesis of Polyetherols with Pyrimidine Ring," *Journal of Applied Polymer Science*, Vol. 106, No. 6, 2007, pp. 4067-4074. doi:10.1002/app.26742

[31] E. Chmiel-Szukiewicz, "Polyetherols Obtained from 6-Aminouracil and Oxiranes," *Journal of Applied Polymer Science*, Vol. 103, No. 3, 2007, pp. 1466-1472. doi:10.1002/app.25100

[32] J. Lubczak, "Reactions in the System (Hydroxymethyl)Melamines-Oxiranes: Addition," *Indian Journal of Chemistry*, Vol. 33B, 1994, pp. 125-131.

[33] K. Schwetlick, "Kinetische Methoden zur Unter-Suchung von Reactionsmechanismen," VEB, Deutscher Verlag der Wissenschaften, Berlin, 1971.

8

Hydroxyalkylation of Cyclic Imides with Oxiranes. Part II. The Mechanism of Reaction in Presence of Triethylamine

Jacek Lubczak
Faculty of Chemistry, Rzeszów University of Technology, Rzeszów, Poland
Email: jml@prz.edu.pl

ABSTRACT

The mechanism of reaction of cyclic monoimides with oxiranes was established based upon kinetic studies and product analysis. It has been established that the reaction proceeds through initial formation of an adduct of imide and triethylamine. The crucial bond in adduct has ionic character; in non-aqueous solvents it is present as ion pair, while in water the adduct dissociate and free ions are present. The adduct enables the proton transfer from imide to oxirane. The rate determining step is reaction of imide and this adduct. Different values of entropy of transition states obtained from thermodynamic calculations suggest slightly different structure of transition state of rate determining step.

Keywords: Imides; Oxiranes; Mechanism; Kinetics; Non-Kinetics Evidences

1. Introduction

The reaction of cyclic monoimides with oxiranes in presence of triethylamine (TEA) leads to N-(hydroxy-alkyl) imides of the general formula (I):

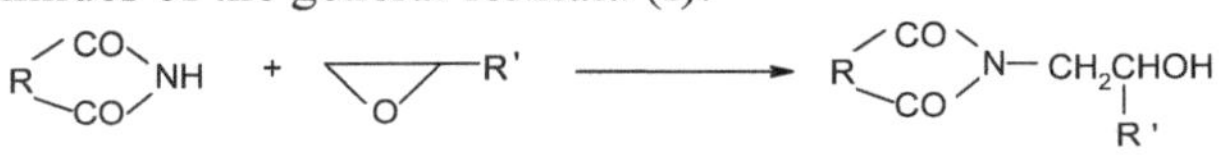

(I)

where R = $(-CH_2-)_2$, $(-CH_2-)_3$, -Ph

R' = -H, $-CH_3$

As it was presented in [1] the reaction follows the rate law (2):

$$V = k''_{1/2} c_K^{1/2} c_{AH}^{3/2} c_B^{1/2} \qquad (2)$$

where c_K, c_{AH}, and c_B mean the catalyst, imide groups, and oxirane concentrations, respectively.

The process can be considered to have the competitive and consecutive character consistent with reaction of N-(hydroxyalkyl)imides, which are the primary products, with oxirane. However the consecutive product is formed with low yield, *ca* 5 - 10 mol%. The rate law describing the reaction of N-(2-hydroxyalkyl)imides with oxiranes obeys another kinetics (first order related to oxirane concentration vs 1/2 for the first step), indicating that the reactions of oxiranes with N-(hydroxyalkyl)imides, which reveals the alcohol character, occur via different mechanism [2-6]. Based upon the rate constants the monoimides were ordered according to lowering reactivity as: PI ≥ SI > GI >HMSI. It has been also found that ethylene oxide (EO) was more reactive than propylene oxide (PO). The free energy of activation ($\Delta G^{\neq}$) was also obtained for those processes. The temperature dependence was linear which provided further evidence that the same mechanism operates for all studied systems.

Here we propose the mechanism of those reactions based on product analysis, intermediate identification and kinetic measurements.

2. Experimental

2.1. Substrates

The substrates were the same as in Part I.

2.2. Conductometric Measurements

The conductivity of imides, TEA, and their mixtures was studied in aqueous solutions at variable molar ratio in thermostated conductometric cells of 20 cm^3 capacity at 20°C (±0.1). The solutions were then diluted twice and conductivity was measured again. No PI solutions were measured due to its low solubility in water.

2.3. Cryoscopic Measurements

The cryoscpic measurements were done for 0.12-0.13-molal solutions of imide and TEA at 1:1 molar ratio (the total concentration of both was 0.25 mol/kg solvent in dioxane ($K_{cryos.}$ = 4.64). The method was standardized by determination of molar mass of imide and TEA. The presence of of imide-TEA adduct was deduced based on following approach. The total concentration of both reagents c_t was the sum of imide (c_{AH}) and catalyst (c_K) concentrations. The actual molal concentration of all components (c_e) present in solution (imide, TEA, and adduct) were found from the measurements of decrease of the freezing point solution related to solvent from (3):

$$-\Delta T = K_k c_e \tag{3}$$

If the adduct is present in solution of mixture the $c_e < c_t$ is expected. If x is the concentration of adduct, the equilibrium:

$$\underset{(c_{oAH}-x)}{AH} + \underset{(c_{oK}-x)}{K} = \underset{x}{A^-, HK^+} \tag{4}$$

takes place and hence the association constant K_a can be obtained from (5):

$$K_a = \frac{x}{(c_{0AH} - x)(c_{0K} - x)} \tag{5}$$

where the association degree β is defined as the ratio of adduct concentration (x) to component concentration (c_o) which is present in lower concentration:

$$\beta = (x/c_o)100\% \tag{6}$$

3. Results and Discussion

3.1. Mechanism of Reaction

The mechanism of the reaction of imides with oxiranes can be deduced based on experimental kinetic law Equation (2). It was assumed that adducts (ion pairs or hydrogen bonded complexes of imides (*AH*) and catalyst (*K*) take part in elementary reactions. The crucial adduct is formed between imide group and nitrogen of TEA. The products are formed upon reaction of imide with adduct of triethylammonium cation and oxirane molecule (*AHK*) according to the scheme (the acronyms under the reaction were introduced to be later used in kinetic equation):

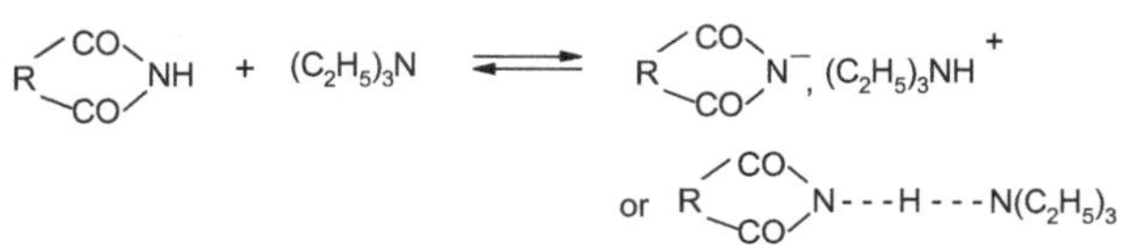

$$AH + K \underset{k_{-1}}{\overset{k_1}{\rightleftarrows}} AHK \tag{7}$$

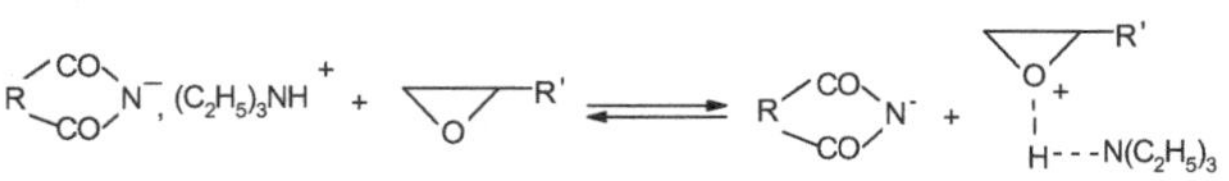

$$AHK + B \underset{k_{-2}}{\overset{k_2}{\rightleftarrows}} A^- + BHK^+ \tag{8}$$

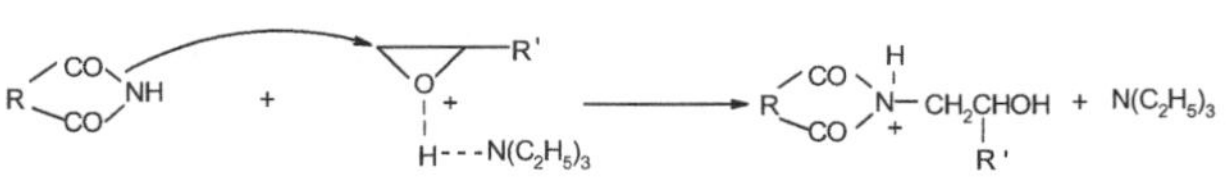

$$AH + BHK^+ \xrightarrow{k_3} AHBH^+ + K \tag{9}$$

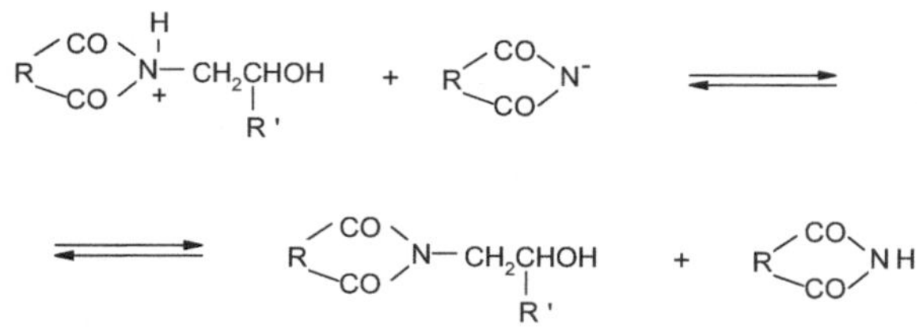

$$AHBH^+ + A^- \underset{k_{-4}}{\overset{k_4}{\rightleftarrows}} ABH + AH \tag{10}$$

Thus the elementary steps are:

1) Formation of AHK adduct from imide and TEA (7), which plays the role of proton transfer agent from imide to oxirane;

2) Reaction of adduct with oxirane with formation of new adduct BHK^+;

3) Reaction of BHK^+ with next molecule of imide resulting in formation of the alcohol protonated on imide group and recovery of catalyst (9);

4) Neutralization of imide anion by proton transfer from protonated alcohol to imide anion (10).

3.2. Kinetic Evidences for Proposed Mechanism

If one assume that:

1) The equilibrium between adduct (*AHK*) and TEA is reached very fast

2) The protonation constant of oxirane (K_2) is low and equilibrium (8) establishes slower that that in reaction (7)

3) The rate determining step is reaction of AH with adduct BHK^+ the rate of oxirane consumption equals the rate of disappearance of BHK^+ adduct, *i.e.*:

$$V_{gen} = -\frac{dc_B}{dt} = -\frac{dc_{BHK^+}}{dt} = k_3 c_{AH} c_{BHK^+} \tag{11}$$

The concentration of c_{BHK} can be found from stationary state condition [7]:

$$\frac{dc_{BHK}}{dt} = k_2 c_{AHK} c_B - k_{-2} c_{A-} c_{BHK^+} - k_3 c_{AH} c_{BHK^+} = 0 \tag{12}$$

Because

$$c_{A-} = c_{BKH^+} \tag{13}$$

then:

$$\frac{dc_{BHK}}{dt} = k_2 c_{AHK} c_B - k_{-2} c^2_{BHK^+} - k_3 c_{AH} c_{BHK^+} = 0 \tag{14}$$

After rearrangement the Equation (14) is:

$$k_{-2} c^2_{BHK^+} + k_3 c_{AH} c_{BHK^+} - k_2 c_{AHK} c_B = 0 \tag{15}$$

The solution for c_{BHK} gives:

$$c_{BHK^+} = \frac{-k_3 c_{AH} + \sqrt{k_3^2 c_{AH}^2 + 4k_{-2} k_2 c_{AHK} c_B}}{2k_{-2}} \tag{16}$$

Taking into account the Assumption 3 the relationship:

$$k_3 c_{AH} \lll 4k_{-2} k_2 c_B c_{AHK}$$

is obvious, which furthermore allows to simplity Equation (16) to:

$$c_{BHK} = \frac{\sqrt{4k_{-2} k_2 c_{AHK} c_B}}{2k_{-2}} = \sqrt{K_2 c_{AHK} c_B} \quad (17)$$

Putting (17) into (11) the rate of reaction becomes:

$$V_{gen} = k_3 \sqrt{K_2} c_{AH} c_{AHK}^{1/2} c_B^{1/2} \quad (18)$$

As before, the concentration of adduct AHK can be obtained from stationary state condition:

$$\frac{dc_{AHK}}{dt} = k_1 c_K c_{AH} - k_{-1} c_{AHK} - k_2 c_{AHK} c_B + k_{-2} c_{BHK^+}^2 = 0 \quad (19)$$

Introducing Equation (17) for c_{BHK^+} the relationship:

$$k_1 c_K c_{AH} - k_{-1} c_{AHK} - k_2 c_{AHK} c_B + k_{-2} K_2 c_{AHK} c_B = 0 \quad (20)$$

is obtained, from which:

$$c_{AHK} = \frac{k_1 c_K c_{AH}}{k_{-1}} \quad (21)$$

or:

$$c_{AHK} = K_1 c_K c_{AH} \quad (22)$$

Thus:

$$V_{gen.} = k_3 \sqrt{K_1 K_2} c_K^{1/2} c_{AH}^{3/2} c_B^{1/2} \quad (23)$$

when

$$c_{AH} \gg c_B \quad (24)$$

the latter equation can be simplified into:

$$V_{gen.} = k_3'' c_K^{1/2} c_B^{1/2} \quad (25)$$

where:

$$k_3' = k_3 \sqrt{K_1 K_2} c_{AH}^{3/2} \quad (26)$$

The obtained kinetic equation is in agreement with the experimental one, which suggests that proposed mechanism operates in the studied systems.

At this step the small differences of reactivity of PI and SI (cf. Part I) with oxiranes can be explained on the basis of proposed mechanism. Higher reactivity of PI is electronic in origin; phtalimide anion is mesomerically stabilized, thus the lifetime of AHK is larger in case of phtalimide, which enables the reactions (8) and (9).

Different values of entropy of transition states obtained from thermodynamic calculations (cf part I, **Table 7**) suggest slightly different structure of transition state of rate determining step. The value of –150 J/mol s can be attributed to transition State (II) which results straightforward from reaction (9), while the values < –200 J/mol s suggest the cyclic transition State (III) exists, in which imide anion participates: participates:

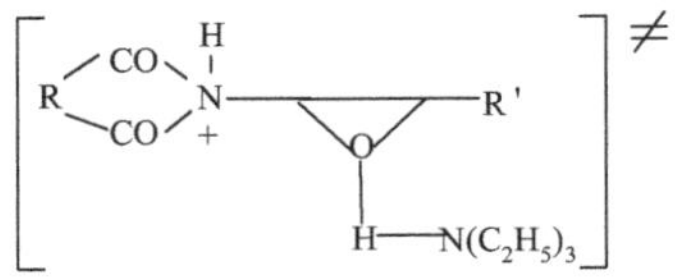

(II)

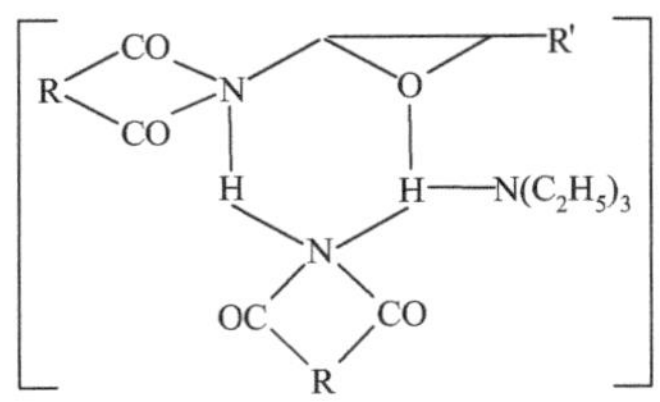

(III)

It has been observed, that the studied reaction are the fastest in DMF in comparison to other solvents (DMSO or dioxane; see Part I., **Table 8**). Presumably the reason is an involvement of solvent molecule in hydrogen cation transfer from imide to oxirane, *i.e.* the same as the catalytic role of TEA in reactions (7) and (8).

3.3. Non-Kinetics Evidences

The crucial points in proposed mechanism are the formation of 1:1 imide-TEA adduct and fast equilibria. Conductivity of aqueous solution of TEA is related to the equilibrium according the to scheme of reaction:

$$(C_2H_5)_3N + H_2O \rightleftarrows (C_2H_5)_3NH^+ + OH^- \quad (27)$$

Introducing the imide or HMSI to the solution of TEA results in considerable increase of conductivity (**Table 1**), which is not additive, *i.e.* it is higher than the sum of conductivities of single components. This suggests that reaction (28) occurs:

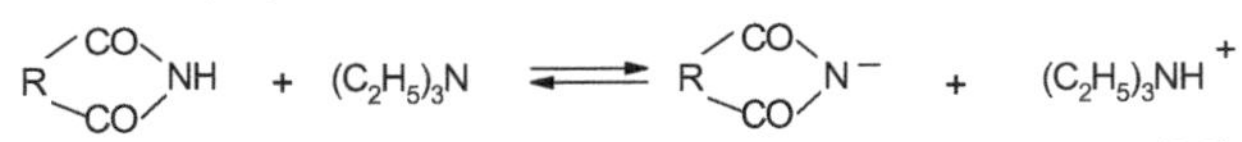

(28)

The measurements performed with variable composition indicated that in imide—TEA mixtures the maximum conductivity is reached immediately at equimolar solution of imide and TEA for the SI-TEA system (**Figure 1**). This was expected result for the stoichiometry of adduct.

Yet another kind of association takes place in non-aqueous solvents. Cryoscopic measurements revealed that all mixtures of imides with TEA follow the non-equality $c_e < c_t$, thus in these conditions the single species (ionic pair or hydrogen bonded adduct) is formed at the 5.2 - 7.7 mol% level in equimolar solutions (**Table 2**). The association is the highest at the PI-TEA system due to mesomeric stabilization of phtalimide anion. This causes shifting of the equilibrium (7) to the right. It should also be noted that association occurs also in the system

Table 1. Changes of conductivity of aqueous solutions of imides, TEA, and equimolar mixtures imide—TEA.

Concentration [mol/dm^3]	Equivalent conductivity [S cm^2] of									
	SI	GI	HMSI	TEA	adduct SI -TEA		adduct GI-TEA		adduct HMSI-TEA	
					additive	experim.	additive	experim.	additive	experim.
0.500	0.018	0.023	-	5.944	5.962	22.729	5.967	22.079	-	-
0.250	0.030	0.026	0.041	6.458	6.488	28.558	6.484	28.038	6.499	16.242
0.125	0.014	0.034	0.053	9.841	9.855	31.117	9.875	31.117	9.394	18.725
$6.25\cdot10^{-2}$	0.073	0.042	0.063	13.427	13.500	36.234	13.469	34.154	13.490	20.643
$3.13\cdot10^{-2}$	0.083	0.062	0.073	17.430	17.513	37.107	17.492	37.107	17.503	21.826
$1.56\cdot10^{-2}$	0.104	0.104	0.133	22.318	22.422	40.934	22.422	39.686	22.451	23.683

Table 2. Analysis of imide—TEA adduct in dioxane by cryoscopic method.

Imide	Molal concentration [mol/kg]		Sum of molal concentrations [mol/kg]		Equilibrium constant	Association degree β [%]
	Imide	TEA	Calculated from weight	Determined cryoscopically		
SI	0.1309	0.1238	0.2547	0.2462	0.60	6.8
PI	0.1286	0.1253	0.2539	0.2441	0.72	7.8
GI	0.1427	0.1457	0.2884	0.2797	0.49	6.1
HESI	0.1180	0.1181	0.2361	0.2300	0.48	5.2

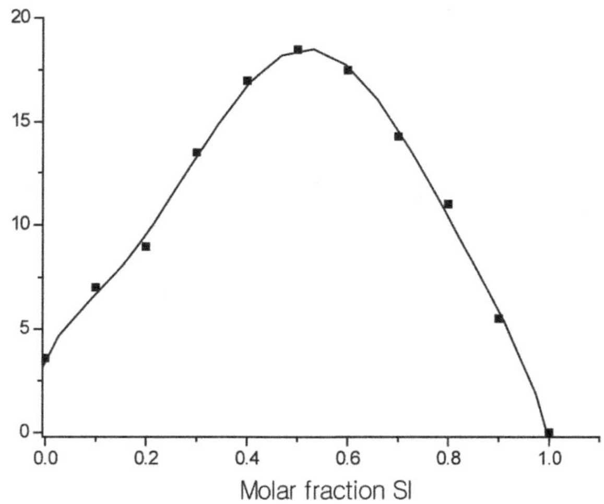

Figure 1. Changes of conductivity of solutions for the SI—TEA system of variable composition.

HESI-TEA. This suggests that proton donor to oxirane can be not only the imide but also its product of reaction, the alcohol, while the TEA catalyst mediates the transfer of proton from imide or alcohol to oxirane.

It has been noted that upon addition of oxirane to HMSI partial release of formaldehyde from N-hydroxymethyl group occurred accompanied by its migration to the end of hydroxyalkyl chain. Based upon cryoscopic data and previous studies it was possible to determine the mechanism of this rearrangement. Previously we have focused on kinetic models for initial steps of oxirane reactions with N-(hydroxymethyl) derivatives of some azacyclic compounds (barbituric, isocyanuric, and uric acids) leading to the formation of polyetherols [8-11]. The percentage of N-hydroxymethyl and groups in HMSI and O-hydroxymethyl groups in the product of its reaction with oxirane was determined as weakly bound formaldehyde. According to the data the HMSI contained 1 mole of formaldehyde per mole of HMSI (%CH_2O_{calc} = 23.3, %CH_2O_{found} = 23.2). It was also found that upon ad- dition of 1 mole of EO to HMSI *ca* 0.3 mole of unstable bonded formaldehyde as O-(hydroxymethyl) groups remains, but after addition of 2 moles EO practically whole formaldehyde is blocked by oxirane. These results suggest, that either O-(hydroxymethyl) groups are formed as the result of partial dissociation of formaldehyde from N-(hydroxymethyl) groups of HMSI and then added to N-(hydroxyalkyl) group ac- cording to the scheme:

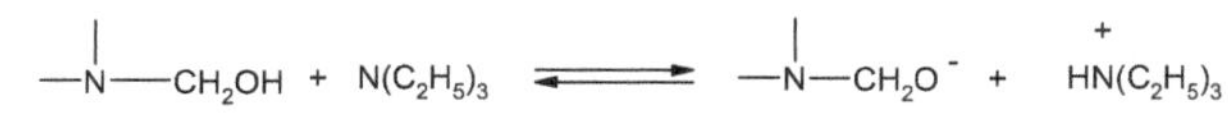

(29)

(30)

(31)

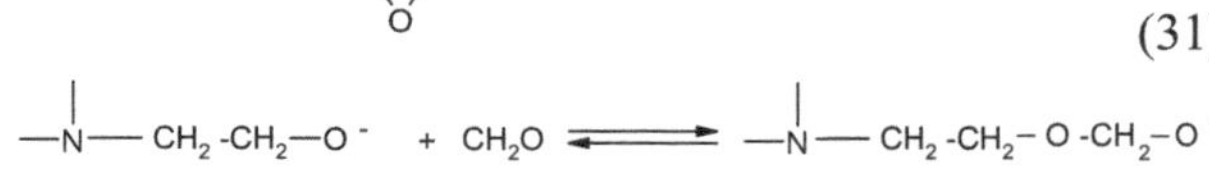

(32)

(33)

or first the oxirane is attached to N-(hydroxymethyl) group with formation of (IV), from which the formaldehyde is released and rearranged into the end of chain according to the scheme:

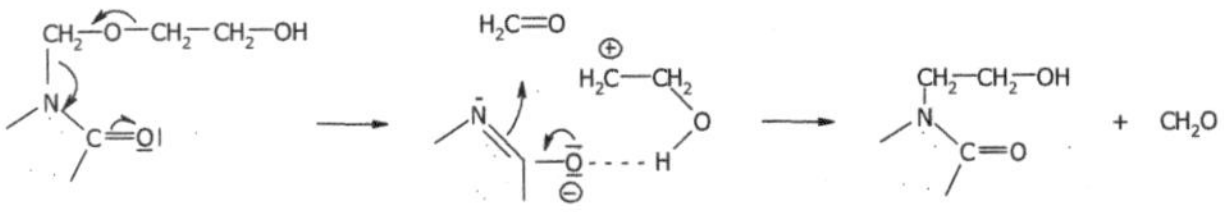

(34)

(35)

Cryoscopic determination of molar mass of HMSI in dioxane and DMSO indicated, that in both solvents the molar mass of HMSI was equal to calculated one (128.9 and 128.5 g/mol, respectively; calc. 129.1). This suggests that no dissociation of formaldehyde from N-(hydroxymethyl) groups occurs and the formaldehyde rearrangement takes place after blocking with oxirane (reaction 34). This corroborate well with our previous studies on the reaction of hydroxymethylated derivatives of uric acid with oxiranes [11]. The system HMSI-TEA was also studied by cryoscopic method (**Table 2**). As in other cases the relation $c_e < c_t$, was kept valid, what confirms that dissociation of formaldehyde from alcoholate did not occur upon TEA.

It is not possible to define the character of the bond in adduct on the basis of cryoscopic measurements. Therefore the IR and ^{1}H-NMR spectra were recorded. The IR spectrum of SI shows the valence band of NH at about 3200 cm^{-1} (**Figure 2(b)**). When 3-fold excess of TEA was added, the band disappeared (**Figure 2(d)**) accom-panied by appearance of weak band from ammonium cation $(C_2H_5)_3NH^+$ in the region 2300 - 2500 cm^{-1}. Undo-ubtedely, the proton transfer to TEA from imide took place. This band was more pronounced in the mixture of HESI and TEA (**Figures 2(c)** and **(e)**). The spectral features were: the disappearance of hydroxyl valence band at 3500 cm^{-1}, and simultaneous appearance of absorption at 2400 cm^{-1}. Thus, the product of reaction can also be the proton donor, which agrees well with cryoscopic data.

IR spectra rather preclude the formation of hydrogen bonded adduct; in such a case the broadening of the N-H

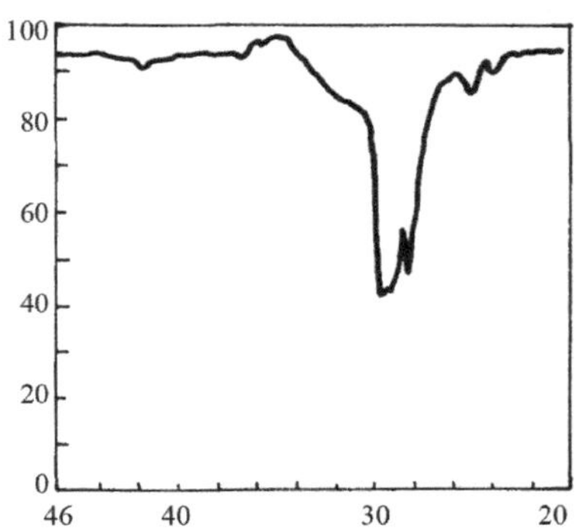

(a)

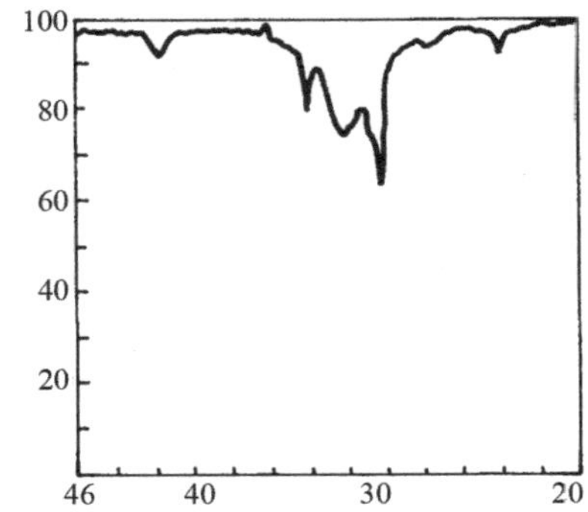

(b)

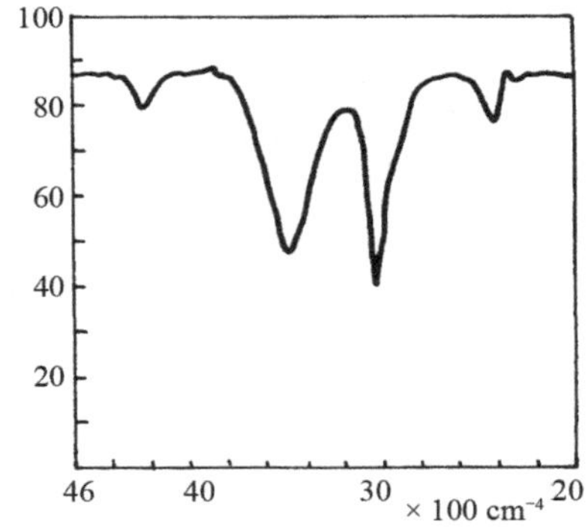

(c)

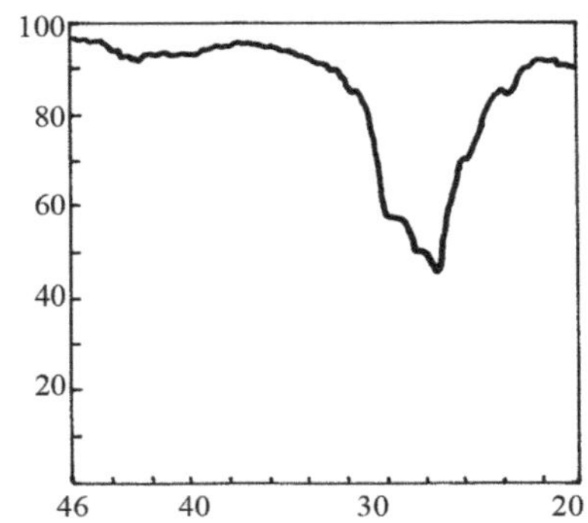

(d)

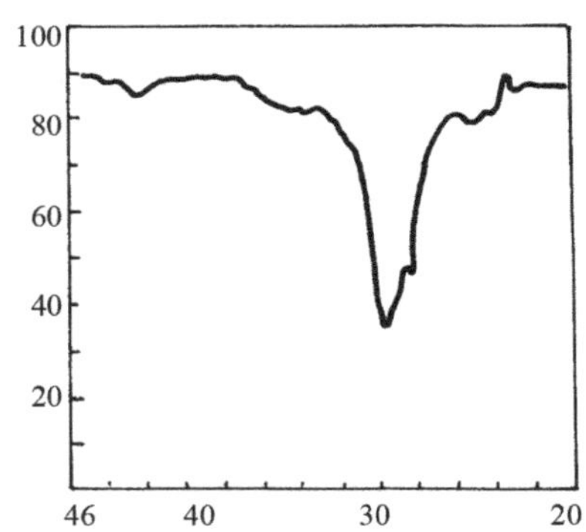

(e)

Figure 2. The relevant region of IR spectrum of TEA (a), SI (b), HESI (c), and mixtures of TEA with SI (d) and HESI (e).

band should be observed, while only the disappearance was the case. ^1H NMR spectra provide further evidences on the presence of ion pair. Addition of TEA to imide solution in DMSO-d_6 resulted in disappearance of imide group proton resonance >NH and simultaneous grow of resonance from $(C_2H_5)_3NH^+$ cation shifted upfield. The resonances are separated by 4.2 ppm in SI/SI-TEA system and by 6.1 ppm PI/PI-TEA. This spectroscopic picture also indicates the formation of ion pair instead of hydrogen bonding. Individual chemical shifts for the system SI/SI-TEA and PI/PI -TEA are as below:

SI (2.8, 4H, CH_2; 11.4, 1H, NH); SI/TEA (1.05, 9H, CH_3; 2.55, 6H, CH_2 from TEA, 2.8, 4H, CH_2 from SI; 7.2, 1H, NH^+).

PI (7.8, 4H, Ph; 11.6, 1H, NH); /PI-TEA (1.05, 9H, CH_3; 2.55, 6H, CH_2 from TEA, 7.8, 4H, Ph; 5.5, 1H, NH^+).

4. Summary and Conclusions

1) Based upon experimental kinetic the mechanism of the reaction of cyclic imides and oxiranes was proposed and verified by instrumental and analytical methods.

2) The initial step of reaction is the formation of 1:1 imide: TEA adduct, which intermediates the proton transfer from imide to oxirane. The adducts could not be isolated, although their presence in solutions was demonstrated experimentally.

3) The crucial bond in adduct has ionic character; in non-aqueous solvents it is present as ion pair, while in water the adduct dissociate and free ions are present.

4) The suggested mechanisms can be applied in expalantion of first stages of obtaining of polyetherols in reaction of oxiranes with cyclic imides containg a couple of imide groups such as isocyanuric acid, parabanic acid and barbituric acid etc.

REFERENCES

[1] J. Lubczak, "Hydroxyalkylation of Cyclic Imides with Oxiranes. Part I. Kinetics of Reaction in Presence of Triethylamine as Catalyst," *Open Journal of Physical Chemistry*, Vol. 2, No. 2, 2002, p. 2.

[2] L. Shechter and J. Wynstra, "Glycidyl Ether Reactions with Alcohols, Phenols, Carboxylic Acids, and Acid Anhydrides," *Industrial and Engineering Chemistry*, Vol. 48, No. 1, 1956, pp. 86-93. doi:10.1021/ie50553a028

[3] S. Blum, P. Walsh and R. Bergman, "Epoxide-Opening and Group-Transfer Reactions Mediated by Monomeric Zirconium Imido Complexes," *Journal of American Chemical Society*, Vol. 125, No. 47, 2003, pp. 14276-14277. doi:10.1021/ja037267t

[4] L.-Z. Dai and M. Shi, "A Gold(I)-Catalyzed Intramolecular Reaction of Propargylic/Homopropargylic Alcohols with Oxirane," *Chemistry—A European Journal*, Vol. 14, No. 23, 2008, pp. 7011-7018. doi:10.1002/chem.200701954

[5] N. Iranpoor, H. Firouzabadi, R. Azadi and F. Ebrahimzadeh, "Regioselective Synthesis of *vic*-Halo Alcohols and Symmetrical or Unsymmetrical *vic*-Dihalides from Epoxides Using Triphenylphosphine *N*-Halo Imides," *Canadian Journal of Chemistry*, Vol. 84, No. 1, 2006, pp. 69-75. doi:10.1139/v05-261

[6] J. F. Larrow and E. N. Jacobsen, "Asymmetric Processes Catalyzed by Chiral (Salen)Metal Complexes," *Topics in Organometallic Chemistry*, No. 6, 2004, pp. 123-152. doi:10.1007/b11772

[7] K. Schwetlick, "Kinetische Methoden zur Un-tersuchung von Reactionsmechanismen", VEB, Deutscher Verlag der Wissenschaften, Berlin 1971.

[8] I. Cisek-Cicirko and J. Lubczak, "Reactions of Hydroxymethyl Derivatives of Uric Acid with Oxiranes. II. an Analysis of Reaction Course and Product Structure," *Journal of Applied Polymer Science*, Vol. 83, No. 9, 2002, pp. 1955-1962. doi:10.1002/app.10108

[9] A. Ślączka and J. Lubczak, "Hydroxyalkylation of Barbituric Acid. II. Synthesis of Polyetherols with Pyrimidine Ring," *Journal of Applied Polymer Science*, Vol. 106, No. 6, 2007, pp. 4067-1074. doi:10.1002/app.26742

[10] M. Kucharski, J. Lubczak and E. Rokaszewski, "Addition of Oxiranes to Hydroxymethyl Derivatives of Isocyanuric Acid," *Chemia Stosowana*, Vol. 27, No. 1-2, 1983, pp. 65-77.

[11] J. Lubczak, "Reactions of Hydroxymethyl Derivatives of Uric Acid with Oxiranes: Recognition of Mechanism Based on Kinetic Studies," *International Journal of Chemical Kinetics*, Vol. 38, No. 5, 2006, pp. 345-350. doi:10.1002/kin.20167

Aminolysis of 1-(1-Hydroxybenzotriazolyl)-2,4-dinitrobenzene and 2-(1-Hydroxybenzotriazolyl)-5-nitropyridine

Sherine N. Khattab[1], Mohamed A. H. Kharaba[2], Amal El-Hawary[2], Ayman El-Faham[1,3,4], Ezzat A. Hamed[1*]

[1]Department of Chemistry, Faculty of Science, Alexandria University, Ibrahimia, Egypt
[2]Department of Chemistry and Physics, Faculty of Education, Alexandria University, Alexandria, Egypt
[3]Institute for Research in Biomedicine, Barcelona Science Park, Baldiri Reixac, Spain
[4]Department of Chemistry, College of Science, King Saud University, Riyadh, KSA
Email: ezzatah2@hotmail.com

ABSTRACT

The reaction 1-(1-hydroxybenzotriazolyl)-2,4-dinitrobenzene 1 and 2-(1-hydroxybenzotriazolyl)-5-nitro-pyridine 2 with amines undergoes amination followed by elimination of the 1-hydroxyl benzotriazolyl anion. The kinetic data for the reaction of 1 and 2 with Mo, CHA and An in MeOH and AN proceeded by uncatalysed mechanism in which the rate limiting step is the leaving group departure, whereas the reaction with Mo in toluene proceeded by uncatalysed mechanism in which the formation of the zwitterionic intermediate is the rate determining step. While the reactions of 1 with CHA and An and the reaction of 2 with CHA in toluene proceeded by SB mechanism in which the rate determining step is the proton transfer process. The reactions of 1 and 2 with Mo in the three solvents and with CHA and An in MeOH and AN is greatly depended on the stability of the zwitterionic intermediate. The effect of ring activation is due to the ground state stabilization and the more efficient delocalization of the negative charge with a nitro group than with a ring-nitrogen in the transition state. The low activation enthalpies $\Delta H^{\#}$ and the highly negative activation entropies $\Delta S^{\#}$ are due to the intramolecular hydrogen bonding with the ammonio hydrogen present in the transition state.

Keywords: Nucleophilic Aromatic Substitution Reactions; Kinetic Measurements, 1-Hydroxybenzotriazole; Solvent Effect; Morpholine; Cyclohexylamine and Aniline; Uncatalyzed and Specific Base Catalyzed Mechanisms

1. Introduction

Nucleophilic aromatic substitution (S_NAr) reaction constitutes an active field of organic chemistry, which is very important from viewpoints of both practical [1-3], computational [4-6] and theory of nucleophilic aromatic substitution reactions [7-10]. Different mechanisms were suggested for unimolecular nucleophilic aromatic substitution [11], aryne mechanism [12], nucleophilic aromatic substitution with rearrangement [13-15], nucleophilic aromatic substitution involving hydrogen substitution and nucleophilic aromatic substitution *via* addition-elimination process (S_NAr). Kinetic studies involving primary and secondary amines as nucleophiles, in solvents of different polarities, play a central role not only in firmly establishing the two or multi-step nature of the mechanism, but also in answering questions regarding the relative rates of intermediate complex formation and decomposition [16-24]. Possible pathway mechanism for this type of reactions are specific base (SB) [22-26], specific base-general acid (SB-GA) [27], or dimer-mechanism [28,29].

In continuation of our studies in the field of nucleophilic aryl and acyl reactions [30-45], we addressed kinetic studies of the leaving group ability of 1-hydroxybenzotriazole (HOBt) in two example: 1-(1-hydroxybenzotriazolyl)-2,4-dinitro-benzene 1 and 2-(1-hydroxy-benzotriazolyl) 5-nitro-pyridine 2. The kinetic studies of 1 and 2 with different amines such as morpholine, cyclohexylamine and aniline in methanol (MeOH), acetonitrile (AN) and toluene (Tol) will be measured. Other parameters such as ring activation, steric effect at the reaction centre, ground state stabilization of substrates 1 and 2 and finally, a plausible mechanism for each substrate with the titled amine and solvent of different polarity will be discussed.

2. Results and Discussion

2.1. Synthesis of the Starting Materials and Reaction Products

1-(1-hydroxybenzotriazolyl)-2,4-dinitrobenzene 1 and 2-(1-hydroxybenzotriazolyl) 5-nitropyridine 2 were readily prepared by the reaction of 1-hydroxy benzotriazole (HOBt) with 1-chloro-2,4-dinitrobenzene and 2-chloro-5-nitropyridine, in the presence of 2 equivalents of triethylamine (Et_3N), **Scheme 1**. The structure of compounds 1 and 2 was confirmed by Elemental analysis, IR and NMR spectroscopy (Exp. section).

The reaction of 1 and 2 with cyclohexylamine (CHA), morpholine (Mo) and aniline (An) in methanol (MeOH), acetonitrile (AN) and toluene (Tol) undergoes amination followed by elimination of the 1-hydroxyl benzotriazolyl anion as indicated from the isolated products namely *N*-cyclohexyl-2,4-dinitroaniline 3 [46], 4-(2,4-dinitrophenyl) morpholine 4 [47], 2,4-dinitro-*N*-phenylaniline 5 [48], *N*-cyclohexyl-5-nitropyridin-2-amine 6 [49], 4-(5-nitropyridin-2-yl) morpholine 7 [50] and 5-nitro-*N*-phenylpyridin-2-amine 8 [51] respectively, **Scheme 1**. The purity of the products was checked by comparing their melting point with the previously reported ones in the literature.

2.2. Kinetics and the Reaction Mechanism

In general, the reaction of 1 and 2 with Mo, CHA, An in MeOH exhibited lower rates than those in AN and has different kinetic behavior in Tol. Also the order of reactivity of amines towards compounds 1 and 2 followed the order Mo > CHA > An while their pK_a values are in the order CHA > Mo > An, **Tables 1** and **4**. This is due to 1) that MeOH molecules reduce the nucleophilicity of the amines by solvation through hydrogen bonding; 2) the formation of less polar transition state which is favored in less polar solvent; 3) the pK_a values of the amines, which may be responsible for the rate differences; 4) the stability of the zwitterionic or the Meisenheimer intermediates involved in the reaction pathway(s). This stabilization arises from the intramolecular and intermolecular interactions; 5) the difference in nature of the electron-withdrawing power of the substituent in the substrate.

2.2.1. Reactions of (1) with Mo, CHA and An in MeOH, AN and Tol

1) In MeOH and AN

The rate of the reactions of 1 with Mo, CHA and An are followed spectrophotomertrically to 90% completion of the products. The condition of measurements ensuring first-order kinetics.

The pseudo-first order rate constants observed (k_{obs}) obeyed Equation (1) with negligible k_o (~0) in MeOH and AN indicate that general base catalysis is absent in the present aminolysis. The apparent second-order rate constants k_A ($l \cdot mol^{-1} \cdot s^{-1}$) are summarized in **Table 1**.

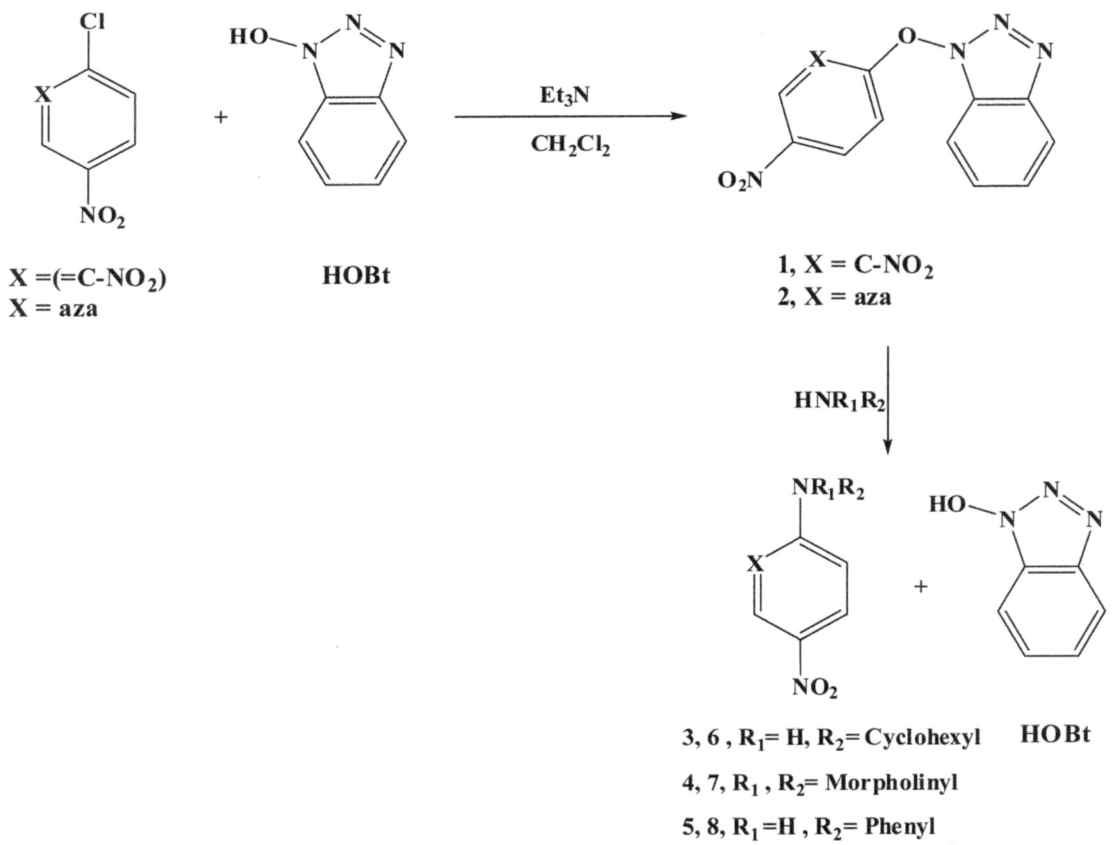

Scheme 1. Synthesis and reaction of 1 and 2 with cyclohexylamine (CHA), morpholine (Mo) and aniline (An).

Table 1. Rate constants[a] for the reactions of 1-(1-hydroxybenzotriazolyl)-2,4-dinitrobenzene 1 with Mo, CHA and An in MeOH, AN and Tol.

Amine/ T°C	15°C	20°C	25°C	30°C	35°C	40°C	45°C	$\Delta H^{\#}$ kcal·mol^{-1}	$-\Delta S^{\#}$ cal·mol^{-1}·K^{-1}
					10^3 k_2 l·mol^{-1}·s^{-1}/Methanol				
Mo[b]		117.5	189.0	259.0	375.0	420.0		6.36	43.24
CHA[b]			35.0	40.0	50.0	60.0	79.0	7.42	41.73
An[b]		5.3	6.5	9.0	10.3	13.0		7.63	42.94
					10^3 k_2 l·mol^{-1}·s^{-1}/Acetonitrile				
Mo			332.9	432.1	519.7	588.7	637.9	5.48	42.30
CHA			193.8	243.2	308.4	350.2	407.1	6.38	40.41
An			19.1	26.2	30.3	33.7	43.0	6.47	44.67
					10^3 k_2 l·mol^{-1}·s^{-1}/Toluene[c]				
Mo	585.5	652.3	765.8	884.7	1080.3			4.79	43.00
					10^3 k_3 l·mol^{-2}·s^{-2}				
CHA			234.0	282.0	388.0	455.0	547.0	7.88	35.07

[a]$k_2 = k_{obs}$/[morpholine]. Rate constants were calculated by the least-squares method and the correlation coefficient was 0.99 in most cases; [b]pK_a values of CHA, Mo and An are 10.567, 8.38 and 4.36 respectively; [c]Reaction of 1 with An in Tol at 50°C shows third-order kinetics with $k_3 = 1 \times 10^{-4}$ l·mol^{-2}·s^{-1}.

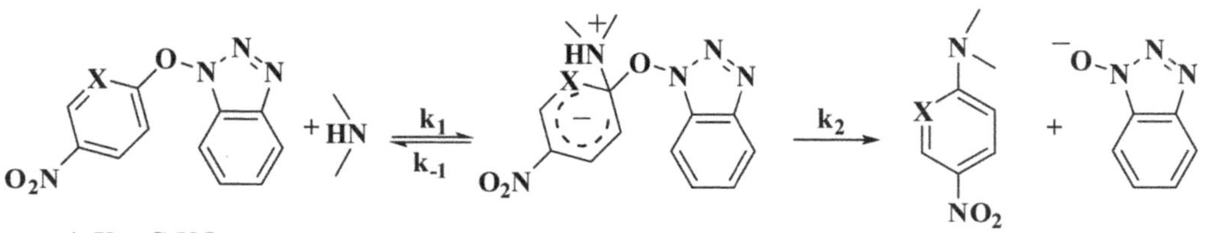

1, X = C-NO$_2$
2, X = aza
Am
I (Zwitterionic Intermediate)
P
$^{-}$OBt (LG)

Scheme 2. The mechanism of the reaction of 1 and 2 with cyclohexylamine (CHA), morpholine (Mo) and aniline (An).

$k_{obs} = k_o + k_A$ [Amine] (Amine = Mo, CHA, An) (1)

It shows that the rate is inconsistent with a stronger nucleophile as normally observed for a typical nucleophilic aromatic substitution reactions.

The second order rate law points out that the rate determining step is either the formation of the zwitterion intermediate I or the uncatalysed leaving group departure step, **Scheme 2**. It involves the reaction of 1 with the titled amines Am (Mo, CHA, An) which is proceeding by the attack of the amine on the *ipso* carbon of the substrate to form the zwitterionic intermediate I, followed by departure of 1-hydroxyl benzotriazolyl anion to give the substitution product.

The differentiation between the two considerations depends greatly on the rate of leaving group departure compared to the rate of the formation of the zwitterionic intermediate I and the pK_a values of the amine used. This will be achieved by changing the nature of the leaving group *i.e. element-effect like* and the use of amines with different pK_a values.

The present kinetic results shows difference in rates with changing in the nature of the leaving group from OBt to Cl, I, O-N = C(COOEt)$_2$, O-N = C(CN)COOEt, O-N = C(CN)$_2$ [36,37,51-53].

This indicates that the rate constants follow the order ON = C(CN)$_2$ > ON = C(CN)COOEt > OBt > I > Cl. As a result, one can conclude that the departure of the leaving group in the reactions of **1** with Mo, CHA and An in MeOH and AN is the rate-determining step, **Scheme 2**, unlike most nucleophilic aromatic substitution reactions [1-7].

The rate enhancement for the reactions of the amines with compound 1 follows the order Mo > CHA > An which is not in harmony with the order of pK_a values of the amines under investigation. Also the small rate constant ratios between two amines compared to the difference between the pK_a values of the corresponding amines is a further indication that the rate determining step is the departure of the $^{-}$OBt group.

The thermodynamic parameters computed in **Table 1** indicate that the reactions in MeOH and AN are not greatly influenced by the nature of the amines. Entropies of activation are negative as expected for bimolecular reactions. The small difference of activation parameters $\Delta H^{\#}$ and $\Delta S^{\#}$ for the reaction of 1 with Mo, CHA, An in MeOH and AN may be due to a slight solvent participation in the transition state.

2) In Toluene

The kinetic of the reaction of 1 with Mo, CHA and An were studied at 15°C - 50°C under pseudo-first-order conditions with respect to 1. The second-order rate constants of the reaction at various concentrations of Mo are shown in **Table 1**. These data show that no significant acceleration in rate constants occurred with the increase in the Mo concentrations. Hence the reaction is not catalyzed by base and it can be suggested that the formation of the zwitterionic intermediate is the rate determining, **Scheme 2**. This suggestion is consistent with the difference in rate constants as well as the activation parameters for the reaction of 1 with the same amine in MeOH and AN. The rate constants for the reaction of 1 with Mo in Tol at 25°C, 30°C and 35°C are higher than those in MeOH by 4, 3.4, 2.8 times and in AN by 2.3, 2.0, 3.5 times. Also, the enthalpy of activation in Tol ($\Delta H^{\#}$ = 20.04 kJ/mol) is smaller than those in MeOH ($\Delta H^{\#}$ = 26.59 kJ/mol) and AN ($\Delta H^{\#}$ = 22.89 kJ/mol) reflecting a higher rate in the former solvent presumably due to the change in the rate determining step in **Scheme 2**.

On the other hand, the values of the second-order rate constants k_2 for the CHA and An increased linearly with amine concentrations, *i.e.* catalyzed reaction. A plot of k_2 (values k_{obs}/[amine]) versus amine (An or CHA) concentrations showed a straight line with an intercept on the *y*-axis indistinguishable from zero. This indicates that the contribution of k_2, the unanalyzed pathway, is negligible, **Scheme 3**.

In these systems, CHA is more efficient than An in catalyzing the reaction as shown from k_2 values, **Table 1**. The dependence of the reaction on two amine molecules for the reactions of 1 with An and CHA in Tol can proceed by a dimer mechanism [28,29], specific base-general acid (SB-GA) [27], or specific base catalysis (SB) [22-26] (**Scheme 3**). The dimer mechanism is rejected on the ground that the plots of k_2 against [amine] for the reactions of 1 with An and CHA do not exhibit a curvilinear response and the plots of quotient k_2 [amine] against [amine] did not gave straight lines [28,29].

The confirmation that the reaction is subjected to SB-GA mechanism is done by measuring the reaction rates in the presence of anilinium hydrochloride and cyclohexylamine hydrochloride at constant An and CHA concentrations respectively and vice versa. Actually, neither aniline hydrochloride or cyclohexylamine hydrochloride are completely soluble in toluene. This excludes catalysis by SB-GA mechanism, which as previously reported, occurs only in dipolar solvents such as acetonitrile and DMSO [54].

The general base catalysis (SB) is likely to involve rate limiting deprotonation of the formed zwitterionic intermediate I, followed by rapid elimination of the leaving group [55]. Thus by adding or omitting bases in the reaction mixture, we may exert considerable control of the relative rate of the product formation and thus on the overall rate-controlling step, **Scheme 3**.

The kinetic expression of SB mechanism when the amine acts as both nucleophile and catalyst base can be expressed by Equation (2).

$$k_A = \frac{k_1 k_2 k_{Am}[Am]}{k_{-1} k_2} = K_1 k_{Am}[Am] \quad (2)$$

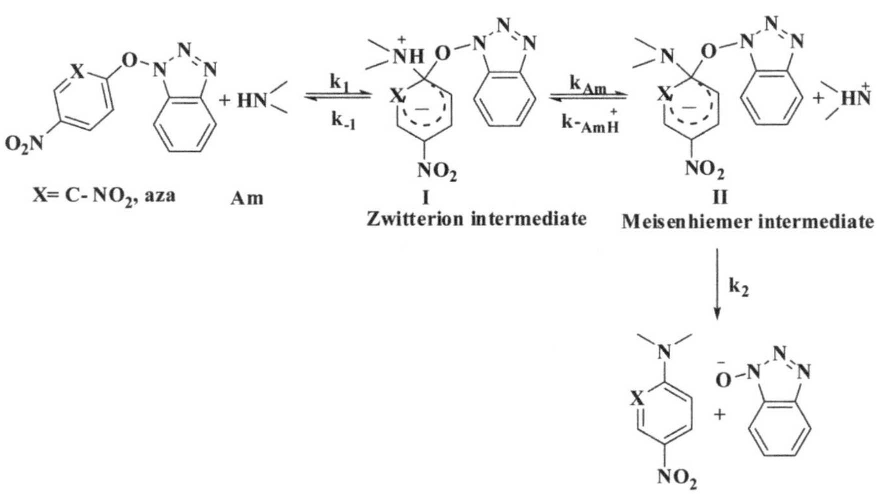

Scheme 3. The mechanism of the reaction of 1 and 2 with cyclohexylamine (CHA) and aniline (An) in Tol by the specific base catalysis mechanism (SB).

where, K_1 is the apparent second-order rate constant. Where,

$$\frac{k_1}{k_{-1}} = K_1$$

Equation (2) is consistent with the amine catalysis for the reaction of 1 with amine in Tol because a linear correlation between the k_A value and [Am] values passing through the origin or intercept on the y-axis indistinguishable from zero is observed. Accordingly, the proton transfer process is the rate-determining step. This confirmed that the reaction is subjected to general base catalysis as shown in **Scheme 3**, where the measurements was carried out in the presence of 1,4-diazabicycle-[2,2,2] octane (DABCO). A complete study of the combined influence of amines was then outlined as follow: the reaction of 1 with CHA in Tol was run at a fixed [amine] (in the range 0.0525 mole·dm^{-3}) in the presence of varying amounts of [DABCO] (up to 0.1049 mole·dm^{-3}). The reactions were also measured at fixed [DABCO] (0.0525 mole·dm^{-3}) in presence of varying [amine]. Results are set forth in **Table 2**. The plot of k_A vs [DABCO] gives straight lines and excellent correlation coefficient. The slopes are also given in the table. The value of $k_{Dabco}/k_{amine} \geq 1.86$ in presence of constant [DABCO] and variable [amine], where it is 2.27 in presence of constant [amine] and variable [DABCO].

In addition, values of K_1k_{DABCO}, where k_{DABCO} represent the k_A rate constants for the reactions of 1 with An DABCO catalyzed route, were obtained from the slopes of the approximately linear plot of the data in **Table 3**.

It is observed that the rate constant in presence of DABCO is larger by about 5 and 2 times than in its absence, (at constant amine concentration and varying DABCO and at constant DABCO concentration and varying An concentration, respectively). Consequently, the catalysis present for the reaction of 1 with both amines in Tol is general base catalysis.

2.2.2. Reaction of 1-(1-Hydroxybenzotriazolyl)-5-nitropyridine 2 with Mo and CHA in MeOH, AN and Tol

All reactions of 2 with CHA and Mo were carried out under pseudo-first-order conditions, with various amine concentrations and obeyed the kinetic law given in Equation (3).

Table 2. Kinetics[a] of the reaction of 1 with CHA in the presence of DABCO in toluene at 40°C.

[Dabco] mol·dm^{-3}	[CHA] mol·dm^{-3}						10^2 slope[b]
	0.0175	0.0350	0.05250	0.0699	0.0874	0.1049	
0.00	4.35	20.58	32.52	36.24	47 .44	58.92	54.80
0.0175			37.00				
0.0350			44.22				
0.0525	5.07	21.11	55.00	65.12	73.77	110.58	1022.80
0.0699			61.11				
0.0874			96.19				
0.1049			133.57				
10^2 slope[b]			1244.7				

[a]10^2 k_A l·mole^{-1}·s^{-1}. [b]the correlation coefficient 0.99.

Table 3. Kinetics[a] of the reaction of 1 with An in the presence of DABCO in toluene at 50°C.

[Dabco] mol·dm^{-3}	[An] mol·dm^{-3}						10^4 slope[b]
	0.549	1.098	1.646	2.195	2.744	3.290	
0.00	0.979	1.76	2.71	4.95	7.68	10.82	1.00
0.549			2.99				
1.098			3.185				
1.646	2.034	2.88	7.178	11.55	16.66	24.83	20.00
2.195			21.65				
2.744			36.58				
3.290			48.79				
10^4 slope[b]			45.00				

[a]10^4 k_A s^{-1}·mole^{-1}. [b]the correlation coefficient 0.99.

$$\text{Rate} = k_2[2][\text{Am}] = k_{obs}[\text{Am}] \quad (3)$$

$$k_{obs} = k_2[\text{Am}] \quad (4)$$

The second-order rate constants, k_A, summarized in **Table 2**, were obtained from a straight-line plot of k_{obs} vs. the concentration of the amine. It should be noted that compound 2 was too unreactive for satisfactory measurement for the reactions with An in all solvents even toluene due to the low reactivity of compound 2 and aniline (An). The kinetic data for the reactions of 2 with Mo and CHA in MeOH and AN are compatible with rate limiting nucleophilic attack and also with an intermediate decomposition without amine catalysis. In a similar way as discussed previously for the reaction of 1 with the same amines, the differentiation between the two considerations depends greatly on the same factors. This leads us to suggest that the rate determining step is the expulsion of the leaving group step, **Scheme 2**. Inspection of the activation parameters, **Table 4**, shows that the activation enthalpy values are slightly variable and the activation entropy values are mainly responsible for the rate difference for the reactions of 2 with Mo and CHA in MeOH and AN.

The same kinetic behavior and mechanism are observed for the reaction of Mo with compound 2 in Tol. The data show that reaction is not catalyzed by a base and the formation of the zwitterionic intermediate is the rate determining, **Scheme 2**. This is corroborated with the difference in rate constants as well as the activation parameters for the reaction of 2 with the same amine in MeOH and AN, **Table 4**.

The kinetic results for the reactions of 2, with CHA in Tol indicate a squared dependence of k_{obs} on amine concentration reflecting the condition $k_{-1} \gg k_{Am}$ so that proton transfer is rate-limiting, **Scheme 3**. Plots of k_2 versus CHA concentration were linear with negligible intercept indicating the uncatalyzed decomposition of the intermediate, the k_2 step. The dimer and SB-GA mechanisms are rejected due to the reasons mentioned previously for the reaction of 1 with the same amine.

The suggested SB mechanism, **Scheme 3** is supported by measurements of the reaction of 2 in the presence of added 1,4-diazabicycle [2,2,2] octane (DABCO). A complete study of the combined influence of amines was then outlined: The reactions were run at a fixed [amine] (in the range 0.525 mole·dm^{-3}) in the presence of varying amounts of [DABCO] (up to 1.049 mole·dm^{-3}). In addition, the reactions were also measured at fixed [DABCO] (0.525 mole·dm^{-3}) in the presence of varying [amine]. Results are set forth in **Table 5**. The plot of k_2 vs [DABCO] gives straight lines and excellent correlation coefficient. The value of k_{Dabco}/k_{amine} is ≥ 2.00 in presence of constant [DABCO] and variable [amine] and it is 5.00 in presence of constant [amine] and variable [DABCO]. Accordingly the kinetic results of the reaction of 2 with CHA is best analyzed in terms of **Scheme 3** in which base catalysis is attributed to rate-limiting proton transfer from the zwitterionic intermediate I to base, followed by rapid expulsion of $^{-}$OBt anion.

Table 4. Rate constants[a] l·mol^{-1}·s^{-1} for the reactions of 2-(1-hydroxybenzotriazol-yl)5-nitropyridine 2 with Mo and CHA in MeOH, AN and Tol.

Amine/T°C	25°C	30°C	35°C	40°C	45°C	ΔH$^{\#}$ kcal·mol^{-1}	−ΔS$^{\#}$ cal·mol^{-1}·K^{-1}
Mo[b]	4.97	7.69	10.30	12.10	13.49	8.68	39.82
CHA[b]	0.50	0.70	0.90	1.20	1.50	9.72	41.07
Mo	34.90	45.40	55.50	76.20	91.10	8.58	36.49
CHA	2.10	3.00	4.20	5.20	6.20	9.66	38.36
Mo	83.10	102.0	128.0	155.0	193.0	7.40	43.29
CHA	1.20	2.10	2.80	3.70	4.30	11.19	34.17

[a]k_2 = k_{obs}/[morpholine]. Rate constants were calculated by the least-squares method and the correlation coefficient was 0.99 in most cases; [b]pK_a values of CHA and Mo are 10.567 and 8.38 respectively.

Table 5. Kinetics[a] of the reaction of 2 with CHA in the presence of Dabco in toluene at 40°C.

[Dabco] mol·dm^{-3}	[CHA] mol·dm^{-3}						10^4 slope[b]
	0.175	0.350	0.525	0.699	0. 874	1.049	
0.00	2.020	4.970	6.023	18.530	26.240	42.300	37.00
0.175			10.440				
0.350			23.100				
0.525	4.330	10.790	40.190	30.210	49.580	68.720	630.00
0.699			60.650				
0.874			74.400				
1.049			86.780				
10^4 slope [b]			810.000				

[a]10^4 k_A·s^{-1}·mole^{-1}; [b]the correlation coefficient 0.99.

2.3. Comparison between the Reactions of 1 and 2 with Mo, CHA and An in MeOH, AN and Tol

2.3.1. Relative Activation of a Ring-Nitrogen and 2-Nitro of Substrates 1 and 2

The bulk of the 2-substituent as well as its activating effect clearly have to be considered. The effect of changing the nature of the ring substituent on the rate constants k_A is given in **Tables 1** and **4**. Kinetic data for the reactions of Mo, CHA and An with 1 and Mo, CHA with 2 in MeOH allow an assumption of the relative activation of a ring-nitrogen and a 2-nitro group expressed in terms of ($k^{2\text{-}NO_2}/k^N$) ratio. Values for MeOH/Mo is 33 - 38; AN/Mo is ≈ 7 - 9 and Tol/Mo is 8 - 9. The results show that the activation effects of the *ortho* substituents is in the order NO_2 > ring aza. Hence, in the absence of steric encumbrance [37,56], the activating power of a 2-nitro group is greater than that of a ring-nitrogen in S_NAr. The low difference in rates can be attributed to the ground state stabilization [24,57] *i.e.* the mesomeric effect of the 4-nitro group, **Figure 1**; (b) the more efficient delocalization of the negative charge with a nitro group than with a ring-nitrogen in the transition state [58].

2.3.2. Effect of Solvent on Kinetic Results and Mechanism

It has long been recognized that the rate of reactions involving the formation of extensively ionic transition state from uncharged molecules are strongly solvent dependent. Methanol is much better than acetonitrile and greater than toluene at solvating charged polarizable species such as the zwitterions. In methanol, hydrogen bond is relatively weak and the intermediates, because they are zwitterionic in character, are extensively solvated. Toluene represents an ideal medium to promote the need for base catalysis for the decomposition of the intermediate. Therefore, aromatic nucleophilic substitution reactions are more prone to base catalysis in aprotic solvents of low relative permittivity than in polar and dipolar solvents [58].

It is observed that the reactions of 1 and 2 with Mo in the three solvents and with CHA and An in MeOH and AN proceeded by uncatalysed nucleophilic aromatic substitution, **Tables 1** and **4**. Therefore, the rate of the reaction is greatly depended on the effect of solvent on the stability of the zwitterionic intermediate.

Therefore, the increase in stability of the zwitterionic intermediate led to an increase in the rate constants, when the formation of intermediate is the rate limiting step and causes a decrease in rate when the departure of the leaving group is the controlling step. This stabilization arises from the intramolecular hydrogen bonding between the ammonium hydrogen and the *ortho*-nitro or the aza ring and the extensive solvation of the intermediate through intermolecular hydrogen bond. We do not neglect the solvation of amine.

Actually, the rate constants in MeOH showed lower values than that in AN which in turn showed different kinetic order in Tol; for the reactions of 1 with Mo, CHA and An and for the reactions of 2 with Mo and CHA. Compound 1 with Mo (Tol:AN:MeOH, 4:1.7:1), with CHA (AN:MeOH, 5:1), with An (AN:MeOH, 2-2.9:1); reaction of compound 2 with Mo (Tol:AN:MeOH 16-14:7-6:1) and with CHA (Tol:AN; 4:1).

Our results postulate that 1) methanol molecules extensively solvate the intermediate I and the amine and accordingly weaken the intramolecular hydrogen bonding in the intermediate I; 2) the dipolar solvent AN slightly solvates the amine and timely solvates the intermediate I by intermolecular hydrogen bond; 3) The higher rate constants for the reactions of compounds 1 and 2 with Mo in toluene compared to those in methanol and AN at all temperatures is due to the destabilization of the zwitterion intermediate which overcomes the maximization of intramolecular hydrogen bonding between the ammonium hydrogen and *ortho*-nitro or aza groups in the intermediate [37,40,41,59]. The higher rate constants of 1 and 2 with Mo and CHA in AN than in MeOH can be explained by the stabilization of the intermediate I by MeOH molecules through intermolecular hydrogen bonding interaction.

The relatively low $\Delta H^{\#}$ values and the absence of amine catalysis in MeOH and AN, **Table 1**, suggest a fast proton transfer, **Scheme 2**. While the large negative $\Delta S^{\#}$ in both solvents perhaps indicates intermolecular (in MeOH, **Figure 2**) and intra as well as intermolecular hydrogen-bond in AN, **Figure 3** (cyclic transition state) and **Figure 4** respectively.

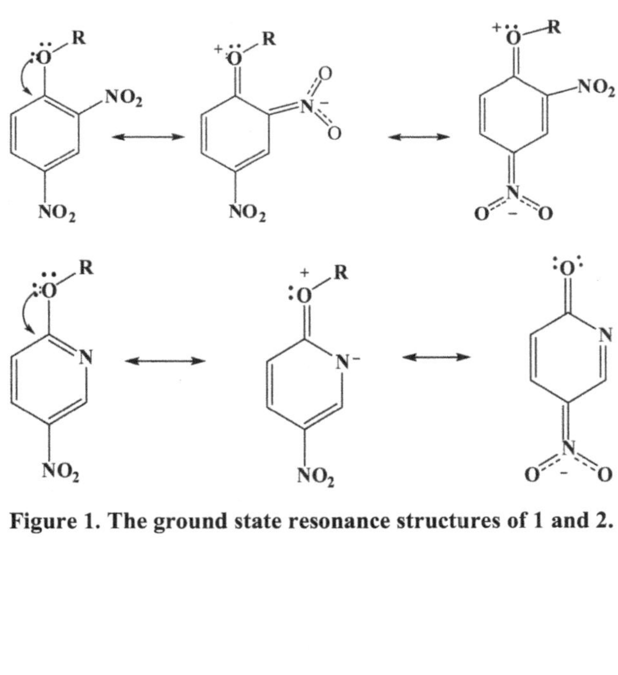

Figure 1. The ground state resonance structures of 1 and 2.

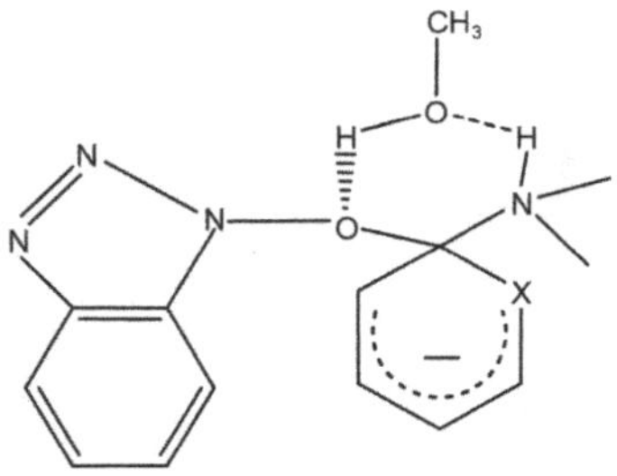

Figure 2. Intermolecular hydrogen bond of compounds 1 and 2 with MeOH.

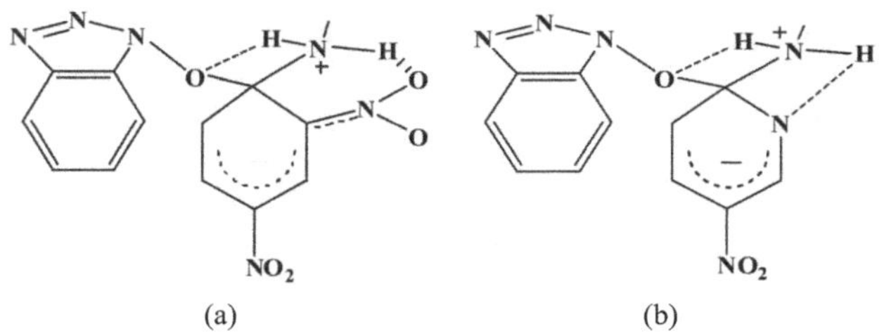

Figure 3. (a) Intramolecular hydrogen-bond of compound 1 in AN; 3(b) Intramolecular hydrogen-bond of compound 2 in AN.

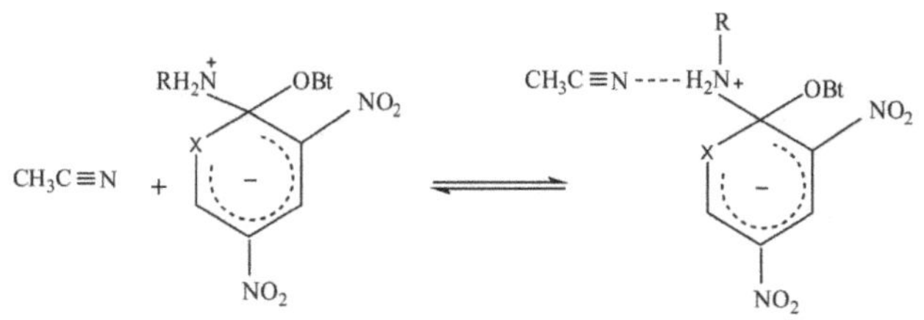

Figure 4. Intermolecular hydrogen-bond of compound 1 in AN.

2.3.3. The Effect of the Nature of Amines (Mo, CHA, An)

In general, base catalysis is more observed in less polar solvents. Unusually, when amines of the same nature and under the same experimental conditions react by different mechanism, there is considerable difference in their basicities which has been ascribed to an increase in the value of k_{-1} with decreasing basicity of the nucleophile [57,59, 60]. pK_a values, measured in MeOH, for the conjugate acids of the amines are given in **Table 1**.

Generally, the rate constants in MeOH showed lower values than that in AN which in turn showed less rate constant values than that in Tol; for the reactions of 1 with Mo, CHA and An; and for the reactions of 2 with Mo and CHA, **Table 1** and **Table 6**. These results indicate that for the uncatalysed reaction, the amine with less pK_a value (Mo) reacts with 1 and 2 faster than the reaction of CHA, that has higher pK_a value, with the same substrates. This is consistent with **Scheme 3** in which the departure of the leaving group is the rate determining step.

2.3.4. The Activation Parameters $\Delta S^{\#}$ and $\Delta H^{\#}$

It seems that the reactions of (1, 2) with Mo, CHA and An proceed via very similar transition states, **Figure 5**. This involves much greater charge separation than exists in the reactants. The more polar nature of the transition state (than of the reactants) means that the change from reactants to transition state will be accompanied by considerable loss of freedom of the solvent and consequent decrease in entropy, **Tables 1** and **4**.

The low activation enthalpies $\Delta H^{\#}$ and the highly negative activation entropies $\Delta S^{\#}$, **Table 1**, are in accordance with the proposed transition state $T_2^{\#}$ for the aza compound 2, whereas compound 1 exhibit $T_1^{\#}$ (hydrogen bond to the *o*-nitro group). Such hydrogen bonding facilitates the attack of the amine (Mo, CHA, An) to form zwitterionic intermediate I. Actually, the energy of formation of the N-H.....O bond is ~2000 cal and that of N-H.....N is ~1930 cal [37]. The expulsion of OBt^- anion in the rate-determining step is assisted by the intramolecular hydrogen-bonding with the ammonio hydrogen present in $T^{\#}_{1,2}$, which accordingly will lower the $\Delta H^{\#}$ value, whereas the structured and rigid transition state reflects the observed large negative $\Delta S^{\#}$ value.

Table 6. Comparison between the amines: Mo, CHA and An.

Reaction of 1 with	Mo	CHA	An
MeOH	189 - 420	35 - 60	6.5 - 13.0
Ratio	1	4.6 - 5.3	29 - 32
AN	332.9 - 637.9	193 - 407	19.1 - 43.0
Ratio	17	10	1
Reaction of 2 with	Mo	CHA	An
MeOH	4.97 - 13.4	0.5 - 1.50	unreactive
Ratio	8.93 - 9.9	1	unreactive
AN	34.9 - 91.1	2.10 - 6.20	unreactive
Ratio	14-16	1	unreactive

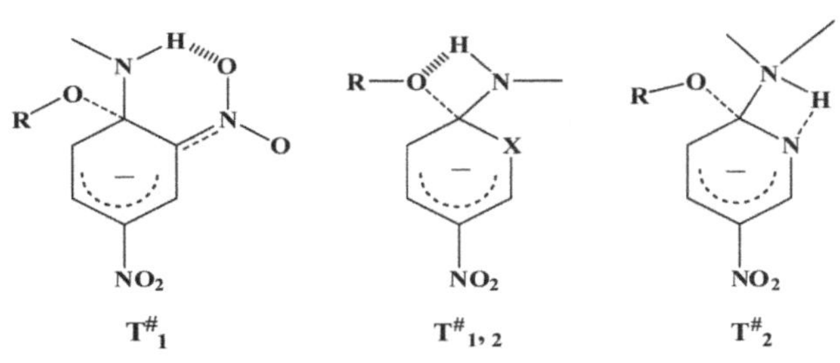

Figure 5. Possible transition states for the reaction of 1 and 2 with amines.

3. Conclusions

The kinetic data for the reactions of compounds 1 and 2 with Mo, An and CHA in MeOH and AN proceeded by uncatalysed mechanism in which the rate limiting step is the leaving group departure, whereas their reactions with Mo in toluene proceeded also by uncatalysed process, in which the formation of zwitterionic intermediate is the rate determining step. On the other hand, the reaction of 1 with CHA and An and the reaction of 2 with CHA in Tol proceeded by SB mechanism in which the rate determining step is the proton transfer process.

The rate for the reactions of (1, 2) with Mo, An and CHA depends on 1) the more efficient delocalization of the negative charge with a nitro group than with a ring-nitrogen in the transition state; 2) the stabilization of the formed activated complex by intramolecular hydrogen bonding in AN and Tol and by intermolecular hydrogen bonding in MeOH and AN; 3) the nucleophilicity of Mo, An and CHA.

4. Experimental Section

4.1. Materials

Methanol, Toluene and acetonitrile were obtained from PROLABO, BDH and were used without further purification. Aniline, cyclohexylamine and morpholine were obtained from Aldrich and were distilled before use. Melting points were determined with a Mel-Temp apparatus and are uncorrected. The UV spectra were carried out on a 160-A UV-VIS recording spectrophotometer Shimadzu. Infrared spectra (IR) were recorded on a Perkin-Elmer 1600 series Fourier Transform instrument as KBr pellets. Proton NMR spectra (^{1}H-NMR) were recorded on a JEOL 500 MHz spectrometers at ambient temperature. Tetramethylsilane (TMS) was used as an internal reference for all ^{1}H-NMR spectra with chemical shifts reported as ppm relative to TMS. The elemental analyses were performed at the Micro analytical Unit, Cairo University, Cairo, Egypt.

4.2. Synthesis of the Starting Materials

4.2.1. Preparation of 1-Hydroxybenzotriazole: (HOBt)

1-Chloro-2-nitrobenzene (0.675 g, 5 mmol) and (1 mL, 15 mmol) of hydrazine hydrate were dissolved in 30 mL ethanol. The reaction mixture was refluxed for 24 hours, after removing the solvent in vacuum, the residue was dissolved in aqueous Na_2CO_3 (10%) solution. The solution was washed with ether to remove the starting material and acidified with concentrated HCl to precipitate the product. The product was washed with water and recrystallized from ether [61].

4.2.2. Preparation of 1-(2,4-Dinitrophenoxy)-1H-benzo[d][1,2,3] Triazole 1

1-Chloro-2,4-dinitrobenzene (1) (2.025 g, 10 mmol) was added to a mixture of 1-hydroxy benzotriazole (1.35 g, 10 mmol) and Et_3N (1.4 mL, 10 mmol) in CH_2Cl_2 (5 mL). Then the reaction mixture was stirred for 3 h at room temperature and then was refluxed on water bath for 2 hours. The crude product was filtered, washed with cold water and recrystallized from ethanol. The purity was checked by TLC (1:4, methanol:chloroform). Yield: 1.5 g (78%); m.p.: 141°C. UV (methanol) λ nm: λ_{max} = 250 nm (ε = 22210); UV (toluene) λ nm: λ_{max} = 291 (ε = 6080); UV (acetonitrile) λ nm: λ_{max} = 297 (ε = 2830) and λ = 236 (ε = 16990). IR (KBr): 1535, 1345, (NO_2 asym and sym. respectively), 1073 (C-O ether) cm^{-1}. ^{1}H NMR [($CDCl_3$)]: δ 6.98 (d, 1H, J = 9.2 Hz, Ar-H), 7.56 (t, 1H, J =7.7 Hz, Ar-H), 7.60 - 7.67 (m, 2H, Ar-H), 8.16 (d, 1H, J =8.4 Hz, Ar-H), 8.39 (dd, 1H, J = 9.2, 2.3 Hz, Ar-H), 9.02 (d, 1H, J =2.3 Hz, Ar-H). $C_{12}H_7N_5O_5$: Requires: C, 49.37; H, 3.19; N, 22.15%. Found: C, 46.07; H, 2.98; N, 22.47%.

4.2.3. Preparation of 1-(5-Nitropyridin-2-yloxy)-1H-Benzo[d][1,2,3]Triazole 2

2-Chloro-5-Nitro pyridine (1.585 g, 10 mmol) was added to a mixture of 1-hydroxy benzotriazole (1.35 g, 10 mmol) and Et_3N (1.4 mL, 10 mmol) in CH_2Cl_2 (5 mL). Then the reaction mixture was stirred for 3 h at room temperature and then was refluxed on water bath for 2 hours. The crude product was filtered, washed with cold water and recrystallized from ethanol. The purity was checked by TLC (2:8 Ethyl acetate: *n*-Hexane). Yield: 1.2 g, 80%; m.p.: 130°C. UV (methanol) λ nm: λ_{max} = 254.5 nm (ε = 18990); UV (toluene) λ nm: λ_{max} = 289.5 (ε = 9580); UV (acetonitrile) λ nm: λ_{max} = 284 (ε = 6230) and λ = 240 (ε = 17190). IR (KBr):1560 (C=N), 1522, 1359 (NO_2 asym and sym. respectively), 1207 (C-O ether) cm^{-1}. ^{1}H NMR [(CD_3Cl)]: δ 7.26 (d, 1H, J = 7.6 Hz, Py-H_3), 7.38 (m, 2H, Ar-H_{OBt}), 7.46 (d, 1H, J = 6.9 Hz, Ar-H_{OBt}), 8.10 (d, 1H, J = 6.9 Hz, Py-H_4), 8.65 (t, 1H, J = 6.1 Hz, Ar-H_{OBt}), 8.90 (d, 1H, J = 2.3 Hz, Py-H_6). $C_{11}H_7N_5O_3$: Requires: C, 54.54; H, 4.23; N, 24.46, O, 16.77%. Found: C, 54.78; H, 4.6; N, 24.11, O, 16.37%.

4.3. Synthesis of the Reaction Products

4.3.1. Preparation of N-cyclohexyl-2,4-Dinitroaniline 3

The reaction of 2,4-dinitrochlorobenzene (0.5 g, 2.5 mmol), with cyclohexylamine (1.5 mL, 13 mmol) in absolute methanol (10 mL) was refluxed for five hours. Work up afforded a yellow precipitate which was filtered and crystallized from benzene-petroleum ether as yellow crystals (0.52 g, 79.42%), m.p. = 212˚C - 213˚C (Lit [46] m.p. 214.1˚C). The purity was checked by TLC (1:9 ethyacetate: n-hexane). UV (methanol) λ nm: λ_{max} = 346 (ε = 14210) and λ = 212 (ε = 14399); UV (toluene) λ nm: λ_{max} = 346 (ε = 17690); UV (acetonitrile) λ nm: λ_{max} = 368 (ε = 14332) and λ = 218 (ε = 14822).

4.3.2. Preparation of 4-(2,4-Dinitrophenyl)morpholine 4

The reaction of 2,4-dinitrochlorobenzene (0.5 g, 2.5 mmol), with morpholine (1.25 mL, 13 mmol) in absolute methanol (10 mL) was refluxed for five hours. Work up afforded a yellow precipitate which was filtered and crystallized from benzene-petroleum ether as yellow crystals (0.6 g, 70%), m.p. = 220˚C - 221˚C (Lit [47] m.p. 221.9˚C). The purity was checked by TLC (1:9 ethyacetate: n-hexane). UV (acetonitrile): λ_{max} = 375 nm (ε = 14160); UV (methanol): λ_{max} = 372 nm (ε = 15980); UV (toluene): λ_{max} = 372 nm (ε = 15170).

4.3.3. Preparation of 2,4-Dinitro-N-phenylaniline 5

The reaction of 2,4-dinitrochlorobenzene (0.5 g, 2.5 mmol) and aniline (1.2 mL, 10.5 mmol) in absolute methanol (10 mL) was refluxed for five hours. Work up afforded a dark yellow precipitate which was filtered and crystallized from benzene-petroleum ether as a dark yellow crystal (0.49 g, 73.4%), m.p. = 158˚C - 160˚C. (Lit [44] m.p. = 159˚C - 161˚C). The purity was checked by TLC (1:9 ethyacetate: *n*-hexane). UV (acetonitrile): λ_{max} = 367 (ε = 1410) and λ = 346 (ε = 17470); UV (methanol): λ_{max} = 365 (ε = 12340) and λ = 346 (ε = 15760); UV (toluene): λ_{max} = 367.5 nm (ε = 14990).

4.3.4. Preparation of N-cyclohexyl-5-nitropyridin-2-amine 6

The reaction of 2-chloro-5-nitropyridine (0.5 g, 2.0 mmol), with cyclohexylamine (1.5 mL, 13 mmol) in absolute methanol (10 mL) was refluxed for five hours. Work up afforded a yellow precipitate which was filtered and crystallized from benzene-petroleum ether as yellow crystals (0.7 g, 71.42%), m.p. = 184˚C - 185˚C. (Lit [49] m.p. = 186˚C). The purity was checked by TLC (1:9 ethyacetate: n-hexane). UV (acetonitrile): λ_{max} = 370 (ε = 11490), UV (methanol): λ_{max} = 370 (ε = 14120). λ = 346 (ε = 14710); UV (toluene): λ_{max} = 367.5 nm (ε = 6710).

4.3.5. Preparation of 4-(5-Nitropyridin-2-yl) Morpholine 7

The reaction of 2-chloro-5-nitropyridine (0.5 g, 2.0 mmol) with morpholine (1.25 mL, 13 mmol) in absolute methanol (10 mL) was refluxed for five hours. Work up afforded a yellow precipitate which was filtered and crystallized from benzene-petroleum ether as yellow crystals (0.6 g, 70%), m.p: 137˚C - 138˚C (Lit [50] m.p. 140˚C - 142˚C). The purity was checked by TLC (1:9 ethyacetate: n-hexane). UV (acetonitrile): λ_{max} = 375 nm (ε = 24720); UV (methanol): λ_{max} = 375nm (ε = 20200); UV (toluene): λ_{max} = 370 nm (ε = 22380).

4.4. Kinetic Measurements

The kinetics of 1-(1-hydroxybenzotriazolyl)-2,4-dinitrobenzene 1,2-(1-hydroxy-benzotriazolyl) 5-nitropyridine 2 with cyclohexylamine, morpholine and aniline in methanol, acetonitrile and toluene was measured spectrophotometrically using a Shimadzu (UV-160A) spectrophotometer in conjunction with a Shimadzu thermo bath (TB-85). Temperature control (±0.1˚C) was attained by circulating water through cell compartments. The kinetic runs were carried out at five temperatures (15˚C - 50˚C) in the same solvents and the absorbance variation with time were recorded at wave length λ = 370 nm, λ = 346 nm, λ = 370 nm for the product of the reaction depending on the nature of the amine. All reactions were carried out under pseudo-first-order conditions, with various concentrations of amine ranges from 1 to 10^{-3} M and final concentration of substrates 1, 2 (1×10^{-4} mol·dm^{-3}) were used. The pseudo-first-order rate constants k_{obs} were estimated by applying Equation (5).

$$\text{Log}\left(A_t - A_\infty\right) = \frac{-k_\psi t}{2.303} + \log\left(A_0 - A_\infty\right) \tag{5}$$

where A_0, A_t and A_∞ are the values of absorbance at zero time, time t and at the end of the reaction, respectively. The A_∞ for each run was taken as the experimentally determined values and k_ψ is the pseudo-first-order rate constant.

Plots of k_{obs} values verses at least six amine concentration gave straight lines passing through the origin with slope equal to the second-order rate constants (k_2) for the reaction in acetonitrile, methanol and toluene. Except for 1 and 2 with CHA in toluene and 1 with An in toluene the plots of k_{obs} versus square amine concentrations gave straight lines passing through the origin with slope equal to the third-order rate constants (k_3). Whereas, plots of k_A values verses amines concentration gave parabolic curve for the reaction of 1 and 2 with CHA in toluene.

REFERENCES

[1] J. Miller, "Aromatic Nucleophilic Substitution," Elsevier, New York, 1969, p. 18.

[2] F. Terrier, "Nucleophilic Aromatic Displacement," The Influence of the Nitro Group, VCH, New York, 1991.

[3] E. Buncel, M. R. Crampton, M. J. Sttrauss and F. Terrier, "Electron Deficient Aromatic and Anionic Sigma Complexes," Elsevier, Amsterdam, 1984.

[4] M. N. Glukhovtsev, R. D. Bach and S. Laiter, "Single-Step and Multistep Mechanisms of Aromatic Nucleo-

philic Substitution of Halobenzenes and Halonitrobenzenes with Halide Anions: *Ab Initio* Computational Study," *Journal of Organic Chemistry*, Vol. 62, No. 12, 1997, pp. 4036-4046. doi:10.1021/jo962096e

[5] A. Patrícia Bento, M. Solà and F. Matthias Bickelhaupt, "*Ab Initio* and DFT Benchmark Study for Nucleophilic Substitution at Carbon (SN2@C) and Silicon (SN2@Si)," *Journal of Computational Chemistry*, Vol. 26, No. 14, 2005, pp. 1497-1504. doi:10.1002/jcc.20261

[6] N. S. Nudelman, "The Chemistry of Amino," Nitroso, Nitro and Related Groups, Wiley & Sons, London, 1996.

[7] S. Park and S. Lee, "Effects of Ion and Protic Solvent on Nucleophilic Aromatic Substitution (S_NAr) Reactions," *Bulletin of the Korean Chemical Society*, Vol. 31, No. 9, 2010, pp. 2571-2573. doi:10.5012/bkcs.2010.31.9.2571

[8] N. Cheron, L. El Kaim, L. Grimaud and P. Fleurat-lessard, "Evidences for the Key of Hydrogen Bonds in Nucleophilic Aromatic Substitution Reactions," *Chemistry: A European Journal*, Vol. 17, No. 52, 2011, pp. 14929-14934. doi:10.1002/chem.201102463

[9] J. Burdon and I. W. Parsons, "A Recent Attempt to Apply Frontier Orbital Theory to Nucleophilic Aromatic Substitution," *Journal of the American Chemical Society*, Vol. 99, No. 23, 1977, pp. 7445-7447. doi:10.1021/ja00465a007

[10] I. Fernández, G Frenking and E. Uggerud, "Rate-Determining Factors in Nucleophilic Aromatic Substitution Reactions," *Journal of Organic Chemistry*, Vol. 75, No. 9, 2010, pp. 2971-2980. doi:10.1021/jo100195w

[11] S. L. Edward and M. I. Joan, "The Reactions of Diazonium Salts with Nucleophiles. X. A Tracer Demonstration of the Reversible Step in Diazonium Ion Hydrolysis," *Journal of the American Chemical Society*, Vol. 86, No. 1, 1964, pp. 32-34. doi:10.1021/ja01055a009

[12] J. F. Bunnett, "Mechanism and Reactivity in Aromatic Nucleophilic Substitution Reactions," *Quarterly Review of the Chemical Society*, Vol. 12, No. 1, 1958, pp. 1-16. doi:10.1039/qr9581200001

[13] F. Pietra, "Mechanisms for Nucleophilic and Photonucleophilic Aromatic Substitution Reactions," *Quarterly Review of the Chemical Society*, Vol. 23, No. 4, 1969, pp. 504-521. doi:10.1039/qr9692300504

[14] M. K. Stern, F. D. Hileman and J. K. Bashkin, "Amination of Nitrobenzene via Nucleophilic Aromatic-Substitution for Hydrogen-Direct Coupling of Aniline and Nitrobenzene," *Journal of the American Chemical Society*, Vol. 114, No. 23, 1992, pp. 9237-9238. doi:10.1021/ja00049a095

[15] N. B. Chapman, J. Shorter and K. J. Toyne, "High Coupled Orbitals," *Journal of the American Chemical Society*, Vol. 98, 1976, pp. 6702-6710.

[16] G. Guanti, G. Petrillo and S. Thea, "Aromatic Nucleophilic Substitution Reactions in the Naphthalene Series. A Kinetic Study of the Reaction of 2,3-Dinitronaphthalene with Piperidine in Benzene," *Tetrahedron*, Vol. 38, No. 4, 1982, pp. 505-511. doi:10.1016/0040-4020(82)80094-X

[17] D. Spinelli and G. Consiglio, "Linear Free Energy Relationships in the Thiophene series. Part I. Leaving Group Effect in Piperidino-Substitution in Methanol of some 2-L-3-Nitro-5-Thiophene," *Journal of the Chemical Society, Perkin Transactions* 2, No. 10, 1975, pp. 989-993. doi:10.1039/p29750000989

[18] J. F. Bunnett and R. E. Zahler, "Aromatic Nucleophilic Substitution Reactions," *Chemical Reviews*, Vol. 49, No. 2, 1951, pp. 273-412. doi:10.1021/cr60153a002

[19] A. J. Parker, "Protic-Dipolar Aprotic Solvent Effects on Rates of Bimolecular Reactions," *Chemical Reviews*, Vol. 69, No. 1, 1969, pp. 1-32. doi:10.1021/cr60257a001

[20] L. Forlani, C. Boga and M. Forconi, "Kinetics and Mechanism of Reactions between 2,4,6-Trinitrofluorobenzene and Alcohols," *Journal of the Chemical Society, Perkin Transactions* 2, No. 7, 1999, pp. 1455-1458. doi:10.1039/a900436j

[21] S. K. Dotterer and R. L. Harris, "Study of Nucleophilic Aromatic-Substitution," *Journal of Organic Chemistry*, Vol. 53, No. 4, 1988, pp. 777-779. doi:10.1021/jo00239a015

[22] A. C. Arvanites and D. W. Boerth, "Modeling of the Mechanism of Nucleophilic Aromatic Substitution of Fungicide Chlorothalonil by Glutathione," *Journal of Molecular Modeling*, Vol. 7, No. 7, 2001, pp. 245-256. doi:10.1007/s008940100032

[23] S. Tascioglu and M. B. Gurdere, "Elucidation of the Mechanism of an Aromatic Substitution Reaction by the Utilization of Micelles as Mechanistic Probes," *Colloids and Surfaces A*, Vol. 173, 2000, pp. 101-107.

[24] M. R. Crampton, T. A. Emokpae, C. Isanbor, A. S. Batsanov, J. A. K. Howard and R. Mondal, "Effects of Ortho- and Para-Ring Activation on the Kinetics of S_NAr Reactions of 1-Chloro-2-nitro and 1-Penoxy-2-nitrobenzenes with Aliphatic Amines in Acetonotrile," *European Journal of Organic Chemistry*, Vol. 2006, No. 5, 2006, pp. 1222-1230. doi:10.1002/ejoc.200500774

[25] C. Isnabor and T. A. Emokpae, "Anilinoysis of Nitro-Substituted Diphenyl Ethers in Acetonitrile: The Effect of Some Ortho-Substituents on the Mechanism of S_NAr Reactions," *International Journal of Chemical Kinetics*, Vol. 42, 2009, pp. 37-49.

[26] R. A. Akpojivi, T. A. Emopkae and J. Hirst, "The Origin of the Dichotomy of Amine Effects in Aromatic Nucleophilic Substitution Reactions," *Journal of the Chemical Society, Perkin Transactions* 2, No. 3, 1994, pp. 443-449.

[27] J. F. Bunnett and G. Davis, "Kinetics of Reactions of 2,4-Dinitrochlorobenzene with Some Familiar Nucleophilic Reagents," *Journal of the American Chemical Society*, Vol. 76, No. 11, 1954, pp. 3011-3015. doi:10.1021/ja01640a044

[28] C. E. S. Alvar and N. S. Nudelman, "Unusual Solvent Effects in the Reactions of 1-Halo-2,4-dinitrobenzenes and Aniline in Aprotic and Dipolar-Aprotic Solvents. Effects of Aggregates," *ARKIVOC*, Vol. X, 2003, pp. 95-106.

[29] C. E. S. Alvaro, M. C. Savini, V. Nicotra, J. S. Yankelevich and N. S. Nudelman, "Reaction of 2,4-Dinitrochlorobenzene with Aromatic Amines in Toluene: Effect of Nucleophile Structure," *Molecules*, Vol. 5, No. 3, 2000,

pp. 401-402. doi:10.3390/50300401

[30] M. F. Fathalla, M. F. Ibrahim and E. A. Hamed, "Kinetics of Alkaline Hydrolysis of 2-Thioaryl-3,5-dinitropyridine Derivatives in 50% v/v DMSO-Water," *Journal of Chemical Research*, Vol. 2004, No. 2, 2004, pp. 150-151.

[31] H. O. H. Al-Howsaway, M. F. Fathalla, A. A. El. Bardan and E. A. Hamed, "Reaction of 4-Chloro-3,5-dinitrobenzotrifluoride with Aniline Derivatives. Substituent Effects," *Journal of Chemical Research*, Vol. 2007, No. 9, 2007, pp. 509-512.

[32] M. F. Fathalla, T. S. Kassem and E. A. Hamed, "Kinetics of the Reaction between Methyl-2,4-dichloro-3,5-dinitro Benzoate and Piperidine: Solvent Effect," *Indian Journal of Chemistry*, Vol. 47A, 2008, pp. 1348-1354.

[33] N. M. El-Mallah, S. A. Senior, G. M. Nabil, M. Sh. Ramadan and E. A. Hamed, "Effect of Acetonitrile-Water Mixtures on the Reaction of Dinitrochlorobenzene and Dinitrochlorobenzotriflouride with Hydroxide Ion," *International Journal of Chemical Kinetics*, Vol. 42, 2010, pp. 453-463.

[34] M. A. H., Kharaba, A. M. G. Nassar, A. A. Youssef and H. Faid-Allah, "A Comparative Kinetic Investigation of Some Nucleophilic Aromatic Displacements," *Polish Journal of Chemistry*, Vol. 65, 1991, pp. 1029-1034.

[35] E. A. Hamed, "Nucleophilic Substitution at the Pyridine Ring. Kinetics of the Reaction of 2-Chloro-3,5-dinitropyridine with Arylthiolates in Methanol," *International Journal of Chemical Kinetics*, Vol. 29, No. 7, 1997, pp. 515-521. doi:10.1002/(SICI)1097-4601(1997)29:7<515::AID-KIN5>3.0.CO;2-X

[36] E. A. Hamed, A. A. El- Bardan, E. F. Saad, G. A. Gohar and G. M. Hassan, "Nucleophilic Substitution at the Pyridine Ring. Conformational Preference of the Products and Kinetics of the Reactions of 2-Chloro-3-Nitro and 2-Chloro-5-nitropyridines with Arenethiolates," *Journal of the Chemical Society*, *Perkin Transactions* 2, No. 11, 1997, pp. 2415-2421. doi:10.1039/a701902e

[37] E. A. Hamed, "Nucleophilic Substitutions at the Pyridine Ring: Kinetics of the Reaction of 2-Chloro-3-nitro and 2-Chloro-5-nitropyridines with Piperidine and Morpholine in Methanol and Benzene," *International Journal of Chemical Kinetics*, Vol. 29, 1997, pp. 599-605. doi:10.1002/(SICI)1097-4601(1997)29:8<599::AID-KIN5>3.0.CO;2-P

[38] N. M. El-Mallah, E. A. Hamed and A. A. El-Bardan, "Kinetics of the Reaction of 2,4-Dinitrophenyl-4'-substituted Sulfones with Morpholine in Methanol," *Egyptian Journal of Chemistry*, Vol. 38, 1995, pp. 319-328.

[39] E. A. Hamed, A. A. El-Bardan, E. F. Saad and M. F. Fathalla, "Kinetics of the Reaction of 2,4-Dinitrophenyl aryl Sulphides Sulphoxides and Sulphones with Hydroxide Ion in 80%(v/v) Ethanol-Water," *Egyptian Journal of Chemistry*, Vol. 43, 2000, pp. 177-185.

[40] A. A. El-Bardan, "Kinetics of the Reaction of 2-Chloro-3-nitro- and 2-Chloro-5-nitropyridines with Aryloxide Ions in Methanol," *Journal of Physical Organic Chemistry*, Vol. 12, No. 4, 1999, pp. 347-353. doi:10.1002/(SICI)1099-1395(199904)12:4<347::AID-POC134>3.0.CO;2-G

[41] F. El. M. El. Hegazy, S. Z. Abdel Fathah, E. A. Hamed and S. M. Sharaf, "Kinetics of the Reaction of 2-Chloro-3,5-dinitropyridine with Meta- and Para-Substituted Anilines in Methanol," *Journal of Physical Organic Chemistry*, Vol. 13, No. 9, 2000, pp. 549-554. doi:10.1002/1099-1395(200009)13:9<549::AID-POC297>3.0.CO;2-5

[42] M. F., Fathalla and E. A. Hamed, "Kinetics of the Reaction between Methyl-2,4-dichloro-3,5-dinitrobenzoate and Piperidine, Piprazine, Morpholine and Thiomorpholine in Methanol and Benzene," *Journal of Chemical Research*, Vol. 7, 2006, pp. 413-416.

[43] Sh. N. Khattab, S. Y. Hassan, E. A. Hamed and A. El-Faham, "Synthesis and Morpholinolysis of *N,N*-Diethyl Carbamate Derivatives of 4-HOAt, 7-HOAt and HOBt," *Journal of Chemical Research*, Vol. 4, 2007, pp. 247-251.

[44] B. H. M. Asghar, M. F. Fathalla and E. A. Hamed, "Solvent and Substituent Effects on the Reaction of 2- and 4-Chloro-3,5-dinitrobenzotrifluorides with Substituted Anilines," *International Journal of Chemical Kinetics*, Vol. 41, No. 12, 2009, pp. 777-786. doi:10.1002/kin.20455

[45] Sh. N. Khattab, S. Y. Hassan, E. A. Hamed, F. Albericio and A. El-Faham, "Synthesis and Aminolysis of *N,N*-Diethyl Carbamic Ester of HOBt Derivatives," *Bulletin of the Korean Chemical Society*, Vol. 31, No. 1, 2010, pp. 75-81. doi:10.5012/bkcs.2010.31.01.075

[46] J. J. Blanksma and G. F. Wilmink, "Action of Cyclohexylamine on Aromatic Halogeno-nitro-compounds," *Recueil des Travaux Chimiques des Pays-Bas*, Vol. 66, No. 7, 1947, pp. 445-453. doi:10.1002/recl.19470660710

[47] C. F. Bernasconi and P. Schmid, "Base Catalysis of the Reaction of Morpholine with 2,4-Dinitrophenyl Phenyl Ether in 10 Percent Dioxane-90 Percent Water," *Journal of Organic Chemistry*, Vol. 32, No. 10, 1967, pp. 2953-2956. doi:10.1021/jo01285a003

[48] R. H. de. Rossi, A. B. Pierini and R. A. Rossi, "Kinetics of the Reaction of Weakly Basic Amines with Activated Aromatic Substrates. Reaction of Imidazole and Aniline with 1-Fluoro-2,4-dinitrobenzene," *Journal of Organic Chemistry*, Vol. 43, No. 15, 1978, pp. 2982-2986. doi:10.1021/jo00409a010

[49] N. S. Nudelman and S. B. Cerdeira, "^{1}H and ^{13}C NMR Studies of Substituted Nitropyridines and Nitrobenzenes," *Magnetic Resonance in Chemistry*, Vol. 24, 1986, pp. 507-511. doi:10.1002/mrc.1260240607

[50] Sh. N. Khattab, E. A. Hamed, F. Albericio and A. El-Faham, "Synthesis and Aminolysis of 2,4-Dinitrophenyl and 5-Nitropyridine *N*-Hydroxy Oxime Derivatives," *Bulletin of the Chemical Society Japan*, Vol. 84, No. 6, 2011, pp. 633-639. doi:10.1246/bcsj.20110015

[51] A. A. El-Bardan, G. M. El-subruiti, F. M. El-Hegazy and E. A. Hamed, "Kinetics and Reactivity of Substituted Anilines with 2-Chloro-5-nitropyridine in Dimethyl Sulfoxide and Dimethylformamide," *International Journal of Chemical Kinetics*, Vol. 34, 2002, pp. 654-650.

doi:10.1002/kin.10053

[52] M. F. Fathalla and Sh. N. Khattab, "Spectrophotometric Determination of pKa's of 1-Hydroxybenzotriazole and Oxime Derivatives in 95% Acetonitrile-Water," *Journal of the Chemical Society of Pakistan*, Vol. 33, 2011, pp. 324-332.

[53] C. F. Bernasconi and P. Schmid, "The Second-Order Coefficient for Formation of 2,4-Dinitrophenol from *N*-(2,4-Dinitrophenyl)morpholine," *Elwood and Gates*, Vol. 32, 1967, pp. 2956-2959.

[54] R. A. Chamberlain and M. R. Crampton, "Mechanism of Base Catalysis in the Reactions of Phenyl Aryl Ethers with Aliphatic Amines in Dimethyl Sulfoxide," *Journal of the Chemical Society*, *Perkin Transactions* 2, No. 10, 1995, pp. 1831-1838. doi:10.1039/p29950001831

[55] R. A. Chamberlain, M. R. Crampton and I. A. Robotham, "Kinetic and Equilibrium Studies of σ-Adduct Formation and Nucleophilic Substitution in the Reactions of Morpholine with 1,3,5-Trinitrobenzene and Some Phenyl Aryl Ethers in Dimethyl Sulphoxide," *Journal of Physical Organic Chemistry*, Vol. 9, No. 3, 1996, pp. 152-158. doi:10.1002/(SICI)1099-1395(199603)9:3<152::AID-POC769>3.0.CO;2-4

[56] M. F. Fathalla, M. F. Ibrahim and E. A. Hamed, "Kinetics of Alkaline Hydrolysis of 2-Thioaryl-3,5-dinitropyridine Derivatives in 50% v/v DMSO-Water," *Journal of Chemical Research*, Vol. 2004, No. 2, 2004, pp. 150-151.

[57] T. A. Emokpae, P. U. Uwakwe and J. Hirst, "The Reactions of Aniline, *N*-Methylaniline, N-Butylamine and Piperidine with Some Nitroaryl Phenyl Ethers in Methanol," *Journal of the Chemical Society, Perkin Transactions* 2, No. 12, 1990, pp. 2191-2195. doi:10.1039/p29900002191

[58] W. Eggiman, P. Schmid and H. Zollinger, "Steric Effects in Nucleophilic Aromatic Substitution Reactions with Aromatic Amines on Nucleophilic Aromatic Substitution," *Helvetica Chimica Acta*, Vol. IV, 1975, pp. 257-268. doi:10.1002/hlca.19750580132

[59] J. F. Bunnett and G. Davis, "The Mechanism of Aminolysis of Esters," *Journal of the American Chemical Society*, Vol. 82, No. 3, 1960, pp. 665-674. doi:10.1021/ja01488a043

[60] J. Hirst and T. O. Bankole, "Kinetics of Some of the Reactions of 2-Fluoro-and 2-Chloro-5-nitropyridine and 1-Fluoro- and 1-Chloro-2,4-dinitrobenzenes with Aniline, Piperidine in Acetone and Methanol," *Journal of the Chemical Society B*, No. 0, 1969, pp. 848-952.

[61] W. König and R. Geiger, "*N*-Hydroxyverbindungen Als Katalysatoren für die Aminolyse Aktivierter Ester," *Chemische Berichte*, Vol. 106, 1973, pp. 3626-3635.

Molecular Simulation of Ion Transport at the Water/Vapor Interface

Jalal Dweik[1,2*], Mahmoud Srour[2], Khaled Karaky[2], Marwan Kobeissi[2], Wissam Joumaa[3], Khalil Abou-Saleh[2]
[1]Institut Européen des Membranes, CNRS UMR 5635, Université Montpellier II, ENSCM, Montpellier, France
[2]Département de Physique et Chimie, Faculté des Sciences, Université Libanaise, Nabatieh, Liban
[3]Laboratoire de Physiotoxcicité Environnementale EDST017, Faculté des Sciences, Université Libanaise, Nabatieh, Liban
Email: kabousaleh@ul.edu.lb, *dweik83@gmail.com

ABSTRACT

Molecular dynamics was used to quantify the role of the size, charge and polarisability of F^-, Cl^-, Br^-, I^- and Na^+ ions in their distribution in the water/vapour interface system. Our results show that the larger polarizable anions I^- and Br^- is attracted to the surface which is traced back to surface-modified ion hydration, while the F^- was repelled from the interface and the Cl^- occupied the total volume of the water slab. Moreover, by artificially increasing the ions charge, anions were localized to the center of the water slab. These results demonstrate that the effect of polarizability cannot be neglected in the transport mechanism.

Keywords: Molecular Dynamics; Ions; Interface; Interaction Potentials; Solvation; Polarization

1. Introduction

The adsorption and distribution of ions at the water/vapor interface is a fundamental process encountered in a wide range of biological, physical, and chemical systems [1,2]. For instance, ions at the vapor/water interface are important for atmospheric chemistry involving ocean surfaces and seawater aerosols [3-5]. Ion specificity plays also an important role in the interactions of between colloidal particles in electrolytes [6]. Ions at the biomolecule/water interface are crucial for protein stability [7,8] to name at least one biological implication. Recently, Tobias and coworkers have used classical Molecular Dynamics techniques and polarizable potential models to examine the equilibrium properties such as density profiles and surface tensions of ions at the water/vapor interface [9-12]. These authors were able to reproduce the experimental surface tensions for different alkali-halide ion-pairs and proposed a molecular model of the distribution of the ions at the water/vapor interface [12,13]. Dang and coworkers [14-16] also reported an extensive study of ion transport across the water/vapor interface using a constrained Molecular Dynamics technique. These authors showed that the dynamics is governed by the free energy profile associated with the transfer of an ion from water to the vapor phase. Recently, Netz and coworkers [17], report a study in which classical molecular dynamics computer simulations were used to examine the distribution of ions at the water/vapor and water/hydrophobic solid surfaces interfaces. We note here, it is of great interest to evaluate the effect of size and polarization on the molecular mechanism of ion binding to the water/vapor interface. The effects of size and polarizability play a crucial role in understanding the equilibrium properties example, orientational structure, surface potentials and surface tensions of the water/vapor interface.

The first physical picture of ions at an aqueous interface was given by Onsager and coworkers [18]. In order to understand the salvation of ions at the water/low-dielectric medium interface is that of the image charge. Within this model the ions are fully characterized by the charge q. It can be shown that an aqueous ion with charge q is effectively repelled from the water/vapor interface by a fictitious image charge $q' = q(\varepsilon - 1)/(\varepsilon + 1)$, *i.e.*, which is of the same sign and practically the same value of charge [12,18,19]. This model doesn't take into account of the size and the polarisability of the ion, so we have carried out classical Molecular Dynamics (MD) simulations, which are particularly well adapted for studying phenomena at the nanoscale especially the roles of the ions specific (charge, size, polarizability). The simulation results show an important effect of hydrophobic and polarisabilty of ion in the water slab.

In this paper, we report on a molecular simulation

*Corresponding author.

study of the thermodynamics and the structure of ions at the water/vapor interface. We address both the effects of size and polarizability of the ions, which are known or expected to play a crucial role on the behavior of such systems by carrying out Molecular Dynamics simulations of the following alkali-halide aqueous electrolytes: Na^+X^- with X = F, Cl, Br, and I. The behavior of the ions at the interface is characterized by calculating density distributions across the interface as well as pair correlation functions. We also determine the ion partition coefficients K_{ij} which are defined as the ratio of the ion concentration in region i to that in region j (i, j are the water phase, the vapor phase or the interface). In order to address the effect of the ionic force, we also vary the absolute value of the charge carried by the ions from 0 (neutral) up to 1 (monovalent cation). The remainder of this paper is organized as follows. The simulation details including the description of the models and the interaction potentials are given in Section 2. The results are presented and discussed in Section 3. The conclusions and suggestions for future work are given in Section 4.

2. Computational Details

2.1. Molecular Dynamics

The molecular dynamics simulations were performed using the Amber9 [20] program package. The simulation box is consisted of a 3 nm thin film containing 864 water molecules and 18 pairs of sodium/halide ions NaX (X = F, Cl, Br, and I). This corresponds to a sodium halide concentration of 1.2 M. The square section of the thin film is 3 nm × 3 nm (**Figure 1**). The size of the parallelepiped box along the direction perpendicular to the film is 10 nm so that each face of the film is in contact with the vapor phase. Periodic boundary conditions were applied along the three dimensions x, y, and z in order to avoid finite size effects. The initial configuration of the simulation was obtained by distributing in a homogeneous way the water molecules and ions in the film. The Verlet leap frog algorithm was used to integrate the equations of motion with a time step of 1 fs. The simulations were performed at a constant temperature of 300 K using the weak coupling method [21] thermostat with a relaxation time step of 3 ps for the thermostat. Each simulation consisted of an equilibration stage of 500 followed by a data collection stage of 1000 ps. The OH bond vibrations were frozen using the Shake algorithm [22,23].

The distribution of ions in the water thin film was investigated by determining both density profiles and partition coefficients. The concentration profiles C(z) for water and ions [11] are measured along the z axis (*i.e.*, perpendicular to the water/vapor interface) by counting the number N(z) of molecules or ions in thin slices of a thickness Δ = 0.2 Å:

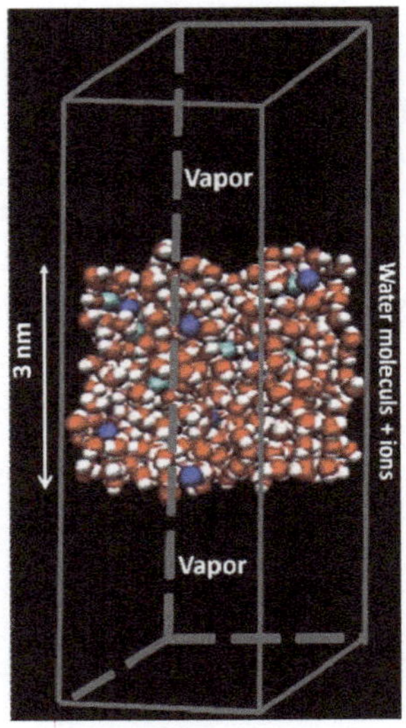

Figure 1. Typical molecular configuration of a 3 nm thin electrolyte film containing 864 water molecules and 18 pairs of sodium/halide ions NaX (X = F, Cl, Br, and I). This corresponds to a sodium halide concentration of 1.2 M. The square section of the thin film along *x* and *y* is 3 nm × 3 nm. The size of the parallelepiped box along the direction perpendicular to the film is 10 nm so that each face of the film is in contact with the vapor phase. Red and white spheres are oxygen and hydrogen atoms of the water molecules, respectively. Green and blue spheres are halide and sodium ions, respectively.

$$C(z) = N(z)/L_x L_y \Delta \tag{1}$$

where L_x and L_y are the size of the parallelepiped simulation box. All of the concentration profiles presented in this paper were averaged over 2000 configurations separated by a time of 0.5 ps. Partition coefficients (K) were determined as the ratio of the average ion concentration in region i to that in region j (i, j are either the water phase, the vapor phase, or the interface):

$$K_{ij}(z) = \langle C_i \rangle / \langle C_j \rangle \tag{2}$$

In the remaining of this paper, the following convention will be adopted: the subscripts V, I, and W denote the vapor, the interface and the water region, respectively. As a convention, we define the interface as the region where the water density is comprised between 1% and 99% of its maximum value. As for the concentration profiles, the partition coefficients were averaged over 2000 configurations and the error was estimated as the square root of the standard deviation.

2.2. Interaction Potentials

The rigid polarizable water model "POL3" was chosen because it takes better account of the polarizability of oxygen atoms (water). [24] was used to describe the water-water interaction. The geometry of the POL3 water molecule is shown in **Figure 2**. This model describes reasonably well the structure and thermodynamic properties of bulk liquid water and the water/vapor interface. The ions are described as rigid polarizable ions using the parameters proposed by Dang and coworkers [14]. The interaction energy E_{pair} between the molecules and ions is the sum of the coulombic interaction and a repulsion and dispersion terms modeled by a Lennard-Jones potential:

$$V_{pair} = \sum_{i<j} \frac{q_i q_j}{r_{ij}} + \sum_{i<j} \varepsilon_{ij} \left[\left(\frac{\sigma_{ij}}{r_{ij}} \right)^{12} - 2 \left(\frac{\sigma_{ij}}{r_{ij}} \right)^{6} \right] \quad (3)$$

where r_{ij} is the distance between the atoms i and j and q_i and q_i are the partial charges. σ and ε are the Lennard – Jones parameters for the repulsion and dispersion interactions. All the parameters used in the present work for the water molecules and the elements F, Cl, Br, I, and Na are listed in **Tables 1-2**. The interaction energy in our model also includes a nonadditive polarization energy term, E_{pol} [25]:

$$V_{pol} = -\sum_{i=1}^{N_{at}} E_i^0 \cdot \mu_i - \frac{1}{2} \sum_{i=1}^{N_{at}} \sum_{j=1, i \neq j}^{N_{at}} \mu_i \cdot T_{ij} \cdot \mu_j + \sum_{i=1}^{N_{at}} \frac{|\mu_i|}{2\alpha} \quad (4)$$

The first term in Equation (3) corresponds to the charge-dipole and dipole-dipole interactions, respectively. The last term in Equation (3) is the energy associated with the generation of the dipole moment $\boldsymbol{\mu}_i = \boldsymbol{\alpha}_i \boldsymbol{E}_i$.

The latter is defined from the local field E_i, which is related to the electric field created at the position of atom i by the fixed charges j in the system $\boldsymbol{E}_i^0$ [14]:

$$\boldsymbol{E}_i = \boldsymbol{E}_i^0 + \sum_{j \neq i} \overline{T}_{ij} \cdot \mu_j \quad \text{with} \quad \boldsymbol{E}_i^0 = \sum_{j \neq i} \frac{q_j \boldsymbol{r}_{ij}}{r_{ij}^3} \quad (5)$$

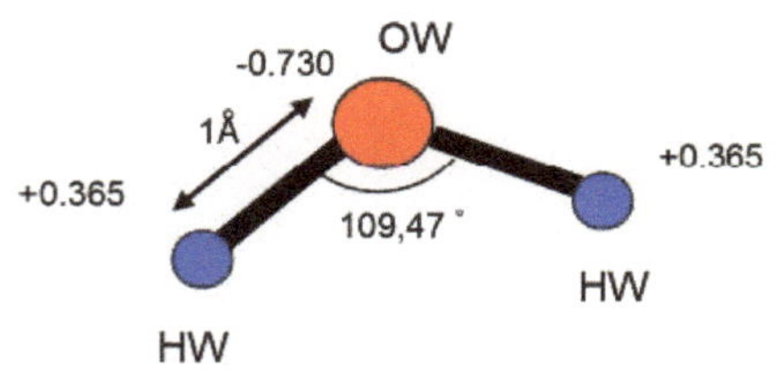

Figure 2. The geometry of rigid polarizable water model "POL3" (taking into account the polarisability of the oxygen atoms) [20] was used to describe the water-water interaction.

Table 1. Potential parameters for the polarizable rigid water model "POL3" [19].

Element	Weight (g/mole)	σ (Å)	ε (kcal/mol)	α_i (Å^3)	Charge
H	1.008	0.000	0.000	0.170	+0.365
O	16.00	3.210	0.156	0.528	–0.730

Table 2. Potential parameters for the different ions considered in this work [10,14].

Element	Weight (g/mole)	σ (Å)	ε (kcal/mol)	α_i (Å^3)	Charge
Na	22.99	2.341	0.1	0.24	+1
F	19.00	3.173	0.1	0.65	–1
Cl	35.45	4.348	0.1	3.25	–1
Br	79.9	4.708	0.1	4.53	–1
I	126.9	5.160	0.1	6.9	–1

where α_i is the atomic polarizability and T_{ij} is the dipole tensor given by:

$$\overline{T}_{ij} = \frac{1}{r_{ij}^3} \left(\frac{3 \boldsymbol{r}_{ij} \boldsymbol{r}_{ij}}{r^2_{ij}} - 1 \right) \quad (6)$$

In the course of the simulation runs, the Car-Parrinello method [26] was used to evaluate and optimize the induced dipoles. The smooth particle mesh Ewald method [27] was used to calculate the electrostatic energy and the van der Waals interactions and the real space part of the Ewald sum were truncated at 12 Å.

3. Results and Discussions

3.1. Effect of the Size of the Anion (Polarizable Water and Non Polarizable Ions)

We first investigate the effect of the size of the anions on their behavior in aqueous electrolyte interfaces by considering the following systems: water with NaX where X = F, Cl, Br, and I. In this part, non polarizable ions were considered (the effect of the ion polarizabily will be discussed in details below). **Figure 3** shows the water, cation and anion concentration profiles across the water/vapor interface at 300 K as a function of the distance z normal to the interface. In order to quantify the degree of exclusion or solvation of halide ions in the electrolyte solutions, we also show in **Table 3** the ion partition coefficient K_{WI} which is defined as the ratio of the ion concentration in water to that at the interface. The anions are strongly repelled from the interface and remain within the water film when small ions such as fluoride and chloride are considered (see **Figures 3(a)** and **(b)**). In this case, the sodium cations follow the anions as they are also located within the water film. These results are confirmed by the fact that the partition coefficient K_{WI} is much larger than 1 for both the anion and cation ($K_{WI} \sim$ 36 for sodium and $K_{WI} \sim 76$ for fluoride in sodium fluoride and $K_{WI} \sim 24$ for sodium and $K_{WI} \sim 27$ for chloride in sodium chloride). The behavior observed above for small electrolytes such as sodium fluoride and sodium chloride confirms the classical Onsager-Samaras model [18], in which ions are repelled from the water/vapor interface due to their image charges on the vapor side. Different behaviors are observed when larger anions are considered such as bromide and iodide. The concentration profiles for sodium bromide shows that these anions also remain located preferentially in the bulk region of the water film (**Figure 3(c)**). Nevertheless, the bromide concentration at the water/vapor interface is non negligible compared to fluoride and chloride. In contrast, the sodium cations remain located far from the interface but the concentration profile splits into two peaks which are located on each side of the main concentration peak for bromide. These results, which departs from what is observed for the smallest anions, suggests that the system evolves towards another equilibrium configuration as the size of the anion increases. This is confirmed by the results for the largest anion, which shows that iodide is preferentially adsorbed at the vapor interface as two marked peaks are observed in the region where the water concentration drops off to zero (**Figure 3(d)**). For this electrolyte solution, the concentration peaks for sodium splits into two peaks which are located within bulk water but close to the peaks for iodide. These results for large anions are consistent with the fact that the ion partition coefficient K_{WI} decreases as the cation gets larger (**Table 3**). An interesting result that summarizes our observations is that the ion partition coefficient K_{WI} decreases for both the anion and cation as one follows the series NaF, NaCl, NaBr, and NaI.

Table 3. Ion partition coefficents K_{WI} between the water phase (W) and the interface (I) for thin electrolyte solutions of a thickness 3 nm when non polarizable ions are considered. K_{WI} is defined as the ratio of the ion concentration in the water phase to that at the interface. The numbers in parentheses indicate the uncertainty of the ion partition coefficient.

Sodium fluoride		Sodium chloride		Sodium bromide		Sodium iodide	
Na	F	Na	Cl	Na	Br	Na	I
36 (2)	76 (2)	24 (1)	27 (1)	17 (2)	7 (2)	4 (3)	2 (1)

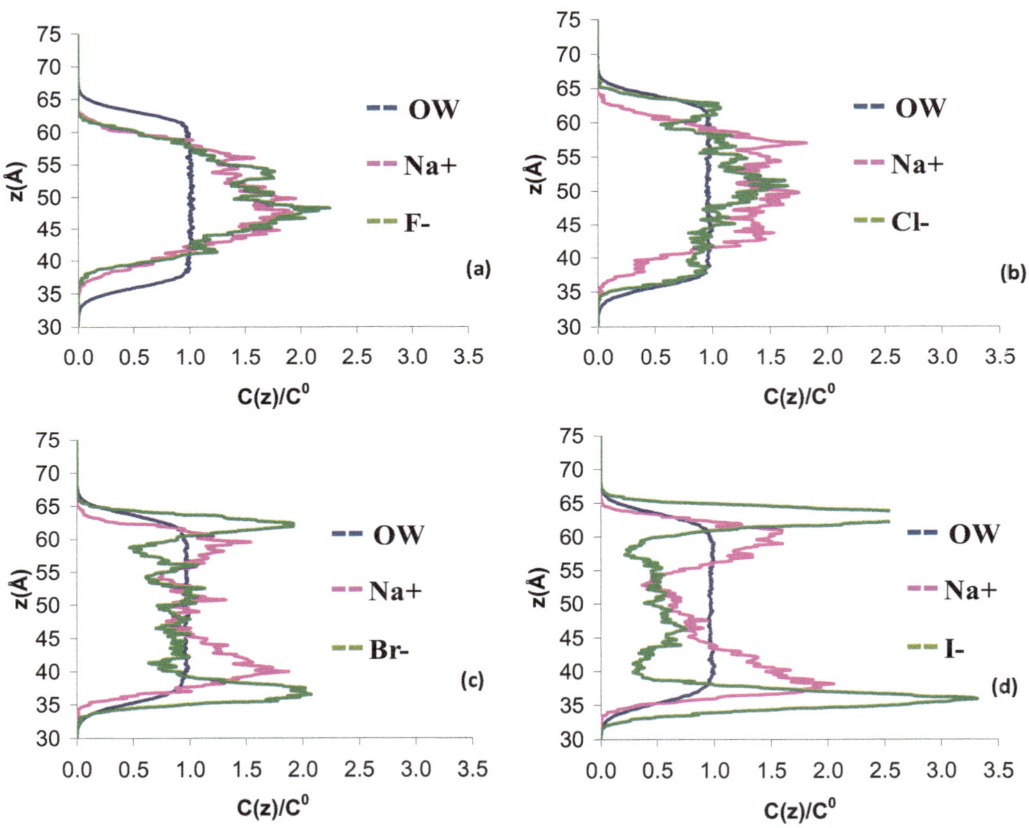

Figure 2. Concentration profile C(z) for a 3 nm thin electrolyte film containing 864 water molecules and 18 pairs of polarizable sodium/halide ions NaX (X = F, Cl, Br, and I). z is the distance normal to the interface (z = 0 is located in the center of the electrolyte film). The blue line corresponds to the data for the water while the pink and red lines correspond to the data for the cation and anion, respectively. The concentration profiles are normalized to the concentration C(0) in the center of the film.

The different regimes observed in this work (anions located in bulk water versus adsorbed at the interface) can be explained in the frame of the so-called "hydrophobic effect" that has been recently reviewed by Chandler [28]. Small anions and cations such as sodium, fluoride, and chloride remain located within bulk water as the volume fraction occupied by these ions is small; the cost in free energy to have these ions in water remains low so that no exclusion is observed. As mentioned above, the classical picture of image charge, which predicts that the ions are repelled from the surface, is relevant in this situation. On the other hand, when larger ions are considered such as bromide and iodide, the volume fraction occupied by the ions becomes significant and leads to high free energy penalties. In this case, the ions tend to move towards the vapor interface. These results show that the hydrophobic effect plays a crucial role as it affects the adsorption and distribution of ions at the water/vapor interface. The origin of this hydrophobic effect is the cost in free energy to solvate the ions as their presence may hinder hydrogen bonding. On the one hand, water molecules can easily solvate small systems (fluoride, chloride) without breaking the hydrogen bonding network so that no exlusion of the ions is observed. On the other hand, large solutes (iodide), which cannot be solvated without drastically hindering hydrogen bonding between the water molecules, tend to be expelled from bulk water towards the water/vapor interface. Depending on the temperature, the free energy penalty responsible for the exclusion of large ions is either enthalpic or entropic. As explained by Chandler [28], a process involving a change in the average number of hydrogen bonds per water molecule will be mainly driven by the enthalpic contribution and the cost in free energy will decrease with increasing temperature. In contrast, a process involving a conformational change in the hydrogen bonding without significant change in the number of hydrogen bonds will be governed by the entropic contribution and the cost in free energy will increase with temperature. In the case of sodium iodide, we found that the fraction of iodide moving from bulk water to the interface increases with increasing temperature (results not shown); this result suggests that solvation and exclusion of large halide ions is driven by the entropic contribution to the free energy.

3.2. Polarizability Effect (Polarizable Water and Ions)

We now discuss the effect of the anion and cation polarizability on the behavior of aqueous electrolyte interfaces by considering the following systems: water with NaX where X = F, Cl, Br, and I. **Figure 3** shows the water, cation, and anion concentration profiles across the water/vapor interface at 300 K as a function of the distance z normal to the interface. In order to quantify the degree of exclusion or solvation of the polarizable ions in the electrolyte solutions, we also show in Table 4 the ion partition coefficient K_{WI} which is defined as the ratio of the ion concentration in the bulk water region to that at the vapor/water interface. As in the case of the non polarizable ions, the two following types of concentration profiles are observed. The anions are strongly repelled from the interface and remain within the bulk region of the water film when small ions such as fluoride and chloride are considered (see **Figures 3(a)** and **(b)**). Again, in this case, the sodium cations follow the anions as they are also located within the water film. These results are confirmed by the fact that the partition coefficient K_{WI} is much larger than 1 for both the anion and cation ($K_{WI} \sim$ 45 for sodium and $K_{WI} \sim 116$ for fluoride in sodium fluoride and $K_{WI} \sim 30$ for sodium and $K_{WI} \sim 4$ for chloride in sodium chloride). In contrast, the concentration profiles for sodium bromide and sodium iodide shows that these anions are preferentially adsorbed at the water/vapor interface (the effect is even more marked for the largest anion). In this case, the concentration peaks for sodium splits into two peaks which are located within bulk water but close to the peaks for bromide or iodide. These results for large anions are consistent with the fact that the ion partition coefficient K_{WI} decreases as the cation gets larger (**Table 4**). As in the case of the non polarizable ions, our results show that the ion partition coefficient K_{WI} decreases for both the anion and the cation as one follows the series NaF, NaCl, NaBr, and NaI.

As in the case of the non polarizable ions, the two regimes observed for the polarizable ions (anions located in bulk water versus adsorbed at the interface) can be explained in the frame of the "hydrophobic effect". Interestingly, the comparison between the ion coefficient partition K_{WI} for the polarizable and non polarizable ions (**Tables 3-4**) suggests that the polarizability strengthens the solvation of the small ions and the exclusion of the large ions. On the one hand, K_{WI} is larger for the small polarizable ions such as fluoride and chloride than for

Table 4. Ion partition coefficents K_{WI} between the water phase (W) and the interface (I) for thin electrolyte solutions of a thickness 3 nm when polarizable ions are considered. K_{WI} is defined as the ratio of the ion concentration in the water phase to that at the interface. The numbers in parentheses indicate the uncertainty of the ion partition coefficient.

Sodium fluoride		Sodium chloride		Sodium bromide		Sodium iodide	
Na	F	Na	Cl	Na	Br	Na	I
45 (2)	116(1)	30 (13)	4 (2)	6 (3)	1 (0.7)	2.2 (0.1)	0.3 (0.2)

their non polarizable counterpart. On the other hand, K_{WI} is smaller for the large polarizable ions such as bromide and iodide than for their non polarizable counterpart. We note here that the simulation results are reproduced by Tobias and co-workers, such as density profiles and surface tensions of ions at the water/vapor interface [9,12]. We can also observe experimentally the surface tension and the distribution of ions polarizability [9,13]. To extend our understanding of polarizable ions, it is interesting to note that the PMF (potential mean field) calculated by Dang and co-workers [14]. The physical origin of these results can be explained as follows: At the water/vapor interface, the salvation is uncompleted (anisotropic salvation); the anion is subjected to a nonzero net dipole. The induced dipole moments of the anions are increased. Note that in the bulk, this net dipole on average equals to zero due to a symmetric salvation of the ion. Consequently, the ion-dipole induction energies between the anion-water are enhanced due to induction effects.

3.3. Effect of the Charge (Polarizable Water and Ions)

We have also examined the dependence on the charge of the polarizable ions on the transport mechanism across the water/vapor interface. Although it does not correspond to any physical situation, the charge of the anion and cation can be arbitrarily changed in order to study the effect of the charge. **Figures 4** and **5** show the water, cation, and anion concentration profiles across the water/vapor interface at 300 K for the system $Na^{x}I^{-x}$ (where x varies from 0 up to 1.66). We also quantify in **Table 5** the exclusion or solvation of the ions in the electrolyte solutions by calculating the ion partition coefficients K_{ij} which are defined as the ratio of the ion concentration in region i to that in region j (i, j are the water phase, the vapor phase or the interface). On the one hand, both the anions and cations are excluded from the water film when the charge is lower or equal than 0.1. On the other hand, the iodide ions are preferentially adsorbed at the water/vapor interface when the charge x varies between 0.33 and 1. In this case, the sodium cations are located within bulk water but close to the peaks for iodide. When x is larger than 1 (strong ionic strength), both the anions and cations are located within the bulk part of the electrolyte film. These results, which show that the type of behavior for the cations and anions is directly related to the charge of the ions, can be rationalized as follows. The

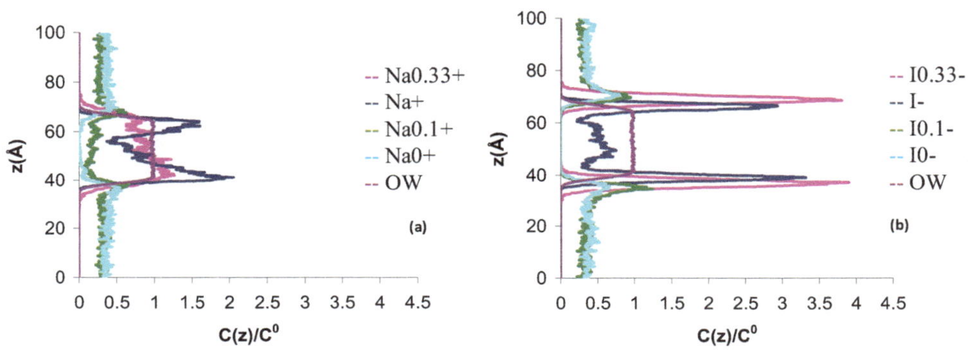

Figure 4. Concentration profile C(z) for a 3 nm thin electrolyte film containing 864 water molecules and 18 pairs of polarizable sodium iodide ions $Na^{x}I^{-x}$ with x varying from 0 to 1. z is the distance normal to the interface (z = 0 is located in the center of the electrolyte film). The concentration profiles are normalized to the concentration C(0) in the center of the film.

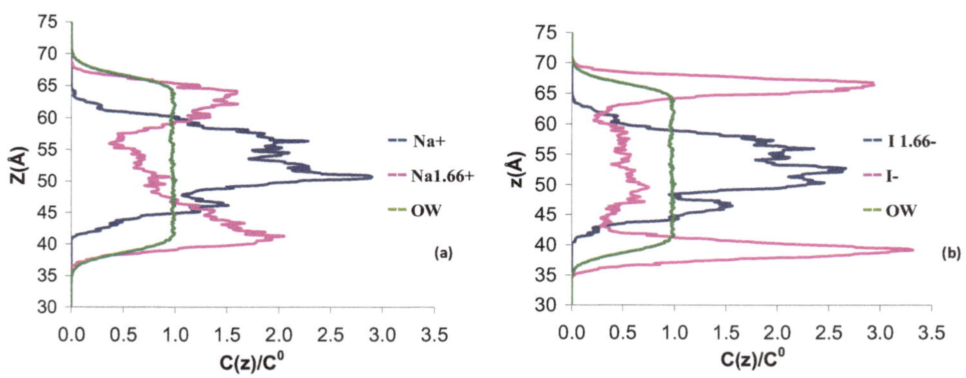

Figure 5. Concentration profile C(z) for a 3 nm thin electrolyte film containing 864 water molecules and 18 pairs of polarizable sodium iodide ions $Na^{x}I^{-x}$ with x varying from 1 to 1.66. z is the distance normal to the interface (z = 0 is located in the center of the electrolyte film). The concentration profiles are normalized to the concentration C(0) in the center of the film.

solvation interaction becomes weak as the ion charge decreases due to the decrease in the coulombic energy. As a result, exclusion of the ions is observed when the charge is very low because the cost in free energy associated with the solvation of the ions is high (the breaking of hydrogen bonds due to the presence of large iodide ions cannot be counterbalanced by the solvation energy). On the other hand, the solvation interaction is strong when the charge is large so that the cost in free energy due to the presence of the iodide ions is compensated by the solvation energy. In the present case of iodide ions having different charges, we note that the solvation/exclusion is driven by the enthalpy contribution to the free energy; indeed, the entropic contribution is independent of the charge x while the enthalpic energy is directly related to x through the solvation energy that depends on the coulombic interaction.

4. Discussion and Conclusions

This paper reports on a Molecular Dynamics simulation study of ions at the water/vapor interface. The fol lowing alkali-halide aqueous electrolytes Na^+X^- with X= F, Cl, Br, and I are considered. In the case of sodium iodide, we also consider the cases of Na^xI^{-x} electrolyte solutions in which the charge x varies from 0 up to 1.66. The cations and anions are found to behave differently in the system as they are either located within the bulk region of the system or preferentially adsorbed at the interface. Our results also show that the size, charge, and polarizability of the ions play a crucial role on their location in the system.

In order to summarize the effect of the size and the polarizability of the ions, we show in **Figure 6** the effect of the polarizability α and size R^* of the anion X on the phase diagram of the thin electrolyte film. The ions are either located within the bulk region *i*), near the interface *ii*), or at the interface *iii*). In order to determine to which type of phase diagram a system belongs, we adopted the following arbitrary convention:

$$\begin{aligned}&\text{Belongs to } (i) \text{ if } K_{WI} > 10\\&\text{Belongs to } (ii) \text{ if } 3 < K_{WI} < 10\\&\text{Belongs to } (iii) \text{ if } K_{WI} < 10\end{aligned} \tag{7}$$

For a given polarizability, the ion partition coefficient K_{WI} decreases for both the anion and cation as one follows the series NaF, NaCl, NaBr, and NaI. This result implies that the ions tend to be repelled towards the interface as the size of anion increases. This is an illustration of the so-called "hydrophobic effect". Small anions and cations remain located within bulk water as the cost in free energy to have these ions in water remains low. In contrast, when larger ions tend to move towards the water/vapor interface they occupy a large volume fraction which leads to a high free energy penalty. The crossover between the different regimes (*i*, *ii*, and *iii*) is also observed when the polarizability increases. As mentioned above, the comparison between the ion coefficient partition K_{WI} for the polarizable and non polarizable ions suggests that the polarizability strengthens the solvation of the small ions and the exclusion of the large ions. The simulations results observed experimentally using the surface tension and the distribution of ions polarizability [9,13]. The simulations correctly reproduce the order of the increase of surface tension $\Delta\gamma$, $I^- < Br^- < Cl^- < F^-$ [9,32] and the pic of density profiles at the water/vapor interface [12]. The presence of the larger, polarizable anions at the water/vapor interface is also consistent with a larger number of experimental and theoretical observations of surface locations of these ions on small to moderately sized water clusters [29-31].

Figure 7 shows the effect of the size $R^*/2$ and charge x of the anion I on the phase diagram of the thin electrolyte film of Na^xI^{-x} (x varies from 0 up to 1.66). Depending on the values taken by these parameters, five different types of phase behavior are observed. The ions are either located within the bulk region *i*), near the interface *ii*), at the interface *iii*), at the interface *iv*), or in the vapor phase *v*). In order to determine to which type of phase behavior a system belongs, we adopted the following arbitrary convention:

$$\begin{aligned}&\text{Belongs to } (i) \text{ if } K_{VW} \text{ and } K_{WI} > 10\\&\text{Belongs to } (ii) \text{ if } K_{VW} = 0 \text{ and } 2 < K_{WI} < 10\\&\text{Belongs to } (iii) \text{ if } K_{VW} = 0 \text{ and } K_{WI} < 2\\&\text{Belongs to } (iv) \text{ if } K_{WI} = 0 \text{ and } K_{VW} < 30\\&\text{Belongs to } (v) \text{ if } K_{WI} = 0 \text{ and } K_{VW} > 30\end{aligned} \tag{8}$$

Table 1. Ion partition coefficents K_{ij} between the water phase (W), the vapor phase (V), and the interface (I) for a thin electrolyte solution NaI when the charge x of the ions varies (Na^xI^{-x}). K_{ij} is defined as the ratio of the ion concentration in the region i to that in the region j. The numbers in parentheses indicate the uncertainty of the ion partition coefficient.

	$x = 0.00$		$x = 0.10$		$x = 0.33$		$x = 1.00$		$x = 1.66$	
	Na	I	Na	I	Na	I	Na	I	Na	I
K_{AI}	1.1 (0.8)	1.5 (1)	0.7 (0.5)	0.8 (0.7)	0.0 (0.0)	1.9 (0.5)	0.0 (0.0)	0.0 (0.0)	0.0 (0.0)	0.0 (0.0)
K_{AW}	18 (1.0)	2406 (1)	1.8 (0.3)	545 (0.2)	0.0 (0.0)	5.6 (0.05)	0.0 (0.0)	0.0 (0.0)	0.0 (0.0)	0.0 (0.0)
K_{WI}	0.06 (0.02)	0.0001 (0.02)	0.4 (0.2)	0.001 (0.06)	1.6 (0.9)	0.003 (0.001)	2.3 (0.1)	0.3 (0.1)	∞	∞

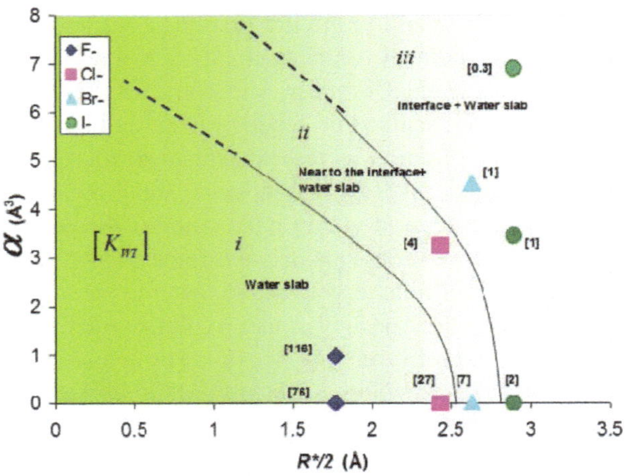

Figure 6. Effect of the polarizability α **and size** R^* **of the anion on the phase diagram of a thin electrolyte film containing 864 water molecules and 18 pairs of polarizable sodium ions NaX (X = F, Cl, Br, and I). The ions are either located within the bulk region, near the interface, or at the interface. The number in parentheses indicates the ion partition coefficient** K_{WI} **defined as the ratio of the ion concentration in water to that at the interface.**

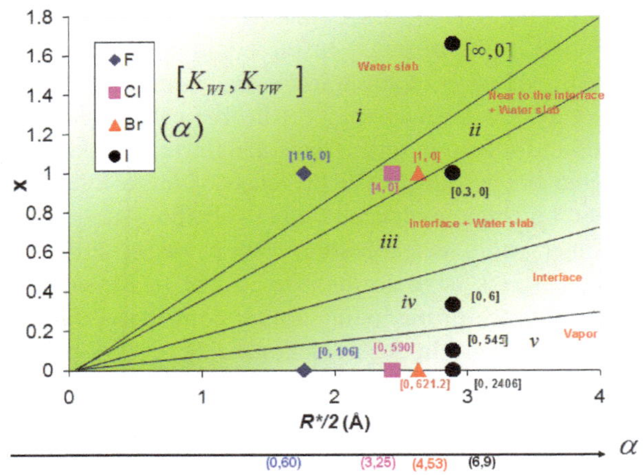

Figure 7. Effect of the charge x **and size** R^* **of the anion on the phase diagram of a thin electrolyte film containing 864 water molecules and 18 pairs of polarizable sodium ions NaX (X = F, Cl, Br, and I). The ions are either located within the bulk region, near the interface, or at the interface. The numbers in parentheses and in brackets indicates the ion polarizability and partition coefficients** K_{WI} **and** K_{AW}**, respectively.**

For a given size of the ions, the phase behavior evolves from *i*, *ii*, *iii*, *iv*, and *v* as the charge x decreases from 1.66 down to 0. For neutral systems, both the cations and anions are expelled from the water film as their presence in liquid water corresponds to a high free energy penalty (no compensation arises from the ion/water interaction as there is no coulombic interaction). As the charge x increases, the ions start to be preferentially adsorbed at the interface and, for large values of x, they are located within the bulk region of the water film. In the latter case, even for large ions such as iodide, the coulombic interaction is large enough to counterbalance the cost in free energy associated with the breaking or reorganization of the water hydrogen bond network (no "hydrophobic effect").

The results reported in the present work underline the crucial role played by the ion charge, size and polarizability on the phase behavior of electrolyte solutions in contact with a vapor phase. In future work, we plan to explore the behavior of similar interfacial electrolyte solutions such as electrolyte solutions confined in nanoporous membranes. These systems, which are known to enhance the diffusion of certain species such as ions and to have a strong selectivity, are very attractive for nanofiltration and therapeutic applications. We will investigate by using the molecular simulation the effect of the size of the nanopores as well as the effect of the charge of the membrane surface.

5. Acknowledgements

We wish to thank the Research Program P-NANO of the French National Agency for Research (ANR) for financial support of this work (Research Project "SIMON-ANOMEM"). Thanks also to Benoit Coasne and John Palmeri for related help and discussions.

REFERENCES

[1] S. Mclaughin, "The Electrostatic Properties of Membranes," *Annual Review of Biophysics and Biophysical Chemistry*, Vol. 18, 1989, pp. 113-136. doi:10.1146/annurev.bb.18.060189.000553

[2] B. Honig, W. L. Hubell and R. F. Flewelling, "Electrostatic Interactions in Membranes and Proteins," *Annual Review of Biophysics and Biophysical Chemistry*, Vol. 15, 1986, pp. 163-193. doi:10.1146/annurev.bb.15.060186.001115

[3] J. H. Hu, Q. Shi, P. Davidovits, D. R. Worsnop, M. S. Zahniser and C. E. Kolb, "Reactive Uptake of Cl_2(g) and Br_2(g) by Aqueous Surfaces as a Function of Br^- and I^- Ion Concentration: The Effect of Chemical Reaction at the Interface," *Journal of Physical Chemistry*, Vol. 99, No. 21, 1995, pp. 8768-8776. doi:10.1021/j100021a050

[4] B. J. Finlayson and J. C. Hemminger, "Physical Chemistry of Airborne Sea Salt Particles and Their Components," *Journal of Physical Chemistry A*, Vol. 104, No. 49, 2000, pp. 11463-11477. doi:10.1021/jp002968n

[5] B. J. Finlayson-Pitts, "The Tropospheric Chemistry of Sea Salt: A Molecular-Level View of the Chemistry of NaCl and NaBr," *Chemical Reviews*, Vol. 103, No. 12, 2003, pp. 4801-4822. doi:10.1021/cr020653t

[6] F. W. Tavares, D. Bratko and J. M. Prausnitz, "The Role of Salt-Macroion van der Waals Interactions in the Colloid-Colloid Potential of Mean Force," *Current Opinion in Colloid & Interface Science*, Vol. 9, No. 1-2, 2004, pp. 81-86. doi:10.1016/j.cocis.2004.05.008

[7] M. G. Cacace, E. M. Landau and J. J. Q. Ramsden, "The Hofmeister Series: Salt and Solvent Effects on Interfacial Phenomena," *Quarterly Reviews of Biophysics*, Vol. 30, No. 3, 1997, pp. 241-277. doi:10.1017/S0033583597003363

[8] R. J. Piazza, "Interactions in Protein Solutions near Crystallization: A Colloid Physics Approach," *Journal of Crystal Growth*, Vol. 196, No. 2-4, 1999, pp. 415-423. doi:10.1016/S0022-0248(98)00867-7

[9] P. Jungwirth and D. J. Tobias, "Molecular Structure of Salt Solutions: A New View of the Interface with Implications for Heterogeneous Atmospheric Chemistry," *Journal of Physical Chemistry B*, Vol. 105, No. 43, 2001, pp. 10468-10472. doi:10.1021/jp012750g

[10] P. Jungwirth and D. J. Tobias, "Chloride Anion on Aqueous Clusters, at the Air-Water Interface, and in Liquid Water: Solvent Effects on Cl^- Polarizability," *Journal of Physical Chemistry*, Vol. 106, No. 2, 2002, pp. 379-383. doi:10.1021/jp012059d

[11] P. Jungwirth and D. J. Tobias, "Surface Effects on Aqueous Ionic Solvation: A Molecular Dynamics Study of NaCl at the Air/Water Interface from Infinite Dilution to Saturation," *Journal of Physical Chemistry B*, Vol. 104, No. 32, 2000, pp. 7702-7706. doi:10.1021/jp000941y

[12] P. Jungwirth and D. J. Tobias, "Specific Ion Effects at the Air/Water Interface," *Chemical Reviews*, Vol. 106, No. 4, 2006, pp. 1259-1281. doi:10.1021/cr0403741

[13] V. Padmanabhan, J. Daillant and L. Belloni, "Specific Ion Adsorption and Short-Range Interactions at the Air Aqueous Solution Interface," *Physical Review Letters*, Vol. 99, No. 8, 2007, Article ID: 086105, pp. 1-4.

[14] L. X. Dang and T. S. Chang, "Molecular Mechanism of Ion Binding to the Liquid/Vapor Interface of Water," *Journal of Physical Chemistry B*, Vol. 106, No. 2, 2002, pp. 235-238. doi:10.1021/jp011853w

[15] L. X. Dang and D. E. Smith, "Molecular Dynamics Simulations of Aqueous Ionic Clusters Using Polarizable Water," *Journal of Chemical Physics*, Vol. 99, No. 9, 1993, pp. 6950-6956. doi:10.1063/1.465441

[16] L. X. Dang, "Computer Simulation Studies of Ion Transport across a Liquid/Liquid Interface," *Journal of Chemical Physics B*, Vol. 103, No. 39, 1999, pp. 8195-8200. doi:10.1021/jp991824+

[17] D. Horinek and R. Netz, "Specific Ion Adsorption at Hydrophobic Solid Surfaces," *Physical Review Letters*, Vol. 99, 2007, pp. 226104-22607. doi:10.1103/PhysRevLett.99.226104

[18] L. Onsager and N. N. T. Samaras, "The Surface Tension of Debye-Hückel Electrolytes," *Journal of Chemical Physics*, Vol. 2, No. 8, 1934, pp. 528-536. doi:10.1063/1.1749522

[19] C. Wagner, "The Surface Tension of Dilute Solutions of Electrolytes," *Physikalische Zeitschrift*, Vol. 25, 1924, pp. 474-477.

[20] D. A. Case, T. A. Darden, T. E. Cheatham, *et al.*, "AM BER9," 2006. http://ambermd.org/doc9/amber9.pdf

[21] H. J. C. Berendsen, J. P. M. Postma, W. F. van Gunsteren, A. DiNola and J. R. Haak, "Molecular Dynamics with Coupling to an External Bath," *Journal of Chemical Physics*, Vol. 81, 1984, pp. 3684-3690. doi:10.1063/1.448118

[22] J. W. Caldwell and P. A. Kollman, "Structure and Properties of Neat Liquids Using Nonadditive Molecular Dynamics: Water, Methanol, and N-Methylacetamide," *Journal of Physical Chemistry*, Vol. 99, No. 16, 1995, pp. 6208-6219. doi:10.1021/j100016a067

[23] L. Perera and M. L. Berkowitz,"Structures of $Cl^-(H_2O)_n$ and $F^-(H_2O)_n$ (n=2,3,...,15) Clusters. Molecular Dynamics Computer simulations," *Journal of Chemical Physics*, Vol. 100, 1994, pp. 3085-3093. doi:10.1063/1.466450

[24] G. Markovich, L. Perera, M. Berkowitz and O. Cheshnovsky, "The Solvation of Cl^-, Br^-, and I^- in Acetonitrile Clusters: Photoelectron Spectroscopy and Molecular Dynamics Simulations," *Journal of Chemical Physics*, Vol. 105, No. 7, 1996, pp. 2675-2685. doi:10.1063/1.472131

[25] J. P. Ryckaert, G. Ciccotti and H. J. C. Berendsen, "Numerical Integration of the Cartesian Equations of Motion of a System with Constraints: Molecular Dynamics of *n*-Alkanes," *Journal of Computational Physics*, Vol. 23, No. 3, 1977, pp. 327-341. doi:10.1016/0021-9991(77)90098-5

[26] W. F. Van Gunsteren and H. J. C. Berendsen, "Algorithms for Macromolecular Dynamics and Constraint Dynamics," *Molecular Physics*, Vol. 34, No. 5, 1977, pp. 1311-1327. doi:10.1080/00268977700102571

[27] U. Essmann, L. Perera, M. L. Berkowitz, T. Darden and L. G. Pedersen, "A Smooth Particle Mesh Ewald Method," *Journal of Chemical Physics*, Vol. 103, No. 19, 1995, pp. 8577-8593. doi:10.1063/1.470117

[28] D. Chandler, "Interfaces and the Driving Force of Hydro Phobic Assembly," *Review*, *Nature*, Vol. 437, No. 29, 2005, pp. 640-647.

[29] R. W. Gora, S. Rosak and Leszczynski, "Properties and Nature of Interactions in $Cl^-(H_2O)_n$, n = 1,6 Clusters: A Theoretical Study," *Journal of Chemical Physics Letters*, Vol. 325, No. 1-3, 2000, pp. 7-14. doi:10.1016/S0009-2614(00)00624-2

[30] S. J. Stuart and B. J. Berne, "Surface Curvature Effects in the Aqueous Ionic Solvation of the Chloride Ion," *Journal of Physical Chemistry A*, Vol. 103, No. 49, 1999, pp. 10300-10307. doi:10.1021/jp991671q

[31] P. Ayotte, G. H. Weddle, J. Kim and M. A. Johnson, " Vibrational Spectroscopy of the Ionic Hydrogen-Bond: Fermi Resonance and Ion-Molecule Stretching Frequencies in the Binary $X^-\cdot H_2O$ (X=Cl, Br, I) Complexes via Argon Predissociation Spectroscopy," *Journal of the American Chemical Society*, Vol. 120, No. 47, 1998, pp. 12361-12362. doi:10.1021/ja981979f

[32] A. P. dos Santos, A. Diehl and Y. Levin, "Surface Tensions, Surface Potentials and the Hofmeister Series of Electrolyte Solutions," *Langmuir*, Vol. 26, No. 13, 2010, pp. 10778-10783. doi:10.1021/la100604k

11

Study of Thermal Conductivity of Porous Silicon Using the Micro-Raman Method

Amaria Ould-Abbas, Mama Bouchaour, Nasr-Eddine Chabane Sari
Research Unit Materials and Renewable Energy (RUMRE), University of Abou Bekr Belkaid of Tlemcen, Tlemcen, Algeria
Email: aouldabbes@yahoo.fr

ABSTRACT

In this work, we are interesting in the measurement of thermal conductivity (on the surface and in-depth) of Porous silicon by the micro-Raman spectroscopy. This direct method (micro-Raman spectroscopy) enabled us to develop a systematic means of investigation of the morphology and the thermal conductivity of Porous silicon oxidized or no. The thermal conductivity is studied according to the parameters of anodization and fraction of silicon oxidized. Thermal transport in the porous silicon layers is limited by its porous nature and the blocking of transport in the silicon skeleton what supports its use in the thermal sensors.

Keywords: Mono-Crystal Silicon; Porous Silicon; Thermal Conductivity; Micro-Raman Spectroscopy

1. Introduction

Porous silicon (PS) is obtained by electrochemical attack of single crystalline silicon [1]. PS is a tuneable material in respect of its pore size distributions depending on its formation conditions. From the macro-porous silicon (with thickness bigger than 0.1 μm or 0.1 nm) to the meso-porous silicon (about 2 to 100 nm of thickness) to the nano-porous silicon (thickness smaller than 10 nm), the range of its applications varies. This material is used in photonics and optoelectronics [2], quantum electronics [3], silicon-on insulator technology [4] and very recently in sensors [5]. Since the last decade, PS has also been investigated in the area of photovoltaic [6] as an antireflection coating and as a sacrificial layer for the layer transfer process (LTP) [7,8].

This material presents a thermal conductivity near to thermal conductivity of silicon dioxide [9]. This material is an excellent candidate to ensure the thermal insulation for the micro sensors on silicon because it ensures the mechanical stability of the microstructure [10]. For this reason, PS layers have been effectively used as material for local thermal isolation on bulk silicon [11,12] and as material for the fabrication of micro-hotplates for low-power thermal sensors [13,14]. Nevertheless, before implementing the PS sensor, it is necessary to understand the thermal transport in such nanostructure and the experimental study of its thermal conductivity, according to its characteristic parameters.

In this work, we are interesting in anodization of silicon to release the principal factors, acting on the morphology in-depth of the layer. We expose a direct method of measurement of the thermal conductivity for PS layers. This parameter is founded on their micro-Raman response to a laser excitation producing a thermal gradient in material. This latter is evaluated by the micro-Raman spectroscopy. The analysis of surface and indepth distribution of the crystallites composing the material is presented.

2. Experimental Procedure

The PS samples were prepared by electrochemical anodic etching of *p*-type, (100)-oriented, 0.01 - 0.025 $\Omega \cdot cm^{-1}$ silicon wafers, in a solution of HF (48%):$C_2H\ddot{O}H$ of 1:1. Current densities range were 5 and 75 mA/cm^2 during 10 min. The current density and the time attack selected permit to obtain different layers with different porosities and different thicknesses. After the electrochemical treatment, samples were rinsed in deionized water and were dried under nitrogen flow.

Samples prepared in this study are described in **Table 1**.

Table 1. Anodization parameters of the samples prepared with two current densities.

Curent density (j) $(mA/cm)^2$	Time (min)	Porosity (P) (%)	Thicknes (e) (μm)
5	10	20	5
75	10	48	40

Thereafter, a thermal oxidation at low temperature (300˚C) under dry atmosphere of O_2 is applied on these samples in order to stabilize the whole structure mechanically and to allow carrying out following treatments [15,16]. We called this stage "pre-oxidation".

Meso-porous layers of 175 μm thickness were obtained by complex methods then were oxidized under dry atmosphere at different temperature (between 300˚C and 700˚C).

The thermal characterization was carried out by micro-Raman spectroscopy in order to determine the size of crystallites of PS as well as thermal conductivity according to the parameters of anodization. **Figure 1** shows a schematic diagram of the micro-Raman method [17].

The laser used for our measurements is an Argon laser with a wavelength of 514 nm. The spectra Raman Stokes are collected in parallel polarization, in retro-diffusion mode using a microscope Olympus BH2, coupled to a monochromator Dilor XY and of a CCD detector (Charge Coupled Detector).

3. Theory

3.1. Thermal Conductivity of Prepared Nano-Porous Silicon

The description of thermal transport of a nanostructured porous material necessarily passes by considerations of a morphological nature

The introduction of a percolation factor g_0 proposed by Gesele *et al.* [10], allows to account for structural microphone-topology and is interpreted like the solid fraction of inter-connected phase contributing to thermal transport.

$$g_0 = (1-P)^2 \quad (1)$$

$$K_{\text{SiP}} = g_0(1-P)\cdot K_{Cr} = (1-P)^3 \cdot K_{Cr} \quad (2)$$

K_{cr} conductivity describes elementary contribution to thermal transport of each quasi-spherical crystallite forming the column-like pseudo network.

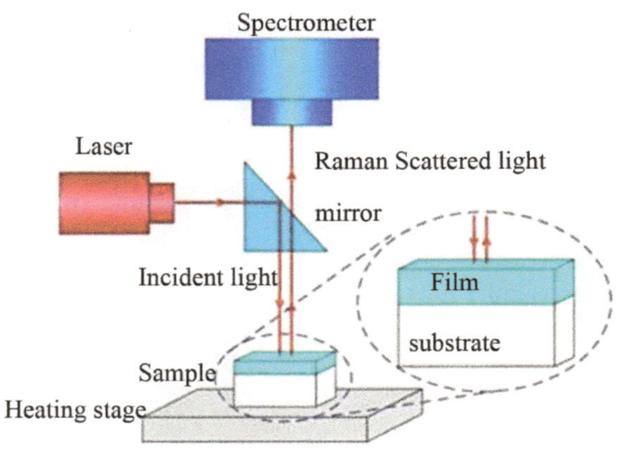

Figure 1. Schematic diagram of the micro-raman method.

Because of their nanometric dimension r_{cr} = 8 - 12 nm lower than 7 times the mean free path phonons in single-crystal silicon at ambient temperature Λ_{Si} = 43 nm [18,19], the transport of heat in a crystallite is not described any more by the traditional model of Fourier.

In this case, the phonons are diffused in crystallites and neither variation in temperature, nor thermal conductivity cannot be defined there. Only, the diffusion of phonons to the interfaces makes it possible to locally restore thermodynamic balance [9]

The transport of phonons intervenes according to a ballistic phenomenon between two interfaces. Their mean free path effective Λ_{Si} is then defined according to their lenght de diffusion in single-crystal silicon Λ_{Si} and of the size of crystallites r_{cr} [16]. Effective thermal conductivity in each crystallite is written [20].

$$K_{cr}^{*} = \frac{K_{\text{Si}}}{1+\frac{4}{3}\frac{\Lambda_{\text{Si}}}{r_{Cr}}} \quad (3)$$

Hence, k_{eff}^{cr} can be determined as:

$$k_{eff}^{cr} = \frac{k_{\text{Si}}}{1+\frac{4}{3}\frac{\Lambda_{\text{Si}}}{r_{cr}}} \quad (4)$$

Replacing the value of $k_{\text{eff}}^{\text{Cr}}$ from Equations (2) and (4) to Equation (1), we obtain the final form of thermal conductivity of nano-PS:

$$k_{eff}^{cr} = \frac{k_{\text{Si}}(1-P)^3}{1+\frac{4}{3}\frac{\Lambda_{\text{Si}}}{r_{cr}}} \quad (5)$$

The variation of thermal conductivity with the porosity of as prepared nano PS layer, for different sizes of non-oxidized Si nanocrystallites is presented in **Figure 2**, *r* is crystallite size of PS.

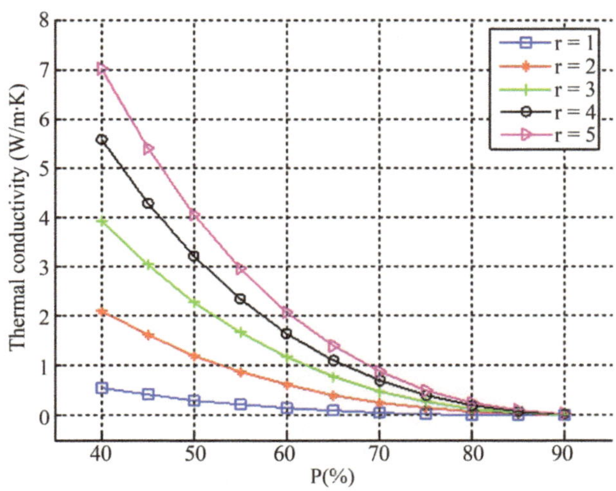

Figure 2. Variation of thermal conductivity with the porosity of as-prepared nano-porous silicon layer.

3.2. Thermal Conductivity of Oxidized Nano-Porous Silicon

After oxidation, each Si crystallite is covered with a SiO_2 sheet and its initial size is reduced. In this case, the thermal conductivity of the remaining Si core of the partially oxidized Si/SiO_2 crystallite k_{cr}^* can be expressed as:

$$k_{cr}^* = \frac{k_{Si}}{1+\frac{4}{3}\frac{\Lambda_{Si}}{r_{cr}\sqrt[3]{(1-\xi)}}} \tag{6}$$

where ξ is the oxidized fraction of the origin Si crystallite and it is defined as:

$$\xi = \frac{\left(V_0^* - V_{Si}^*\right)}{2.27\, V_0} \tag{7}$$

where V_0 is the initial volume of Si nano-crystallite, V_0^* is the total volume of the crystallite after oxidation procedure and V_{Si}^* is the volume of Si core of the oxidized crystallite. Coefficient 2.27 [21] considers extension of the crystallite volume after oxidation.

It is easily proved that total thermal conductivity of oxidized nano-PS $k^*_{\text{nano-PS}}$ can be written as:

$$k^*_{\text{nano-PS}} = \frac{(1-P)^3\,(1+1.27\xi)^4}{\dfrac{(1-\xi)\left(1+\frac{4}{3}\frac{\Lambda_{Si}}{r_{cr}\sqrt[3]{(1-3)}}\right)}{k_{Si}} + \dfrac{2.27\xi}{k_{SiO_2}}} \tag{8}$$

k_{SiO_2} is the thermal conductivity of the dioxide silicon.

The variation of thermal conductivity with oxidized fraction of nano-porous silicon layer, for three different sizes of the original non-oxidized Si nanocrystallites is presented in **Figure 3**.

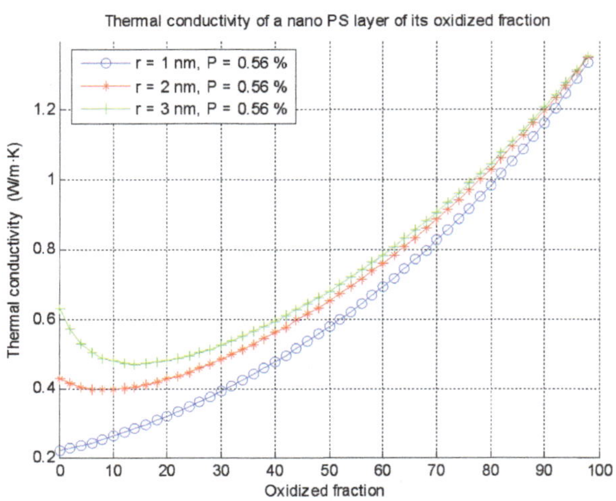

Figure 3. Dependence of the thermal conductivity of a nano PS layer of its oxidized fraction for three different sizes (r) of silicon nanocrystallites porosity of the layer is 56%.

The function $k^*_{\text{nano-PS}} = f(\xi)$ decreases in the range of small ξ values and has a minimum value (for r_{cr} of 2 - 3 nm). It corresponds to heat transport preferably through the decreased Si cores. Then, $k^*_{\text{nano-PS}}$ increases along the increase of oxidized fraction, because of the heat transport occurring through the growing oxide sheet, and comes up to the thermal conductivity value of silicon dioxide (1.4 $Wm^{-1}\cdot K^{-1}$) when the porous silicon layer is completely oxidized and transformed to a SiO_2 layer [22,23].

3.3. Determination of the Crystallites Size

In the determination of crystallites the size, we use Campbell and Fauchet method [24]. It consists in correlating position and/or width with middle height of the Raman response of the semiconductor to crystallites size. In order to control morphology on the surface and in-depth of SiP layers, it is necessary to measure the size of crystallites forming material. Moreover, the thermal properties of the layer depend directly on their value. Raman Stokes intensity, obtained with low laser power, is modeled by the following equation:

$$I(\omega) = \frac{d^3q\left|C(q_0,q)\right|^2}{\left(\omega-\omega(q)\right)^2+\left(\Gamma_0/2\right)^2} \tag{9}$$

with $C(q_0,q)$ Fourier coefficients resulting from the decomposition of wave function in material in Fourier series, $\omega(q)$ the curve of dispersion of phonons and Γ_0 the width with middle height of the Raman peak of crystalline material. In the case of mesoporous silicon, we choose $C(q_0,Q)$ with a function of containment of phonons in quasi-spherical crystallites such as:

$$C(0,q) = \frac{Ar}{(2\pi)^{3/2}}\exp\left(-\frac{1}{2}\frac{q^2r^2}{4}\right) \tag{10}$$

with r crystallites size, A constant and Q wave vector.

4. Results and Discussion

The results presented in **Table 2** on the size of crystallites are obtained from the measurement of the width middle height of Raman peak function.

The exploitation of Raman spectra can lead to the determination of crystallites size only when the incidental power laser is low for not provokes the heating of the surface (which would modify the position and the shape

Table 2. Crystallites sizes obtained from the raman analyses.

Porosity %	Crystallites sizes (r) Nm
20	9.5
48	8.7

of the spectra) (**Figure 4**).

By the method of micro characterization Raman, presented in a preceding study [24]. The method of measurement that we developed is founded on two effects complementary to the micro-Raman spectroscopy. The first is the use of the laser beam localised on the surface of material like local source of heating. The rise in the temperature resulting depends mainly on thermal conductivity from the sample. The second effect is the displacement of the peak Raman with the temperature of the sample, which makes it possible to deduce the local rise in the temperature from material.

By using the linear model suggested by Nonnenmacher [25], the thermal conductivity of a material in thin layer can be written in the form:

$$K_s = \frac{2P}{\pi a\left(T_j - T_s\right)} \tag{11}$$

with P the power of heating, T_j the temperature of surface and T_s the temperature of the substrate, has the diameter of the laser spot, We obtained the following results. The porosity is directly, related to the nano-structure of material. It is the principal parameter acting on thermal conductivity, **Figure 5** shows that thermal conductivity decreases when porosity increases. This is explained by the effect of reduction of the effective size of the nano-crystals silicon; when porosity increases, crystallites size decreases as well as surfaces of interconnection between crystallites what more reduces the thermal way by conduction in the silicon skeleton.

The variations of thermal conductivity with the temperature of porous silicon are presented in **Figure 6** for three porosities. Each sample has a thickness of 50 µm. Their thermal conductivity is measured after anodization and after oxidation during 1 hour, in dried O_2 at 150˚C, 300˚C and at last at 450˚C.

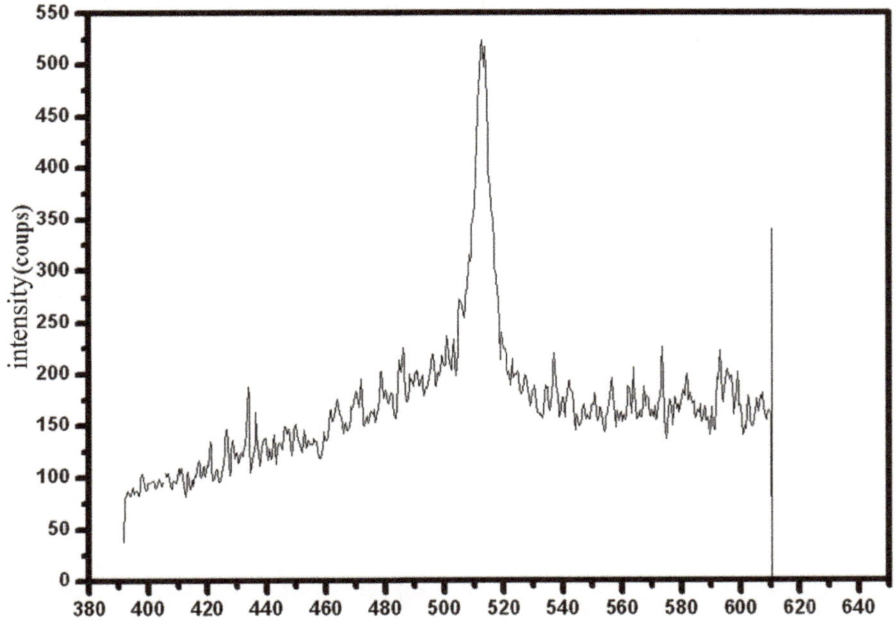

Figure 4. Raman spectra of the sample (J = 5 mA/cm², t = 10 mn).

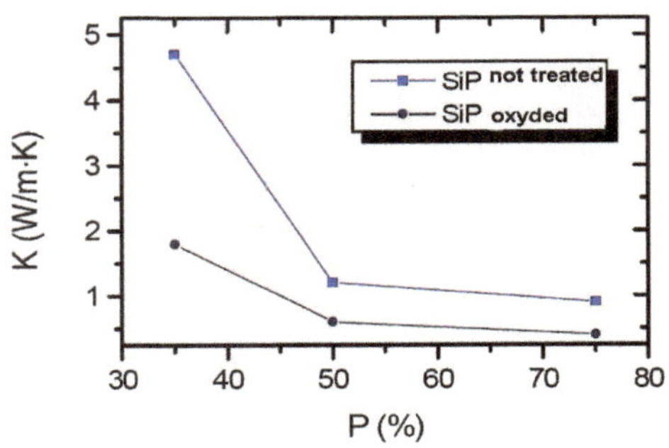

Figure 5. Thermal conductivity of meso PS samples oxidized 1h under dry O_2.

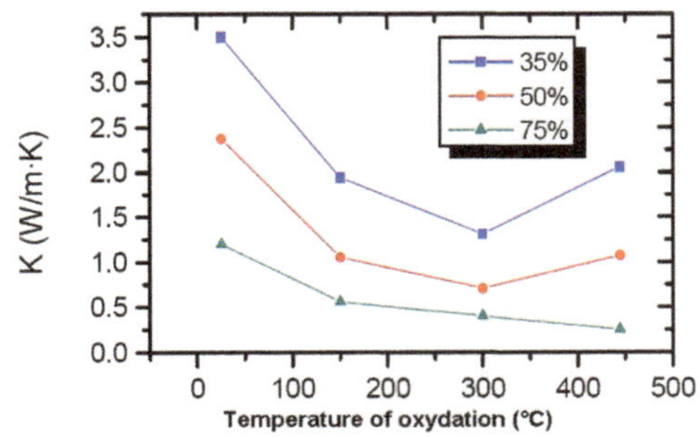

Figure 6. Thermal conductivity of meso-PS obtained according to oxydation temperature at 150˚C, 300˚C and 450˚C.

Whatever their porosity, the thermal conductivity of the samples follows the same evolution. Conductivity decrease initially until reaching a minimum then increases slightly

5. Conclusions

The micro-Raman spectroscopy enabled us to develop a systematic means of investigation of morphology and thermal conductivity of porous silicon. The thermal conductivity of porous silicon was studied according to the parameters of anodization (initial doping and current density and rate oxidation).

One will retain the principal following tendencies:

- Thermal conductivity strongly decrease when the layer porosity increases.
- Oxidation under dried O_2 induced a strong decrease of conductivity to reach a minimum.

The thermal conductivity of porous silicon is quite lower than that of mono-crystal silicon. Thermal transport in the porous silicon layers is limited by its porous nature and the blocking of transport in the silicon skeleton what supports its use in the thermal sensors.

6. Acknowledgements

The authors wish to thank S. Perichon and D. Barbier for Raman manipulations and for their fruitful discussions

REFERENCES

[1] L. T. Canham, "Silicon Quantum Wire Array Fabrication by Electrochemical and Chemical Dissolution of Wafers," *Applied Physics Letters*, Vol. 57, No. 10, 1990, pp. 1046-1049. doi:10.1063/1.103561

[2] A. G. Cullis, L. Canham and P. D. J. Calcott, "The Structural and Luminescence Properties of Porous Silicon," *Journal of Applied Physics*, Vol. 82, No. 3, 1997, pp. 909-962. doi:10.1063/1.366536

[3] A. J. Read, R. J. Needs, K. J. Nash, L. T. Canham, P. D. J. Calcott and A. Qteish, "First-Principles Calculations of the Electronic Properties of Silicon Quantum Wires," *Physical Review Letters*, Vol. 69, No. 8, 1992, p. 1232. doi:10.1103/PhysRevLett.69.1232

[4] G. Bomchill, A. Halimaoui and R. Herino, "Porous Silicon: The Material and Its Applications to SOI Technologies," *Microelectronic Engineering*, Vol. 8, No. 3-4, 1988, pp. 293-310. doi:10.1016/0167-9317(88)90022-6

[5] A. Foucaran, B. Sorli, M. Garcia, F. Pascal-Delannoy, A. Giani and A. Boyer, "Porous Silicon Layer Coupled with Thermoelectric Cooler: A Humidity Sensor," *Sensors & Actuators*, Vol. 79, No. 3, 2000, pp. 189-193. doi:10.1016/S0924-4247(99)00285-X

[6] R. Bilyalov, L. Stalmans, G. Beaucarne, R. Loo, M. Cayman, J. Poortmans and J. Nijs, "Porous Silicon as an Intermediate Layer for Thin-Film Solar Cell," *Solar Energy Materials and Solar Cells*, Vol. 65, No. 1-4, 2001, pp. 477-485. doi:10.1016/S0927-0248(00)00130-6

[7] R. B. Bergmann, "Crystalline Si Thin-Film Solar Cells: A Review," *Applied Physics A*, Vol. 69, No. 2, 1999, pp. 187-194. doi:10.1007/s003390050989

[8] R. Brendel, "A Novel Process for Ultrathin Monocrystalline Silicon Solar Cells on Glass," *Proceeding of the 14th European Photovoltaic Solar Energy Conference*, Barcelona, 6-10 June 1997, p. 1354.

[9] S. Huang, X. D. Ruan, J. Zou, X. Fu and H. Y. Yang, "Thermal Conductivity Measurement of Submicrometer-Scale Silicon Dioxide Films by an Extended Micro-Raman Method," *Microsystem Technologies*, Vol. 15, 2009, pp. 837-842

[10] G. Gesele, J. Linsmeier, V. Drach, J. Fricke and R. Arens-Fischer, "Temperature-Dependent Thermal Conductivity of Porous Silicon," *Journal of Physics D*, Vol. 30, No. 21, 1997, p. 2911. doi:10.1088/0022-3727/30/21/001

[11] A. G. Nassiopoulou and G. Kaltsas, "Porous Silicon as an Effective Material for Thermal Isolation on Bulk Crystalline Silicon," *Physica Status Solidi A*, Vol. 182, No. 1, 2000, pp. 307-311. doi:10.1002/1521-396X(200011)182:1<307::AID-PSSA307>3.0.CO;2-#

[12] V. Lysenko, S. Perichon, B. Remaki and D. Barbier, "Thermal Isolation in Microsystems with Porous Silicon," *Sensors and Actuators A*, Vol. 99, No. 1-2, 2002, pp. 13-24. doi:10.1016/S0924-4247(01)00881-0

[13] P. Maccagnani, R. Angelucci, P. Pozzi, A. Poggi, L. Dori, G. C. Cardinali and P. Negrini, "Thick Oxidised Porous Silicon Layer as a Thermo-Insulating Membrane for High-Temperature Operating Thin- and Thick-Film Gas Sensors," *Sensors and Actuators B*, Vol. 49, No. 1-2, 1998, pp. 22-29. doi:10.1016/S0925-4005(97)00337-7

[14] C. Tsamis, A. Tserepi and A. G. Nassiopoulou, "Thermal Properties of Suspended Porous Silicon Micro-Hotplates for Sensor Applications," *Sensors and Actuators B*, Vol. 95, No. 1-3, 2003, pp. 78-82. doi:10.1016/S0925-4005(03)00409-X

[15] S. M. Sze, "Physics of Semiconductor Devices," John Wiley and Sons, New York, 1981, pp. 42-43.

[16] J. J. Yon, K. Barla, R. Herino and Bomchil, "The Kinetics and Mechanism of Oxide Layer Formation from Porous Silicon Formed on P-Si Substrates," *Journal of Applied Physics*, Vol. 62, No. 3, 1987, pp. 1042-1048. doi:10.1063/1.339761

[17] Y. Arita and K. Kuranari, "Thermal Behavior of Porous Silicon," *Japanese Journal of Applied Physics*, Vol. 15, 1976, pp. 1655-1664. doi:10.1143/JJAP.15.1655

[18] S. Huang, X.-D. Ruan, X. Fu, H.-Y. Yang and A. Majumdar, "Transient Ballistic and Diffusive Phonon Heat Transportin Thin Films," *Journal of Heat Transfer*, Vol. 115, 1993, p. 7.

[19] A. A. Joshi and A. Majumdar, "Transient Ballistic and Diffusive Phonon Heat Transportin Thin Films," *Journal of Applied Physics*, Vol. 74, No. 1, 1993, pp. 31-39.

[20] G. Chen, "Nonlocal and Nonequilibrium Heat Conduc-

tion in the Vicinity of Nanoparticles," *Journal of Heat Transfer*, Vol. 118, No. 3, 1996, p. 539. doi:10.1115/1.2822665

[21] L. Roussel, B. Remaki, G. Delhomme, A. Dittmar and D. Barbier, "Study of Nano-Porous Silicon with Low Thermal Conductivity as Thermal Insulating Material," *Journal of Porous Materials*, Vol. 7, No. 1-3, 2000, pp. 177-182. doi:10.1023/A:1009626518619

[22] Charles Populaire, These, lyon, 2005.

[23] A. Ould-Abbas, M. Bouchaour, M. Madani and N.-E. C. Sari, "Thermal Analysis and Morphological of Porous Silicon Oxidized or Not by Raman Spectroscopy," *Applications to Mems Thermal*, *Proceeding of Thermal French Congres*, *SFT* 2006, Île de Ré, 16-19 Mai 2006.

[24] I. H. Campbell and P. M. Fauchet, "The Effects of Microcrystal Size and Shape on the One Phonon Raman Spectra of Crystalline Semiconductors," *Solid State Communications*, Vol. 58, No. 10, 1986, pp. 739-741. doi:10.1016/0038-1098(86)90513-2

[25] M. Nonnenmacher and H. K. Wickramasinghe, "Scanning Probe Microscopy of Thermal Conductivity and Subsurface Properties," *Applied Physics Letters*, Vol. 61, No. 2, 1992, pp. 168-170. doi:10.1063/1.108207

12

A Graph-Theoretical Approach to Calculate Vibrational Energies of Atomic and Subatomic Systems

Jorge Galvez
Department of Physical Chemistry, Faculty of Pharmacy, University of Valencia, Valencia, Spain
Email: jorge.galvez@uv.es

ABSTRACT

One of the challenges still pending in string theory and other particle physics related fields is the accurate prediction of the masses of the elementary particles defined in the standard model. In this paper an original algorithm to assign graphs to each of these particles is proposed. Based on this mapping, we demonstrate that certain indices associated with the topology of the graph (graph theoretical indices) are very effective in predicting the masses of the particles. Specifically, the spectral moments of the graph adjacency matrix weighted by edge degrees play a key role in the excellent correlations found. Moreover, the same topological pattern is found in other well known quantum systems such as the particle in a box and the vibrational frequencies of diatomic molecules, such as hydrogen. The results shown here open a suggestive pathway for the use of graph-theoretical approaches in predicting properties of elementary particles and other physical systems, which seem to match similar topological patterns.

Keywords: Graph Theory; Physical Systems; Vibrational Energy

1. Introduction

Among the most important achievements generally recognized to string theory is its potential capability to predict the masses of elementary particles, specifically of the 12 particles that make up the three families of the standard model [1]. The process, however, require very complex mathematical calculations which are still far from complete.

There is a well known precedent in particle physics to the study of interactions between particles, which are the Feynman diagrams [2]. **Figure 1** shows the Feynman diagram for electron-positron annihilation.

It is interesting that, although representing a space-time event, Feynman diagrams are ultimately graphs. A graph is a set of points called vertices (or nodes) connected by lines called edges. The studies of graphs constitute a discipline known as graph theory. It was L. Euler (1707-1783) the one who introduced the notion of graph [3], which was developed later on by A. Cayley [4] and J. J. Sylvester [5] along the nineteenth century. In the 20th century, graph theory became an essential tool in different areas of science and technology where connectivity plays a role. Think, for instance, of the optimization of communication and transport networks [6], the design of electrical circuits (e.g. in computers) [7], the synchronization of interacting oscillators with different topologies [8], the analysis of social networks [9], among others.

Although graphs in general are also widely employed in theoretical physics, however graph-theoretical indices are rarely used [10]. Our goal here is just using graph-theoretical indices to predict the masses of the elementary particles referred above as well as of other physical systems such as molecular vibrational energy.

2. Assignation of Graphs to Some Vibratory Physical Systems. Theoretical Framework

Starting by the elementary particles, we must remind the origin of the Feynman diagrams. It is well known that the interactions between particles are expressed as an integral, which is typically too difficult to do, so that Feynman developed a perturbation theory which was based upon an expansion in terms of graphs. The type of such graphs depends on the interactions.

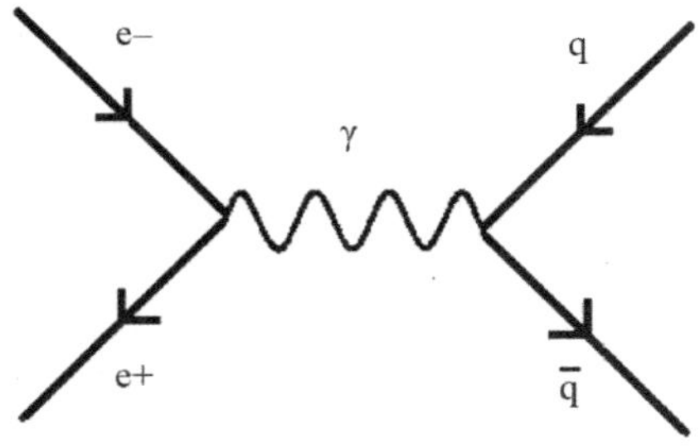

Figure 1. Feynman diagram for electron-positron annihilation. e: electron (–) or positron (+) q: quark γ = gamma emission.

A similar approach can be also found in the so called force-based algorithms [11], in which the entire graph is then simulated as if it were a physical system, for example an elementary particle. Moreover, the use of graphs and matrices to study the interactions between elementary particles, for example quarks, has been also carried out by other authors [12].

On string theory, the way to explain that an elementary particle, say an electron, has a given mass is based on the different modes of vibration of the strings. Indeed, in the perturbative approach to string theory, the strings interact by joining and splitting [13].

Each vibrational mode has an associated energy according to Einstein's equation:

$$E = m \times c^2 \quad (1)$$

This energy is transformed into the elementary particle mass. In this sense we can say that elementary particles are made of strings.

In the perturbative string theory, the influence of each incoming string results from adding together the influences of graphs with ever more loops.

The algorithm proposed here is similar, *i.e.* assigning each perturbative term to each one of the particles so that each graph simply adds a loop over the next particle with increasing mass; in other words we propose assigning one-to-one the perturbative terms to the particles.

The process of assigning individual graphs to each elementary particle proposed here is done as follows: First, the elementary particles are sorted in increasing order of mass ranging from the electron-neutrino ($<10^{-8}$) up to top-quark (189), both in GeVs. **Table 1** reflects the values, in increasing order of energies, for these particles. On the first column is the allocation of families for every one of the particles. A data taken from reference [13].

Table 1. Elementary particles, according to the standard model, ordered in increasing values of energy. F1, F2 and F3 represent the three families in which they are classified.

Family (Standard Model)	Elementary Particle	Mass (GeV)[a]
F1	Electron-neutrino	0.00000001 (aprox)
F2	Muon-neutrino	0.0003 (aprox)
F1	Electron	0.00054
F1	Up-quark	0.0047
F1	Down-quark	0.0074
F2	Muon	0.11
F2	Strange-quark	0.16
F3	Tau-neutrino	0.33 (aprox)
F2	Charm-quark	1.6
F3	Tau	1.9
F3	Bottom-quark	5.2
F3	Top-quark	189

For the first particle, namely electron-neutrino, we assign a simple graph that is equivalent to the interaction electron-positron described before. **Figure 2** illustrates the Feynman-like 3D diagram associated to the strings interaction, *i.e.* the string/antistring pair, together with the simple (no loops) graph assigned to it:

The following larger energy particles are represented by pseudographs which take into account different topologies with 1, 2, 3, ...11 holes. For example, for the muon-neutrino, the second lowest energy particle, we would have the following equivalence between the strings' interaction and the graph (**Figure 3**):

In this case, the two strings interact creating a hole whose graph theoretical equivalent is the one-loop graph on the right (B) of **Figure 3**. In short, the two diagrams in **Figure 3** correspond to the Feynman representations for string (A) and quantum field theory (B) [14]. The advantage of this approach is that there is just one diagram for each order of perturbation and that each diagram avoid the drawback of short-distance infinities, *i.e.* contrary to quantum field theory, the graphs in **Figures 2(a)** and **3(a)** have not singularities (nodes). It is also curious that the topological pattern that we meet here is the same as the one of harmonic vibrations of a macroscopic string (such as a violin string), in which each frequency of the overtones (n = 2, 3, 4…) is an integer multiple of the first one (n = 1, fundamental), as shown in **Figure 4**.

Note that the graph-theoretical equivalent of each mode of vibration would be graphs with a progressive increase in the number of loops.

Figure 5 shows the graphs allocated to the different particles, according to our algorithm.

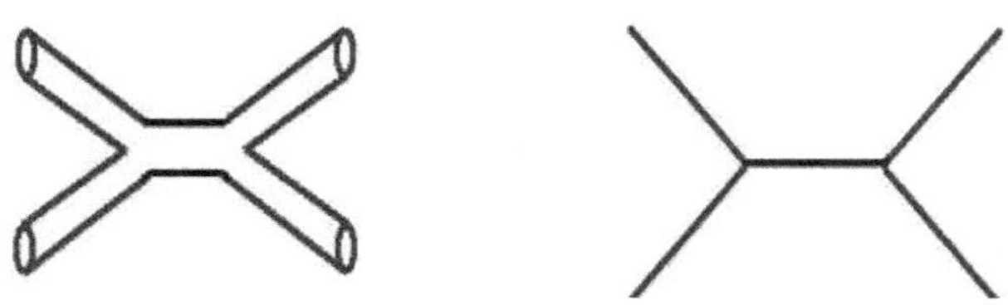

(a) (b)

Figure 2. Allocation graph to electron-neutrino. Left (a) Feynman-like graph for the springs interaction and right; (b) corresponding simple graph.

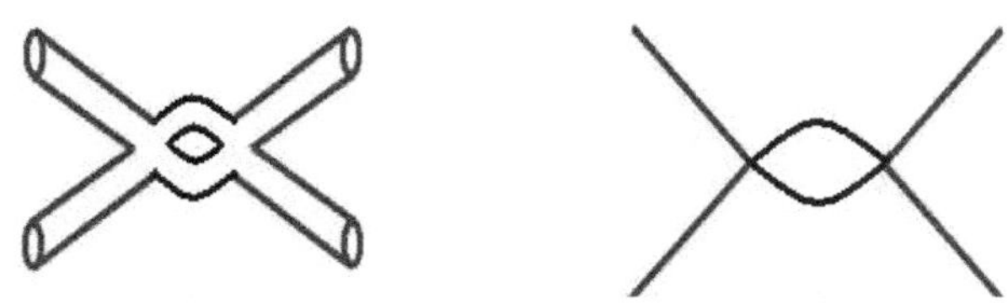

(a) (b)

Figure 3. Allocation of graph to muon-neutrino, the second particle with lower mass. To the left (a) Feynman graph and on the right (b) corresponding pseudograph.

Vibrational mode	n	Frequency
	1	f_1
	2	$f_2=2f_1$
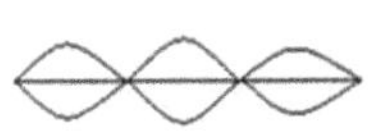	3	$f_3=3f_1$
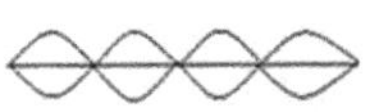	4	$f_4=4f_1$
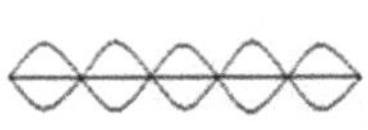	5	$f_5=5f_1$

Figure 4. Simple modes of vibration of a string. The harmonic frequencies shown are multiples of the first one (fundamental vibration).

Particle	Graph	Graph's label	n	Ψ
electron-neutrino		G1	0	0
muon-neutrino	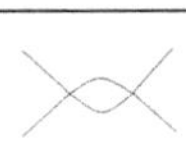	G2	1	$\Psi(x)=\pm\sin(\frac{\pi x}{2})$
electron	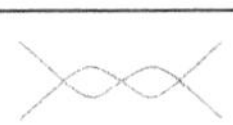	G3	2	$\Psi(x)=\pm\sin(\pi x)$
up-quark	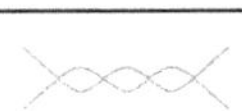	G4	3	$\Psi(x)=\pm\sin(\frac{3\pi x}{2})$
down-quark	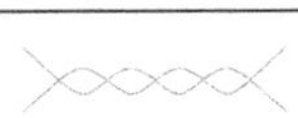	G5	4	$\Psi(x)=\pm\sin(2\pi x)$
muon	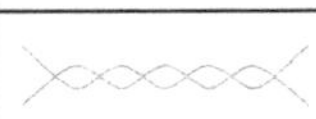	G6	5	$\Psi(x)=\pm\sin(\frac{5\pi x}{2})$
strange-quark		G7	6	$\Psi(x)=\pm\sin(3\pi x)$
tau-neutrino	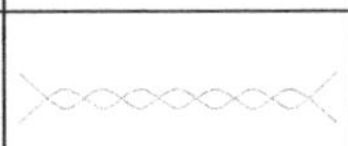	G8	7	$\Psi(x)=\pm\sin(\frac{7\pi x}{2})$
charm-quark	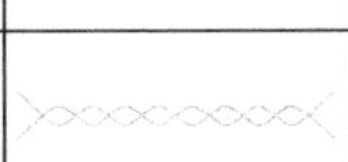	G9	8	$\Psi(x)=\pm\sin(4\pi x)$
tau		G10	9	$\Psi(x)=\pm\sin(\frac{9\pi x}{2})$
bottom-quark	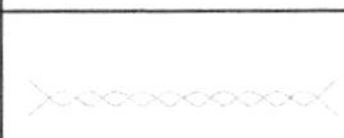	G11	10	$\Psi(x)=\pm\sin(5\pi x)$
top-quark	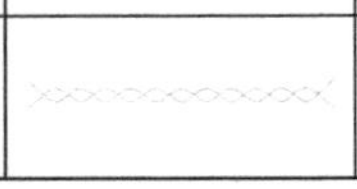	G12	11	$\Psi(x)=\pm\sin(\frac{11\pi x}{2})$

Figure 5. Allocation of graphs to different particles. On the right, the real part of wavefunctions for particle in a box for different quantum levels (n).

Except for the less massive particle, the electron-neutrino, which is assigned a simple graph, the rest are assigned on pseudographs with increasing number of adjacent loops.

Furthermore, the same graph allocation can be done on other quantum systems such as the particle in a box, whose wave functions (the real part) are also outlined in **Figure 5** for the different quantum levels (n). These wave functions show a shape that is equivalent to the graphs assigned to each particle. This approach relating graphs to wavefunctions has been proposed previously by the author [15].

In fact, the proposed allocation of graphs can be based on very simple theoretical terms considering the vibrating strings as particles in a one-dimensional box and giving each particle a quantum number (n). Indeed, considering only the real part of the wave function associated with a one-dimensional box of length "*a*", we have [16]:

$$\Psi(x)=\sqrt{2/a}\,\sin(\frac{n\pi x}{a}) \qquad (2)$$

Which normalized for $a = 2$ leads to:

$$\Psi(x)=\pm\sin(\frac{n\pi x}{2}) \qquad (3)$$

It is clear from Equation (3) that the number of maxima-minima associated with the wave function is equal to the quantum number n, which can take the values 0, 1, 2, 3... The value 0 is associated to the ground state which corresponds to annihilation.

A similar result could be reached using the harmonic oscillator model, in which, for the one-dimensional case, the number of interior nodes on the boundary points is zero for the ground state and increases by one for each successive excited state, according to:

$$\Psi_v(x)=\left(2^v v!\right)^{-1/2}(\alpha/\pi)^{1/4}\,e^{-\alpha x^2/2}H_v\left(\alpha^{1/2}x\right) \qquad (4)$$

where:

$\alpha = 4\pi^{2vm/h}$

v = Vibrational quantum number

H_v = Hermite polinomials .

In this case there is a residual vibrational energy that would correspond to the lowest mass particle (electron-neutrino), so that the association would be between the number of nodes and the number of loops in the graph. This way we can see that, regardless the model used, if we translate into graph-theoretical terms the wave functions, each graph has the same number of loops as number of singular points (maxima-minima or nodes) has the wave function. The relationship between wave functions and graphs is implicit in Heisenberg's matrix formulation of quantum mechanics, because for each array or matrix a graph of connections can be assigned. Moreover, Filk has solidly demonstrated the equivalence between wave functions and graphs [17].

Within this theoretical framework, it is to be expected that the same allocation of graphs we have proposed for elementary particles, may also work in other physical systems, as for example, the vibrational energies of a diatomic molecule. In concrete we have applied this approach to predict the vibrational levels energy for the molecule of hydrogen. Given that in this case we have a residual vibrational energy for the lowest quantum level ($v = 0$), we assign the graphs G2, G3, G4, … to the first, second, third, ...vibrational levels, respectively.

3. Results and Discussion

After the allocation of the different graphs, the next step is to calculate the topological descriptors associated with each graph and try to establish a correlation with the energy of the particles. For calculation of topological descriptors, the software Dragon (a pull of some 330 indices) was used [18] and to obtain the regression equations, the BMDP statistical package [19] was run on the data.

The results clearly demonstrate that spectral moments are the variables that contribute most to the variance as for the correlation with the elementary particle masses. Thus, **Figure 6** illustrates the representation of the parameter F (Fisher-Snedecor) as a function of the spectral moments' ranges (ordinals). As can be seen, in almost all cases the value is greater than 100, so demonstrating that the spectral moments play a key role in the predictive equation of the particle masses, being the #7 to #9 the major contributors to the variance (F around 500).

The best regression equations with one and two variables were:

$$\log M = -56.69 + 4.780 \times ESpm08x \quad (5)$$

N = 12 R = 0.99095 SE = 0.3739 F = 545.1

$$\log M = -53.99 + 4.537 \times ESpm08x + 4.028 \times EEi \quad (6)$$

N = 12 R = 0.99684 SE = 0.2333 F=708.5

The symbols in Equations (5) and (6) are:

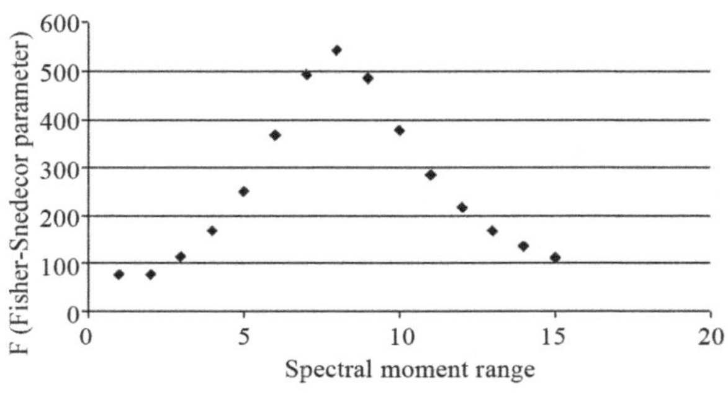

Figure 6. Representation of the values of Fisher-Snedecor parameter (F) as a function of the spectral moments.

M = Particle mass (GeV)

$ESpm08x$ = Spectral moment 08 from edge adjacency matrix weighted by edge degrees

$EEig11x$ = Eigenvalue 11 from edge adjacency matrix weighted by edge degrees.

N = Number of data

R = Regression coefficient

SE = Standard error of estimate

F = Fischer-Snedecor parameter

The spectral moments of the edge adjacency matrix are defined as the traces, *i.e.*, the sum of the main diagonal entries, of the different powers of the matrix [20].

The k-th spectral moment of a graph G, $\mu_k(G)$, with eigenvalues $\lambda_1, \lambda_2 \ldots, \lambda_n$, is defined as:

$$ESpmkx = \mu_k(G) = \sum_{t=1}^{n} (\lambda_t)^k \quad (7)$$

It is to be noted that the spectral moment μ_k represents the number of weighted closed walks of length k existing in the graph, what makes the outcome from Equations (5) and (6) all the most reasonable considering the allocation of graphs done here. Moreover, both equations showed to be non-random and stable under the cross validation (leave-one-out) procedure. Interestingly, although both equations exceed 0.99 of correlation coefficient, the larger value of the parameter F (Fisher-Snedecor) as well as the significant drop in the standard error (from 0.37 to 0.23) in Equation (6), clearly shows that the two-variable equation is more significant.

Table 2 illustrates the values of the two topological indices from the selected regression equation ($ESpm08x$ and $EEig11x$) for each particle, as well as the comparison between the experimental and the calculated values from the two-variable equation in both, logarithmic and direct scale. As can be seen, the variable explaining most of the variance is $ESpm08x$, whereas the other, $EEig11x$, only improves the prediction for the two most massive particles (bottom and top quarks).

Furthermore, there is a good matching between the actual and the calculated values, what is to be emphasized given the simplicity of the procedure followed and the ease of calculation. It is also noteworthy the low standard errors achieved, which, except for the lowest and the highest value, are always below 0.1, *i.e.* 10%, what can be considered a significant achievement taking into account the wide range of magnitude of the particle energies.

Figure 7 illustrates the plot of the experimental versus the calculated values. It is noteworthy the excellent agreement.

Interestingly, in both Equations (5) and (6), the indices therein are those derived from the edge adjacency matrices weighted by edge degrees, what accounts for the importance of interconnections between the nodes rather

Table 2. Values of the topological indices together to the observed and calculated (Equation (6)) values of particle masses.

Particle	*ESpm08x*	*EEig11x*	log mass (experimental value, GeV)	log mass (calculated by Equation (6))	Standard error of pred. value	Mass experimental value (GeV)	Mass calculated by Equation (6)
Electron-neutrino	10.2	0	−8	−7.848	0.195	10^{-8}	1.4×10^{-8}
Muon-neutrino	11.1	0	−3.523	−3.661	0.089	0.0003	0.0002
Electron	11.2	0	−3.268	−3.244	0.082	0.0005	0.0006
Up-quark	11.3	0	−2.328	−2.536	0.074	0.0047	0.0029
Down-quark	11.5	0	−2.131	−1.892	0.072	0.0074	0.0128
Muon	11.6	0	−0.959	−1.329	0.075	0.1100	0.0469
Strange-quark	11.7	0	−0.796	−0.826	0.081	0.1600	0.1494
Tau-neutrino	11.8	0	−0.482	−0.377	0.088	0.3300	0.4202
Charm-quark	11.9	0	0.204	0.036	0.096	1.5999	1.0874
Tau	12.0	0	0.279	0.413	0.104	1.9002	2.5876
Bottom-quark	12.1	0.1	0.716	1.089	0.097	5.2000	12.2603
Top-quark	12.1	0.8	2.277	2.164	0.224	189.02	145.85

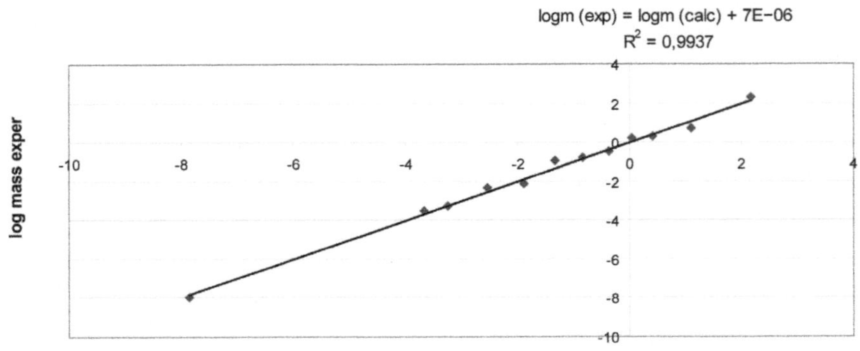

Figure 7. Plot of the observed versus the calculated values of the particle masses according to Equation (6).

than that of the nodes themselves. Of course, this implies that, as expected, the topology associated with each particle plays a key role, beyond geometric characteristics related to size or shape. In particular the presence of successive adjacent loops stands as an essential element of the predictive, which is interpreted as successive stages of annihilation-splitting of the interacting strings.

It is also compelling that the great difference between the less massive particle (electron-neutrino) and the next one (muon-neutrino) is a pure "topological" gap because it is the difference between no-loop and one-loop graphs.

As pointed above, no one theory has been yet capable to exact prediction of the elementary particle mass-energy, although some interesting attempts have been made [21]. The problem remains far to be solved and this is why any step forward is of great interest.

Regarding the results of prediction of the vibrational energy levels for hydrogen molecule, the best regression equation obtained for the first twelve levels was

$$Ev = -1773.80 + 450.23 \times ESpm03r \qquad (8)$$

N =12 R =0.9998 SE = 2.6275 F = 26602.74

The symbols in Equation (8) are

Ev = Vibrational energy (KJ/mole).

ESpm03r = Spectral moment 03 from edge adjacency matrix weighted by resonance integrals.

N = Number of data.

R = Regression coefficient.

SE = Standard error of estimate.

F = Fischer-Snedecor parameter.

Table 3 illustrates the comparison between the experimental values and those calculated from the Equation (8). As can be observed, in this case the graph G2 is assigned to the first quantum state (v = 0), because of the existence of the vibrational residual energy in the anharmonic oscillator.

As seen, another spectral moment, namely *ESpm03r*,

accounts for more than 99% of the variance, what is important because we are in presence of a completely different physical system.

The preponderant role of spectral moments as a measure of the energy in different systems has enough precedents in the literature, although little application has been done so far on particle physics. For example, Broote and Körner have posed arguments on the role of spectral moments in perturbation theory, particularly in the study to the low-energy hadron production associated to electron-positron annihilation [22].

The spectral moments have been related elsewhere to the concept of "energy of graphs" [23]. Moreover, Estrada [24] has demonstrated in a very solid way that the spectral moments can be related to the energy of any system, according to the equation:

$$\begin{aligned} E(t) &= \int_{-\infty}^{\infty} \varepsilon(t)\Gamma\,^{\wedge}[\varepsilon(t)]\,d\varepsilon(t) \\ &= \int_{-\infty}^{\infty} \varepsilon(t)\sum_i c_i[\varepsilon(t)]\mu_i[E(t)]\,d\varepsilon(t) \\ &= \sum_i c_i[\varepsilon(t)]\mu_i[E(t)] \end{aligned} \quad (9)$$

where $E(t)$ are the different types of energies involved and $\varepsilon(t)$ are the discrete values of the energy levels for energy of type t. Γ is the spectral density function and $\mu_i[E(t)]$ are the corresponding spectral moments associated to that type of energy.

Based on Equation (9), for a given spectral moment value, for example μ_k, and given that the spectral den- sity function is continuous and differentiable, the energy of the system can be expressed as a series expansion by powers of μ_k in the form:

$$E(t) = \alpha + \beta\mu_k + \gamma\mu_k^2 + \cdots = c\sum_{j=0}^{\infty}\frac{(c'\mu_k)^j}{j!} = c\,e^{c'\mu_k} \quad (10)$$

where c and c' are just two proportionality constants. Taking logarithms and considering that in our case we have a vibrational energy, Ev, it results:

$$\log Ev = \log c + c'\mu_k \quad (11)$$

Since the vibrational energy is the mass of the particle (in GeV units), Ev may be replaced by m (mass of the particle), so coming finally to the equation:

$$\log m = \log c + c'\mu_k \quad (12)$$

Equation (12) stands for a possible theoretical interpretation of Equations (5) and (6), however the definitive interpretation of the role of the topological indices in Equations (5), (6) and (8) remains an open question that might perhaps encourage other fellow experts to take over the issue under the framework of theoretical physics.

The application of graphs in various fields, particularly in molecules, has been extremely productive, demonstrating to be a very effective tool for predicting molecular properties [25-27].

Furthermore, our research group has cover a long way in the topological design of new drugs [28-30].

Table 3. Allocation of graphs and comparison between observed and calculated values of vibrational energies for the molecule of hydrogen.

Vibrational quantum number	Graph allocated	*ESpm03r*	Vibrational energy observed* (KJ/mole)	Vibrational energy calculated through Eq. (8) (KJ/mole)	Standard error of predicted value
0	G2	3.99	26.1	22.2	1.58
1	G3	4.11	76.0	77.1	1.29
2	G4	4.22	123.1	126.2	1.06
3	G5	4.32	167.3	169.9	0.90
4	G6	4.41	208.6	210.4	0.79
5	G7	4.49	247.1	247.3	0.76
6	G8	4.56	282.7	281.1	0.79
7	G9	4.63	315.4	313.0	0.86
8	G10	4.70	345.2	342.3	0.95
9	G11	4.76	372.1	370.2	1.07
10	G12	4.82	396.2	396.3	1.18
11	G13	4.87	417.4	421.1	1.30

*Data taken from J.M. Hernando. Problemas de Quimica Fisica. Graficas A. Martin S.A. ISBN: 84-400-6995-2.

Not surprisingly, the topology associated to graphs can also be useful in predicting properties in other areas, such as that of atomic and subatomic systems. As previously mentioned, the role of graphs in the interactions between particles (Feynman diagrams) and in string theory, is well known. However, it is noteworthy that the contribution done here is significant because it shows how the graph-theoretical algorithms can perform simple calculation of atomic and sub-atomic properties, what in turn might suggest new approaches to solve the much more complex problems associated with the theoretical formalisms of particle physics.

4. Conclusion

The use of graph-theoretical indices for the prediction of properties of atomic and subatomic particles is a rather unexplored field. The results outlined here demonstrate that this approach can be useful on this goal. Indeed, although there must be other alternative graph-theoretical algorithms as the one illustrated here, it seems clear that it points toward a new framework that can yield interesting results in the near future.

5. Acknowledgements

The author thanks the Spanish Ministry of Science and Innovation by its Grant project SAF2009-13059-C03-02. He is also grateful to Maria Galvez-Llompart for her invaluable help in preparing this work.

REFERENCES

[1] Smolin and J. Harnad, "The Trouble with Physics: The Rise of String Theory, the Fall of a Science, and What Comes Next," *The Mathematical Intelligencer*, Vol. 30, No. 3, 2008, pp. 66-69. doi:10.1007/BF02985383

[2] R. P. Feynman, "The Theory of Positrons," *Physical Review*, Vol. 76, No. 6, 1949, pp. 749-759. doi:10.1103/PhysRev.76.749

[3] F. Harary, "Proof Techniques in Graph Theory," Academic Press Inc., Burlington, 1969.

[4] M. C. Heydemann and B. Ducourthial, "Cayley Graphs and Interconnection Networks," *Physics, Science & Math*, Vol. 497, No. 497, 1997, pp. 167-226.

[5] J. J. Sylvester and F. Franklin, "A Constructive Theory of Partitions, Arranged in Three Acts, an Interact and an Exodion," *American Journal of Mathematics*, Vol. 5, No. 1, 1882, pp. 251-330. doi:10.2307/2369545

[6] T. M. Cover, "Comments on Broadcast Channels," *IEEE Transactions on Information Theory*, Vol. 44, No. 6, 1998, pp. 2524-2530. doi:10.1109/18.720547

[7] N. Deo, "Graph Theory with Applications to Engineering and Computer Science," PHI Learning Ltd., New Delhi, 2004.

[8] W. Ren, "Synchronization of Coupled Harmonic Oscillators with Local Interaction," *Automatica*, Vol. 44, No. 12, 2008, pp. 3195-3200. doi:10.1016/j.automatica.2008.05.027

[9] A. W. Wolfe, "Social Network Analysis: Methods and Applications," *American Ethnologist*, Vol. 24, No. 1, 1997, pp. 219-220. doi:10.1525/ae.1997.24.1.219

[10] F. Harary, "Graph Theory and Theoretical Physics," Academic Press, New York, 1967.

[11] T. M. J. Fruchterman and E. M. Reingold, "Graph Drawing by Force-Directed Placement," *Software: Practice and Experience*, Vol. 21, No. 11, 1991, pp. 1129-1164. doi:10.1002/spe.4380211102

[12] N. Cabibbo, "Unitary Symmetry and Leptonic Decays," *Physical Review Letters*, Vol. 10, No. 12, 1963, pp. 531-533. doi:10.1103/PhysRevLett.10.531

[13] B. Greene, "The Elegant Universe," Vintage Books, New York, 2000.

[14] R. Penrose, "Applications of Negative Dimensional Tensors," Academic Press Inc., Burlington, 1971.

[15] J. Galvez, R. Garcia-Domenech and J. V. de Julian-Ortiz, "Assigning Wave Functions to Graphs: A Way to Introduce Novel Topological Indices," *Communications in Mathematical and in Computer Chemistry*, Vol. 56, No. 3, 2006, pp. 509-518.

[16] I. Levine, "Quantum Chemistry," 5th Edition, Prentice Hall, New Jersey ,1999.

[17] T. Filk, "Relational Interpretation of the Wave Function and a Possible Way Around Bell's Theorem," *International Journal of Theoretical Physics*, Vol. 45, No. 6, 2006, pp. 1166-1180. doi:10.1007/s10773-006-9125-0

[18] Dragon, "Talete Srl," Milano, 2006. http://www.talete.mi.it/

[19] W. J. Dixon, M. B. Brown, L. Engelman and R. I. Jennrich, "7M Package," University of California Press, San Francisco, 1990.

[20] E. Estrada, "Spectral Moments of the Edge Adjacency Matrix of Molecular Graphs. 1. Definition and Applications to the Prediction of Physical Properties of Alkanes," *Journal of Chemical Information and Computer Sciences*, Vol. 36, No. 4, 1996, pp. 844-849. doi:10.1021/ci950187r

[21] G. N. Shah and T. A. Mir, "Are Elementary Particle Masses Related," *The 29th International Cosmic Ray Conference*, Pune, 3-10 August 2005, pp. 219-222.

[22] S. Groote and J. G. Körner, "Spectral Moments of Two-Point Correlators in Perturbation Theory and Beyond," *Physical Review*, Vol. 65, No. 3, 2002, 30 p. doi:10.1103/PhysRevD.65.036001

[23] B. Zhou, I. Gutman, J. A. de la Peña, J. Rada and L. Mendoza, "On Spectral Moments and Energy of Graphs," *Communications in Mathematical and in Computer Chemistry*, Vol. 57, No. 1, 2007, pp. 183-191.

[24] E. Estrada, "Quantum-Chemical Foundations of the Topological Substructure Molecular Design," *The Journal of Physical Chemistry*, Vol. 112, No. 23, 2008, pp. 5208-5217. doi:10.1021/jp8010712

[25] L. B. Kier, W. J. Murray, M. Randić and L. H. Hall, "Molecular Connectivity V: Connectivity Series Concept Applied to Density," *Journal of Pharmaceutical Sciences*, Vol. 65, No. 8, 1976, pp. 1226-1230. doi:10.1002/jps.2600650824

[26] R. Garcia-Domenech, J. Galvez, J. V. de Julian-Ortiz and L. Pogliani, "Some New Trends in Chemical Graph Theory," *Chemical Reviews*, Vol. 108, No. 3, 2008, pp. 1127-1169. doi:10.1021/cr0780006

[27] S. C. Basak, D. R. Mills, A. T. Balaban and B. D. Gute, "Prediction of Mutagenicity of Aromatic and Heteroaromatic Amines from Structure: A Hierarchical QSAR Approach," *Journal of Chemical Information and Computer Sciences*, Vol. 41, No. 3, 2001, pp. 671-678. doi:10.1021/ci000126f

[28] P. Jasinski, B. Welsh, J. Galvez, D. Land, P. Zwolak, *et al.*, "A Novel Quinoline, MT477: Suppresses Cell Signaling through Ras Molecular Pathway, Inhibits PKC Activity, and Demonstrates *in vivo* Anti-Tumor Activity against Human Carcinoma Cell Lines," *Investigational New Drugs*, Vol. 26, No. 3, 2008, pp. 223-232. doi:10.1007/s10637-007-9096-x

[29] P. Jasinski, P. Zwolak, R. Isaksson, V. Bodempudi, K. Terai, *et al.*, "MT103 Inhibits Tumor Growth with Minimal Toxicity in Murine Model of Lung Carcinoma via Induction of Apoptosis," *Investigational New Drugs*, Vol. 29, No. 5, 2011, pp. 846-852. doi:10.1007/s10637-010-9432-4

[30] J. Galvez, J. Llompart, D. Land and G. M. Pasinetti, "Compositions for Treatment of Alzheimer's Disease Using Abeta-Reducing and/or Abeta-Anti-Aggregation Compounds," US Patent No. 2010114636, 2010.

13

Steady-State and Time-Resolved Fluorescence Studies of Anthrylacrylic Ester

Anil B. Naik[1], L. R. Naik[1*], V. Jayathirtha Rao[2], Haridas Pal[3]
[1]Department of Studies in Physics, Karnatak University, Dharwad, India
[2]Organic Division II, Indian Institute of Chemical Technology, Hyderabad, India
[3]Radiation and Photochemistry Division, Bhabha Atomic Research Centre (BARC), Mumbai, India
Email: [*]naik_36@rediffmail.com

ABSTRACT

The photophysical properties of Anthrylacrylic ester were investigated in various solvents using steady-state and time-resolved fluorescence techniques. The dual fluorescence was observed in various solvents of different polarity functions. The emission bands observed are around 400 nm (band I) and 480 nm (band II) for 350 nm and 390 nm excitation at low concentrations (10^{-5} mol·dm^{-3}) but at higher concentrations (10^{-3} mol·dm^{-3}) only the emission (band II) at 480 nm alone was observed. The emission band observed around 400 nm shifts from 390 nm to 420 nm (shift 30 nm) when the excitation wavelength was varied from 300 nm to 380 nm and the emission band at 480 nm shifts from 430 nm to 510 nm (shift 80 nm) for the change of excitation wavelength from 380 nm to 400 nm. However, both emissions were observed for 380 nm excitation at lower concentrations. Solvatochromic shifts of electronic absorption and fluorescence emissions (band I & II) as a function of polarity functions were analyzed and the changes in dipole moment of the molecule was estimated ($\Delta\mu$ = 3.36 D & 5.98 D). Fluorescence decays monitored over each emission maxima showed bi-exponential behavior, and yielded two lifetime components respectively in the range 0.97 - 7.1 ns (at λ_{em} = 400 nm) and 0.34 - 7.23 ns (at λ_{em} = 480 nm). Based on the steady-state and time-resolved emission measurements emission band I (400 nm) and emission band II (480 nm) are respectively assigned as due to locally excited state (Trans-forms) and due to isomers (Cis-form).

Keywords: Solvent Effect; Concentration Effect; Isomers

1. Introduction

Photophysical processes involved in the excited state, Intra/Inter-Molecular Charge (ICT) transfer reaction is an interesting area for many researchers in the field of chemical physics and physical chemistry. For the last few decades both experimental and theoretical studies started from the dual fluorescence of the 4-N, N-dimethyl amino benzonitrile (DMABN) and ascribed to Intramolecular Charge Transfer (ICT) process [1]. Later on several researchers designed new molecules and studied the possibility of photoinduced ICT reactions [2-10]. Subit kumar Saha *et al.* have reported excited state isomerizations, effect of viscosity and temperature dependent torsional relaxation on TICT fluorescence of trans-2[4-(dimethylamino) styryl] benzothiazole [2]. Amrita Chakraborty *et al.* have reported photophysical properties of Trans-3-(4-monomethylamino-phenyl)-acrylonitrile and discussed about the Twisted Intramolecular Charge Transfer (TICT), process [3]. Rupashree Balia Singh and others have reported spectroscopic and theoretical evidence for the photoinduced twisted intramolecular charge transfer state formation in N, N-dimethylaminonaphthyl-(acrylo)-nitrile and compared experimental results with the different theoretical models [4]. The three different theoretical models used to explain the observed anomlous emission of the molecules are the Twisted Intramolecular Charge Transfer (TICT), Planarized Intramolecular Charge Transfer (PICT) and hybridized intramolecular charge transfer process [5-10]. V. Raj Gopal *et al.* have reported results on wavelength dependent Trans to Cis and quantum chain isomerizations of Anthrylethylene derivatives [11]. But they studied wavelength dependent Trans to Cis photostationary composition by preferential light absorption and excitation of the Trans-isomer at longer wavelength only. This is only a qualitative description of the Trans-Cis photoisomerization of the Anthrylethylene derivatives but it is necessary to study the kinetics of photoisomerization and formation of photostationary state in terms of time-resolved fluorescence studies.

In the present work, ground and excited state proper-

[*]Corresponding author.

ties of Anthrylacrylic ester were investigated to explore its photophysical properties in various conditions using steady-state absorption and time-resolved fluorescence measurements. The ground and excited state geometric isomers (*i.e.*, structural change of Trans to Cis forms) of this molecule is shown in **Figure 1**. Orientation of the ester group with respect to double bond is different, it can be S-Cis or it can be S-Trans. Cis-Trans isomers are due to rotation around double bond. The energy difference between Cis and Trans-isomers is much higher at room temperature and only the more stable Trans isomers predominate. If dynamic equilibrium between Trans and Cis-isomers is of prime importance in describing the photophysical behaviour of Anthrylacrylic ester, this should be reflected in the decay parameters. Such studies were not reported in previous work and we have thus performed the experiments in various solvents to confirm the kinetic behaviour of Anthrylacrylic ester in various conditions. Fluorescence decays monitored over each emission maxima showed bi-exponential behavior, and yielded two lifetime components and are in favor of the existence of different isomers having two excited electronic sates. Bi-exponential behavior might also be interpreted in terms of aggregates, S_2 emission twisted intramolecular charge transfer states or even the presence of an impurity but the detailed study on this molecule as a function of concentration and excitation energy supports the conversion of Trans into Cis form of the molecule. Anthrylacrylic ester exhibits dual emissions as a function of concentration and excitation energy. The high-energy emission is assigned due to locally excited state or Franck-Condon state (Trans-isomer) and the lower energy emission due to charge transfer state/Cis-isomer.

2. Experimental

Anthrylacrylic ester was obtained from Indian Institute of Chemical Technology, (IICT) Hyderabad [11], and was used without further purification. All solvents were of spectroscopic grade (Fluka) and were used as received. The purity of the solvents was checked before use by HPLC-MS method [12]. The absorption spectra in various solutions at 1×10^{-5} mol·dm^{-3} were recorded using Hitachi Model U-3010/U-3310 spectrophotometer and fluorescence measurements were recorded using Hitachi model F-7000 fluorescence spectrophotometer. Time-resolved fluorescence measurements were carried out using a picosecond operated time-correlated single photon counting unit, which is described elsewhere [13]. In the present work, 373 nm (Nano LED, 1.2 ns, 1 MHz) was used as the excitation light source and a TBX4 de tection module (IBH) coupled with a special Hamamatsu PMT was used for fluorescence detection. For the present set up, the instrument response was ~1.2 ns (~240 ps) at

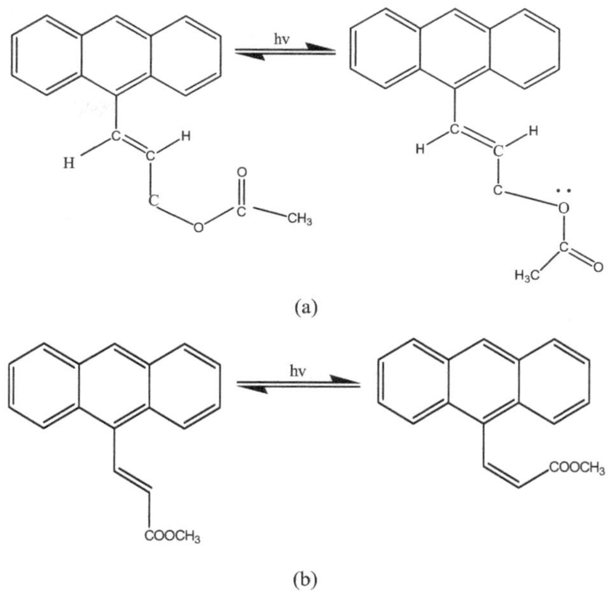

Figure 1. (a) Ground state conformers at room temperature (at very low energy) [Orientation of the carbonyl group with respect to double bond is different, it can be S-Cis or it can be S-Trans]; (b) Excited state geometric isomers (Cis-Trans isomers are due to rotation around double bond. This energy difference is much higher at room temperature only the more stable Trans isomers predominates).

Full Width at Half Maxima (FWHM). Fluorescence decays were recorded with a vertically polarized excitation beam and fluorescence was collected at magic angle 54.7° and was analyzed by a deconvolution method at room temperature.

3. Results and Discussion

3.1. Steady State Absorption and Fluorescence Excitation Studies

Figure 2 shows the electronic absorption spectrum of Anthrylacrylic ester in various solvents of different polarity functions at low concentration (1×10^{-5} mol·dm^{-3}). The peak positions of the absorption spectrum in all solvents are around 384 - 386 nm and not much change in the absorption spectrum were observed by change of solvent polarity. However in case of methanol and toluene solutions the absorption maxima are located around 382 nm and 388 nm. But no change in absorption band is observed except little broadening of the spectra as a function of concentration (1×10^{-5} to 1×10^{-3} mol·dm^{-3}) (not shown in **Figure 2**) in all solvents used.

The fluorescence excitation maxima for emissions 400 nm and 480 nm recorded at low concentration (at 1×10^{-5} mol·dm^{-3}) in almost all solvents located respectively in the regions 345 nm and 385 nm indicates the existence of two independent absorbing species (**Figure 3**). Comparison of steady-state absorption with the fluorescence

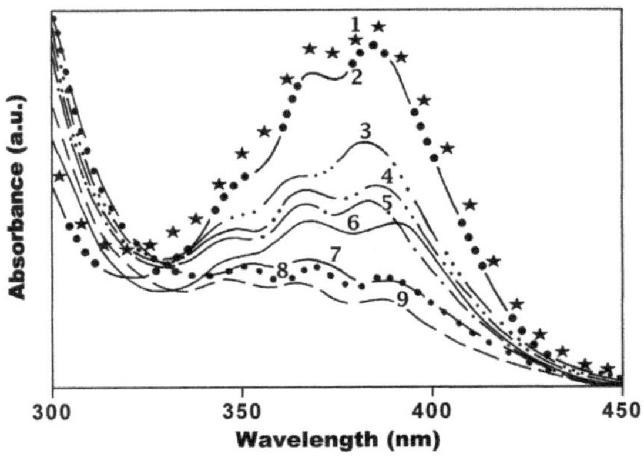

Figure 2. Absorption spectra of Anthrylacrylic ester at 1 × 10^{-5} mol·dm^{-3} in (1) Acetonitrile; (2) n. Propanol; (3) Butanol; (4) Tetrahydrofuran; (5) Methanol; (6) Toluene; (7) Dioxane; (8) Benzene; (9) Cyclohexane.

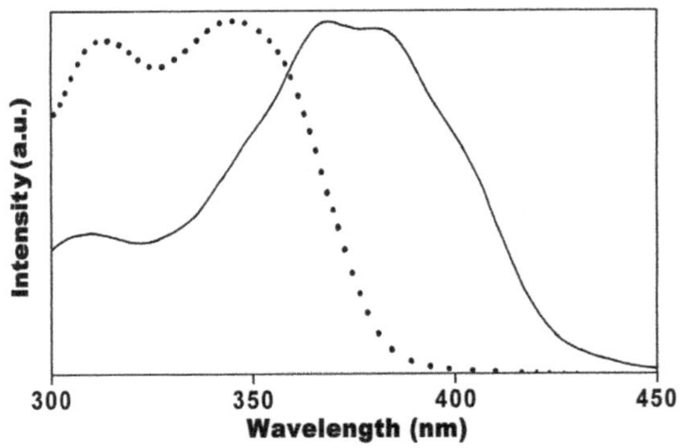

Figure 3. Excitation spectra of Anthrylacrylic ester in Cyclohexane at concentration 1 × 10^{-5} mol·dm^{-3} for λem 400 nm (● ● ● ● ●) and 480 nm (———).

excitation (not shown) in all solvents as a function of concentration inferred the existence of different iso-mers/conformers.

3.2. Fluorescence Emission Studies

The dual emission of the dye was observed at low concentration (1 × 10^{-5} mol·dm^{-3}) in all solvents used. Thus in order to understand the existence of dual emissions of the dye in different solutions the effect of concentration and excitation wavelength has been investigated and the result shows concentration and excitation wavelength dependence. The dramatic change in fluorescence behavior was observed along with the increase in the Stoke's shift as a function of polarity of the solvents. At low concentration (1 × 10^{-5} mol·dm^{-3}) the dye exhibits two emission bands in all solvents with their peak positions ranging from 390 - 420 nm (I band) and 430 - 510 nm (II band) for λ_{ex} 350 nm and 400 nm (**Figures 4** and **5**).

The intensity contribution of the emission band I and band II are respectively (94% - 92%) and (6% - 8 %). The peak positions of emission bands are solvent dependent and are red shifted with increasing polarity of the solvent, indicating the increased basicity of the solute in the excited state. The emission band II shifts more towards (red shift) higher wavelength region (436 - 500 nm) as compared to the emission band I (390 - 430 nm) due to the change of solvent from non-polar to polar and moreover, Anthrylacrylic ester is highly fluorescent in non-polar solvent as compared to polar solvent. Intensity of shorter wavelength emission band around 400 nm (I band) goes on decreases and merges with the second emission band around 480 nm (band II) in all solvents as the concentration of the dye is increased from 1 × 10^{-5} to 1 × 10^{-3} mol·dm^{-3}. The emission maxima observed above 450 nm at higher concentrations (1 × 10^{-3} mol·dm^{-3}) is assigned to the charge transfer state and the emission maxima observed around ~400 nm at lower concentrations (10^{-6} mol·dm^{-3}) is assigned to the locally excited state or Franck Condon state. Concentration dependent fluorescence emission spectra of Anthrylacrylic ester in benzene solutions are shown in **Figures 4** and **5** respectively for λ_{ex} 350 nm and 390 nm.

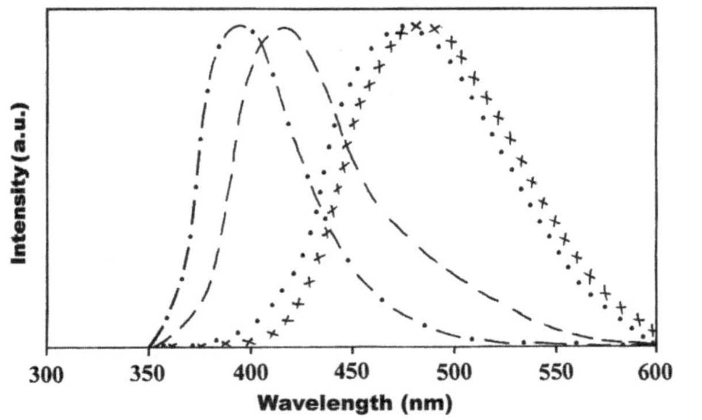

Figure 4. Emission spectra of Anthrylacrylic ester in Benzene as a function of concentration 1×10^{-6} (– · – · –), 1×10^{-5} (– – –), 1 × 10^{-4} (•••••) and 1 × 10^{-3} (x x x x x) [mol·dm^{-3}] for λ_{ex} = 350 nm.

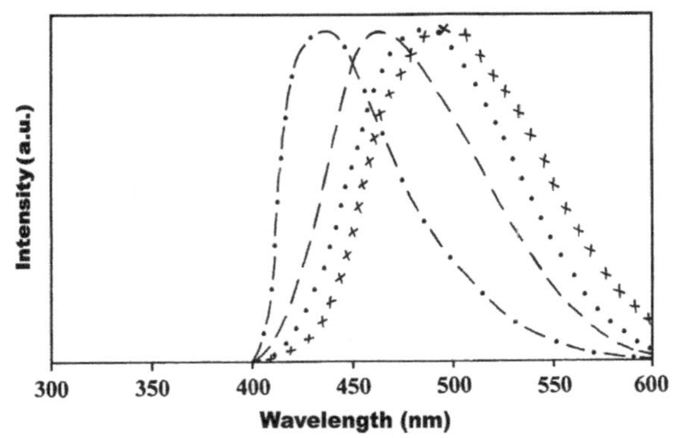

Figure 5. Emission spectra of Anthrylacrylic ester in Benzene as a function of concentration 1 × 10^{-6} (– · – · –), 1 × 10^{-5} (– – –), 1 × 10^{-4} (● ● ● ● ●) and 1 × 10^{-3} (x x x x x) (mol·dm^{-3}) for λ_{ex} = 400 nm.

With the increase of concentration from 1×10^{-5} to 1×10^{-3} mol·dm^{-3}, there is a continuous shift of emission maxima towards higher wavelength region (red shift) both in polar and non-polar solvents (for both λ_{ex} 350 nm and 390 nm) which reveals that the longer wavelength band is associated with isomers/charge transfer state. The emission spectra recorded as a function of excitation wavelength ranging from 300 nm to 370 nm in various concentrations shows that the peak position of the emission band I (390 - 420 nm) was slightly red shifted (**Figure 6**). However, fluorescence emission for excitation wavelength 390 nm and above the longer wavelength band II (430 - 510 nm) alone was observed. But at lower concentrations both the emissions (band I and band II) were observed for 380 nm excitation. This finding clearly indicates that two emissions are originated through different emitting species. At lower concentration (1×10^{-5} mol·dm^{-3}) the fluorescence emission bands observed around 400 nm (I band) and 480 nm (II band) in non-polar and polar solvents at excitation wavelength about 380 nm was assigned as Trans (52%) and Cis (48%) photoisomerization. However, at the excitation wavelength 390 nm and above leads to enhanced accumulation of Cis-isomer to over 94% from Trans isomer (6%) and hence our experimental results are in agreement with the earlier researchers [11]. Consequently at higher concentrations (1×10^{-3} mol·dm^{-3}) the conversion of Trans to Cis photoisomerizations are respectively 6% and 94% (I band and II band). At lower concentrations the band at shorter wavelength region (band I) is found to be predominant over the band at higher wavelength region (band II). With increase of concentration there is a continuous shift of emission band II towards red region with increase in intensity (~94%) while there is a little shift of fluorescence emission band I towards the red region with suppressed intensity (~6%) reveals the charge transfer character of the state. This phenomenon has been observed in all polar and non-polar solvents. Thus, fluorescence characteristics of Anthrylacrylic ester studied as a function of concentration and excitation wavelength reveals the existence of two emitting species. In addition the experimental results may also be explained in terms of the possible resonance structure of the molecules as shown in **Figures 1(a)** and **1(b)**. **Figure 1(a)** shows the ground-state conformers; at room temperature the orienttation of the ester group with respect to double bond is different , it can be S-Cis or it can be S-Trans. **Figure 1(b)** shows excited state geometric isomers; Cis-Trans isomers are due to rotation around double bond.

It is evident from experimental observation that the energy difference between Cis-Trans isomers is much higher at room temperature therefore more stable Trans isomers predominates. However, at very low concentration and at excitation wavelength around 385 nm the S-Cis form and S-Trans forms of the molecule are in equilibrium. Based on these observations, the fluorescence band observed at longer wavelength region (band II) is assigned to isomers/charge transfer state (in particular Cis-isomers). However, the possibility of the occurrence of fluorescence emission band towards longer wavelength side due to excimers/dimers, is less probable because in which case there should be a regular increase in intensity of higher wavelength emission band while the intensity of the lower wavelength emission band should decrease with increasing concentrations [12,14].

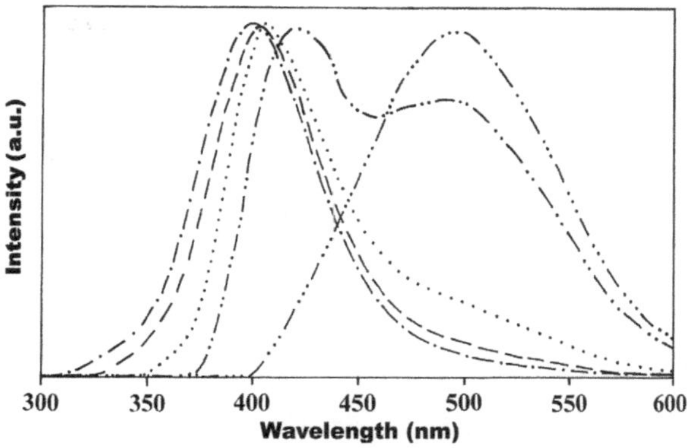

Figure 6. Emission spectra of Anthrylacrylic ester in Acetonitrile as a function of excitation wavelength at 1×10^{-5} mol·dm^{-3} (1) 320 nm (– · – · –); (2) 340 nm (– – –); (3) 360 nm (······); (4) 380 nm (– ·· – ·· –) and (5) 400 nm (– ··· –).

3.3. Solvatochromic Measurements

The electronic absorption and fluorescence emission characteristics of Anthrylacrylic ester in various solvents of different polarity functions were investigated in order to study the role of polarity in modifying both ground and excited states of the molecule and the changes in dipole moment of the molecule were estimated by using Lippert-Mataga's relation [15-20].

$$\Delta\bar{\nu} = \left(\bar{\nu}_a - \bar{\nu}_f\right) = 2F\left(\mu_e - \mu_g\right)^2 / hca^3 \qquad (1)$$

where $\bar{\nu}_a$ and $\bar{\nu}_f$ are the absorption and fluorescence emission maxima wave numbers in cm^{-1}, μ_g and μ_e are the ground and excited state dipole moments.,

$$F(\varepsilon, n) = \left[\frac{\varepsilon - 1}{2\varepsilon + 1} - \frac{n^2 - 1}{2n^2 + 1}\right]$$

where ε and n are respectively dielectric constant and refractive index of the solvents and were taken from literature [21]. The value of the solute cavity radius a, was calculated by using the Suppan relation [22].

$$a = \left(3M/4\pi\delta N\right)^{1/3} \qquad (2)$$

where δ is the density of solute molecule, M the molecular weight of the solute and N is the Avogadro's number. Using Equation (2) the cavity radius a, estimated for Anthrylacrylic ester were found to be $3.71A^0$. The solute-solvent properties and solvent shift data are collected in **Tables 1** and **2**.

Figures 7(a) and **7(b)** shows the plot of Stoke's shift (in·cm^{-1}) versus solvent polarity functions (F) with good correlation coefficients R = 0.97 and R = 0.96 for I and II emission bands (emissions bands I and II are due to S-Trans and S-Cis-isomers). Using the corresponding slopes the change in dipole moments for S-Trans and S-Cis-isomers were estimated to be $\Delta\mu = 3.36$ D and $\Delta\mu = 5.98$ D with errors 3% and 4%, respectively. These change in dipole moments for emission bands I and II are in favor of the existence of two isomers called S-Trans and S-Cis-isomers. The higher excited state dipole moments of S-Trans and S-Cis-isomers than that of the ground state dipole moment indicates that the excited state electronic charge distribution should be substantially different from that of the ground state charge distribution [23-25].

Dielectric constant and refractive index were taken from reference [21] and E_T^N values were taken from references [18] and [19].

For understanding polarization dependence/hydrogen bonding effect on spectral characteristic, the change in dipole moment was estimated by using E_T^N function, which is the empirical measure of solvent polarity and is based on the solvatochromic behaviour of a betaine dye as a probe solute. On this scale each solvent gives the quantitative expression for polarity. The correlation between the spectral shifts with E_T^N has been developed by Ravi *et al.* [24] theoretically and accordingly the excited state dipole moment was estimated by using the following expression.

$$\Delta\bar{\nu} = \bar{\nu}_a - \bar{\nu}_f = 11307.6\left[\left(\frac{\Delta\mu}{\Delta\mu_D}\right)^2\left(\frac{a_D}{a}\right)^3\right]E_T^N + \text{constant} \qquad (3)$$

where, E_T^N is the microscopic solvent polarity function proposed by Reinhardt,

$$\Delta\mu_D(9\ D) \text{ and } a_D\left(6.2\ A^0\right)$$

are respectively the change of dipole moment and Onsager cavity radius of a standard Betaine dye in the solvent and $\Delta\mu$ and a are the corresponding quantities of the molecule of interest. The errors involved in the estimation of Onsager radius is reduced to some extent because of the ratio a_D/a in Equation (2). Equation (2) illustrates that the Stoke's shift changes linearly with the solvent polarity function E_T^N. A plot of Stoke's shift versus E_T^N gives a straight line with correlation coefficient of 0.92 (**Figure 8**) and from the slope of this plot the change in dipole moment was estimated to be about $\Delta\mu = 7.12$ D. A good correlation coefficient and the higher value of change in dipole moment obtained in this case are in favour of Trans to Cis form of the molecule after excitation.

Table 1. Solvent properties and polarity functions of Anthrylacrylic ester.

Solvents	Dielectric const (ε) at 25°C	Refractive index (n) at 25°C	F(ε,n)	E_T^N
n-hexane	1.880	1.372	−0.0038	0.006
Dioxane	2.210	1.420	0.0212	0.164
Benzene	2.284	1.498	0.0039	0.117
Toluene	2.379	1.494	0.0140	0.110
Butanol	17.51	1.397	0.2642	0.506
Propanol	20.21	1.383	0.2748	0.546
Methanol	32.63	1.326	0.3094	0.762
Acetonitrile	35.94	1.342	0.3054	0.472

Table 2. Solvent shift data for Anthrylacrylic ester in various solvents at 1 × 10^{-5} mol·dm^{-3}.

Solvents	Absorption maxima (nm)	Emission maxima (nm) Trans (I)	Cis (II)	$(\bar{\nu}_a)$ cm^{-1}	$(\bar{\nu}_f)$ cm^{-1} Trans (I)	Cis (II)	$(\bar{\nu}_a-\bar{\nu}_f)$ cm^{-1} Trans (I)	Cis (II)
n-hexane	384	392	436	26,042	25,510	22,936	532	3106
Dioxane	386	395	459	25,906	25,316	21,786	598	4120
Benzene	386	396	460	25,906	25,252	21,739	654	4167
Toluene	388	398	462	25,773	25,125	21,645	648	4128
Butanol	384	402	490	26,042	24,805	20,408	1167	5634
Propanol	386	403	493	25,906	24,813	20,284	1093	5622
Methanol	382	404	498	26,178	24,752	20,080	1426	6098
Acetonitrile	386	405	495	25,906	24,691	20,202	1215	5704

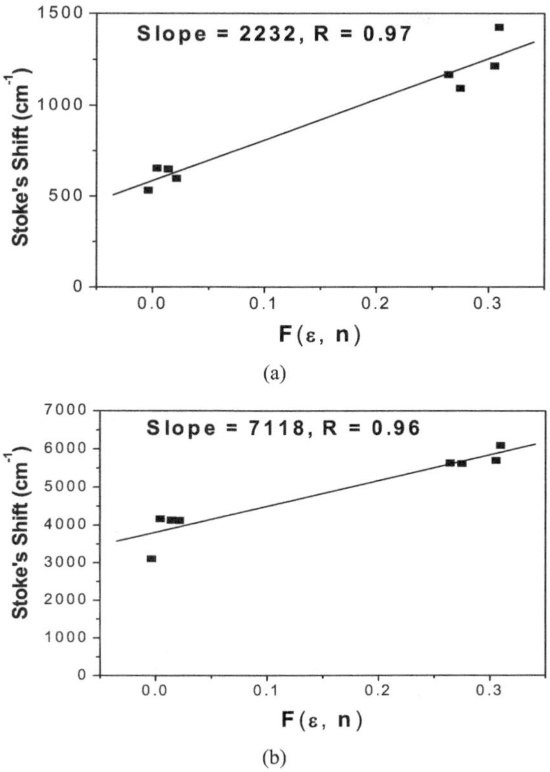

Figure 7. (a) Plot of Stoke's shift versus F(ε, n) according to Lippert-Mataga's relation for λ_{em} = 400 nm (Trans-form); (b) Plot of Stoke's shift versus F(ε, n) according to Lippert-Mataga's relation for λ_{em} = 480 nm (Cis-form).

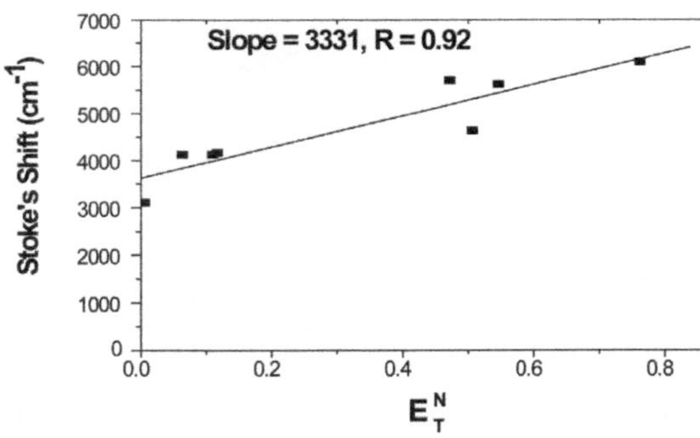

Figure 8. Plot of Stoke's shift versus E_T^N.

3.4. Fluorescence Decays

The fluorescence decays of Anthrylacrylic ester in various solvents (1 × 10^{-5} mol·dm^{-3}) were recorded to understand the mechanism of isomerization and the possible involvement of charge transfer excited states. Fluorescence decay in various solvents for each emission was fitted to bi-exponential function [26].

$$I(t) = a_1 \exp(-t/t_1) + a_2 \exp(-t/t_2) \quad (4)$$

where τ_1, τ_2 are the lifetimes of the two decay components and a_1, a_2 are their respective amplitudes such that $a_1 + a_2 = 1$.

The fluorescence lifetimes estimated in different solvents are listed in **Table 3**. As can be seen from the **Table 3**, the decay parameters depend on the emission wavelength and nature of the solvent. From the **Table 3** it is to be noted that lifetime (τ_1) and amplitudes (a_1) of the first component in all solvents is less than the second component ($\tau_1 < \tau_2$) and ($a_1 < a_2$) for emission 400 nm. However the lifetime (τ_1) of the first component in all solvents is less than the second components ($\tau_1 < \tau_2$) with ($a_1 > a_2$) the amplitude of the first component greater than the second component *i.e.*, the relative contribution is reversed for emission 480 nm indicating the conversion of Trans-isomer to Cis-isomers. However, the average lifetime lies between 1.64 ns to 2.76 ns except dioxane and benzene (7.06 ns & 5.30 ns) for λ_{em} 480 nm and 3.61 ns to 6.30 ns for λ_{em} 400 nm.

In other words the large difference between two decay components indicates the presence of two emitting species or a single species from two different excited states [27]. From the fluorescence decay measurements the fluorescence emission maxima at 400 nm was assigned to as normal species (Trans-isomer) and that of at 480 nm as due to the formation of isomers/CT states (Cis-isomers). It is observed that at lower concentrations the band I of Anthrylacrylic ester due to the monomer

Table 3. Fluorescence lifetimes of Anthrylacrylic ester in various solvents at 1 × 10^{-5} mol·dm^{-3} (at λ_{ex} = 373 nm at room temperature).

Solvents	Λem (nm)	τ_1 (ns)	τ_2 (ns)	τ_{av} (ns)	χ^2
n-Hexane	400	1.45[35.52]	4.59[64.48]	3.61	1.10
	480	1.06[80.45]	6.45[19.55]	2.12	1.21
Cyclohexane	400	1.24[25.59]	5.81[74.41]	4.64	1.14
	480	1.01[86.56]	6.39[13.44]	1.72	1.10
1,4-Dioxane	400	1.89[16.76]	6.96[83.24]	6.11	1.05
	480	2.77[36.46]	7.23[63.54]	5.62	1.28
Benzene	400	1.91[19.30]	6.31[80.70]	5.46	1.08
	480	1.13[29.45]	7.04[70.55]	5.30	1.34
Toluene	400	1.73[17.80]	5.98[82.20]	5.22	1.18
	480	0.89[54.45]	6.69[45.55]	3.52	1.17
Methanol	400	0.97[49.16]	6.16[50.84]	3.61	1.31
	480	0.66[62.34]	5.98[37.68]	2.66	1.19
n-Propanol	400	1.81[51.84]	6.84[48.16]	4.23	1.43
	480	0.72[66.42]	6.44[35.58]	2.76	1.06
Butanol	400	1.98[43.26]	7.10[56.74]	4.88	1.18
	480	0.42[36.79]	6.46[63.21]	4.23	1.31
THF	400	2.46[15.08]	6.98[84.92]	6.30	1.07
	480	0.68[63.81]	6.33[36.19]	2.72	1.31
Acetonitrile	400	1.46[9.63]	6.51[90.37]	6.02	1.21
	480	0.34[76.82]	5.98[23.18]	1.64	1.10

predominates over the fluorescence emission due to isomers/charge transfer states. In dilute solutions the average separation of solute molecules is too large to allow an appreciable proportion of the excited molecules to encounter a second solute molecule within the lifetime of the excited state. The formation of conformers/charge transfer state in dilute solutions is small however, as concentration increases; the average separation between solute molecules becomes too small to allow the interaction between solute/solvent molecules leading to the formation of charge transfer state. In order to study the existence of dynamic equilibrium between Trans and Cis-isomers at low concentration and at excitation wavelength around 380 nm the fluorescence decays were measured at λ_{ex} 373 nm as it is the best suited wavelength available for us at the time of experimental measurements. The fluorescence decay profiles are shown in **Figure 9** in Acetonitrile solutions at λ_{em} 400 nm and 480 nm along with weighted residuals at λ_{ex} 373 nm. Fluorescence decay profiles (**Figures 9(a)** and **(b)**) show the clear cut difference between Trans and Cis-isomers. In our opinion, the observation of bi-exponential fluorescence decay for Anthrylacrylic ester itself offers unique interpretation in terms of Cis and Trans-isomers.

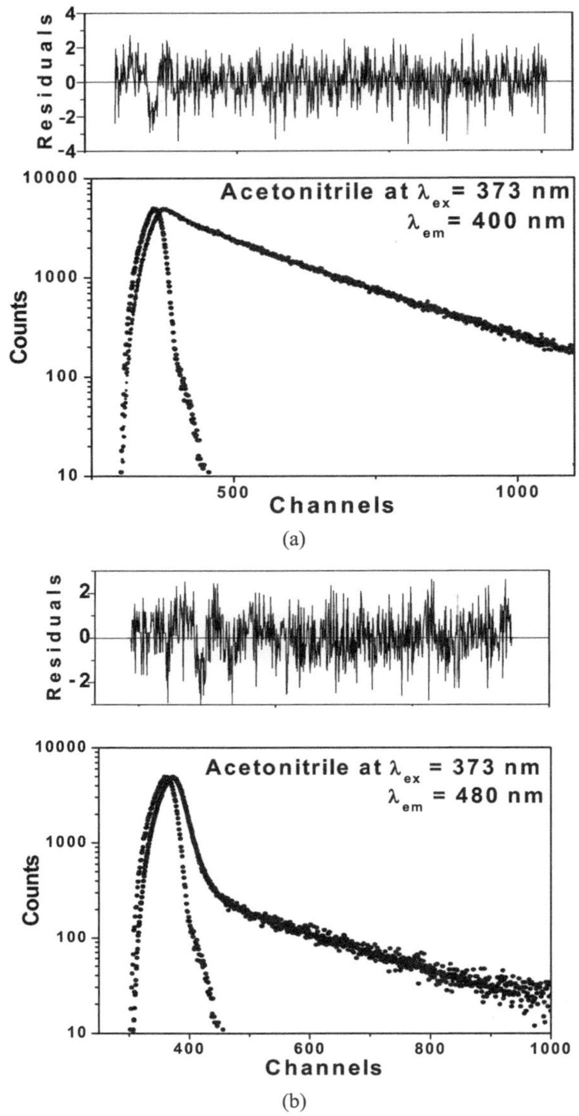

Figure 9. (a) Fluorescence decay of Anthrylacrylic ester at λ_{em} 400 nm; (b) Fluorescence decay of Anthrylacrylic ester at λ_{em} 480 nm.

4. Conclusions

The electronic absorption and fluorescence emission characteristics of Anthrylacrylic ester in various solvents of different polarity were investigated in order to study the role of polarity in modifying both ground and excited states of the molecule. It is observed that the excited state dipole moment of Anthrylacrylic ester is greater than its ground state dipole moment indicating the more polar character of the excited state.

The fluorescence emission exhibits two emission bands one at 400 nm (band I) and another at 480 nm (band II) for λ_{ex} 350 nm and 390 nm at low concentration. The change of solvent, concentration and excitation wavelength induces dramatic changes in the photoproperties such as emission spectra and fluorescence lifetimes. In addition the estimated values of change in dipole moments for emission bands I and II are respectively $\Delta\mu = 3.36$ D and $\Delta\mu = 5.98$ D shows the wide separation between the existence of S-Trans and S-Cis-isomers of the molecule. Thus the existence of more than two species monomer (Trans) and charge transfer state isomers (Cis-isomers) are inferred from the concentration and excitation wavelength dependence of fluorescence emission. The ratio of formation of Trans to Cis-isomers is also concentration and excitation wavelength dependent.

Fluorescence decays monitored over each emission maxima showed bi-exponential behavior, and yielded two lifetime components and are in favor of the existence of different isomers having two excited electronic sates. Based on the experimental results of steady state and time-resolved emission studies, two different types of electronic states of Anthrylacrylic ester in the excited state inferred are the LE state (Trans-isomers) and charge transfer state/isomers (Cis-isomer) having average lifetimes between 3.61 ns to 6.3 ns & 1.64 ns to 5.6 ns respectively.

5. Acknowledgements

One of the authors (Anil B. Naik) thanks to CSIR, New Delhi for awarding Senior Research Fellowship (SRF). The authors would like to thank Prabhat K. Singh from Molecular Photochemistry Section, Radiation and Photochemistry Division, Bhabha Atomic Research Centre (BARC), Mumbai, India for technical help during fluorescence decay measurements. Thanks to Prof. M. V. Kulkarni for fruitful discussions.

REFERENCES

[1] E. Lippert, W. Ludar and H. Boss., "Advances in Molecular Spectroscopy," Pergamon Press, Oxford, 1962.

[2] S. K. Saha, P. Purkayastha, A. B. Das and S. Dhara, "Excited State Isomerization and Effect of Viscosity- and Temperature-Dependent Torsional Relaxation on TICT Fluorescence of Trans-2-[4-(Dimethylamino)Styryl] Benzothiazole," *Journal of Photochemistry and Photobiology A: Chemistry*, Vol. 199, No. 2-3, 2008, pp. 179-187.

[3] A. Chakraborty, S. Kar and N. Guchhait, "Photophysical Properties of Trans-3-(4-Monomethylamino-Phenyl)-Acrylonitrile: Evidence of Twisted Intramolecular Charge Transfer (TICT) Process," *Chemical Physics*, Vol. 324, No. 2-3, 2006, pp. 733-774. doi:10.1016/j.chemphys.2005.12.032

[4] R. B. Singh, S. Mahanta, S. Kar and N. Guchhait, "Spectroscopic and Theoretical Evidence for the Photoinduced Twisted Intramolecular Charge Transfer State Formation in N,N-Dimethylaminonaphthyl-(Acrylo)-Nitrile," *Journal of Luminescence*, Vol. 128, No. 9, 2008, pp. 1421-1430. doi:10.1016/j.jlumin.2008.01.015

[5] Z. R. Grabowski, K. Rotkiewicz and A. Siemiarezak, "Dual Fluorescence of Donor-Acceptor Molecules and the Twisted Intramolecular Charge Transfer (TICT) States," *Journal of Luminescence*, Vol. 18-19, 1979, pp. 420-424. doi:10.1016/0022-2313(79)90153-4

[6] J. Catalan, C. Diaz, V. lopez, P. Perez and R. M. Claramunt, "The TICT Mechanism in 9,9'-Biaryl Compounds: Solvatochromism of 9,9'-Bianthryl, N-(9-Anthryl) Carbazole, and N,N'-Bicarbazyl," *The Journal of Physical Chemistry*, Vol. 100, No. 47, 1996, pp. 183-192.

[7] C. Rulliere, Z. R. Grabowski and J. Dobkowski, "Picosecond Absorption Spectra of Carbonyl Derivatives of Dimethylaniline: The Nature of the Tict Excited States," *Chemical Physics Letters*, Vol. 137, No. 5, 1987, pp. 408-413. doi:10.1016/0009-2614(87)80224-5

[8] K. A. Zachariasse, "Comment on Pseudo-Jahn-Teller and TICT-Models: A Photophysical Comparison of Meta-and Para-DMABN Derivatives: The PICT Model for Dual Fluorescence of Aminobenzonitriles," *Chemical Physics Letters*, Vol. 320, No. 1-2, 2000, pp. 8-13. doi:10.1016/S0009-2614(00)00230-X

[9] A. Sobolewski and W. Domcke, "Charge Transfer in Aminobenzonitriles: Do They Twist," *Chemical Physics Letters*, 250, No. 3-4, 1996, pp. 428-436. doi:10.1016/0009-2614(96)00014-0

[10] L. R. Naik, A. B. Naik and H. Pal, "Steady-State and Time-Resolved Emission Studies of Thioflavin-T," *Journal of Photochemistry and Photobiology A: Chemistry*, Vol. 204, No. 2-3, 2009, pp. 161-167. doi:10.1016/j.jphotochem.2009.03.016

[11] V. R. Gopal., M. Reddy and V. J. Rao, "Wavelength Dependent Trans to Cis and Quantum Chain Isomerizations of Anthrylethylene Derivatives," *The Journal of Organic Chemistry*, Vol. 60, No. 24, 1995, pp. 7966-7973. doi:10.1021/jo00129a043

[12] L. R. Naik and N. N. Math, "Photo Physical Properties of 8-Hydroxy Quinoline," *Indian Journal of Pure & Applied Physics*, Vol. 43, 2005, pp. 743-749.

[13] M. Kumbhakar, S. Nath, T. Mukherjee and H. Pal, "Effect of Temperature on the Dynamics of Electron Transfer in Heterogeneous Medium: Evidence for Apparent Marcus Inversion," *Journal of Photochemistry and Photobiology A: Chemistry*, Vol. 182, No. 1, 2006, pp. 7-12. doi:10.1016/j.jphotochem.2006.01.005

[14] S. A. Rani, J. Sobhanadri and T. A. P. Rao, "Solvent and Concentration Effects on the Steady State Fluorescence of Fluorenone," *Journal of Photochemistry and Photobiology A: Chemistry*, Vol. 94, No. 1, 1996, pp. 1-5. doi:10.1016/1010-6030(95)04114-1

[15] M. Ravi, A. Samanta and T. P. Radhakrishnan, "Excited State Dipole Moments from an Efficient Analysis of Solvatochromic Stokes Shift Data," *The Journal of Physical Chemistry*, Vol. 98, No. 37, 1994, pp. 9133-9136. doi:10.1021/j100088a007

[16] C. Antonious, M. S. Aaron, J. J. Buna, M. Tine, A and L. Cisse, "Determination of the First Excited Singlet State Dipole Moments of Coumarins by the Solvatochromic Method," *Spectroscopy Letters*, Vol. 27, No. 4, 1994, pp. 439-449.

[17] B. Acemioglu, M. Arik, H. Efeoglu and Y. Onganer, "Solvent Effect on the Ground and Excited State Dipole Moments of Fluorescein," *Journal of Molecular Structure: Theochem*, Vol. 548, No. 1-3, 2001, pp. 165-171. doi:10.1016/S0166-1280(01)00513-9

[18] M. P. Haas and J. M. Warman, "Photon-Induced Molecular Charge Separation Studied by Nanosecond Time-Resolved Microwave Conductivity," *Chemical Physics*, Vol. 73, No. 1-2, 1982, pp. 35-53. doi:10.1016/0301-0104(82)85148-3

[19] A. Kawaski, "On the Estimation of Excited-State Dipole Moments from Solvatochromic Shifts of Absorption and Fluorescence Spectra," *Zeitschrift für Naturforschung*, Vol. 57A, 2002, pp. 255-262.

[20] L. R. Naik and N. N. Math, "Estimation of Ground and Excited State Dipole Moments of Coumarin 450 by Solvatochromic Shift Method," *Journal of Photosceince*, Vol. 12, No. 2, 2005, pp. 57-61.

[21] R. C. Weast, "Hand book of Chemistry," 68th Edition, CRC Press, Boca Raton, 1987.

[22] P. Suppan, "Excited-State Dipole Moments from Absorption/Fluorescence Solvatochromic Ratios," *Chemical Physics Letters*, Vol. 94, No. 3, 1983, pp. 272-275. doi:10.1016/0009-2614(83)87086-9

[23] S. Kumar, V. C. Raoand and R. C. Rastogi, "Excited-State Dipole Moments of Some Hydroxycoumarin Dyes Using an Efficient Solvatochromic Method Based on the Solvent Polarity Parameter, E_T^N," *Spectrochimica Acta Part A: Molecular and Biomolecular Spectroscopy*, Vol. 57, No. 1, 2001, pp. 41-47.

doi:10.1016/S1386-1425(00)00330-9

[24] M. Ravi, T. Soujanya, A. Samantha and T. P. Radhakrishna, "Excited-State Dipole Moments of Some Coumarin dyes from a Solvatochromic Method Using the Solvent Polarity Parameter, E_T^N," *Journal of the Chemical Society, Faraday Transactions*, Vol. 91, 1995, pp. 2739-2742. doi:10.1039/ft9959102739

[25] R. Giri, S. S. Rathi, M. K. Machwe and V. V. S. Murti, "Solvent and Substituent Effects: Depolarization of Fluorescence of Solutions of 4-Phenyl-6,7-Dimethoxy Coumarin," *Indian Journal of Pure & Applied Physics*, Vol. 28, No. 6, 1988, pp. 445-448.

[26] L. R. Naik, H. M. Suresh Kumar, S. R. Inamdar and N. N. Math, "Steady-State and Time-Resolved Emission Studies of 6-Methoxy Quinoline," *Spectroscopy Letters*, Vol. 38, No. 4-5, 2005, pp. 645-659. doi:10.1081/SL-200062818

[27] A. B. Naik, L. R. Naik, J. S. Kadadevarmath, P. Haridas and V. J. Rao, "Fluorescence Quenching of Anthrylvinyl Acetate by Carbon Tetrachloride," *Journal of Photochemistry and Photobiology A: Chemistry*, Vol. 214, No. 2-3, 2010, pp. 145-151.

14

Thermodynamic Properties of Li-LiH (LiD, LiT) Systems. The Phase Diagram

Konstantin A. Yakimovich[1], **Tatiana Biryukova**[2]
[1]Joint Institute for High Temperatures Russian Academy of Sciences, Moscow, Russia
[2]The Institute of Informatics Problems, Russian Academy of Sciences, Moscow, Russia
Email: teplo1930@mail.ru, yukonta@mail.ru

ABSTRACT

The paper presents a comprehensive critical review and detailed analysis of the literature, published in the field of the phase diagram solution Li-LiH (LiD, LiT). Special attention is paid to the completeness of the source information on the paper's topic resulting in an extended reference list. Particularly, paper reviews rarely used sources including little-known publications from research centers, proceedings of the international scientific meetings and dissertations. These publications were more thoroughly analyzed in order to make the information available to the scientific society.

Keywords: Diagram; Phase Equilibrium; Lithium Hydride; Deuteride

1. Introduction

The phase diagram of the Li-LiH system is complex. Many investigators have studied its individual parts by different methods such as plotting thermograms, analysis of the electrical resistance as a function of temperature and composition of the solution, and measurement of the partial pressure of hydrogen or of its isotopes when the condensed phase is in equilibrium with the vapor. These studies have identified several systematic features in the phase diagram of the Li-LiH system.

2. The Phase Diagram

The temperature-composition diagram for the systems under consideration is shown schematically in **Figure 1** (x is the mole fraction of lithium hydride in the condensed phase). The α-phase (on the left-hand side), which is enriched with lithium, is arbitrarily distinguished from the β-phase (on the right-hand side), which is enriched with hydride. The subscript l or s denotes the liquid or the solid phase, respectively. This system has eutectic properties at point c as well as monotectic properties (the line dfg).

Above the monotectic temperature, the liquid phase has a broad immiscibility zone. Experimental data show that the boundaries of the immiscibility zone (the curve dkf) are essentially the same for the Li-LiH and Li-LiD systems. At the same time, the experiment shows that the position of point c with respect to temperature and concentration is slightly different in these systems.

The characteristic points in the diagram for a Li-LiH system according to [1-3] are given in **Table 1** (the concentrations at points f and g are rough estimates).

Figure 2 shows the behavior of the isotherms in the pressure-composition diagram of the Li-LiH system above the temperature of the monotectic. The concentration

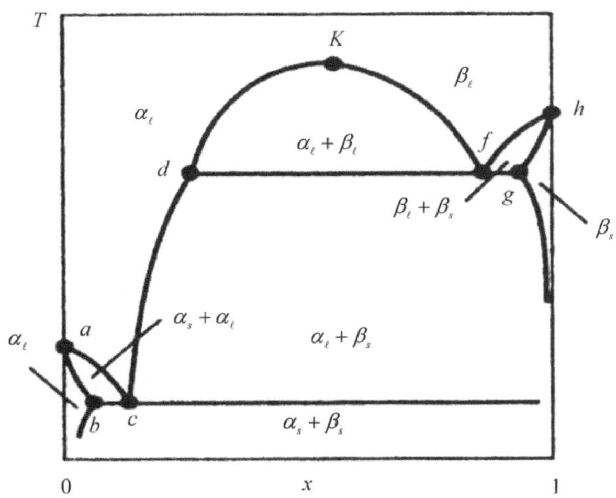

Figure 1. The phase diagram solution type Li-LiH (schematic).

Table 1. Characteristic points of the temperature composition diagram for the Li-LiH system.

Coordinates	a	b	c	d	κ	f	g	h
T, K	453.65	453.55	453.55	958	1273	958	958	970
x	0	0.56×10^{-4}	1.6×10^{-4}	0.26	0.60	0.98	0.99	1

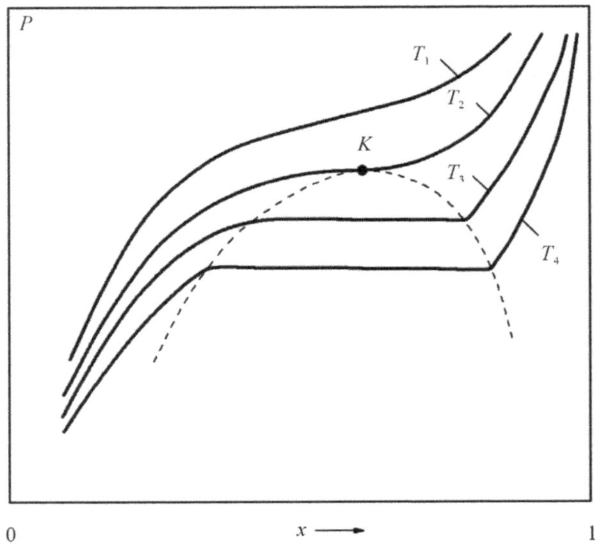

Figure 2. The pressure-composition phase diagram of the Li-LiH system (schematic); $T_1 > T_2 > T_3 > T_4$**.**

interval inside the immiscibility zone (the dashed curve), in which the vapor pressure remains constant, is called a "plane" region, and the vapor pressure in this region is called a plane pressure.

It is interesting to note the behavior of the isotherms as the condensed-phase composition approaches the stoichiometric composition of lithium hydride. The fact that P goes to infinity as $x \to 1$ means that the stoichiometric composition of lithium hydride is actually an unattainable hypothetical state. This isotherm behavior applies not only to the Li-LiH system but also to many other systems containing metal hydrides.

3. Behavior of Isotherm

Analysis of the phase diagram of the Li-LiH system (**Figure 1**) showed that in a certain temperature region there is an immiscibility zone, in which the vapor pressure and, in particular, the partial hydrogen pressure remain constant as the concentration of the components in the condensed phase changes. At the same time, the hydrogen pressure increases sharply beyond the immiscibility zone as the concentration of the components in the condensed phase approaches the stoichiometric composition of LiH.

The following conventional line of reasoning can be used to explain this phenomenon. Let us assume that the closed system in the liquid phase originally contained a pure lithium hydride of stoichiometric composition which evaporates in the first stage of the process. If there were no chemical reaction in the vapor phase, the amount of evaporated lithium hydride would be such that the vapor pressure would correspond to the saturation vapor pressure of LiH at the temperature T. In the vapor phase, however, there is a dissociation reaction

$$\mathrm{LiH} \leftrightarrow \mathrm{Li} + \frac{1}{2}\mathrm{H_2}. \qquad (1)$$

As a result, the number of LiH molecules in the vapor decreases, causing the partial pressure of the LiH vapor to drop (if it is assumed that the degree to which the vapor deviates from the ideal vapor before the dissociation is the same as the degree to which it deviates after the dissociation). At the same time, the partial pressure of one of the products of the dissociation reaction—atomic lithium—is higher than the saturation pressure of lithium at the temperature T under consideration (if all the produced lithium remains in the vapor state). Since we are dealing with a closed system, the excess lithium must condense. If there were no mixing of lithium with condensed lithium hydride, a new portion of lithium hydride would have to be converted to the vapor phase in order to restore the initial vapor pressure of LiH, which would be contingent in this case solely upon the temperature being kept constant. The evaporation and decomposition would evidently continue until all condensed lithium hydride would be converted to vapor and only pure lithium would remain in the condensed phase.

This situation, in fact, does not exist, which suggests that condensed lithium hydride in a pure form cannot coexist in equilibrium with the vapor. Coexistence can occur only if a certain amount of lithium dissolves in the lithium hydride. The partial vapor pressure of LiH drops to a value at which the condensed Li-LiH phase is in equilibrium with the vapor of a corresponding composition.

Since a solution of lithium in lithium hydride forms in the liquid phase, the partial pressure P_1 of lithium is proportional to the concentration of lithium in the liquid phase if the liquid is in a phase equilibrium with the vapor:

$$P_1 = P_1^0 \gamma_1 (1-x) \qquad (2)$$

where P_1 is the saturation pressure of pure lithium, γ_1 is the activity coefficient of lithium in a binary Li-LiH solution, and x is the mole fraction of lithium hydride in the solution.

The equilibrium constant of dissociation reaction (1)

$$K_p = P_1 \cdot P_3^{1/2} / P_2 \qquad (3)$$

is a function of only the temperature (here P_2 is the partial pressure of lithium hydride, and P_3 is the partial pressure of hydrogen). Since at T = const, the partial pressure of lithium $P_1 \to 0$ as $x \to 1$, and since the pressure of lithium hydride, P_2, cannot tend to zero in this case, the partial hydrogen pressure must tend to infinity in this process. Although these arguments are based on the assumption that the vapor phase is an ideal phase, a similar regularity observed experimentally confirms these conclusions and indicates that at thermodynamic equilibrium the purity of a lithium hydride sample (*i.e.*, the admixture with lithium) is proportional to the equilibrium hydrogen pressure.

The curve *acdfh* (see **Figure 1**) is the solidification line of the Li-LiH system (the liquidus line). This curve was studied in detail for the first time by Messer *et al.* [4]. In analyzing the behavior of the Li-LiH system when the condensed phase is in equilibrium with the vapor, they found that upon the removal of hydrogen from the vapor phase, the solidification temperature of the melt falls to the monotectic temperature of 685°C ± 1°C, remains constant at the molar concentrations of LiH between 98% and 26%, and then falls to 624°C at 13% LiH. This part of the curve is described by Messer *et al.* in terms of the functional dependence

$$\log\left[x/(1+x)\right]=-(3381/T)+2.835, \qquad (4)$$

where x is the mole fraction of LiH in the solution. The heat of solution of solid lithium hydride in liquid lithium was found to be 64.72 kJ/mole from of this equation.

The first data on the liquidus line in Li-LiH and Li-LiD systems below 624°C were published by a group of British scientists headed by Adams and Hubberstey [1, 5-9]. The solubility was determined from the change in the electrical resistance during the dissolution of H_2 or D_2 in lithium and from the precipitation at the time of saturation.

This method is based on a strong increase in the resistance with increasing concentration of H_2 (D_2) in the solution and on the appearance of a discontinuity in the ρ —x diagram (ρ is the resistivity, and x is the composition of the solution) at the saturation point of the solution (the point the hydride precipitates out).

The typical isotherms in **Figure 3** show that the resistivity of the solution increases linearly with increasing hydrogen concentration up to the saturation line and then remains nearly constant. The change in resistivity is clearly defined, which accounts for the high sensitivity in determining the solubility. The experimental data on the solubilities of LiH and LiD in lithium obtained in the studies underconsideration are presented in **Table 2**. These results were approximated by the authors of the cited studies in the form of the equations

$$\log x_H = 3.523-\frac{2308}{T} \quad (523\text{K}<T<775\text{K}); \qquad (5)$$

$$\log x_D = 4.321-\frac{2873}{T} \quad (549\text{K}<T<724\text{K}). \qquad (6)$$

Later they published [10] a little other coefficients:

$$\ln x_H = 3.507-\frac{5314}{T} \quad (523\text{K}<T<775\text{K}); \qquad (7)$$

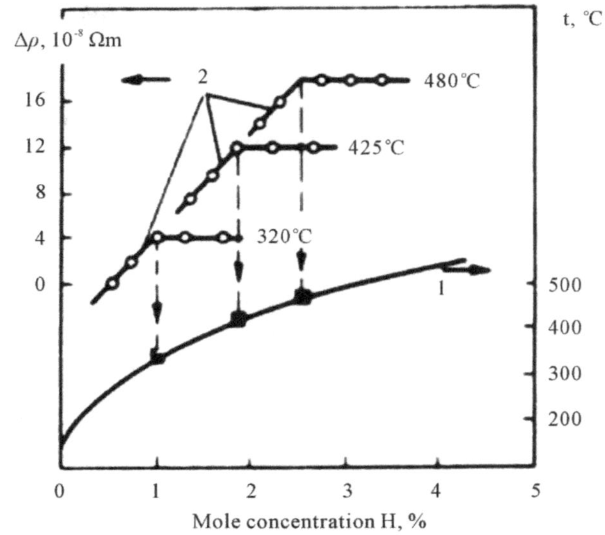

Figure 3. The curve for the solubility (curve 1) of the Li-LiH system plotted according to the data on the change in the resistivity of the solution (curve 2).

Table 2. Experimental data on the solubilities of hydrog en and deuterium in lithium (*x* is the atomic concentrat ion of *H* or *D*, at%).

t, °C	x_H, at%	t, °C	x_D, at%
221	0.037	551	5.680
227	0.063	276	0.122
257	0.140	305	0.229
275	0.212	328	0.330
296	0.301	340	0.527
326	0.462	350	0.493
344	0.596	365	0.691
361	0.753	375	0.773
376	0.892	389	0.990
395	1.148	397	0.990
397	1.195	402	1.095
411	1.420	417	1.393
441	1.930	451	2.350
499	3.480	456	2.631
525	4.540	500	3.900
551	5.680	-	-

$$\ln x_D = 3.101-\frac{5082}{T} \quad (472\text{K}<T<771\text{K}). \qquad (8)$$

In Equations (5)-(8) x is the atomic concentration of H or D in the solution (%).

These equations yield the values 44.2 and 54.8 kJ/mole for the partial molar enthalpies and 29.2 and 44.2 J/(mole deg) for the partial molar entropies of LiH and LiD, respectively. A comparison of the hydrogen and deuterium solubilities led the authors to conclude that at low temperatures deuterium is less soluble than hydrogen, but at $T = 435^{\circ}C$ and $x = 1.83\%$, the situation changes.

The solubility of lithium deuteride in liquid lithium between the eutectic and monotectic temperatures was also determined experimentally by Veleckis *et al.* [3]. The experimental method used by them consisted of three stages: 1) preparation of the equilibrium solution of Li-LiD, 2) removal of the filtered solution into metallic tubes, and 3) analysis of the samples of the solution to determine whether they contain deuterium.

The deuterium content of the sample was determined by thermal decomposition. Each sealed tube containing the sample was inserted into a container made from silica and then heated to 840°C for 10 days in order to completely remove the dissolved deuterium. The evolved gas was collected into a container of specified volume, and the deuterium content in it was determined by mass spectrometry. The data obtained by this method are given in **Table 3**.

Analysis of the available experimental data on the solubility of hydrogen and deuterium in the temperature interval between the eutectic and monotectic temperatures, respectively, in Li-LiH and Li-LiD systems, and the additional evaluation of these data have shown that the experimental data are best described by an equation of the type

$$\ln x = A + B\left(\frac{1}{T} - \frac{1}{T_M}\right) + C\ln\left(\frac{T}{T_M}\right) \tag{9}$$

where x is the mole fraction of LiH or LiD in a solution with lithium and T_M is the temperature of the monotectic.

The coefficients in Equation (9) and the mean-square deviations of the experimental data, σ, are summarized in **Table 4**.

Table 3. Solubility of lithium deuteride in the Li-LiD system from the data of Ref. [3] (*x* is the molar concentration of LiD, %).

t, °C	x	t, °C	x
198.9	0.0514	351.0	0.633
221.1	0.0768	375.1	0.866
246.3	0.129	397.2	1.14
271.5	0.181	397.2	1.17
303.6	0.322	451.4	2.08
323.3	0.427	498.0	3.32

Table 4. Results of analysis of the experimental data on the solubility of hydrogen and deuterium in the Li-LiH and Li-LiD systems. Analysis of the data was based on Equation (9).

System	A	B, K^{-1}	C	T_M, K	σ, %
Li-LiH	−1.517	−3998	3.004	961	1.9
Li-LiD	−1.517	−1197	6.83	963	1.9

Decrease in the solidification temperature of lithium due to the dissolution of hydrogen and deuterium in it (curve *ac* in **Figure 1**) was studied in Refs. [1,9,11]. The measurements were carried out by the thermal-analysis method (thermography).

Small portions (1 cm^3 under normal conditions) of soluble gas were added to pure lithium (30 g of 99.98% pure lithium). The solution was cooled slowly at the rate of 10 deg/h, which made it possible to reliably determine the solidification temperature. Continuous mixing of the solution kept it homogeneous. The melting point of pure lithium was assumed to be 180.49°C ± 0.02°C. With increasing atomic concentration of hydrogen or deuterium, the depression of the solidification temperature, ΔT, increased to the maximum value of 0.082°C at 0.016% H and to 0.075°C at 0.013% D, which corresponded to the eutectic points of the Li-LiH and Li-LiD systems (**Figure 4**).

The solidus lines (curve *ab* in **Figure 1**) were described under the assumption that the solid and liquid phases in Ref. [11] are ideal phases:

$$x_l - x_s = \frac{(T_f - T)\cdot \Delta H_f}{T_f \cdot T \cdot R} \tag{10}$$

Here x_l and x_s, are the atomic concentrations of hydrogen or deuterium on the liquidus and solidus curves; T_f and ΔH_f are, respectively, the temperature and the heat of fusion of pure lithium; and R is the universal gas constant. The concentration at point b (see **Figure 1**) for the Li-H system ($x_b = 0.002\%$) and for the Li-D system ($x_b = 0.001\%$) was determined on the base [12].

4. The Monotectic

The temperature of the monotectic T_M of the Li-LiH system was determined for the first time by Messer *et al.* [4] by the thermographic method. The value $T_M = 685^{\circ}C \pm 1^{\circ}C$ obtained by them, however, turned out to be too low, as was indicated in the subsequent experiments, e.g., by Smith and Webb [13]. Analysis of the behavior of the pressure isotherm in the Li-LiH system, when the condensed phase is in equilibrium with the vapor, as a function of the composition of the condensed phase, shows that in the temperature region of the monotectic the isotherm plotted as $P = f(x)$ is horizontal (see **Figure 2**)

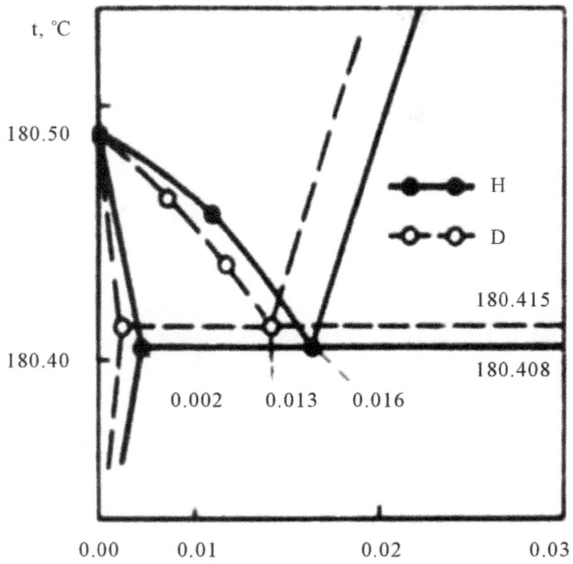

Figure 4. The decrease in the solidification temperature of lithium as a result of dissolution of hydrogen and deuterium.

in the region *df* (see **Figure 1**).

The temperature of the monotectic can therefore be determined from the point of intersection of the functions $P = f(x)$ measured above and below the temperature of the monotectic.

At the temperature of the monotectic, there are four phases in a thermodynamic equilibrium: the vapor phase, a liquid phase with the composition corresponding to point *d* (see **Figure 1**), a liquid phase with the composition at point *f*, and the solid phase with composition at pointy *g*.

In the temperature region slightly above T_M, the vapor phase and the two liquid phases are in equilibrium, whereas in the temperature region slightly below T_{My} the vapor phase, the liquid phases and the solid phase are in equilibrium. The heat and entropy of dissociation in the reaction

$$LiH_{cond} = Li_{cond} + \frac{1}{2} H_{2\,gas} \quad (11)$$

which determines the pressure of the vapor phase, at $T > T_M$ and $T < T_M$ will therefore differ from the heat and entropy of fusion of lithium hydride. This accounts for the discontinuity in the function $P = f(T)$ at the monotectic temperature.

A consistent analysis of Smith and Webb [13] data for the Li-LiH (LiD) systems near temperatures of ±100 K relative to T_M in the form of the relation

$$\ln P = A + B/T \quad (12)$$

yields (the coefficients *A* and *B* are assumed to be constant because of the narrow temperature interval) the following values: $T_M = 961 \pm 2$ K for the Li-LiH system and $T_M = 963 \pm 2$ K for the Li-LiD system.

For the Li-LiT system, the analysis, which took into account the experimental data of Refs. [14,15] for the entire temperature interval studied (because of their limitation, all data were taken into account), yielded the value $T_M = 964 \pm 3$ K.

Smith and Webb [13] and Veleckis [15] also determined the temperatures of the monotectic in Li-LiH (LiD, LiT) systems by analyzing their data on the vapor pressure in the plane region over a broad temperature range. However, the values obtained by them (the data of Ref. 11 are enclosed in parentheses), T_M = 694°C (699.6°C) for Li-LiH, T_M = 690°C (687.7°C) for Li-LiD, and T_M = 688°C for Li-LiT, do not seem to be legitimate princeipally because of their isotopic sequence.

The data on the melting point of the isotopic compounds found in the literature show that the heavier isotopic modifications have a higher melting point. In **Table 5** are several melting points at 10^5 Pa, taken from the data presented in Refs. [16,17].

Table 5. Melting point of the isotopic compounds.

Substance	H_2	D_2	CH_4	CD_2	H_2O	D_2O
T_y K	13.95	18.65	90.64	98.78	273.15	276.95

Although at the monotectic point the system is, strictly speaking, a solution rather than a pure substance, the isotopic dependence of the melting point (of the monotectic) must obviously reflect this trend.

The use by the authors of Refs. [13,15] of a linear approximation $\ln P = f\ (1/T)$ [see Equation (12)] over a broad temperature range clearly is a possible source of error in determining the monotectic temperature. The temperature dependence of the heat and entropy of dissociation of LiH (LiD, LiT) in this case could affect the calculation results.

REFERENCES

[1] P. Hubberstey, R. J. Pulham and A. F. Thunder, *Transactions of the Faraday Society*, Vol. 72, No. 12, 1973, pp. 432-436.

[2] E. Veleckis, E. H. Van Deventer and M. Blander, "Lithium-Lithium Hydride System," *The Journal of Physical Chemistry*, Vol. 78, No. 19, 1974, pp. 1933-1940. doi:10.1021/j100612a013

[3] E. Veleckis, R. M. Yonco and V. A. Maroni, "Solubility of Lithium Deuteride in Liquid Lithium," *Journal of the Less Common Metals*, Vol. 55, No. 1, 1977, pp. 85-92. doi:10.1016/0022-5088(77)90263-6

[4] C. E. Messer, E. B. Damon, P. C. Maybury, *et al.*, "Solid-Liquid Equilibrium in the Lithium Lithium Hydride System," *The Journal of Physical Chemistry*, Vol. 62, No. 2, 1958, pp. 220-222. doi:10.1021/j150560a018

[5] P. F. Adams, P. Hubberstey and R. J. Pulham, "Review of the Solubility of Non-Metals in Liquid Lithium," *Journal of the Less Common Metals*, Vol. 42, No. 1, 1975, pp. 1-11. doi:10.1016/0022-5088(75)90014-4

[6] P. F. Adams, M. G. Down, P. Hubberstey, *et al.*, "Solubilities, and Solution and Solvation Enthalpies, for Nitrogen and Hydrogen in Liquid Lithium," *Journal of the Less Common Metals*, Vol. 42, No. 3, 1975, pp. 325-334. doi:10.1016/0022-5088(75)90052-1

[7] P. Hubberstey, P. F. Adams, R. J. Pulham, In: *Proc. Inter. Conf. on Radiation Effects and Tritium Technology*, Oak Ridge National Lab., Oak Ridge, 1975, pp. 117-124.

[8] R. J. Pulham, P. F. Adams, P. Hubberstey, *et al.*, In: *Proc. Intern. Conf. Radiation Effects and Tritium Tech. Fusion Reactors*, Vol. 4, Report CONF-750989, Oak Ridge National Lab., Oak Ridge, 1976, pp. 144-157.

[9] P. Hubberstey, P. F. Adams, R. J. Pulham, *et al.*, "Hydrogen in Liquid Alkali Metals," *Journal of the Less Common Metals*, Vol. 49, No. 3, 1976, pp. 253-269. doi:10.1016/0022-5088(76)90039-4

[10] P. Hubberstey, A. T. Dadd, P. G. Roberts, *Proceedings of International Conference on Material Behavior and Physical Chemistry in Liquid Metal Systems*, Plenum Press, New York, 1982. pp. 445-454. doi:10.1007/978-1-4684-8366-6_46

[11] P. F. Adams, P. Hubberstey, R. J. Pulham, *et al.*, "Solubilities, and Solution and Solvation Enthalpies, for Nitrogen and Hydrogen in Liquid Lithium," *Journal of the Less Common Metals*, Vol. 42, No. 3, 1975, pp. 285-289. doi:10.1016/0022-5088(75)90052-1

[12] E. Fromm and H. Jehn, "Bull," *Alloy Phase Diagrams*, Vol. 5, No. 3, 1984, pp. 324-326.

[13] H. M. Smith and R. E. Webb, "Equilibrium Dissociation Pressures of Lithium Hydride and Lithium Deuteride," Oak Ridge, 1977.

[14] F. K. Heuman and O. N. Salmon, "The Lithium Hydride, Deuteride and Tritide Systems," Knolls Atomic Power Lab., Schenectady, 1956.

[15] E. Veleckis, *Journal of Nuclear Materials*, Vol. 79, No. 1, 1979, pp. 20-27.

[16] A. R. Ubbelohde, "Melting and Cristal Structure (Clarendon," Oxford, 1965, 420 p.

[17] I. B. Rabinovich, V. Izotopii na Fiziko-Khimicheskie and S. Zhidkostei, "Effect of Isotopy on the Physicochemical Properties of Liquids," Nauka, Moscow, 1968, p. 308.

15

Solar Water Splitting by Semiconductor Nanocomposites and Hydrogen Storage with Quinoid Systems

Thorsten Wilke, Daniela Schricker, Josefin Rolf, Karl Kleinermanns*
Institute of Physical Chemistry, Heinrich-Heine-University, Duesseldorf, Germany
Email: *kleinermanns@uni-duesseldorf.de

ABSTRACT

Photocatalytic splitting of water was carried out in a two-phase system. The efficiencies of different types of nanocrystalline semiconductors were investigated and compared with commercialised TiO_2 nanopowder. Generated hydrogen was chemically stored by use of a quinoid system, which seems to be useable for fuel cells. Solar light sensitive nanocomposites of $CdSe/TiO_2$ and CdS_xSe_y/TiO_2 type were prepared and their good photocatalytic performance was demonstrated. In the visible range of 400 - 600 nm CdS_xSe_y/TiO_2 composites show comparable good results as in the UV range, which is very promising for their use as solar light water splitters. The concept of sensitising TiO_2 with different kind of semiconductor nanoparticles, which is already known from quantum dot sensitised solar cells (QDSC), was demonstrated here for water splitting as well. Furthermore the kinetics of the storage reaction was investigated by UV-Vis spectroscopy and found to proceed via a consecutive reaction with an 1:1 charge transfer complex of quinone and hydroquinone as intermediate. The electron transfer process via a Fe^{2+}/Fe^{3+} redox couple was investigated by UV-Vis spectroscopy as well as by a dye reaction on the TiO_2 surface. A light microscopic view of the surface of larger aggregates of TiO_2 nanoparticles indicated different areas of photocatalytic activity with photocatalysis preferentially at catalyst edges. The global electron transfer process could be traced by following the dye colour in real time.

Keywords: Water Splitting; Nanocomposites; Cadmium Selenosulphide; Titanium Dioxide; Hydrogen Storage

1. Introduction

By the year 2050 the total energy consumption is expected to double, as the world's population is steadily increasing. Fossil fuels are not able to meet this energy demand in the long term, therefore, renewable energy resources will come more sharply into focus. The most promising alternative is solar light, because the amount of energy that arrives on earth every hour from the sun is greater than the amount that is required by the entire humanity in one year [1]. Yet, there is no practical way to transform and store this huge energy reserve efficiently, because the widely used silicon solar cells are of limited use here due to their high costs. Therefore it is necessary to look for less expensive and in sufficient quantities available alternatives to high-purity silicon. Storage of solar energy is possible for example by batteries and capacitors, but, compared to chemical bonds, these storage systems feature a low energy density. In this regard hydrogen is a good energy reservoir. It can be used as fuel for vehicles or can be converted into electrical energy by the use of fuel cells. The generation of hydrogen by electrolysis requires electrical energy, which could be obtained by use of solar cells, but the effectiveness is just approximately 8% for large-scale facilities [2]. Thus, direct photolytic water splitting by the use of suitable and inexpensive nanocrystalline semiconductors would be an promising alternative. Here the water is split without use of electricity and with high efficiency by solar light [1,3].

The semiconductor titanium dioxide has a band gap of about 3 eV and its conduction band potential is high enough for water splitting [4]. By absorption of photons electrons can be promoted to the excited state and electron-hole pairs are generated, which diffuse separately on the surface of the TiO_2 particles:

$$TiO_2 + h\nu \rightarrow e^- + h^+ \qquad (1)$$

The formed holes in the valence band are able to oxidize molecules, for example water:

$$H_2O + 2h^+ \rightarrow \frac{1}{2}O_2 + 2H^+ \qquad (2)$$

The electrons in the conduction band can reduce species, for example H^+ to hydrogen because the reduction potential of TiO_2 is sufficiently negative (–0.65 V [5]):

$$2H^+ + 2e^- \rightarrow H_2 \qquad (3)$$

For water splitting by TiO_2 we finally obtain the fol-

*Corresponding author.

lowing overall reaction [6]:

$$H_2O + 2h\nu \rightarrow \frac{1}{2}O_2 + H_2 \tag{4}$$

Depending on the particle size, TiO_2 nanoparticles (NPs) absorb at <350 - 380 nm [7]. Larger particles show smaller band gaps so that the absorption is red-shifted. Pure titanium dioxide can thus only use the UV part of the solar spectrum. However, most UV radiation is already absorbed by the ozone layer, which means that its fraction on earth is low. Therefore, extension of TiO_2's light absorption to the visible region is of great interest. Depending on particle size cadmium selenide nanoparticles have a band gap of about 2.1 eV [8] and absorb in the range up to 600 nm [9]. So by sensitization of titanium dioxide with cadmium selenide nanoparticles, a much better use of the solar spectrum is possible. An efficient electron injection from the photoexcited nanoparticle into the TiO_2 conduction band requires a close contact, which can not be achieved by simple mixing of both semiconductors. Therefore we prepared nanocomposites by annealing a homogenous mixture of TiO_2 and CdSe nanoparticles at 450°C.

Cadmium selenosulphide CdS_xSe_y are a novel class of photocatalysts. Their good ability as sensitizer in Quantum Dot Solar Cells (QDSC) has already been shown by our group. Their absorption depends strongly on the composition. The best results have been achieved with CdS_6Se_1 with an absorption up to 800 nm [7]. In principle, organic dyes can be used for sensitisation as well [10-12] and results obtained with this class of sensitisers for water splitting with solar light will be published elsewhere [13].

For the chemical storage of the generated hydrogen quinoid systems are suitable. Their use mimics natural processes, e.g. photosynthesis, which also use quinoid systems like plastoquinone for hydrogen transfer [6].

2,3-Dichloro-5,6-dicyano-1,4-benzoquinone (DDQ) is known from literature as a good hydrogen acceptor [14]. An advantage of this benzoquinone derivate is that DDQ and DDHQ are well distinguishable by UV-VIS spectroscopy allowing rather easy quantitative analysis. The constraint of using DDQ is its high reactivity towards water resulting in the evolution of hydrocyanic acid. Because of this behaviour the water splitting must be carried out in a two-phase system.

In 1999 Matsumura *et al.* presented a photocatalytic reaction system for splitting water by use of TiO_2 nanoparticles [14]. The experimental setup consists of two compartments connected by an oil phase, which consists of DDQ and DDHQ dissolved in an organic solvent. The photocatalytic water splitting takes place in the first compartment containing TiO_2 as photocatalyst in an aqueous iron(III) solution, where gaseous oxygen is generated. In the second compartment hydrogen is generated by use of platinum-loaded TiO_2 nanoparticles in an aqueous bromine/bromide solution. DDQ/DDHQ serves as a hydrogen transporter between both compartments. The group of Matsumura has given a proof of principle; however, their hydrogen yields were very low and could be achieved only by very long and intense irradiation.

In this paper we present a further development of the scheme used in Ref. [14] and its extension to use visible solar light for photocatalysis. We compared TiO_2, CdSe sensitised TiO_2 and CdS_6Se_1 sensitised TiO_2 for their efficiency of DDQ reduction by visible light. We also report investigations of the kinetics of the hydrogen storage reaction based on time-resolved UV-Vis investigations of DDQ conversion.

For a closer look at the processes on the surface of the nanocatalysts we recorded successive light microscopic images of TiO_2 aggregates in an aqueous suspension. For visual detection of the electron transfer process a simple dye reaction was used, which verified a fast conversion of Fe^{3+} to Fe^{2+} on the surface of TiO_2 aggregates. A magnification of the micrographs clearly visualised high active areas as bright spots on the surface of TiO_2 nanoparticle aggregates.

2. Experiments

This article is dealing with the investigation of titanium dioxide, cadmium selenide and cadmium selenosulphide nanoparticles as well as nanocomposites of TiO_2 with both semiconductors for their ability to split water photocatalytically.

2.1. Preparation of the Photocatalysts

Titanium dioxide nanopowder (Aerosil TiO_2 P-25) was purchased by Degussa Evonik and used without any treatment.

Cadmium selenide nanoparticles are synthesised as follows [15]: $Cd(ClO_4)_2 \cdot 6\ H_2O$ (985 mg, 2.35 mmol, Aldrich) was dissolved in 125 mL water. Afterwards 5.7 mmol of thioglycolic acid (TGA, Aldrich) as thiol-stabilizer were added during stirring and the pH of the solution was adjusted between 11.2 and 11.8 by dropwise addition of 1 M NaOH (Aldrich). A three neck flask with Al_2Se_3 lumps (134 mg, 0.46 mmol, Aldrich) was connected and flushed with N_2 for about 30 minutes. The cadmium solution was deaerated in the same way. Subsequently 10 - 20 mL of 0.5 M H_2SO_4 (Aldrich) was added to the Al_2Se_3 lumps via a septum and the generated H_2Se gas was passed to the solution via a slow nitrogen flow for about 20 minutes. The colour of the solution changed to orange due to the formation of CdSe particles. The CdSe precursors grew to CdSe nanoparticles under reflux (100°C, 24 hours). For precipitation of the CdSe NCs, isopropanol (Fluka) was added. The suspension was centrifuged for 5 minutes at 13000 rmp, the aqueous phase was discarded and the residual solid was dried in an exsiccator for 1 week. This route gave particles with a size of 2 - 4 nm.

As precursor for the preparation of cadmium selenosulphide nanoparticles, cadmium sulphide nanoparticles were prepared with a size distribution of 2 - 4 nm [16]. 1-thioglycerol (TG, 1.87 mL, 22 mmol, Aldrich) was added to 30 mL of a 80 mM aqueous cadmium sulphate solution ($CdSO_4 \cdot 8/3\ H_2O$, Fluka). During constant stirring for 5 minutes ammonium sulphide (12.4 ml, 36 mmol, 20 % aqueous solution, Aldrich) was added. An intense yellow solution with yellow precipitation was obtained thereafter, which was centrifuged and dried for 1 week in an exsiccator.

Cadmium selenosulphide nanocomposites were prepared as follows: In a specific molar ratio of 6:1 (S:Se) cadmium sulphide nanoparticles and selenium powder (99.5 %, Aldrich) were suspended in methanol. After 1 hour of stirring the methanol was evaporated at 80°C. The resulting CdS:CdSe solid was sintered at 550°C for 15 minutes. The colour of the remaining CdS_6Se_1 nanocomposite is intensive red.

CdS_6Se_1 is not soluble in common solvents, which precludes their direct TEM investigation. The CdS_6Se_1 nanoparticles were absorbed on the surface of small TiO_2 nanoparticles by mixing the finely grinded solid with a clear and colourless solution of TiO_2 nanoparticles. The resulting nanocomposites could be investigated by TEM and elementary analysis by EDX showed their composition.

The clear and colourless solution of TiO_2 nanoparticles was prepared as follows [17]: titanium (IV) butoxide (8.5 mL, 0.025 mmol, Aldrich) was dissolved in ethanol (19 mL, Aldrich) and diethanolamine (2 mL, 20.8 mmol, Fluka). Subsequently 19 mL ethanol and 0.5 mL H_2O were added very slowly. The solution was stirred over night.

$CdSe/TiO_2$ respectively CdS_6Se_1/TiO_2 nanocomposites were prepared by mixing with a suitable TiO_2 paste. TiO_2 nanopowder (5 g Aearosil TiO_2 P25, Evonik) was suspended in nitric acid (1 N, Fluka). This suspension was heated at 80°C for 24 hours. Afterwards the nitric acid was evaporated and the TiO_2 solid was dried for 3 days at 100°C. Finally the TiO_2 solid was treated with 25 ml water, acetylacetone (2.5 g, 24.97 mmol, Merck), Triton-X-100 (1.25 g, 1.92 mmol, Avocado) and polyethylene oxide (M.W. 100000, Alfa Aesar) to give a homogenous paste. 40 mg of CdSe or CdS_6Se_1 were mixed with 1 mL of this TiO_2 suspension in an agate mortar to produce a smooth and uniformly coloured paste, which was dried in a ceramic crucible at 80°C for 2 hours. Afterwards the solid was sintered at 450°C in a muffle furnace for about 30 minutes and ground to a fine powder.

2.2. Photocatalytic Water Splitting

2,3-Dichloro-5,6-dicyano-1,4-benzoquinone (98%, Aldrich), iron(III) chloride (>97%, Aldrich) and n-butyronitrile (purum, ≥99.0%, Fluka) was used without purification.

The water splitting experiments were carried out in a quartz cuvette of 2 × 4 × 4 cm size. 30 mg of the photocatalyst were dispersed in 16 mL of an aqueous iron (III) chloride solution (8.0 mM). 5 mL of a DDQ solution in n-butyronitrile (1.9 mM) was placed carefully on top of the aqueous phase to form the organic phase. In order to avoid an evaporation of the organic solvent, the cuvette was covered by a glass plate of adequate size. To avert heating of the reaction system during irradiation and to reduce diffusion of DDQ into the aqueous phase, the vessel was cooled to 15°C by a water flushed aluminium block, connected to the cuvette by a heat-conductive paste.

The irradiation was carried out for 90 minutes by a 80 W mercury-vapor lamp (Oriel, Germany), which was mildly focused to give 100 mW/cm^2.

To prevent irradiation of the organic phase and photodegradation of DDQ and DDHQ an aluminium mask of adequate size was used.

For investigation of the efficiency of the photocatalysts in the range of the visible solar spectrum, the water splitting experiment was performed by irradiation of the aqueous phase with the unfiltered Hg lamp spectrum and, for comparison, with a filtered lamp spectrum (400 - 600 nm). For this purpose a suitable cut-off filter (Edmund optics GmbH, Karlsruhe) was placed between lamp and reaction vessel.

To exclude contributions to the spectra, which were not due to the process of water splitting, a reference solution of the organic phase was kept in the dark without contact to the aqueous phase during irradiation.

After 90 min irradiation a sample of the organic phase was taken, diluted 1:25 with n-butyronitrile and analysed by UV-Vis spectroscopy.

2.3. Kinetics of the Hydrogen Storage Reaction

The water splitting experiment described above was carried out with 30 mg of P-25 TiO_2. Irradiation of the aqueous phase was performed over 100 minutes with the unfiltered Hg lamp spectrum (100 mW/cm^2). Every 10 minutes a 100 μL sample of the organic phase was taken at exactly the same position via a rigidly adjusted stainless steal cannula. The samples were diluted 1:25 with n-butyronitrile and analysed by UV-Vis spectroscopy. The recorded spectra were processed by a peak analysing software (OriginPro 8G, OriginLab Cooperation, USA) to determine the exact absorption of the involved species and the change of their concentrations with reaction time.

2.4. Tracking of the Electron Transport Process via UV-Vis Spectroscopy

For tracking the electron transport process via Fe^{3+}/Fe^{2+} the aqueous phase was investigated by UV-Vis spectroscopy as well. A suspension of titanium dioxide nanoparticles in water shows strong light scattering, so that the kinetic analysis can not be performed with TiO_2 as photocatalyst. Therefore CdSe in a 10 times lower concentration (1 mg, dissolved in 16 mL of a 8.0 mM aqueous $FeCl_3$ solution) was used, which also photocatalytically splits water. Irradiation of the aqueous phase was performed for 90 minutes with an unfiltered Hg lamp spectrum (100 mW/cm^2). Every 15 minutes a 100 µL sample of the aqueous phase was taken at exactly the same position via a rigidly adjusted stainless steal cannula. The samples were again analysed by UV-Vis spectroscopy to follow the reduction of Fe^{3+} to Fe^{2+} quantitatively.

2.5. Visual Detection of the Electron Transfer Process via a Dye Reaction

About 1 mL of a suspension of 30 mg P-25 TiO_2 nanoparticles in 16 mL of an aqueous solution of iron (III) chloride (8 mM) and potassium ferrocyanide(III) (8 mM) was placed on a glass slide. The slide was positioned under a light microscope and consecutive light microscopic images were taken at 500-times magnification while the sample was irradiated for 5 minutes by a 80 W Hg-Lamp. Bright spots from Prussian blue K $[Fe^{III}Fe^{II}(CN)_6]$ indicate the positions of Fe^{3+} reduction on the surface of the TiO_2 aggregates.

2.6. Experimental Equipment

The particle size distribution of the photocatalyst nanoparticles was investigated by Transmission Electron Microscopy (TEM) measurements, which were performed with a HITACHI TEM 7500 microscope equipped with a Mega View II camera (Soft Imaging System) at the Max-Planck institute for coal research (MPI Mülheim a.d. Ruhr).

TiO_2 nanopowder has been obtained as a commercialised product from Degussa Evonik and was used without treatment. To verify the given properties Transmission Electron Microscopy (TEM) images has been taken. They are shown in **Figure SI1**. An average particle size of 21 nm and a homogeneous size distribution have been found.

Optical absorption spectra of the organic phase were recorded by using a Cary 300 UV-Vis spectrophotometer operated at a resolution of 1 nm.

Light microscopic images have been taken by using an Olympus BX41M reflected-light microscope and an Aiptec AHD 720P digital camcorder.

3. Results and Discussion

Our experimental setup is shown in **Figure 1**. A solution of DDQ in n-butyronitrile forms the organic layer and is located above the aqueous phase, in which the water splitting takes place. A Fe^{2+}-Fe^{3+} redox system transports the electrons between the two phases. The holes in the valence band of titanium dioxide oxidize water, while the excited electrons in the conduction band reduce the Fe^{3+} to Fe^{2+} ions. The reduced iron ions transport electrons to the interface, where quinone is subsequently reduced to semihydroquinone and hydroquinone by accepting electrons from Fe^{2+} and protons from water. Because of the spatial separation of reduction and oxidation processes in two different phases, electron-hole recombination can be minimized and the oxidation of DDHQ by TiO_2-holes can be prevented. By this, a quantitative DDQ to DDHQ conversion can be realized.

The redox scheme in **Figure 2** confirms the energetic feasibility of this approach.

DDQ shows a strong absorption band with a maximum at 278 nm, while DDHQ shows absorption bands at 248 and 347 nm (**Figure 3**).

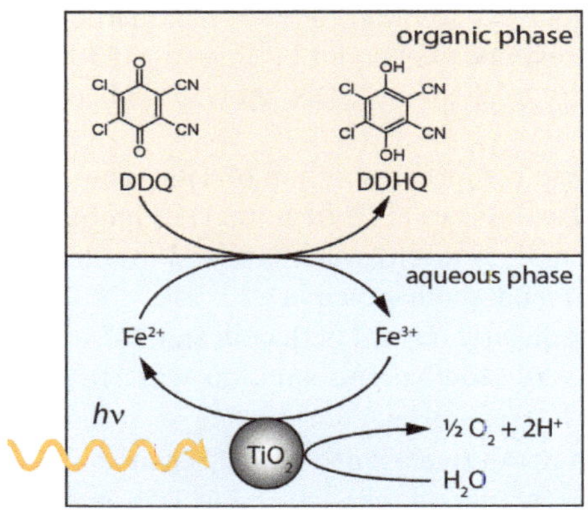

Figure 1. Experimental setup of the two-phase system. The photocatalyst nanopowder, e.g. TiO_2, is suspended in a solution of $FeCl_3$ in water which forms the aqueous phase. It is covered with a layer of the organic phase, a solution of 2,3-Dichloro-5,6-dicyano-1,4-benzoquinone (DDQ) in n-butyronitrile [14]. The reaction cell is tempered by a water cooling block to approximately 15°C to reduce diffusion of DDQ into the aqueous phase. The aqueous phase is irradiated by a Hg lamp, which is mildly focused to give 100 mW/cm^2. Aluminium foil prevents irradiation of the organic phase. The TiO_2 nanoparticles are coated with CdSe or CdS_6Se_1 nanoparticles for sensitisation to solar light (not shown in the figure).

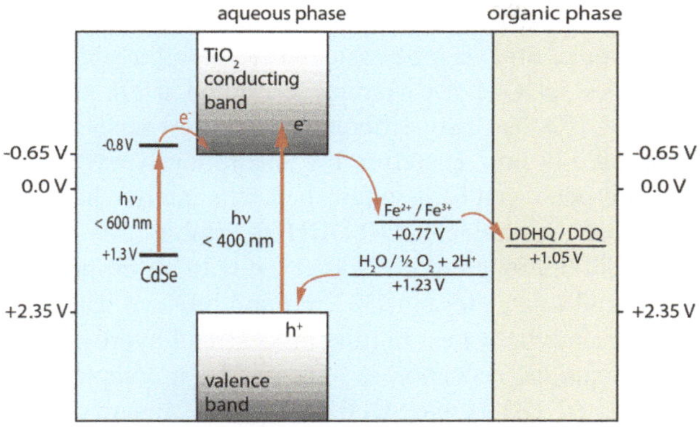

Figure 2. Energy scheme of the two-phase water splitting experiment. Redox potentials are given in Volts [V] against Standard Hydrogen Potential [SHE]. The working principle is based on excitation of the photocatalyst, e.g. TiO_2. The excited electron (e^-) is stored by reducing Fe^{3+} to Fe^{2+} and transported to the interphase by diffusion. The electron transporter is regenerated to Fe^{3+} by reducing DDQ. Holes (h^+) in the valence band are filled by oxidation of water leading to O_2 and H^+. DDHQ is formed at the interface by reaction of H^+ with the reduced DDQ.

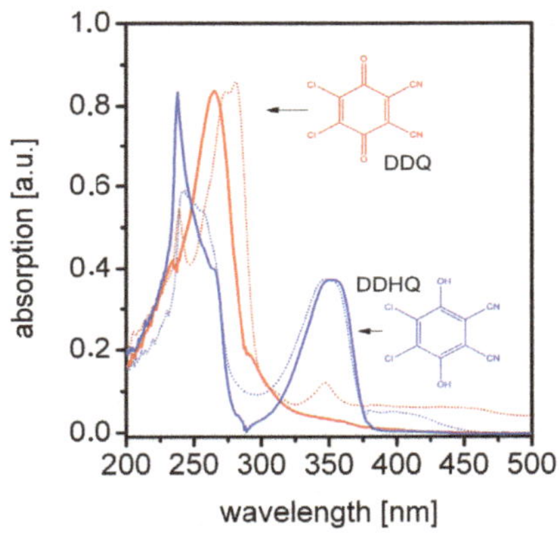

Figure 3. UV-Vis spectra of DDQ (red) and DDHQ (blue) measured in THF (solid curves) and n-butyronitrile (dotted curves). The spectra were obtained using a suprasil quartz cell with 1 cm path length and a double-beam UV-Vis spectrophotometer operated at a resolution of 1 nm.

Figure 4(a) displays a TEM image of synthesized CdSe nanoparticles stabilised with thioglycolic acid (TGA). A narrow size distribution has been found and is presented in the inset. Typical size ranges are 2 - 4 nm. Based on our TEM images we propose Ostwald ripening to be responsible for the formation of the nanoparticles.

A TEM image of the CdS_6Se_1 nanocomposite in contact with TiO_2 is displayed in **Figure 4(b)**. The composite particles show an average size of 50 nm. Elemental analysis by local Energy Dispersive X-ray spectroscopy (EDX) indeed showed that the composite particles contain Ti, Cd, S and Se.

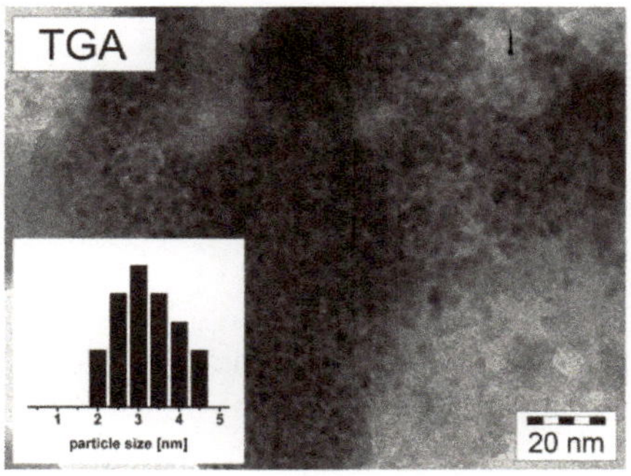

(a)

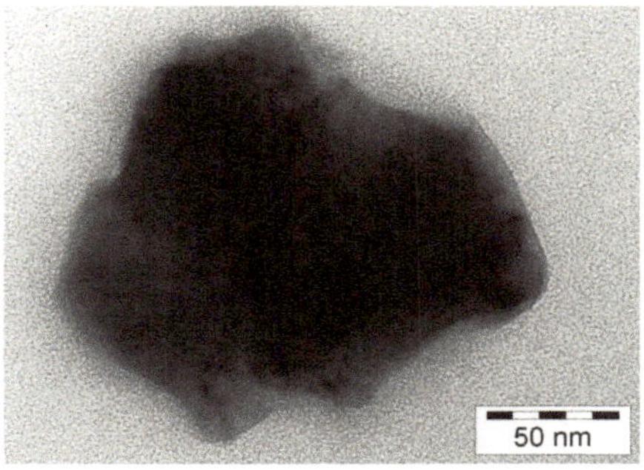

(b)

Figure 4. (a) TEM images of CdSe nanoparticles stabilised with thioglycolic acid (TGA). The synthesis is described in the experimental section. The insets indicate the size distribution as obtained from diameter measurements of 20 - 40 particles. Typical size ranges are 2 - 4 nm; (b). TEM image of CdS_xSe_y/TiO_2 nanocomposites obtained from dispersion of CdS_xSe_y particles in a colourless, clear TiO_2 solution.

The results of the water splitting experiments are presented in **Figures SI2-SI4** and **Figure 5(a)**. Compared are the UV-Vis spectra of the diluted organic phase after 90 minutes irradiation with the unfiltered Hg lamp spectra and a filtered Hg lamp spectrum (400 - 600 nm). Irradiation for <90 minutes shows a significant slower conversion of DDQ to DDHQ when using the filtered Hg lamp and TiO_2, TiO_2/CdSe or CdSe photocatalyst. The conversion of DDQ to DDHQ proceeds via an intermediate 1:1 complex of DDQ and DDHQ. The reaction process can be described as follows:

DDHQ is formed by reduction of DDQ and acceptance of two protons.

$$DDQ + 2e^- + 2H^+ \rightarrow DDHQ \quad (5)$$

A 1:1 charge-transfer complex is formed with remaining DDQ.

$$DDQ + DDHQ \rightarrow [DDQ \cdot DDHQ] \quad (6)$$

The complex is degraded by reducing and accepting two further protons and electrons to obtain two DDHQ molecules.

$$[DDQ \cdot DDHQ] + 2e^- + 2H^+ \rightarrow 2DDHQ \quad (7)$$

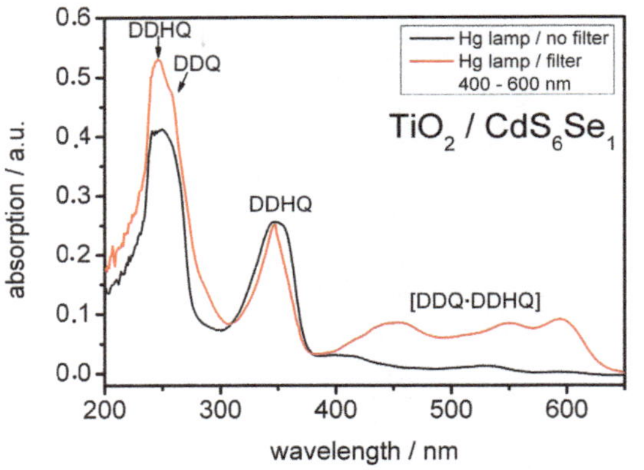

(a)

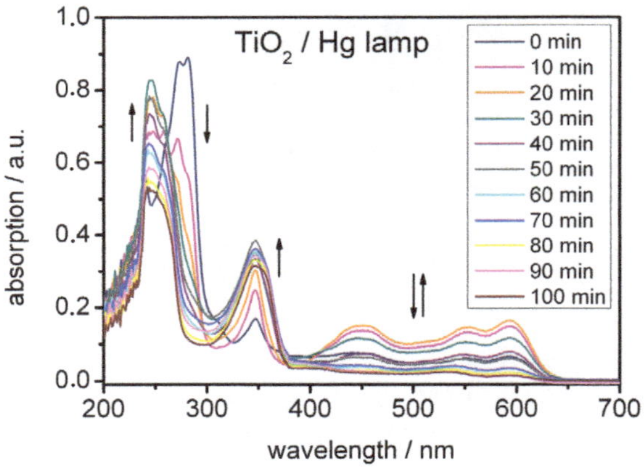

(b)

Figure 5 (a) Water splitting by TiO_2/CdS_6Se_1 nanocomposites as photocatalyst. Shown are the UV-Vis spectra of the organic phase after 90 minutes reaction time. The black line shows the conversion of DDQ to DDHQ after 90 minutes irradiation of the aqueous phase with the unfiltered Hg lamp spectrum. The red line shows the result after 90 minutes irradiation with a filtered lamp spectrum (400 - 600 nm) by using an optical glass filter. The recorded absorption bands can be assigned to three different species: DDQ (absorption at 278 nm), DDHQ (absorption at 248 nm, 347 nm) and complexed [DDQ·DDHQ] (absorption at 448 nm, 509 nm, 552 nm, 593 nm). The remaining DDQ and [DDQ·DDHQ] indicates the achieved conversion of DDQ and is a measure of the efficiency of the used photocatalyst for water splitting and hydrogen storage; (b) Temporal progress of the conversion reaction DDQ to DDHQ. Shown are the UV-Vis spectra of the organic phase measured over 100 minutes. The recorded absorption bands can be assigned to three different species: DDQ (absorption at 278 nm), DDHQ (absorption at 248 nm, 347 nm) and complexed [DDQ·DDHQ] (absorption at 448 nm, 509 nm, 552 nm, 593 nm). Increase and decrease of absorption bands are indicated by arrows. A decrease of the DDQ concentration and a simultaneous increase of the DDHQ concentration with time can be observed. Complexed [DDQ·DDHQ] is formed as an intermediate, whose absorption quickly emerges, increases within 20 minutes, afterwards decreases and completely vanishes after 80 minutes. The absence of isosbestic points indicates at least one further intermediate in the course of the reaction.

The observed absorption bands, shown in **Figure SI2-SI4** and **Figure 5(a)**, can be assigned to these three different species. DDQ shows a strong absorption at 278 nm, whereas DDHQ has two different absorption bands at 248 nm and 347 nm. Therefore the absorption bands of educt and product can be distinguished. The intermediate charge-transfer complex [DDQ·DDHQ] exhibits a wide absorption in the range of 400 - 650 nm with absorption maxima at 448 nm, 509 nm, 552 nm and 593 nm. Efficient photocatalytic water splitting process and hydrogen storage by quinine reduction is indicated by a complete conversion of DDQ into DDHQ without remaining [DDQ·DDHQ] complex. The observed absorption of remaining free and complexed DDQ is a measure of the efficiency of the photocatalyst and the hydrogen storage system. Considering that the irradiation of the earth by the sun has only a small UV component, the performance of the catalyst using the visible part of the sun spectrum is of particular interest. This has been investigated by using a filtered Hg lamp spectrum in the range of 400 - 600 nm. The resulting spectrum is displayed as red line in **Figure SI2-SI4** and **Figure 5(a)**. The black line shows the absorption of the organic phase after irradiation of the aqueous phase with the unfiltered lamp spectrum.

Figure SI2 summarises the performance of TiO_2 nanoparticles without sensitiser. It can clearly be seen that the efficiency of TiO_2 in the visible range is very low. A strong absorption at 278 nm indicates a large amount of unconverted DDQ while the observed absorption bands of DDHQ at 248 nm and 347 nm are, compared with the result of the unfiltered irradiation, rather small. Weak absorptions at 448 nm, 509 nm, 552 nm and 593 nm are referring to a small [DDQ·DDHQ] complex concentration.

The performance of CdSe nanoparticles under the same conditions is presented in **Figure SI3**. For irradiation with a strong UV component a comparably good performance as TiO_2 could be demonstrated. Due to its absorption up to 600 nm the efficiency in the visible range is significantly higher compared to TiO_2. Nevertheless a complete DDQ conversion could not be achieved, which is clearly evident from the observed absorption of unconverted DDQ and complexed DDQ.

Quite contrary to these results, nanocomposites of titanium dioxide and cadmium selenide, characterised by a very close contact between both semiconductors, behave quite differently. Our results are given in **Figure SI4**. Even though a strong absorption in the range of 400 - 650 nm indicates a quite high complex concentration, the conversion of DDQ to DDHQ is higher compared to the exclusive use of CdSe.

This strategy to combine TiO_2 with its excellent properties for water splitting with a sensitising semiconductor for good absorption in the visible range is very promising. The efficiency can be further enhanced by using CdS_6Se_1 in close contact with TiO_2. The performance of this TiO_2/Cd_6Se_1 nanocomposite is shown in **Figure 5(a)**.

Compared to TiO_2/CdSe the efficiency is significantly higher. Even though [DDQ·DDHQ] can still be observed, its concentration is lower than the concentration, which was found by using TiO_2, CdSe or TiO_2/CdSe for visible light catalysis. Taking the high concentration of formed DDHQ in account it can be clearly concluded that the conversion reaction is already far proceeded.

The progress of the conversion reaction from DDQ to DDHQ with time was studied via UV-Vis investigations of the organic phase as function of time for 100 minutes. The UV-Vis spectra are shown in **Figure 5(b)**. The observed absorption bands can be assigned to the three species DDQ (278 nm), DDHQ (248 nm, 347 nm) and [DDQ·DDHQ] (448 nm, 509 nm, 552 nm, 593 nm). The mechanism of the conversion reaction is described above. The spectra indicate a fast degradation of DDQ and a corresponding increase of [DDQ · DDHQ] as intermediate. After 20 minutes the absorption of complexed DDQ starts to decrease for about 1 hour until it is completely reduced to DDHQ. The absorption of DDHQ is found to increase simultaneously with the decrease of DDQ.

Based on these UV-Vis measurements the kinetics of the conversion reaction was assigned. The kinetic curves are presented in **Figure 6**. The temporal course is typical for a consecutive reaction. Contrary to our expectations a small decrease of the DDHQ concentration was observed at the end of the reaction. This can be explained as the result of a number of minor side reactions, which leads to loss of DDHQ, for example diffusion of DDHQ into the aqueous phase, photochemical degradation due to light scattering by the photocatalyst particles into the organic phase and/or hydrolysis of DDHQ at the interphase. Therefore we do not obtain a 100% mass balance. A small quantity of DDHQ and [DDQ · DDHQ] was observed before irradiation which seems to be formed by a side reaction. This portion was assumed to be constant during irradiation and therefore subtracted from the measured concentrations.

The results of our UV-Vis studies of the aqueous phase during the irradiation process are shown in **Figure SI5**. The electron transport via the redox couple Fe^{3+}/Fe^{2+} from the surface of the excited photocatalyst through the aqueous phase to the interface was tracked over 90 minutes. Due to an overlap of the absorption bands of Fe^{3+} and Fe^{2+} the absorption of Fe^{2+} can just be found as a shoulder in the absorption band of Fe^{3+}. For comparison the absorption maxima of Fe^{3+} and Fe^{2+} are indicated by arrows. A rapid decrease of the Fe^{3+} absorption was observed with irradiation time.

At ~90 min the Fe^{2+}/Fe^{3+} concentration ratio does not change anymore with time (concentration equilibrium) because Fe^{2+} is not consumed anymore for DDQ reduction (nearly complete conversion to DDHQ).

Figures 7 and SI6 gives a more detailed, spatially resolved view of the electron transfer at the surface of photoexcited TiO_2. The transfer process was followed by a simple dye reaction. After irradiation an electron is transferred from the surface of an excited TiO_2 particle to an iron(III) ion to form an iron(II) ion, which is immediately trapped by potassium ferrocyanide(III). The so formed complex $K[Fe^{III}Fe^{II}(CN)_6]$ is well-know as a stable and intensive blue coloured dye called Prussian blue. In our experiment it was used as a visual detector for the formation of Fe^{2+} at the surface of photoexcited TiO_2 particles. The dye reaction was followed by use of a light microscope. As single TiO_2 nanoparticles are much too small to be observable by light microscopic magnification, aggregates of TiO_2 nanoparticles with a size of 10 - 15 μm were used. The aggregates were formed by using a particularly highly concentrated TiO_2 suspension. Within the first two minutes a green and turquoise colour appears, which is due to a mixture of the yellow-orange

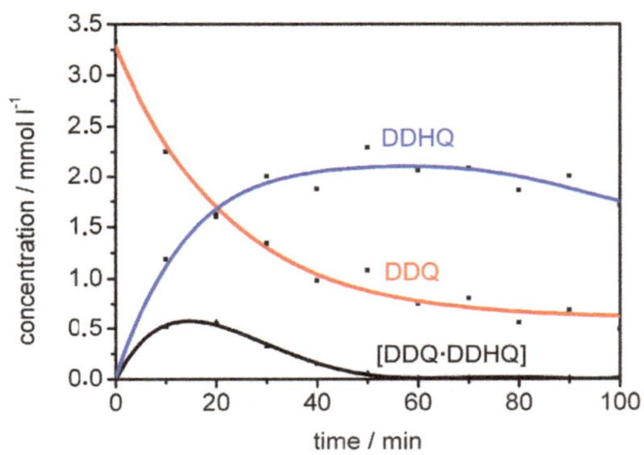

Figure 6. Kinetics of the hydrogen storage reaction. Illustrated are the time-depenent changes in concentration of DDQ (red), DDHQ (blue) and [DDQ·DDHQ] complex (black). The shown kinetic curves are based on UV-Vis measurements of the organic phase for 100 minutes; see Figure 5(b). The kinetics complies with a typical consecutive reaction.

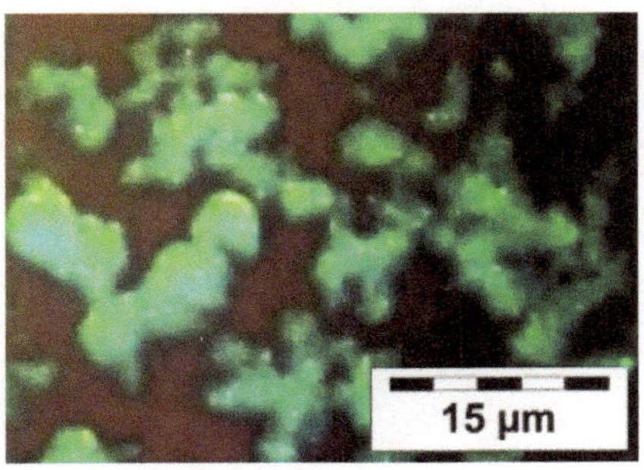

Figure 7. Enlarged detail of the light optical micrograph, see Figure SI6. Bright spots on the surface of the aggregated TiO_2 nanoparticles are clearly visible and indicate highly active areas.

colour of remaining $FeCl_3$ and the blue colour of already formed Prussian blue. After 5 minutes of irradiation the surface of the TiO_2 aggregates appears intensive blue coloured, see Supplementary Information.

It can be assumed that the surface of the TiO_2 aggregates is rough with different kind of edges. An enlargement of the light optical micrograph, which is presented in **Figure 7**, indicates high active reaction areas, which are visible as bright spots on the surface of the TiO_2 aggregates, mostly near edges. Edges therefore appear to be the areas of higher photoreduction activity.

4. Conclusion

We presented an experimental setup to investigate the performance of nanocrystalline semiconductors for photocatalytic water splitting. Chemical storage of the generated hydrogen by quinoid systems seems to be a good alternative to evolution of gaseous H_2 as it is easy to handle and can be use as recyclable fuel for fuel cells, as we show in a further publication [18].

5. Acknowledgements

The authors gratefully acknowledge Dr. Christian W. Lehmann and his group (Max-Planck Institut für Kohlenforschung, Mülheim/Ruhr) for the implementation of the TEM measurements and Dr. Daniel Ogermann (Institute of Physical Chemistry, Heinrich-Heine-University, Düsseldorf) for his assistance in preparation of the photocatalyst samples.

REFERENCES

[1] R. Pike and P. Earis, "Powering the World with Sunlight," *Energy & Enviromental Science*, Vol. 3, No. 2, 2010, p. 173. doi:10.1039/b924940k

[2] N. Kelly, T. Gibson and D. Ouwerkerk, "Generation of High-Pressure Hydrogen for Fuel Cell Electric Vehicles Using Photovoltaic-Powered Water Electrolysis," *International Journal of Hydrogen Energy*, Vol. 36, No. 24, 2011, pp. 15803-15825. doi:10.1016/j.ijhydene.2011.08.058

[3] E. Durgun, S. Ciraci, W. Zhou and T. Yildirim, "Transition-Metal-Ethylene Complexes as High-Capacity Hydrogen-Storage Media," *Physical Review Letters*, Vol. 97, No. 22, 2006, pp. 1-4. doi:10.1103/PhysRevLett.97.226102

[4] M. Grätzel, "Photoelectrochemical Cells," *Nature*, Vol. 414, No. 6861, 2001, pp. 338-344. doi:10.1038/35104607

[5] A. Hagfeldt and M. Grätzel, "Light-Induced Redox Reactions in Nanocrystalline Systems," *Chemical Reviews*, Vol. 95, No. 1, 1995, pp. 49-68. doi:10.1021/cr00033a003

[6] M. Kaneko and I. Okura, "Photocatalysis—Science and Technology," Springer, Heidelberg, 2002.

[7] D. Ogermann, T. Wilke and K. Kleinermanns, "CdS_xSe_y/TiO_2 Solar Cell Prepared with Sintered Mixture Deposition," *Open Journal of Physical Chemistry*, Vol. 2, No. 1, 2012, pp. 47-57. doi:10.4236/ojpc.2012.21007

[8] D. R. Cooper and N. M. Dimitrijevic, "Photosensitization of CdSe/ZnS QDs and Reliability of Assays for Reactive Oxygen Species Production," *Nanoscale*, Vol. 2, No. 1, 2010, pp. 114-121. doi:10.1039/b9nr00130a

[9] I. Robel, M. Kuno and P. V. Kamat, "Size-Dependent Electron Injection from Excited CdSe Quantum Dots into TiO_2 Nanoparticles," *Journal of the American Chemical Society*, Vol. 129, No. 14, 2007, pp. 4136-4137. doi:10.1021/ja070099a

[10] M. Grätzel, "Dye-Sensitized Solar Cells," *Journal of Photochemistry and Photobiology C*, Vol. 4, No. 2, 2003, pp. 145-153. doi:10.1016/S1389-5567(03)00026-1

[11] T. Meyer, D. Ogermann, A. Pankrath, K. Kleinermanns and T. J. J. Müller, "Phenothiazinyl Rhodanylidene Merocyanines for Dye-Sensitized Solar Cells," *The Journal of Organic Chemistry*, Vol. 77, No. 8, 2012, pp. 3704-3715. doi:10.1016/10.1021/jo202608w

[12] R. Menzel, D. Ogermann, S. Kupfer, D. Weib, H. Görls, K. Kleinermanns, L. González and R. Beckert, "4-Methoxy-1,3-thiazole Based Donor-Acceptor Dyes: Characterization, X-Ray Structure, DFT Calculations and Test as Sensitizers for DSSC," *Dyes and Pigments*, Vol. 94, No. 3, 2012, pp. 512-524. doi:10.1016/j.dyepig.2012.02.014

[13] T. Wilke and K. Kleinermanns, to be published, 2012.

[14] T. Ohno, K. Fujihara, K. Sarukawa, F. Tanigawa and M. Matsumura, "Splitting of Water by Combining Two Photocatalytic Reactions through a Quinone Compound Dissolved in an Oil Phase," *Zeitschrift für Physikalische Chemie*, Vol. 213, No. 2, 1999, pp. 165-174. doi:10.1524/zpch.1999.213.Part_2.165

[15] A. L. Rogach, A. Kornowski, M. Gao, A. Eychmüller and H. Weller, "Synthesis and Characterization of a Size Series of Extremely Small Thiol-Stabilized CdSe Nanocrystals," *The Journal Of Physical Chemistry B*, Vol. 103, No. 16, 1999, pp. 3065-3069. doi:10.1021/jp984833b

[16] G. A. Martínez-Castañón, M. G. Sánchez-Loredo, J. R. Martínez-Mendoza and F. Ruiz, "Synthesis of CdS Nanoparticles: A Simple Method in Aqueous Media," *Azojomo*, Vol. 1, No. 1, 2005, pp. 1-2. doi:10.2240/azojomo0170

[17] J. Yu, X. Zhao, J. Du and W. Chen, "Preparation, Microstructure and Photocatalytic Activity of the Porous TiO_2 Anatase Coating by Sol-Gel Processing", *Journal of Sol-Gel Science and Technology*, Vol. 17, No. 2, 2000, pp. 163-171. doi:10.1023/A:1008703719929

[18] T. Wilke, M. Schneider and K. Kleinermanns, "1,4-Hydroquinone is a Hydrogen Reservoir for Fuel Cells and Recyclable via Water Splitting," 2012.

Appendix

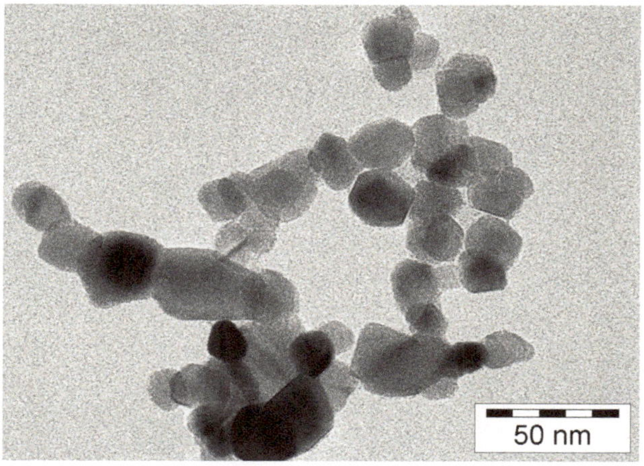

Figure SI1. TEM image of TiO_2 nanoparticles with an average size of 21 nm obtained from Evonik (Aerosil TiO_2 P-25). It consists of a mixture of 80% anastase and 20% rutile with a specific surface area of approximately 50 m^2/g.

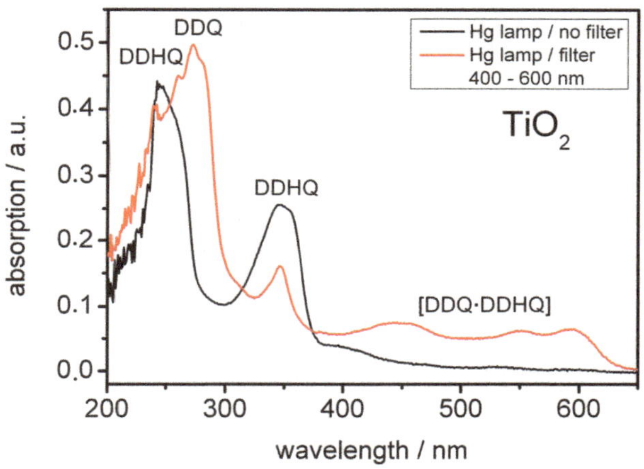

Figure SI2. Water splitting by TiO_2 nanoparticles as photocatalyst. See Figure 5(a) for detailed explanation.

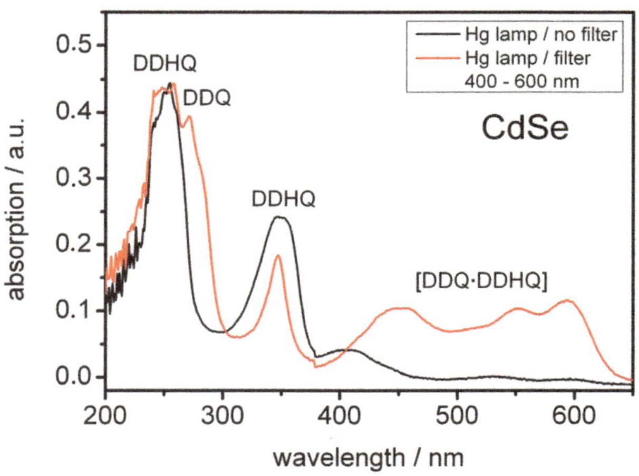

Figure SI3. Water splitting by CdSe nanoparticles as photocatalyst. See Figure 5(a) for detailed explanation.

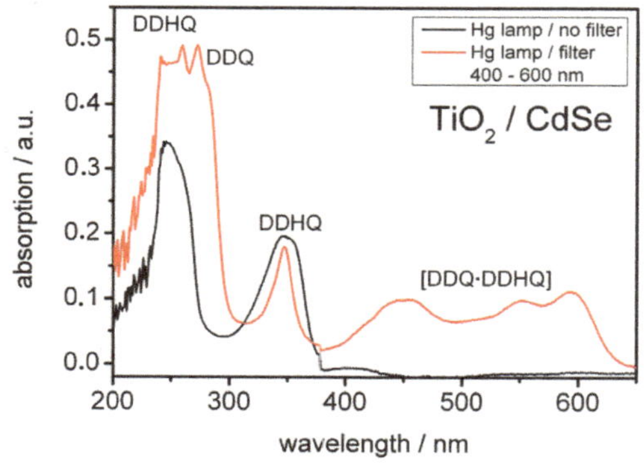

Figure SI4. Water splitting by TiO_2/CdSe nanocomposits as photocatalyst. See Figure 5(a) for detailed explanation.

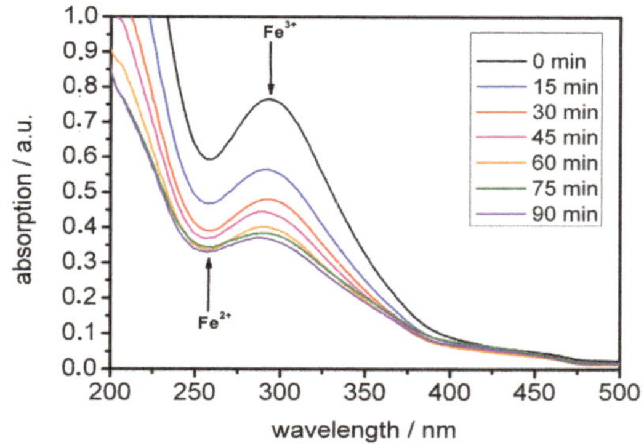

Figure SI5. Tracking of the electron transport process for TiO_2 irradiated with the unfiltered Hg lamp. Presented are UV-Vis spectra of the aqueous phase registered over 90 min. The absorption range of Fe^{3+} and Fe^{2+} are indicated by arrows. A rapid decrease of the Fe^{3+} concentration can be observed while the increasing absorption of Fe^{2+} is an evidence of the electron transport process.

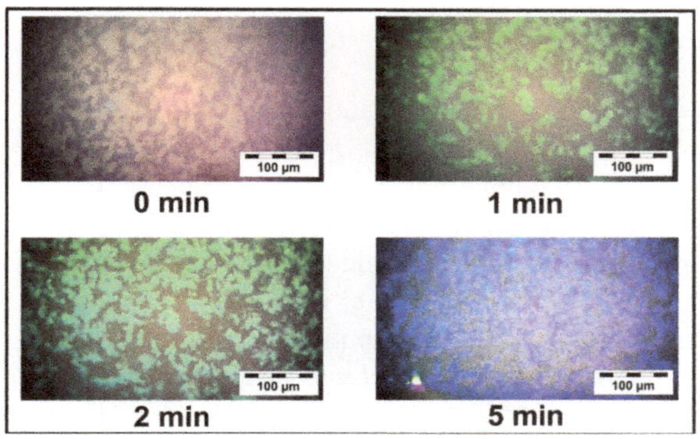

Figure SI6. Visual detection of the electron transfer process via a simple dye reaction. By addition of potassium ferrocyanide(III) to a suspension of TiO_2 nanopowder in water the Fe^{2+} formed during the irradiation is trapped by the ferrocyanide and intense coloured Prussian blue $K[Fe^{III}Fe^{II}(CN)_6]$ is formed on the surface of the TiO_2 aggregates. The green and turquoise colour visible after the first two minutes is based on the mixture of the yellow-orange colour of remaining $FeCl_3$ and the blue colour of already formed Prussian blue.

A Computational Evaluation of Bond Order and Charge Distributions in Isomeric Aminotroponiminiums and Their Benzo-Fused Derivatives

S. Natarajan Balasubrahmanyam[1,2*], Irishi Narayanan Namboothiri[3], Kuruvilla Pius[4]
[1]Department of Organic Chemistry, Indian Institute of Science, Bangalore, India
[2]Jawaharlal Nehru Centre for Advanced Scientific Research, Bangalore, India
[3]Department of Chemistry, Indian Institute of Technology, Mumbai, India
[4]Department of Chemistry, Mahatma Gandhi University, Kottayam, India
Email: *snb@orgchem.iisc.ernet.in

ABSTRACT

This paper reports the results on the nature of bond-order and net charge distributions predicted by *ab initio* Hartree-Fock procedures for 1-amino-2-iminio-, 1-amino-3-iminio- and 1-amino-4-iminiotropylium cations that incorporate, in order, the 1,7-, 1,3- and 1,5-diazapentadienium (vinamidinium) elements. There appears to be very little contribution from tropylium-type charge distribution, the positive charges residing largely in the nitrogen atoms. The partial bond fixations and charge distributions show interesting variation in the three isomers. The 1,3-isomer in which the 1,3-diazapentadienium element is preserved in the favoured zigzag conformation appears to be relatively the best stabilized. The six isomeric benzo-fused derivatives arising from the three amino-iminiotropylium cations show similar differences in patterns of behaviour. Interestingly, the isomer in which a zigzag 1,3-diazapentadienium element is conjugated with a styrene moiety receives the deepest stabilization. While showing that the element largely contributes to the relative stabilization among the systems studied, contribution from certain stereochemical destabilizing factors may not be insignificant.

Keywords: Vinamidinium Element; Amino-Iminiotropylium Cations; Benzo-Fused Amino-Iminiotropylium Derivatives; *Ab initio* Hartree-Fock Procedures; Partial Bond Fixations; Net Charge Distributions; Stereochemical Factors

1. Introduction

An evaluation of bond localizations and charge redistributions by theoretical methods when the amino (poly) vinyliminium (vinamidinium) structural element **1** (for the wide application of this element in chemistry see [1-11]) is made part of a cyclic system (e.g. aminotropon-2-, -3- & -4-iminiums **2a-c**; n = 3, 1 & 2 in **1**; the numbering indicated for **2a** applies to all structures **2**) was of interest because differences in energy content among these possibilities can arise for a number of reasons: While it is possible to conceive of a formal "diaminotropylium"-type charge distribution (**a**, **b**, **c**) for all three cases, only **2a**, which has the N *ipso*-carbons directly linked, can participate as a whole in a vinamidinium-type delocalization (**2'a**). In the cases of its analogues **2b** and **2c** passage of π-information between the nitrogens for delocalization of that type would be, apparently, confined to their "odd" sides (**2'b** and **2'c**) even while both have alternate vinylogous connectivity between the nitrogen *ipso* carbons.

- Available experimental evidence [12] has shown that vinamidinium chains **1**, including those with 1,3-disubstitution, generally prefer *anti*-periplanar (*i.e.* zigzag) conformations. While such a conformation (**2'b**) can be maintained in the vinamidinium part of **2b**, *syn*-periplanar conformations are forced on **2a** and the vinamidinium part of **2c** (**2'c**). Differences in the degree of destabilization may, therefore, be expected to depend on the delocalization that comes into effect on cyclization into forms **2a-c**.
- Angle strain would be introduced when the internal angles open out beyond 120° to accommodate a planar 7-membered cyclic structure. Relief of strain can be expected to depend on the specific partial bond fixations that come about in **2a-c**.
- Alternatively, bond localization could be such as would permit the 7-membered cycles to assume specific puckered conformations, minimizing angle strain

*Corresponding author.

while affecting the nature of the delocalization at the same time.

- The N-C-C-N array may not be uniplanar in the 1,2-system **2a** in order that *syn*-periplanar interaction be relieved, causing it to be different from the 1,3- and 1,4-systems **2b** and **2c** both of which may prefer to be entirely planar.

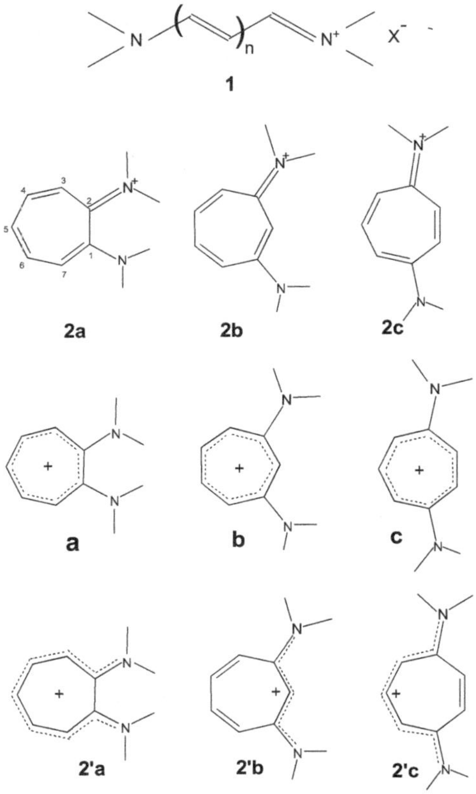

2. Structures with Aminotroponiminium Elements

Attempts to synthesize the aminotroponimines structured like **2a-c** were made at a time when interest in comparing them, in order, with α-, β- and γ-tropolones was running high [13-17]. Ionic representation **2a/2'a** would definitely have been that of the N,N'-tetramethylated primary product when one of the methods developed for the synthesis of the 1,2-free base N,N'-dimethyl derivative **3a** (R=Me) and salts thereof was suitably adapted [13]. The products were found to be highly unstable compared with **3a** which shows unusual stability and appreciable "aromatic" character: it undergoes facile nuclear electrophilic substitution and shows IR absorption bands for the presence of strong intramolecular N-H---N bridging. However, it exhibits its ring proton resonances upfield of those of α-tropolone (**4**) and its dipole moment is in the opposite direction. An initial interpretation of these pieces of evidence was that they were consistent with the establishment of a 10π-electron circuit (**3a'**) involving hybridization of non-bonding electrons of the NHR substituent. (*cf.* the case of N-*tert.* butyl-3-(*tert.* butylamino)-2-nitroprop-1-en-1-amine for which X-ray evidence [18] can be said to accord with "through-resonance" of this type.) It should be noted, however, that a canonical structure (akin to **4**) where negative charge resides mostly on oxygen and the positive charge is distributed in a "tropylium" manner is not a large contributor to resonance in tropone. X-ray diffraction measurements and analysis of dipole moments and NMR spectra have disclosed that tropones and tropolones display appreciable partial bond fixation and must be regarded essentially as non-aromatic although some aromatic character can be accorded because of a susceptibility to aromatic-type substitution. The ^{1}H and ^{13}C spectra of the N,N'-isopropyl derivative of corresponding to **3a** (R=*i*Pr), have been stated to show better consistency with the presence of rapid tautomeric equilibrium in the solution state. However, X-ray crystallographic data have disclosed that the two C-N bonds differ significantly in length (N=C: 1.314 Å; N-C: 1.342 Å), showing that in the solid state the structure is not symmetric on the whole even while the seven-membered ring and the two nitrogen atoms are nearly uniplanar.

Structures **3b** and **3c** can only be just presumed to have been those of the primary products in their attempted synthesis [16,17]. The instability of **3b** and **3c** is in accord with the absence of possibility in them of intramolecular N-H---N bridging. While a hydrochloride of the 1,3-system **3b** is hygroscopic and unstable, both the 1,3- and 1,4-systems (**3b** and **3c**) are said to give stable, crystlline picrates associated with waters of crystallization. The authors of reference [16] managed to record the ^{1}H NMR spectrum of the 1,3-system **3b** and found that the ring protons appear at fields higher than the "aromatic region", the shifts being even less than those of the protons of β-tropolone. While this may be seen as according well with an absence of possibility of establishing a H-bond mediated peripheral circuit the authors do not say explicitly whether the ^{1}H NMR spectrum is or is not consistent with a symmetric structure [16,17]. The ethylene bridged free base **5**, where N-H---N bridging is not possible, has been synthesized [19]. The cycloheptimidazolonium salt **6**, where the two nitrogen atoms that bear partial positive charges are bonded to a carbonyl, is an interesting precursor in a synthesis of N,N'-disubstituted 1-amino-2-imino-cycloheptatrienes [20]. This system and the cycloheptaquinoxalinium salts **7** [15], are among those that formally retain the 1-aminotropon-2-iminium cationic element **2a** *per se*. The electronic structure of cation **7a** has spawned an interesting debate. It was first put forward as an example of a stable anti-aromatic with a peripheral 16π-electron circuit (form **7'a**) [21]. However, an estimation of resonance and circuit resonance energies and ring currents (employing graph theoretical methods [22-25]) showed that it would be more realistic to regard it and its other hetero-analogues **7b-d** as 6π-tropylium-6π-benzo systems **7"a**. π-Electron distributions calculated by MNDO and ^{13}C NMR chemical shifts [26-28] have been found to accord with this structure.

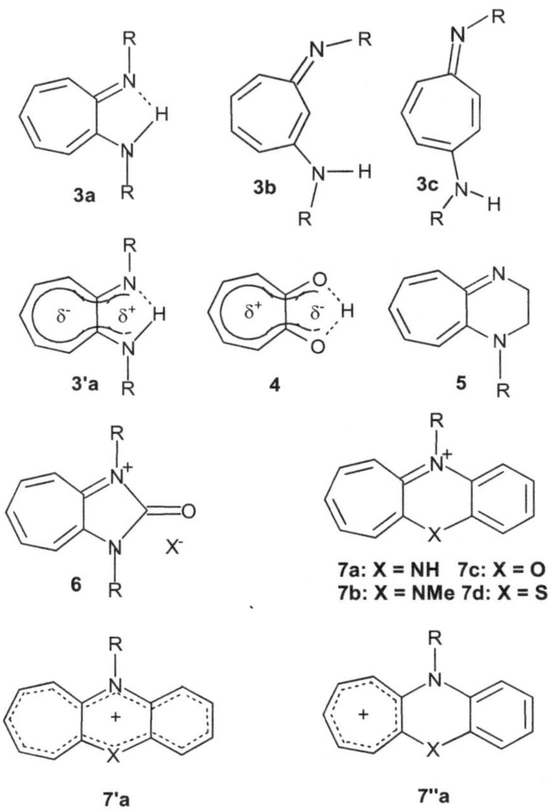

3. Computational

3.1. Aminotroponiminiums

Ab initio MO procedures were applied to systems **2** [with amino (-NH_2) and iminium (= NH_2^+) groups] in the attempt to obtain information on the possibilities raised above. Algorithms of the well-known GAUSSIAN procedures of various levels HF/3-21G, HF/6-31G* and MP2/6-31G* (including one that takes into account electron correlation, all available in our laboratories) were used for implementing the calculations. The MO procedures yield somewhat different relative energies for each case (**Table 1**) for the possible reason that they take into account electron correlation in different ways. MO procedures are generally not regarded as suitable for direct comparison of stabilities on the basis of (small) differences in energy content. We hope to show, at least by implication, that the on-the-average lesser stabilization of the 1,2-system **2a** has significance.

3.2. Benzo Fusion

A strategy adopted in the investigation of mechanisms of reactions of tropolones has been predicated on a possibility that double bond character can be localized at certain positions within the 7-cycle by benzo-fusion [29].

Benzo-fusion to systems **2a-c** yields six isomeric benzo-analogues, formally represented by particular sets of canonical structures **8a-f** shown in **Figure 1** below, that may be expected to differ in the modes of electron delocalization.

Table 1. ***Ab initio*** **relative energies (kcal/mol) of aminotroponiminiums 2a-2c.**

Species	HF/3-21G	HF/6-31G*	MP2/6-31G*
2a	0.5	11.1	2.5
2b	0.0	0.0	0.0
2c	2.6	3.2	–0.2

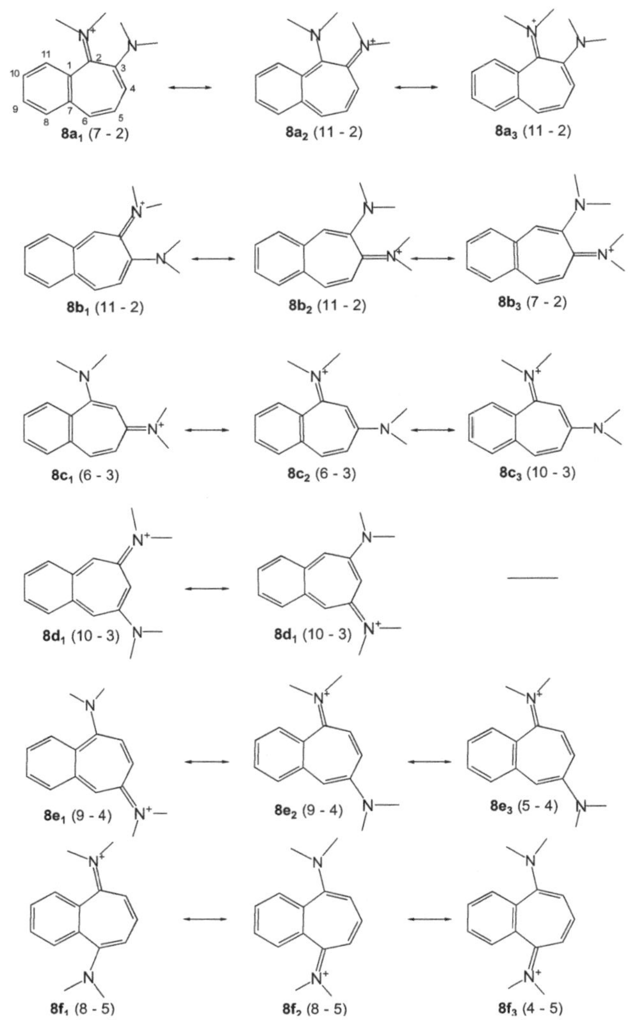

Figure 1. Structures resulting from benzo-fusion of 1, 2-, 1, 3- and 1,4-amino-iminiocycloheptatrienes. Odd numbers in parentheses denote possible "vinamidic" pathways. The numbering indicated for 8a applies also to 8b-f.

It was of interest to examine in what way the MO procedures would reflect the electron redistributions and structural features. The relative energies of the isomeric systems **8** are assembled in **Table 2**.

We discuss possible factors that may underlie why the non-symmetric 1,3-diamino system **8c**, derived from the diaminotropylium **2b**, the most stabilized among the monocyclic systems, is predicted also to be the most stabilized one. The structures were geometry optimized in the two sets of cases. Net atomic charges and HF calculated bond lengths (Å; in italics), shown in the structures in **Figures 2** and **3** below, were chosen to serve as the parameters relevant to making qualitative comparisons among each set of the systems.

Table 2. *Ab initio* relative energies (kcal/mol) of Aminobenzotroponiminiums 8a-8f.

Species	HF/3-21G	HF/6-31G*	MP2/6-31G*
8a	11.6	12.8	8.6
8b	5.0	5.6	4.6
8c	0.0	0.0	0.0
8d	13.6	12.5	10.3
8e	12.4	13.4	9.5
8f	12.5	12.6	8.0

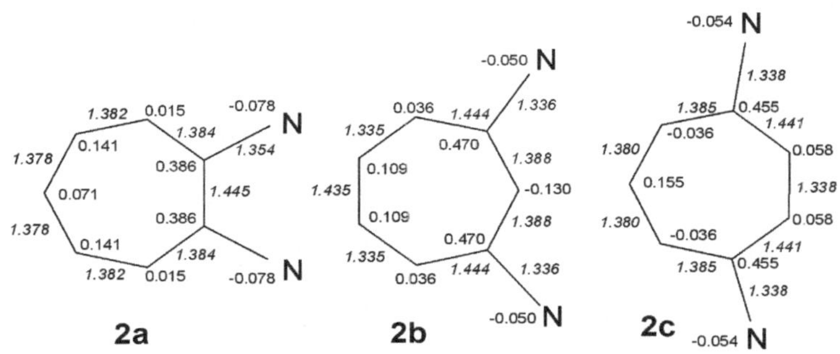

Figure 2. HF bond lengths and net charge distributions in aminotroponiminiums.

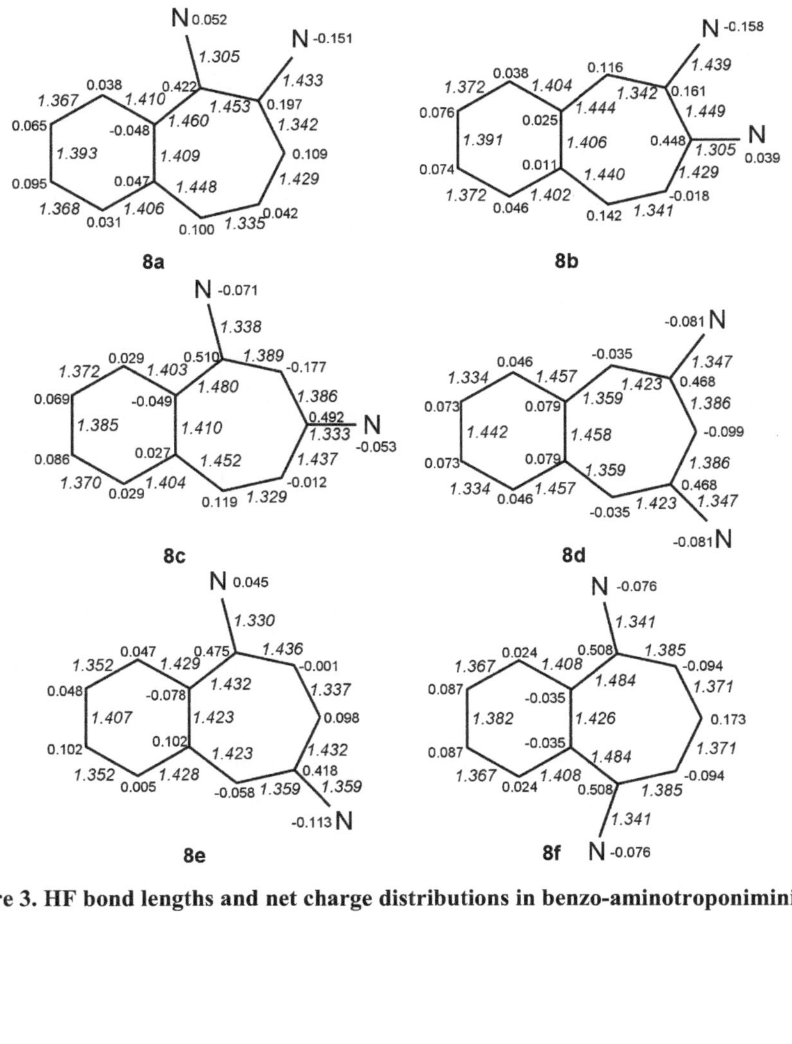

Figure 3. HF bond lengths and net charge distributions in benzo-aminotroponiminiums.

4. Results and Discussion

4.1. The Aminotroponiminiums 3

There seems to be good indication that stabilization correlates inversely with the length of the vinamidinium chain (**2'a** > **2'c** > **2'b**). Differences in the energies of formation of systems **2** apparently depend on maintenance of the zigzag conformation, as is the case in **2b**. Energy changes on the account that one and two *syn*-periplanar conformations are forced on the vinamidinium chains in **2c** and **2a** may not be strictly additive, however.

Factors of further interest are:

- Protonation of nitrogen draws electron density away from the nitrogen-*ipso* carbons to appreciable extents in all these aminotroponiminiums. High positive charges at the nitrogen-*ipso* carbons and build-up of electron density at the α carbons are seen.
- Consequent "vinyl cation" delocalization leads to charge alternation on the vinamidinium parts, the "odd" sides of aminotroponiminiums **2b** and **2c**.
- High bond orders are found at adjacent bonds in **2b** and **2c**.
- Vinamidinium character is well-preserved in **2b** and **2c**.
- Partial bond fixation is clearly seen on the "even" sides of **2b** and **2c** in the form of appreciable shortening of the C(4)-C(5) & C(6)-C(7) bonds in **2b** (1.335 Å) and the C(2)-C(3) bond in **2c** (1.338 Å).
- The pairs connecting the vinylic part with the vinamidic part [C(3)-C(4) & C(7)-C(1) in **2b** (1.444 Å) and C(1)-C(2) and C(3)-C(4) in **2c** (1.441 Å)] remain long (>1.40 Å), emphasizing suppression of delocalization on the "even" sides.

These characteristics are not strictly shared by the vinamidinium system that encompasses the whole of **2a**. While the bond connecting the nitrogen-*ipso* carbons [C(1) & C(2)] is long (1.445 Å) and there is a progresssive decrease in bond lengths along the half cycle, there is no charge alternation but only an alternate placement of low and high residual positive charges, in the form of equal partial positive charges on the symmetrically placed carbons. A similar polarization is found on the "even" sides of **2b** and **2c**: high positive charges at the nitrogen-*ipso* carbons cause a build-up of electron density at the α carbons but successive carbons bear decreasing amounts of positive charges; there is no charge alternation.

"Steric interactions" between the two vicinal nitrogens seem clearly present: atom co-ordinates show out-of-plane movements of the two nitrogen atoms (**Figure 4**). Puckering of the ring preserves C_{2v} symmetry. The C-N bonds are lengthened compared with the values in **2b** or **2c** (1.395 Å in **2a** vs 1.363 Å in **2b** and 1.365 Å in **2c**). The increase in the negative charge at the nitrogens

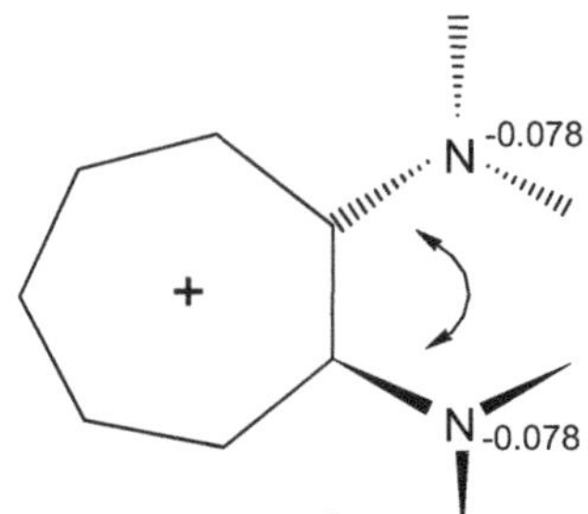

2a

Figure 4. Steric interaction between vicinal amino groups in 1,2-aminotroponiminium.

indicates a more amine-like and pyramidal than iminio-like and planar character.

While these factors can account for some of the differences in the energies of formation among systems **2** it appears that the roughly inverse correlation of stabilization with the length of the vinamidinium chain (**2b** ≅ **2c** > **2a**) must derive from what was anticipated in the "Introduction" section. While the zigzag conformation is maintained in **2b**, one and two *syn*-periplanar conformations are forced on the vinamidinium chains when they take the cyclic forms, respectively **2c** and **2a**, the proviso being energy changes on this account may not be strictly additive, as already noted.

4.2. The Benzo-Aminotroponiminiums

For carrying through qualitative comparisons conveniently features of systems **8** salient from the MO calculations have been assembled in **Table 3** overleaf.

- System **8c**, the nonsymmetric one of the two 1,3-systems, emerges as the most stabilized. All systems **8** share the feature of having the C(8)-C(9) and C(10)-C(11) bonds short and C(7)-C(1) bonds long making it appear that partial bond fixation in the benzo parts accords with a large contribution from *o*-quinonoid canonical structures, including the cases **8c** and **8f** for which such structures cannot be set down without charge separation.
- That the fixation is really not *o*-quinonoid in any of the cases except **8d** is clear from the increase in the lengths of both C(1)-C(2) and C(6)-C(7) bonds.
- While N(A) in **8a** and N(B) in **8b** bear partial positive charges, the partial charges at N(B) in **8a** and N(A) in **8b** are negative.
- The negative charges are much larger than at the negatively charged nitrogens of the other systems. The C-N bond lengths associated with the negative nitrogens are much longer than the ones associated with the positive nitrogens C(A)-N(A)—1.305 & C(B)-N(B)—1.433 Å in **8a** and C(A)-N(A)—1.439 & C(B)-N(B)—1.305 Å in **8b**, betokening amino and

Table 3. Features of systems 8 qualitatively compared.

System	Position of Benzannulation and Minimum Length of Vinamidinium	Sign of Charge on Nitrogen and C-N Bond Length	Specific C-C Bond Lengths	Residual Positive Charges on Nitrogen-*ipso* Carbons	Signs of Charges at *α*-Carbons	Vertical (Y) Component of Dipole Moment[a]
8a	Vin.[b] (7C)	+N(A) – N(B) short long	(A)-C-C-N-(B) long	Large Difference	Negative at C(*α*)-C-N(A) Positive at C(*α*)-C-N(B)	Non-zero
8b	Vin. (7C)	–N(A) + N(B) long short	(A)-C-C-N-(B) long	Large Difference	Positive at C(*α*)-C-N(A) Negative at C(*α*)-C-N(B)	Non-zero
8c[c]	Even (3C)	–N(A & B) short, small diff.	Vin. part isolated by long bonds	Small Difference	Negative; small difference	Zero[d]
8d	Even (3C)	–N(A & B) short, equal	Vin. part isolated by long bonds	Equal	Negative; equal	Zero
8e	Vin. (5C)	–N(A & B) short, small diff.	Unequal pairs on Vin. side	Small Difference	Negative; larger at C(*α*)-C-N(A)	Zero[d]
8f	Even (5C)	–N(A & B) short, equal	Long bonds on even side	Equal	Negative; equal	Zero

[a]See **Figure 5**. [b]Vin. = Vinamidinium. [c]The most stabilized system **8c** is styrene-like and vinamidic character appears best preserved. [d]No evidence of N---(benzo)H *peri*-interaction.

iminio C-N partial bond fixation in the respective cases.

- The ionic positive charge appears redistributed over the entirety of systems **8**. However, the largest residual positive charges are concentrated at the nitrogen-*ipso* carbons. While these charges are equal in the symmetric cases (**8d** & **8f**) and not very different in their non-symmetric analogues **8c** & **8e**, systems **8a** & **8b** are exceptional in that the charge differences are large, those at carbons *ipso* to the negative (amino-type) nitrogens, C(3) in **8a** and C(4) in **8b**, being much lower. The gathering of positive charges at both the nitrogen and carbon of the iminio-type C-N groups could be a destabilizing factor in these systems.
- Carbons *α* to the nitrogen-*ipso* carbons bear small negative charges in all cases. A part of this feature could be normal to the vinamidinium-type charge redistribution but there are three exceptions: carbons α to those bearing low positive charges in systems **8a** and **8b** and C(4) on the "even" side of **8e** bear positive charges.
- Neighbouring carbons on the "even" side of **8f** [C(1) & C(7)] bear negative charges.
- Low bond order (length > 1.45 Å) of the N(A)-C-C-N(B) bond, found in the precursor **2a**, is maintained in both of the benzo systems **8a** and **8b** derived from it. The lengthening arises, presumably, from an accumulation of high residual positive charges at vicinal carbons, C(2) & C(3) in **8a** and C(3) & C(4) in **8b**. It is manifest in a different form in systems **8c-8f**, leading to virtual isolation of the "odd" vinamidinium part from the "even" part—bonds C(1)-C(2) & C(4)-C(5)in **8c**, C(2)-C(3) &C(5)-C(6) in **8d**, C(2)-C(3) & C(4)-C(5) in **8e** and C(1)-C(2) & C(6)-C(7) in **8f** are long.
- The lengthening of the C(1)-C(7) bridging bond in systems **8**, coupled with partial bond fixation within the benzo ring, may or may not always accord with the bond order distribution natural to vinamidinium in those cases where benzo-substitution is on the vinamidinium side, as in **3e**.
- This factor is important if systems **8c** and **8e** or **8e** and **8f** are compared. **8c** is stabilized relative to **8d** despite the cyclizing entity being styrene-like in **8c** and *o*-quinonoid-like in **8d**.
- Absence of an iminio-type C-N group, with its in-plane hydrogens is presumably necessary for any out-of-plane movement of the constituent atoms or puckering of the 7-membered cycle. Systems other than **8a** and **8b** show zero orthogonal component of dipole moment (**Table 4**) even though the possibility of *peri*-interaction exists [30,31], one in each of **8c** and **8e** and two in **8f**.
- A component of dipole moment orthogonal to the general planes is present in both **8a** and **8b** and absent in the others (**Figure 5**). Atom co-ordinate data have indicated that the component arises both from a puckering of

Table 4. Dipole moment components (debyes).

	X	Y	Z
8a[a]	−0.4707404	−0.3161992	0.9706146
8b[a]	−0.5218589	0.0001089	1.2733497
8c	−0.6376398	0.0000000	−1.2426583
8d[b]	0.0000000	0.0000000	−1.3905561
8f	0.1818975	0.0000000	1.3918996
8e[b]	0.0000000	0.0000000	−0.1339071

[a]Non-zero component of dipole moment orthogonal to the general molecular plane (see **Figure 5**) in **8a** & **8b** could be evidence of out-of-plane placements of the nitrogens and/or puckering of the 7-membered ring. [b]C_{2v} symmetry is present.

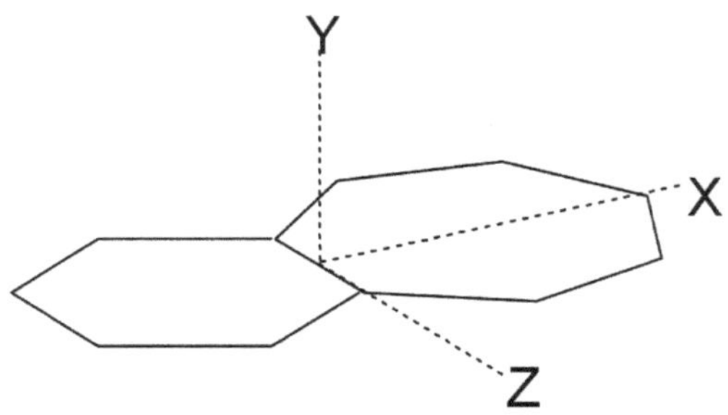

Figure 5. Co-ordinate system for the benzo-aminotropo niminiums.

the 7-membered ring and from out-of-plane placement of the nitrogens. The greater magnitude of the component in **8a** could arise from a larger out-of-plane movement of N(A) from a need to minimize not only *peri*-interaction of the *syn* iminio hydrogen on planar N(A) with the C(11) hydrogen but also the interaction of the *anti* iminio hydrogen with the amine-like N(B). However, the need to maintain uniplanarity of H-N(A)-H with C(1)-C(2)-C(3) in the system seems met partly by puckering of the ring whereby N(A) is moved above the general plane and partly by lengthening of the C(11)-C(1) bond.

- A point of similarity exists between systems **3f** and **3d** even though *o*-quinonoid bond order distribution, available to **3d**, is not very likely for **3f**. The symmetrically placed bonds between the vinamidinium and the rest of the system are lengthened; the latter are the longest bonds encountered in all of systems **8**. This symmetric distortion of the 7-membered cycle may prevent the accumulation of angle strain.

5. Conclusion

The three formalisms employed for the computations HF/3-21G, HF/6-31G* and MP2/6-31G* are in agreement in predicting system **2b** as relatively the best stabilized among the three isomeric aminotroponiminiums compared and system **8c** as the most stabilized among their six positional isomeric benzo analogues. It appears safe to state that the 3-carbon length of the vinamidinium moiety and the freedom it has to assume the preferred zigzag conformation are paramount among the stabilizing factors. The destabilization of **2a** and **8a** relative to the others of its kind likely arises from twists around bonds of low bond order. The lengthening of the C(1)-C(7) bond in all of the systems **8**, coupled with the implied partial bond fixation within the benzo ring, may or may not always accord with the bond order distribution natural to the vinamidinium chain in cases where benzo-substitution is on the vinamidinium side. This appears to be a factor when system **8c** is compared with system **8e** or **8e** with **8f**. The stabilization of **8c** relative to **8d** is of particular interest because the cyclizing entity is vinyl benzene-like (styrene-like) in the former whereas it is *o*-quinonoid in the latter. Of equal interest is a point of similarity between systems **8f** and **8d** even though bond order distribution of the *o*-quinonoid type, available to the latter, is not very likely for the former. The implied symmetric distortion of the 7-membered cycle seems to prevent either system from accommodating angle strain in a manner that prevents them from reaching the level of energy minimization possible for **8c** on this account.

REFERENCES

[1] D. Lloyd and H. McNab, "Vinamidines and Vinamidinium Salts—Examples of Stabilized Push-Pull Alkenes," *Angewandte Chemie International Edition*, Vol. 15, No. 8, 1976, pp. 459-468. doi:10.1002/anie.197604591

[2] S. S. Malhotra and M. C. Whiting, "Researches on Polyenes. VII. Preparation and Electronic Absorption Spectra of Homologous Series of Simple Cyanines, Merocyanines, and Oxonols," *Journal of the Chemical Society*, Vol. 756, 1960, pp. 3812-3822. doi:10.1039/jr9600003812

[3] C. Jutz, "C-C Bond Forming Reactions of Chloromethinium Ions," In: H.-G. Viehe and H. Bohm, Eds., *Iminium Salts in Organic Chemistry*, Wiley, New York, Vol. 9, 1976, pp. 225-343.

[4] J. Zemlicka and Z. Arnold, "Synthetic Reactions of Dimethylformamide. XI. Polyformylation of Ketones of Other Types and the Problem of the Reaction Course," *Collection of Czechoslovak Chemical Communications*, Vol. 26, 1961, pp. 2852-2864.

[5] Z. Arnold, "Synthetic Reactions of Dimethyl Formamide. XII. Formylation of Some Carboxylic Acids and Their Derivatives," *Collection of Czechoslovak Chemical Communications*, Vol. 26, 1961, pp. 3051-3057.

[6] I. M. Mallick and S. N. Balasubrahmanyam, "σ-Correlations and Steric Effects," *Organic Magnetic Resonance*, Vol. 19, No. 3, 1982, pp. 160-163. doi:10.1002/mrc.1270190311

[7] S. N. Balasubrahmanyam, B. Jeyashri and I. N. N. Namboothiri, "Selectivities in the Formation of Pyridines and

Pyrimidines by Ammonia-Induced Cyclocondensations of Vinamidiniums," *Tetrahedron*, Vol. 50, No. 27, 1994, pp. 8127-8142. doi:10.1016/S0040-4020(01)85295-9

[8] R. Gompper and C. Schneider, "New (Dialkylamino)Allenes, -Thiophenes, and -Pyrroles from Vinamidinium Salts," *Synthesis*, No. 3, 1979, pp. 213-215. doi:10.1055/s-1979-28624

[9] J. T. Gupton, D. K. Krolikowski, R. H. Yu, S. W. Riesinger and J. A. Sikorski, "Application of 2-Substituted Vinamidinium Salts to the Synthesis of 2,4-Disubstituted Pyrroles," *Journal of Organic Chemistry*, Vol. 55, No. 15, 1990, pp. 4735-4740. doi:10.1021/jo00302a047

[10] K. Hafner, K. F. Bangert and K. Orgafos, "1,3-Bis(dimethylamino)pentalene," *Angewandte Chemie International Edition*, Vol. 6, No. 5, 1967, pp. 451-452. doi:10.1002/anie.196704511

[11] J. T. Gupton, J. Coury, M. Moebius and S. Fitzwater, "The Reaction of Organometallic Reagents with Unsymmetrically Substituted Vinamidinium and Chloropropeniminium Salts," *Synthetic Communications*, Vol. 16, No. 12, 1986, pp. 1575-1586. doi:10.1080/00397918608056411

[12] M.-L. Filleux-Blanchard and D. le Botlan, "Conformational Isomerism about the Carbon-Carbon Bond in Propenylylidene Ammonium Salts," *Organic Magnetic Resonance*, Vol. 9, 1977, pp. 618-620. doi:10.1002/mrc.1270091103

[13] W. R. Brasen, H. E. Holmquist and R. E. Benson, "Novel Analogs of Tropolone," *Journal of the American Chemical Society*, Vol. 82, No. 4, 1960, pp. 995-996. doi:10.1021/ja01489a055

[14] W. R. Brasen, H. E. Holmquist and R. E. Benson, "N,N'-Disubstituted 1-Amino-7-imino-1,3,5-cyclo Heptatrienes, a Non-Classical Aromatic System," *Journal of the American Chemical Society*, Vol. 83, 1961, pp. 3125-3135. doi:10.1021/ja01475a031

[15] N. Soma, J. Nakazawa, T. Watanabe, Y. Sato and G. Sunagawa, "Seven-Membered Ring Compounds. XV. Preparations of Troponimine Derivatives," *Chemical & Pharmaceutical Bulletin*, Vol. 13, 1965, pp. 457-464. doi:10.1248/cpb.13.457

[16] M. Cavazza and F. Pietra, "Synthesis of 3-Aminocycloheptatrienylidenamines," *Journal of the Chemical Society, Chemical Communications*, Vol. 11, 1976, 412-413. doi:10.1039/c39760000412

[17] M. Cavazza and F. Pietra, "A Synthetic Approach to 4-Amino-1-imino-2,4,6-cycloheptatrienes," *Synthesis*, No. 3, 1977, 204-205. doi:10.1055/s-1977-24326

[18] D. L. Ostercamp, L. M. Preston and K. D. Onan, "The Synthesis and Crystal Structure of N-Tert-butyl-3-(tert-butylimino)-2-nitropropen-1-amine," *Zeitschrift für Naturforschung B*, Vol. 48, 1993, pp. 1138-1142.

[19] T. Nozoe, S. Ishikawa and K. Shindo, "Synthesis and Reactions of 2,3-Dihydro-cyclohepta[b][1,4] Thiazines and 2,3-Dihydro-1H-cyclohepta [b] Pyrazines," *Heterocycles*, Vol. 28, 1989, pp. 733-744.

[20] H. Nakao, "Seven-Membered Ring Compounds. XVIII. Reactions of Quaternary Ammonium Salts of Cycloheptimidazole Derivatives," *Chemical & Pharmaceutical Bulletin*, Vol. 13, No. 7, 1965, pp. 810-818. doi:10.1248/cpb.13.810

[21] T. Nozoe, "Cyclohepta[b][1,4]benzoxazine and Related Compounds—Some Novel Aspects in Troponoid Chemistry," *Pure and Applied Chemistry*, Vol. 54, 1982, pp. 975-986. doi:10.1351/pac198254050975

[22] J. Aihara, "A New Definition of Dewar-Type Resonance Energies," *Journal of the American Chemical Society*, Vol. 98, No. 7, 1976, pp. 2750-2758. doi:10.1021/ja00426a013

[23] J. Aihara, "Resonance Energies of Nonbenzenoid Hydrocarbons," *Bulletin of the Chemical Society of Japan*, Vol. 51, No. 12, 1978, pp. 3540-3543. doi:10.1246/bcsj.51.3540

[24] J. Aihara, "Graph-Theoretical Formulation of London Diamagnetism," *Journal of the American Chemical Society*, Vol. 101, No. 20, 1979, pp. 5913-5917. doi:10.1021/ja00514a009

[25] J. Aihara and T. Horikawa, "Graph-Theoretical Formula for Ring Currents Induced in a Polycyclic Conjugated System," *Bulletin of the Chemical Society of Japan*, Vol. 56, No. 6, 1983, pp. 1853-1854. doi:10.1246/bcsj.56.1853

[26] T. Kurihara, K. Shindo, S. Ishikawa and T. Nozoe, "Benzo[b]cyclohepta[e][1,4]thiazines and Their Diazine Analogues. III. π-Electronic Structures of 5H,11H+- Cyclohepta[b]quinoxalinium Ions and Their O- and S-Analogues," *Bulletin of the Chemical Society of Japan*, Vol. 64, No. 9, 1991, pp. 2668-2676. doi:10.1246/bcsj.64.2668

[27] T. Nozoe, T. Kitahara, K. Takase and M. Sasaki, "Syntheses of Benzo[b]tropazines," *Proceedings of the Japan Academy*, Vol. 32, No. 5, 1956, pp. 349-352.

[28] J. Aihara, "A New Definition of Dewar-Type Resonance Energies," *Journal of the American Chemical Society*, Vol. 98, No. 10, 1976, pp. 2750-2758. doi:10.1021/ja00426a013

[29] F. Pietra, "Revival of Troponoid Chemistry," *Accounts of Chemical Research*, Vol. 12, 1979, pp. 132-138. doi:10.1021/ar50136a004

[30] I. N. N. Naboothiri and S. N. Balasubrahmanyam, "Transformations in Bromo- and Alkoxybenzotropones," *Journal of the Indian Institute of Science*, Vol. 74, 1994, pp. 473-486.

[31] I. N. N. Namboothiri and S. N. Balasubrahmanyam, "On the Temperature-Dependence of Regioselectivity in the Amination of Bromobenzotropones," *Indian Journal of Chemistry—Section B*, Vol. 32, 1994, pp. 1029-1034.

17

A Review of the Aqueous Aerosol Surface Chemistry in the Atmospheric Context

Kalliat T. Valsaraj
Department of Chemical Engineering, Louisiana State University, Baton Rouge, USA
Email: Valsaraj@lsu.edu

ABSTRACT

In this review the surface chemistry and properties of aqueous atmospheric aerosols are explored. Water plays a major role in scavenging pollutants. Reactions occur on thin water films in atmospheric aerosols. The study of the aerosol water surface is important to properly account for chemical transformations in the troposphere. The thermodynamics of adsorption of organic molecules and oxidant species on the aqueous surface and, the techniques employed to quantify the adsorption isotherms are summarized. Experimental techniques for elucidating the reactions on the water surface are described. Field and laboratory data for oxidation reactions of compounds at the air-water interface are summarized. The Langmuir-Hinshelwood reaction mechanism is useful in quantifying the reaction rate on the aqueous aerosol surface. A hypothesis for the large heterogeneous reaction rate on the water surface over the homogeneous bulk aqueous phase reaction is presented.

Keywords: Aerosols; Water Film; Adsorption; Reaction

1. Introduction

Atmospheric aerosols are colloidal systems with one phase (solid or liquid) dispersed in another phase (gas). Solid aerosols are generally organic or inorganic particles dispersed in air ranging from sub-micron to a few microns in diameter; they may either be dry or wet. Liquid aerosols comprise of rain, fog, mist, cloud droplets, snow and ice. **Table 1** provides the essential characteristic dimensions and properties of atmospheric aerosols. It is noteworthy that they have lifetimes ranging from a few minutes to days. The presence of surfaces such as cloud, fog and fine particulates can influence atmospheric processes with characteristic times of the order of hours.

Table 1. Typical properties of atmospheric aerosols [1,2].

Typeof aerosol	Size range/ μm	Surface area/ $m^2 \cdot m^{-3}$	Average life time
Aerosol particles	10^{-2} - 10	$\sim 1 \times 10^{-3}$	4 - 7 d
Fog droplets	1 - 10	$\sim 8 \times 10^{-4}$	3 - 6 h
Cloud droplets	10 - 10^2	$\sim 2 \times 10^{-1}$	7 - 10 h
Rain	10^2 - 10^3	$\sim 5 \times 10^{-4}$	3 - 15 min
Snow	10^3 - 10^5	~0.3	15 - 50 min

Aerosols play a large role in tropospheric chemistry and climate. Aerosols take part in numerous heterogeneous reactions with gases (organic and inorganic) and atmospheric oxidants (ozone, hydroxyl radical, singlet oxygen, nitrate radical, chlorine). Aerosols absorb and reflect sunlight and contribute to direct and indirect effects on climate forcing. The 2007 IPCC assessment [3] on global climate change stated that the largest uncertainty is the effect of aerosols on climate forcing. Over the last few decades an impressive array of data has been accumulated on the role of atmospheric aerosols.

Atmospheric aerosols consist of a core of inorganic or organic nucleus (e.g., soluble salts, insoluble black carbon, soil or soot) that forms the condensation site for water as well as organic and inorganic molecules. Water effects the aerosol properties in various ways: 1) it determines the nature of aerosol (solid, liquid, or glassy/amorphous), 2) the hygroscopic growth and size of the aerosol, 3) uptake of gases and oxidants, and 4) multiphase reactions in aerosol droplets. The water content of the aerosol is determined by environmental factors such as the type of condensation nucleus (CN), the relative humidity, temperature, and oxidation state of molecules on the surface of the aerosol. The water either plays the role of a competitive adsorbate, or acts as a substrate to provide the surface area for reaction. If the water content is substantially less than that required for a monomolecular layer, then the individual water molecules compete for adsorption sites on the surfaces of CN. If, on the other hand, the water forms a thin layer (1 nm to 50 μm) of high specific surface area it provides the site for adsorption and reaction on the aerosol. For water films of ~1 - 2 nm on silica, the surface-induced modifications to the adsorbed film propagate to ~1 - 4 nm from the solid-water interface [4]. When the water layer thickness exceeds few monolayers (>100 μm), the bulk water properties predominate.

Thin layers of water are important from the atmospheric chemistry perspective. The thin water layer in aerosols may have properties different from those of bulk water. Questions that are relevant in this regard include the following: 1) Is there any difference between "bound" and "unbound" water? 2) Is the H-bonding characteristic of water near surfaces different from bulk water? 3) Are the activity coefficients of solutes in thin layers of water (γ_i) different from their infinite dilute activity coefficients ($\gamma_{i\infty}$)?

Figure 1 is the schematic of an aqueous aerosol. The aqueous phase consists of dissolved inorganic ions, small particles and dissolved organic species, some of which are surface active and of large molecular weight that form a distinct phase on the water surface. The adsorption and reaction of trace gases and oxidants in the atmospheric aerosol will be influenced by the nature and composition of the aqueous phase. There have been several reports that describe the relevance of the aqueous surface in atmospheric droplets. For example, one of the earliest reports involved the conversion of sulfur dioxide in fog and cloud droplets to sulfates in the context of the so-called "acid rain" issue [2]. More recently, aqueous aerosol surface reaction was invoked for the depletion of ozone near sea surface aerosols [5]. There is also evidence for the real-time production of secondary organic aerosols in fog droplets [6]. We have provided evidence for the formation of both oxygenated PAHs and other compounds in fog sampled in the United States [7]. A variety of questions remain on the mechanisms of transport, adsorption and reaction on aqueous aerosol droplets [8].

In the following review I will summarize the various physico-chemical aspects pertaining to the adsorption and reaction of organic molecules on aqueous aerosol droplets and water thin films.

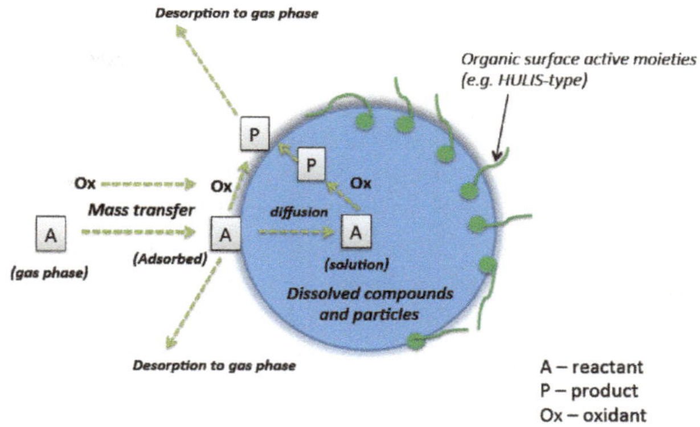

Figure 1. A pictorial representation of an atmospheric aqueous aerosol and its interactions with the gas-phase molecule and oxidant species.

2. The Air-Water Interface

The surface of water is unique in its energy characteristics. The surface tension of water (72 $mN \cdot m^{-1}$) is one of the highest for any liquid. Hence, the air-water interface is considered "the most hydrophobic, non-polar surface known" [9]. The "hyper hydrophobicity of the air-water interface" attracts hydrophobic moieties to the surface [10]. The hydrophobic interactions are the basis for a number of well-known phenomena such as [11]: 1) the low aqueous solubility of hydrophobic molecules in water, 2) the ability of proteins to fold in water, 3) the formation of micelles by surfactants in water, 4) the inability of water to spread on hydrophobic surfaces, 5) the ability to separate hydrophobic particles and compounds by attachment to air bubbles (foam flotation) and, 6) dimerization of hydrophobic chains in water.

The separation of hydrophobic organic compounds (HOC) from water by adsorption on air bubbles, viz. solvent sublation, was developed in our laboratory for wastewater treatment [12]. It was observed that the inverse process, viz., water droplets in air (as in a spray column), was able to remove high concentrations of HOC by adsorption [13]. The latter process was also invoked to explain the high concentrations of HOC observed in fog and dew waters collected from urban areas [14]. Further there has been several observations regarding the unique reactions of gas-phase HOC in fog and cloud droplets leading to the formation of products that are precursors to secondary organic aerosols (SOA) in the aqueous phase [15]. Field data on real-time SOA formation in a London fog event was recently reported [6].

Two aspects concerning trace gas interactions with atmospheric water surfaces need study [16]. Firstly, we need data on the thermodynamics of adsorption at the air-water interface. Secondly, we need information on the reactivity of adsorbed molecules towards gas-phase and liquid-phase oxidant species. In the following discussion, we focus on both aspects.

2.1. Thermodynamics of Adsorption at the Air-Water Interface

Utilizing the thermodynamic criteria of equal fugacities between the air and the air-water interface, we can derive the following equation for the free energy of partitioning for an HOC [17]

$$\Delta_{ads}G^0 = -RT\ \ell n\left(K_{\sigma a}/\delta_0\right) \quad (1)$$

where $K_{\sigma a}$ is the Kemball-Rideal standard state surface thickness (=6 × 10^{-10} m). Experimental values of $K_{\sigma a}$ can be obtained from a variety of techniques. These include direct and indirect methods and are summarized in **Table 2**. The first two methods in **Table 2** are direct methods that are suitable for compounds that have high vapor pressure. The last two indirect methods are suitable for compounds with low vapor pressure. They are especially useful for obtaining thermodynamic adsorption parameters (free energy, enthalpy and entropy) on thin water films of a few microns in thickness.

The experimental methods described above are complimentary to correlations with compound properties such as sub-cooled liquid vapor pressure, octanol-air partition constant, and molecular properties (e.g., molecular connectivity index, molecular surface area, molar volume) [23,24]. Further, the correlations were extended to include polar and non-polar compounds via the use of H-bond acceptor and donor indices [25,26]. Ab-initio and quantum chemistry based calculations of the partition constants using a Universal Surface Area and Solvation Model [27] and the use of the Conductor-Like Screening Model (COSMOTherm) [28] were also reported to provide good comparisons to experimental data.

Molecular dynamics (MD) and Monte Carlo (MC) simulations have been performed for the transfer of HOCs from air to water to determine the free energy of adsorption and solvation of molecules [29-33]. MD simulations provide good estimates of the basic thermodynamic properties, which can be compared to the experimental values. These calculations have been made not only for the HOCs but also for various gas-phase oxidant species (OH, O_3, H_2O_2) at the interface [34]. MD simulations for the air-to-water transfer of a series of PAHs showed that a free energy minimum exists at the air-water interface which is conducive to surface adsorption. The free energy difference was larger between the bulk air and the interface than between the bulk air and bulk water phases, indicating a large energy for complete solvation of the molecule. The free energy values obtained from the calculations agree with the experimental values obtained (**Table 3**). The presence of surfactants at the interface increased the free energy minimum [33]. The MD and MC simulations also allowed an understanding of the differences in surface orientation of the HOC in the presence of the surfactant [33]. It is clear that the adsorption at the air-water interface is a favorable process, but that the complete solvation (solution) of the molecule involves considerable energy penalty for HOCs. However, for the small oxidant species the two free energy values are similar, although there is a slight preference for the interfacial adsorption.

Table 2. Methods suitable for determining the partition constants of HOC from the gas phase to the aqueous surface.

Method	Description	Reference
Surface tension-static	Wilhelmy plate method	[18]
Surface tension-dynamic	Axisymmetric drop shape analysis method	[19]
Gas chromatography	Inverse gas chromatography method	[20,21]
Flow reactor	Adsorption on a thin water film	[22]

Table 3.Experimental and MD simulated values of free energy of adsorption and solution for HOC and Oxidant species [29,35,36].

Compound	$\Delta_{ads}G^0$/kJ·mol^{-1}		$\Delta_{soln}G^0$/kJ·mol^{-1}	
	MD simulation	Experimental	MD simulation	Experimental
Benzene	−15	−16.3 ± 0.4	−3	−4 to −3
Naphthalene	−24	−26.5 ± 0.1	−11	−11 to −7
Phenanthrene	−32	−44.2	−15	−17 to −11
Anthracene	−33		−15	−17 to −10
Ozone	−5		+3	+2 to +3
$^{\bullet}OH$	−24		−18	−17 to −16
$HO_2^{\bullet}$	−31		−28	−31 to −25
H_2O_2	−44		−40	−37 to −36

2.2. Reactions of HOCs at the Air-Water Interface

As described above, both gas-phase oxidants and gaseous HOCs show a surface free energy minimum. We conclude that surface interactions between them should be feasible. The question is whether "surface" water is different from "bulk" water in supporting surface interactions. For example, theoretical work show that there is an increase in energy of surface water molecules of approximately 10 - 15 $kcal \cdot mol^{-1}$ over bulk water molecules [37] and, hence surface contains, on average, far more reactive sites than the bulk.

A number of techniques have been employed to study the reactions of gaseous HOC molecules with typical gas-phase oxidants at the air-water interface, which are summarized in **Table 4**. The air-water interface is difficult to probe compared to a gas-solid interface due to features peculiar to the former. For example, techniques suitable to study reactions of molecules on solids under vacuum are not useful since water has a high partial pressure. Unlike a solid surface, a water surface is constantly renewed and highly fluxional. The mobility of molecules adsorbed on water surfaces is larger than on solid surfaces with sub-nanosecond time scales for surface to bulk exchange of molecules. The delineation of a true surface phase for water is problematic since composition and density are not necessarily sharp and extend over a significant depth. Thus, the probing of water surfaces had to wait till suitable non-linear techniques were available to selectively amplify the molecular signals. One such technique is the sum frequency generation (SFG) method. This technique has been utilized to study the orientation of molecules at the water surface as well as the potential for configurational changes upon adsorption [46,47]. SFG has also been used to explore the dynamics of adsorption of molecules on the water surface [48].

A useful method to simultaneously study the adsorption and reaction involves the tracking of a molecular signal from the adsorbed species. For example, the adsorption and reaction of PAHs can be explored by following their characteristic fluorescence signals. Glancing angle laser induced fluorescence (LIF) of PAHs adsorbed at the air-water interface and its changes as a result of exposure and reaction with gas-phase oxidants (e.g., ozone) can be followed to obtain the progress of the reaction at the interface [40]. LIF can be also used to directly sample other reactions at the surface such as singlet oxygen with PAH molecules [48]. A recent innovation is the use of droplet electrospray mass spectroscopy [40], which involves the introduction of droplets of water containing the probe molecule into an electron spray mass spectrometer. The droplets are then exposed for a very short burst of an oxidant and the surface concentration directly ascertained. The principle of exposing water droplets in a falling droplet reactor to gas phase PAH and ozone and liquid analysis by HPLC/MS was used in our laboratory to study the surface reactions as a function of droplet size [43]. The reaction of aqueous surface adsorbed halide species to gas phase chlorine was reported recently using glancing angle Raman spectroscopy [44]. The pseudo-first order reaction rate constants for various molecules with gas phase oxidants have been explored using the above methods. Two examples are shown in **Figures 2** and **3**. An interesting observation is the saturation-type rate behavior. This can be justified by invoking a Langmuir-Hinshelwood (L-H) type surface reaction mechanism. **Figure 3** shows the schematic of the same. The mechanism involves the reaction of adsorbed organic molecules with the oxidant on the surface of the water. The overall reaction rate is given by

Table 4. A sampling of various techniques used for studying the adsorption and reaction at the water surface.

Technique used	Reference
Vibrational sum frequency generation (VSFG)	[38]
Glancing angle laser induced fluorescence	[39]
Droplet Electrospray ionization mass spectroscopy (ESI/MS)	[40]
Reflection absorption infrared spectroscopy (RAIRS)	[41]
Glancing angle Raman spectroscopy	[42]
Falling droplet reactor with HPLC/MS	[43]
Heterodyne-detected electronic sum frequency generation (HD-ESFG)	[44]
Reactive scattering technique	[45]

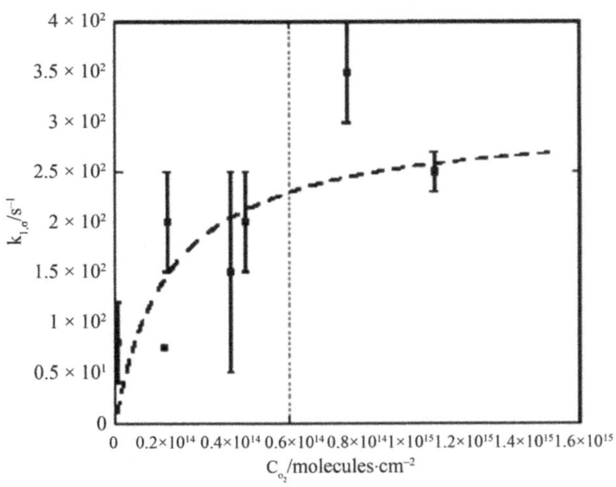

Figure 2. The pseudo first order reaction rate of naphthalene with gaseous ozone on aqueous droplets of 91 μm diameter in a falling droplet reactor. Data obtained from Raja *et al* [43]. The data fit gave k_{max} = 303 s^{-1} and $C_{Ox,1/2}$ = 1.92 × 10^{14} molecules.cm^{-3}. Note the asymptotic approach to a constant rate constant at high ozone concentration in the gas phase.

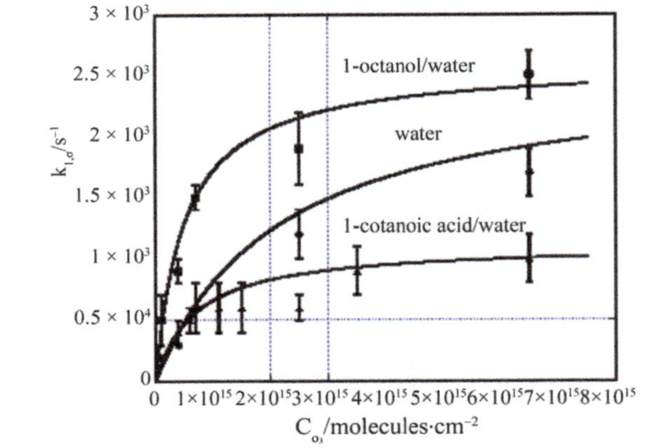

Figure 3. Pseudo first order reaction rate constants for gaseous anthracene reaction with gaseous ozone on different types of aqueous surfaces, viz., 2.5 mM aqueous solution of octanol, pure uncoated water, and 3.8 mM aqueous solution of 1-octanoic acid. Data obtained from Mmereki et al [39]. Note that in each case the rate constant reaches an asymptotic value at high concentrations of gaseous ozone. Values of kmax and $C_{ox,1/2}$ are given in Table 5.

$$r_i^s = k_{2,\sigma} \cdot \left(\Gamma_{ox}^{\max} C_{ox}(g) / \{ C_{ox,1/2} + C_{ox}(g) \} \right) \cdot \Gamma_i^{\sigma} \quad (2)$$

Thus, a pseudo first-order reaction rate constant can be defined

$$k_{1,\sigma} = k_{\max} \cdot \left(C_{ox}(g) / \{ C_{ox,1/2} + C_{ox}(g) \} \right). \quad (3)$$

The above equation explains the saturation-type behavior for the reaction rate constant with oxidant concentration as shown in **Figures 2** and **3**. The values of the two constants are given below in **Table 5** for different water surfaces. Note that $C_{ox,1/2}(g)$ relates to the binding energy of ozone with the water surface and $k_{\max}$ depends on the different bimolecular reaction rate constants on specific surfaces. Thus, these values change depending on the nature of the water surface [49]. A problem with the above treatment is that molecular O_3 has only a small residence time (~10^{-9} s) on surfaces. The molecular mechanism leading to the L-H surface reaction of ozone with PAH can be reconciled only by invoking the presence of long-lived (>100 s) reactive oxygen intermediates (ROI) on the surface [50].

The overall pseudo-first order reaction rate constant should depend on the surface thickness since there are two parallel channels in which reactions occur in a water film once the organic molecules adsorb (and solvate) on the surface. This is shown in **Figure 4**. The overall reaction rate is given by the following equation [22]

$$k_1 = k_{\text{bulk}} + \left(k_{1,\sigma} / \delta \right) \quad (4)$$

where k_{bulk} and $k_{1,\sigma}$ are respectively the homogenous bulk aqueous phase and heterogeneous surface reaction rate constants. It is evident that as the film thickness becomes smaller, the heterogeneous reaction rate will make a larger contribution towards the overall reaction rate. This fact was demonstrated in our laboratory with regard to the UV photooxidation of two gaseous PAHs (naphthalene and phenanthrene). In both cases the major product concentrations were followed in the aqueous films and the first order rate constants for the degradation of PAH into respective products were obtained [36,51]. For example, the reaction rate constants for phenanthrene conversion to major products are given in **Figure 5**. It is clear that for all the three major products of oxidation the reaction rate increases as the film thickness decreases. Similar behavior was also observed for naphthalene degradation.

Table 5. Pseudo first order rate constant parameters for the reaction of anthracene with gas phase ozone on various water surfaces.

Surface	$k_{\max}/s^{-1}$	$C_{ox,1/2}$/molecule·m^{-3}
Water	$(2.55 \pm 0.17) \times 10^{-3}$	$(21.4 \pm 0.4) \times 10^{-20}$
1-octanol on water	$(2.54 \pm 0.14) \times 10^{-3}$	$(5.0 \pm 0.9) \times 10^{-20}$
Octanoic acid on water	$(1.11 \pm 0.14) \times 10^{-3}$	$(6.8 \pm 2.9) \times 10^{-20}$
Hexanoic acid on water	$(0.48 \pm 0.07) \times 10^{-3}$	$(11.8 \pm 3.6) \times 10^{-20}$

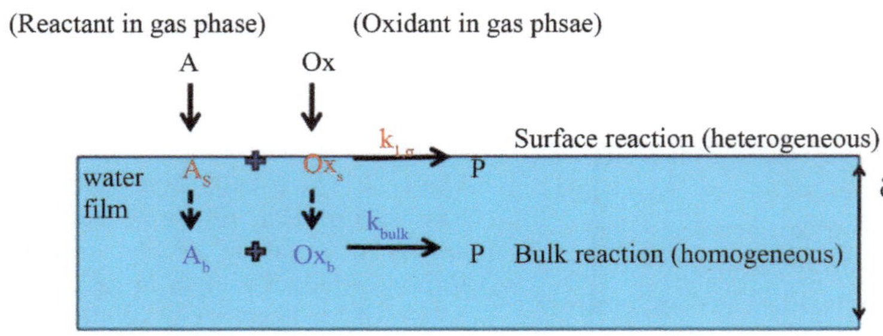

Figure 4. Schematic of the two reaction channels (heterogeneous and homogeneous) for gaseous organic species in water films.

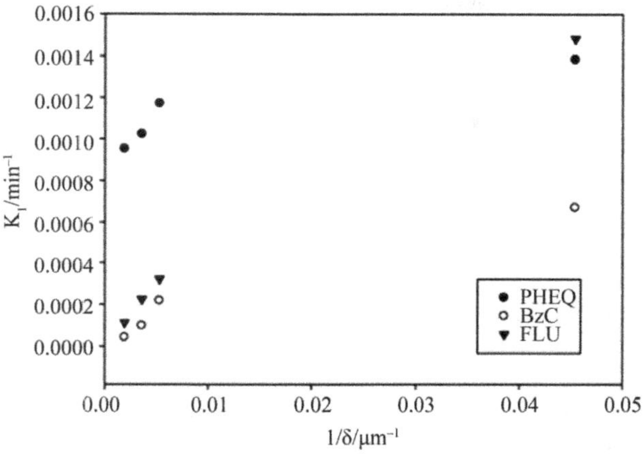

Figure 5. The photooxidation rates of gaseous phenanthrene on water films as a function of film thickness. The formation rate constants for three different compounds (PHEQ-Phenanthraquinone, BzC-Benzocoumarin, FLU-Fluoranthone) from phenanthrene are given. Data obtained from Chen *et al.* [36].

Others have also confirmed the surface reactions on water [52]. The reaction of $Cl_2^{-\bullet}$ with ethanol in water films was investigated using UV diffuse reflectance laser flash photolysis technique. It was demonstrated that the surface reaction rate is different from the bulk reaction rate. The surface reaction rate varied as the square root of the ethanol concentration and reached an asymptote at high concentrations. However, the bulk reaction rate was much slower and varied linearly with ethanol concentration.

Experimental field observations also lend support to the importance of surface reactions. For example, in analyzing the loss of ozone in coastal regions, it was observed that the surface reaction of ozone on the air-water interface of sea salt particles had to be invoked in models to correctly explain the observations [5]. This observation was confirmed with laboratory work and from molecular dynamics and *ab-initio* calculations. Fog and cloud droplets can process organic trace gases and particulates via aqueous phase photochemistry. Field observations have shown that fog droplets provide a micro-reactor environment for the formation of secondary organic aerosols (SOA) by photochemical transformations of organic compounds [6]. Real-time formation of secondary aerosol was detected using on-line aerosol mass spectrometry immediately upon the detection of fog. It was observed that the post-fog atmospheric environment contained particles and gases that were produced by evaporation of water from the fog droplet.

In the organic chemistry literature there is the term "on-water" heterogeneous chemistry that refers to orders of magnitude increase in the rates of catalysis of organic reactions at oil-water interfaces compared to homogeneous reactions in organic solvents [53]. This has been attributed to the breaking of an existing H-bonding network in homogeneous aqueous solution in order to permit catalysis, but not for the "on-water" reactions [54].

In order to explain the preference of surface reactions over bulk reactions for gaseous species on aerosols in the atmosphere, a hypothesis has been proposed [55,56]. For bulk aqueous phase reactions to occur the molecular species that are fully solvated (in a H-bonded water network) have to break their solvation shells. For surface species that are only partly solvated [57] and such an energy penalty does not exist. Diffusion also limits the encounter of the fully solvated species in the case of the bulk solution. There is evidence that the adsorbed molecules at the interface display faster orientational dynamics than in the bulk [58]. Moreover, even when reactions occur within the bulk solution, the diffusion of molecules out of the solvent cage is in competition with recombination reactions. At the surface recombination of reactants is inhibited as shown in photolysis reactions at the surface compared to the bulk [59].

3. Summary

It is evident from principal observations that "surface" water is different from "bulk" water in reactivity. Trace gases (HOC) and atmospheric oxidant species adsorb and react on aqueous surfaces. These processes are of significance for aqueous atmospheric aerosols. Experimental work and molecular simulations support the above conclusion. It is evident that heterogeneous reactions deserve special attention in tropospheric chemistry and is important for thin films of water existing on aqueous aerosols.

4. Acknowledgements

The US National Science Foundation is gratefully acknowledged for support of this work through research grants (AGS 0907261 and AGS 1106569). This material was also presented in part as a Plenary Lecture at the "Recent Advances in Chemical Engineering and Technology" meeting organized by the Indian Institute of Chemical Engineers, Kochi Regional Centre, in Kochi, Kerala, India on 10-12 March 2011.

REFERENCES

[1] J. H. Seinfeld and S. N. Pandis, "Atmospheric Chemistry and Physics," John Wiley and Sons, Inc., New York, 2006.

[2] K. T. Valsaraj, "Elements of Environmental Engineering," 3rd Edition, Taylor and Francis Publishers, New York, 2009.

[3] S. Solomon, S. Qin, D. Chen, Z. Marquis, K. B. Averyt, M. Tignor and H. L. Miller (Eds.), "Climate Change 2007: The Physical Basis, Contributions of Working Group I to

the Fourth Assessment Report of the Intergovernmental Panel on Climate Change," IPCC, Cambridge University Press, Cambridge, 2007.

[4] S. R.-V. Castrillon, N. Giovambattista, I. A. Aksay and P. Debenedetti, "Structure and Energetics of Thin Water Films," *Journal of Physical Chemistry C*, Vol. 115, 2011, pp. 4624-4635. doi:10.1021/jp1083967

[5] E. M. Knipping, M. J. Lakin, K. L. Foster, P. Jungwirth, D. J. Tobias, R. B. Gerber, D. Dabdub and B. J. Finlayson-Pitts, "Experiments and Simulations of Ion-Enhanced Interfacial Chemistry on Aqueous NaCl Aerosols," *Science*, Vol. 288, No. 5464, 2000, pp. 301-306. doi:10.1126/science.288.5464.301

[6] M. Dall'Osto, R. M. Harrison, H. Coe and P. Williams, "Real-Time Secondary Aerosol Formation during a Fog Event in London," *Atmospheric Chemistry and Physics*, Vol. 9, No. 7, 2009, pp. 2459-2469. doi:10.5194/acp-9-2459-2009

[7] S. Raja, R. Ravikrishna, X.-Y. Yu, T. Lee, J. Chen, R. R. Kommalapati, K. Murugesan, X. Shen, K. T. Valsaraj and J. L. Collett, "Fog Chemistry in the Texas-Louisiana Gulf Coast Corridor," *Atmospheric Environment*, Vol. 42, No. 9, 2008, pp. 2048-2061. doi:10.1016/j.atmosenv.2007.12.004

[8] I. J. George and J. P. D. Abbatt, "Heterogeneous Oxidation of Atmospheric Aerosol Particles by Gas-Phase Radicals," *Nature Chemistry*, Vol. 2, No. 9, 2010, pp. 713-722. doi:10.1038/nchem.806

[9] C. J. van Oss, "Interfacial Forces in Aqueous Media," Taylor and Francis, New York, 2006.

[10] E. E. Meyer, K. J. Rosenberg and J. N. Israelachvili, *Proceedings of the National Academy of Sciences*, Vol. 103, 2006, pp. 15739-15746.

[11] J. N. Israelachvili, "Intermolecular and Surface Forces," 3rd Edition, Academic Press, New York, 2010.

[12] J. S. Smith and K. T. Valsaraj, "Solvent Sublation for Industrial Wastewater Treatment," *Chemical Engineering Progress*, Vol. 94, No. 5, 1998, pp. 69-76.

[13] H. F. Rafson (Ed.), "Odor and VOC Control Handbook," McGraw Hill Pub Co., New York, 1998.

[14] K. T. Valsaraj, G. J. Thoma, D. D. Reible and L. J. Thibodeaux, "On the Enrichment of Hydrophobic Organic Compounds in Fog Droplets," *Atmospheric Environment*, Vol. 27A, No. 2, 1993, pp. 203-210. doi:10.1016/0960-1686(93)90351-X

[15] J. D. Blando and B. J. Turpin, "Secondary Organic Aerosol Formation in Cloud and Fog Droplets: A Literature Evaluation of Plausibility," *Atmospheric Environment*, Vol. 34, No. 10, 2000, pp. 1623-1632. doi:10.1016/S1352-2310(99)00392-1

[16] D. J. Donaldson and K. T. Valsaraj, "Adsorption and Reaction of Trace Gas-Phase Organic Compounds on Atmospheric Water Film Surfaces: A Critical Review," *Environmental Science & Technology*, Vol. 44, No. 3, 2010, pp. 865-873. doi:10.1021/es902720s

[17] K. T. Valsaraj, "Trace Gas Adsorption Thermodynamics at the Air-Water Interface: Implications in Atmospheric Chemistry," *Pure and Applied Chemistry*, Vol. 81, No. 10, 2009, pp. 1889-1901. doi:10.1351/PAC-CON-08-07-06

[18] D. J. Donaldson and D. Anderson, "Adsorption of Atmospheric Gases at the Air-Water Interface. 2. C_1-C_4 Alcohols, Acids, and Acetone," *Journal of Physical Chemistry A*, Vol. 103, No. 1, 1999, pp. 871-876. doi:10.1021/jp983963h

[19] R. Braunt and M. J. Conklin, "Dynamic Determination of Vapor/Water Interface Adsorption for Volatile Hydrophobic Organic Compounds (VHOCs) Using Axisymmetric Drop Shape Analysis: Procedure and Analysis of Benzene Adsorption," *Journal of Physical Chemistry B*, Vol. 104, No. 47, 2004, pp. 11146-11152. doi:10.1021/jp001140y

[20] A. Hartkopf and B. L. Karger, "Study of the Interfacial Properties of Water by Gas Chromatography," *Accounts of Chemical Research*, Vol. 6, 1973, pp. 209-221. doi:10.1021/ar50066a006

[21] S. Raja, F. S. Yaccone, R. Ravikrishna and K. T. Valsaraj, "Thermodynamic Parameters for the Adsorption of Aromatic Hydrocarbon Vapors at the Gas-Water Interface," *Journal of Chemical & Engineering Data*, Vol. 47, No. 5, 2002, pp. 1213-1219. doi:10.1021/je025520j

[22] J. Chen, F. S. Ehrenhauser, K. T. Valsaraj and M. J. Wornat, "Uptake and UV-Photooxidation of Gas-Phase PAHs on the Surface of Atmospheric Water Films. 1. Naphthalene," *Journal of Physical Chemistry A*, Vol. 110, No. 29, 2006, pp. 161-916. doi:10.1021/jp062560b8

[23] K. T. Valsaraj, "On the Physico-Chemical Aspects of Partitioning of Non-Polar Hydrophobic Organics at the Air-Water Interface," *Chemosphere*, Vol. 17, No. 5, 1988, pp. 875-887. doi:10.1016/0045-6535(88)90060-4

[24] K. T. Valsaraj, "Binding Constants for Non-Polar Hydrophobic Organics at the Air-Water Interface: Comparison of Experimental and Predicted Values," *Chemosphere*, Vol. 17, No. 10, 1988, pp. 2049-2061. doi:10.1016/0045-6535(88)90015-X

[25] K.-U. Goss and R. P. Schwarzenbach, "Linear Free Energy Relationships Used to Evaluate Equilibrium Partitioning of Organic Compounds," *Environmental Science & Technology*, Vol. 35, No. 1, 2001, pp. 1-9. doi:10.1021/es000996d

[26] K.-U. Goss, "Conceptual Model for the Adsorption of Organic Compounds from the Gas Phase to Liquid and Solid Surfaces," *Environmental Science & Technology*, Vol. 31, No. 12, 1997, pp. 3600-3605. doi:10.1021/es970361n

[27] C. P. Kelly, C. J. Cramer and D. G. Truhlar, "Predicting Adsorption Coefficients at Air-Water Interfaces Using Universal Solvation and Surface Area Models," *Journal of Physical Chemistry B*, Vol. 108, No. 34, 2004, pp. 12882-12897. doi:10.1021/jp037210t

[28] K.-U. Goss, "Predicting Adsorption of Organic Chemicals at the Air-Water Interface," *Journal of Physical Chemistry A*, Vol. 113, No. 44, 2009, 12256-12259. doi:10.1021/jp907347p

[29] R. Vacha, P. Jungwirth, J. Chen and K. T. Valsaraj, "Adsorption of Polycyclic Aromatic Hydrocarbons at the Air-Water Interface: Molecular Dynamics Simulations

and Experimental Atmospheric Observations," *Physical Chemistry Chemical Physics*, Vol. 8, No. 38, 2006, pp. 4461-4467. doi:10.1039/b610253k

[30] R. Vacha, K. Cwiklik, J. Rezac, P. Hobza, P. Jungwirth, K. Valsaraj, S. Bahr and V. Kempter, "Adsorption of Aromatic Hydrocarbons and Ozone at Environmental Aqueous Surfaces," *Journal of Physical Chemistry A*, Vol. 112, No. 22, 2008, pp. 4942-4950. doi:10.1021/jp711813p

[31] T. Somasundaram, R. M. Lyndon-Bell and C. H. Patterson, "The Passage of Gases through the Liquid Water/Vapour Interface: A Simulation Study," *Physical Chemistry Chemical Physics*, Vol. 1, No. 1, 1999, pp. 143-148. doi:10.1039/a805067h

[32] C. D. Wick, B. Chen and K. T. Valsaraj, "Computational Investigation of the Influence of Surfactants on the Air-Water Interfacial Behavior of Polycylic Aromatic Hydrocarbons," *Journal of Physical Chemistry C*, Vol. 114, No. 34, 2010, pp. 14520-14527. doi:10.1021/jp1039578

[33] J. Chen, F. Ehrenhauser, T. Arachi, F. Hung, K. Valsaraj and M. Wornat, "Adsorption of Gas-Phase Phenanthrene on Atmospheric Water and Ice Films," *Polycyclicarom Atichydrocarbons*, Vol. 31, 2011, pp. 1-26. doi:10.1080/10406638.2011.585370

[34] R. Vacha, P. Slavaicek, M. Mucha, B. Finlayson-Pitts and P. Jungwirth, "Adsorption of Atmospherically Relevant Gases at the Air/Water Interface: Free Energy Profiles of Aqueous Solvation of N_2, O_2, O_3, OH, H_2O, HO_2, and H_2O_2," *Journal of Physical Chemistry A*, Vol. 108, 2004, pp. 11573-11579.

[35] R. Vacha, P. Slavicek, M. Mucha, B. J. Finlayson-Pitts and P. Jungwirth, "Adsorption of Atmospherically Relevant Gases at the Air/Water Interface: Free Energy Profiles of Aqueous Solvation of N_2, O_2, O_3, OH, H_2O, HO_2, and H_2O_2," *Journal of Physical Chemistry A*, Vol. 108, No. 52, 2004, pp. 11573-11579. doi:10.1021/jp046268k

[36] J. Chen, F. Ehrenhauser, K. T. Valsaraj and M. J. Wornat, "Adsorption and UV Photooxidation of Gas-Phase Phenanthrene on Atmospheric Films," *ACS Symposium Series*, Vol. 1005, 2009, pp. 127-146. doi:10.1021/bk-2009-1005.ch009

[37] I.-F. W. Kuo and C. J. Mundy, "An *Ab initio* Molecular Dynamics Study of the Aqueous Liquid-Vapor Interface," *Science*, Vol. 303, No. 5658, 2004, pp. 658-662. doi:10.1126/science.1092787

[38] X. D. Zhu, H. Suhr and Y. R. Shen, "Surface Vibrational Spectroscopy by Infrared-Visible Sum Frequency Generation," *Physical Review B*, Vol. 35, No. 6, 1987, pp. 3047-3050. doi:10.1103/PhysRevB.35.3047

[39] B. T. Mmereki, D. J. Donaldson, J. B. Gilman, T. L. Eliason and V. Vaida, "Kinetics and Products of the Reaction of Gas-Phase Ozone with Anthracene Adsorbed at the Air-Aqueous Interface," *Atmospheric Environment*, Vol. 38, 2004, pp. 6091-6103. doi:10.1016/j.atmosenv.2004.08.014

[40] E. S. Enami, M. R. Hoffmann and A. J. Colussi, "Ozonolysis of Uric Acid at the Air/Water Interface," *Journal of Physical Chemistry B*, Vol. 112, No. 14, 2008, pp. 4153-4157. doi:10.1021/jp712010k

[41] R. Vacha, L. Cwiklik, J. Rezac, P. Hobza, P. Jungwirth, K. Valsaraj, S. Bahr and V. Kempter, "Adsorption of Aromatic Hydrocarbons and Ozone at Environmental Aqueous Surfaces," *Journal of Physical Chemistry A*, Vol. 112, No. 22, 2008, pp. 4942-4947. doi:10.1021/jp711813p

[42] S. N. Wren and D. J. Donaldson, "Glancing-Angle Raman Spectroscopic Probe for Reaction Kinetics at Water Surfaces," *Physical Chemistry Chemical Physics*, Vol. 12, No. 11, 2010, pp. 2648-2654. doi:10.1039/b922254e

[43] S. Raja and K. T. Valsaraj, "Heterogeneous Oxidation by Ozone of Naphthalene Adsorbed at the Air-Water Interface of Micron-Size Water Droplets," *Journal of the Air & Waste Management Association*, Vol. 55, No. 9, 2005, pp. 1345-1355.

[44] I. V. Stiopkin, H. D. Jayathilake, A. N. Bordenyuk and A. V. Benderskii, "Heterodyne-Detected Vibrational Sum Frequency Generation Spectroscopy," *Journal of the American Chemical Society*, Vol. 130, No. 7, 2008, pp. 2271-2275. doi:10.1021/ja076708w

[45] C. Waring, P. A. J. Bagot, M. L. Costen and K. G. McKendrick, "Reactive Scattering as a Chemically Specific Analytical Probe of Liquid Surfaces," *The Journal of Physical Chemistry Letters*, Vol. 2, No. 1, 2011, pp. 12-18. doi:10.1021/jz1013032

[46] Y. R. Shen and V. Ostroverkhov, "Sum-Frequency Vibrational Spectroscopy on Water Interfaces: Polar Orientation of Water Molecules at Interfaces," *Chemical Reviews*, Vol. 106, No. 4, 2006, pp. 140-154. doi:10.1021/cr040377d

[47] K. Harper, B. Minofar, M. R. Sierra-Hernandez, N. N. Casillas-Ituarte, M. Roeselova and H. C. Allen, "Surface Residence and Uptake of Methyl Chloride and Methyl Alcohol at the Air/Water Interface Studied by Vibrational Sum Frequency Spectroscopy and Molecular Dynamics," *Journal of Physical Chemistry A*, Vol. 113, No. 10, 2009, pp. 2015-2024. doi:10.1021/jp808630v

[48] K. B. Eisenthal, "Equilibrium and Dynamic Processes at Interfaces by Second Harmonic and Sum Frequency Generation," *Annual Review of Physical Chemistry*, Vol. 43, 1992, pp. 627-661. doi:10.1146/annurev.pc.43.100192.003211

[49] N.-O. A. Kwamena, M. G. Staikova, D. J. Donaldson, I. J. Goerge and J. P. D. Abbatt, "Role of the Aerosol Substrate in the Heterogeneous Ozonation Reactions of Surface-Bound PAHs," *Journal of Physical Chemistry A*, Vol. 111, No. 43, 2007, pp. 11050-11058. doi:10.1021/jp075300i

[50] M. Shiraiwa, Y. Sosedova, A. Ronviere, H. Yang, Y. Zhang, J. P. D. Abbatt, M. Ammann and U. Poschl, "The Role of Long-Lived Reactive Oxygen Intermediates in the Reaction of Ozone with Aerosol Particles," *Nature Chemistry*, Vol. 3, 2011, pp. 291-295. doi:10.1038/NCHEM.988

[51] J. Chen and K. Valsaraj., "Uptake and UV-Photooxidation of Gas-Phase PAHs on the Surface of Atmospheric Water Films. 1. Naphthalene," *Journal of Physical Chemistry A*, Vol. 110, 2006, pp. 9161-9166. doi:10.1021/jp062560b

[52] R. S. Strekowski, R. Remorov and Ch. George, "Direct Kinetic Study of the Reaction of Cl_2 Radical Anions with Ethanol at the Air-Water Interface," *Journal of Physical Chemistry A*, Vol. 107, 2003, pp. 2497-2504. doi:10.1021/jp026174f

[53] S. Narayan, J. Muldoon, M. G. Finn, V. V. Fokin, H. C. Kolb and K. B. Sharpless, "On Water: Unique Reactivity of Organic Compounds in Aqueous Suspension," *Angewandte Chemie International Edition*, Vol. 44, No. 21, 2005, pp. 3275-3279. doi:10.1002/anie.200462883

[54] Y. Jung and R. A. Marcus, "On the Theory of Organic Catalysis 'on Water'," *Journal of the American Chemical Society*, Vol. 129, No. 17, 2007, pp. 5492-5502. doi:10.1021/ja068120f

[55] K. T. Valsaraj, "Trace Gas Adsorption Thermodynamics at the Air-Water Interface: Implications in Atmospheric Chemistry," *Pure and Applied Chemistry*, Vol. 81, No. 10, 2009, pp. 1889-1901. doi:10.1351/PAC-CON-08-07-06

[56] P. Nissensson, C. J. X. Knox, B. J. Finlayson-Pitts, L. F. Phillips and D. Dabdub, "Enhanced Photolysis in Aerosols: Evidence for Important Surface Effects," *Physical Chemistry Chemical Physics*, Vol. 8, No. 40, 2006, pp. 4700-4710. doi:10.1039/b609219e

[57] H. Watanabe, S. Yamaguchi, S. Sen, A. Morita and T. Tahara, "'Half-Hydration' at the Air/Water Interface Revealed by Heterodyne-Detected Electronic Sum Frequency Generation Spectroscopy, Polarization Second Harmonic Generation, and Molecular Dynamics Simulation," *Journal of Chemical Physics*, Vol. 132, No. 14, 2010, Article ID: 144701. doi:10.1063/1.3372620

[58] M. L. Johnson, C. Rodriguez and I. Benjamin, "Rotational Dynamics of Strongly Adsorbed Solute at the Water Surface," *Journal of Physical Chemistry A*, Vol. 113, 2009, pp. 2086-2091. doi:10.1021/jp808842k

[59] P. Nissenson, D. Dabdub, R. Das, V. Maurino, C. Minero, D. Vione, "Evidence of the Water-Cage Effect on the Photolysis of and $FeOH^{2+}$. Implications of This Effect and of H_2O_2 Surface Accumulation on Photochemistry at the Air-Water Interface of Atmospheric Droplets," *tmospheric Environment*, Vol. 44, 2010, pp. 4859-4866. doi:10.1016/j.atmosenv.2010.08.035

18

CdS_xSe_y/TiO_2 Solar Cell Prepared with Sintered Mixture Deposition

Daniel Ogermann, Thorsten Wilke, Karl Kleinermanns*

Institute of Physical Chemistry, Heinrich Heine University, Duesseldorf, Germany

Email: *kleinermanns@uni-duesseldorf.de

ABSTRACT

Grätzel cells were prepared by using CdSe- and CdS_xSe_y-nanoparticles as sensitizer. The quantum dots were incorporated in various sizes and concentrations in a TiO_2 nanoparticle layer by a simple mixing procedure. The advantage of this method compared to anchoring of nanoparticles to TiO_2 by linker molecules or chemical bath deposition is that we are able to control the ratio between TiO_2 and CdSe or CdS_xSe_y more precisely and over a larger range of concentrations. TiO_2 solar cells sensitized by this technique achieved photon-to-current conversion efficiencies (IPCE) of ~40% in the range of 300 - 500 nm with a maximum IPCE of ~70% at 400 nm (sulphide/sulphate electrolyte). The best results at wavelengths above 500 nm were achieved with CdS_xSe_y/TiO_2 cells at a molar ratio of 6:1 (S:Se) with IPCE of 40% at 500 nm and still 15% at 800 nm. Quantum efficiencies obtained with iodine/iodide electrolyte were lower and lead to an overall efficiency of 0.32%. The CdS_xSe_y sensitized solar cells show enhanced stability compared to CdSe based systems and the use of the iodine/iodide electrolyte increases cell endurance further.

Keywords: Solar Cells; Nanocomposites; Cadmium Selenosulphide; Titanium Dioxide

1. Introduction

About 20 years ago Grätzel and O'Regan demonstrated their concept of dye sensitized solar cells (DSC), as a promising alternative to p-n junction silicon cells [1]. While the growth of the photovoltaic market over the last decade has been impressive, the cost of the photovoltaic electricity production is still too high to be competitive with fossil fuels. Dye-sensitized nano-porous TiO_2 photoelectrochemical cells with high IPCE and low cost are possible alternatives to silicium solar cells in a solar powered future [2]. The working cycle of this cell type consists of dye excitation, electron injection into the conduction band (CB) of titanium dioxide-a wide-bandgap semiconductor used for electron transport-and fast reduction of the oxidized dye by a redox couple. Ruthenium dye sensitized TiO_2 cells reached overall solar (standard AM 1.5) to current conversion efficiencies of 10% with stabilities of several months [1,3]. Long term cell degradation due to corrosion caused by the redox electrolyte could be reduced using for example ionic liquid electrolytes [4].

During the last years different types of quantum dots (QD) like InP [5], PbS [6,7], Bi_2S_3 [6,8], InAs [9], CdS [6,10] and CdSe [11,12] have been investigated for their possible use as photo sensitizer [13]. Semiconductor QDs such as CdSe have several advantages as solar cell sensitizers. The working principle of the QD-sensitized solar cell (QDSC) is explained in **Figure 1** and very similar to DSC. The energy gap of the QDs can in principle be controlled by the change of quantum confinement at different quantum dot size so that the absorption

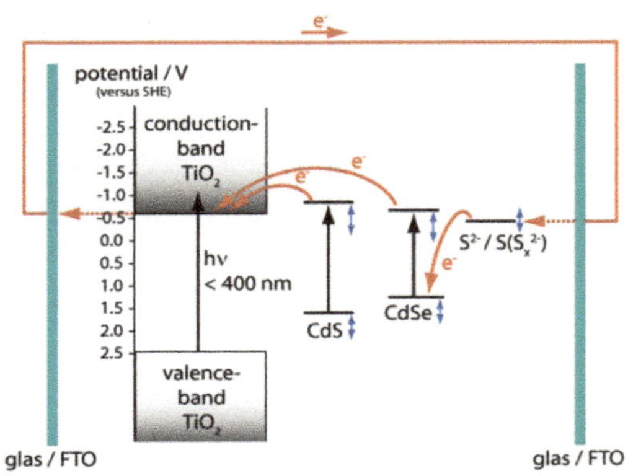

Figure 1. Scheme of the quantum dot solar cell. The working cycle consists of CdS or CdSe excitation, electron (e^-) injection into the conduction band (CB) of titanium dioxide and fast reduction of the oxidized sensitiser h^+ by a Na_2S/Na_2SO_4 redox couple. Redox potentials are given in Volts [V] against Standard Hydrogen Potential [SHE]. Variation of redox potentials due to NC size and preparation conditions as well as electrolyte concentration are indicated by vertical arrows.

*Corresponding author.

spectra can be tuned to match the spectral distribution of sunlight [14]. QDs have large absorption cross sections accompanied by large dipole moments upon electronic excitation facilitating ultrafast charge separation. Quantum yields higher than unity are possible due to hot electron impact ionisation leading to secondary electrons [15-18].

Efficient electron injection from the photoexcited QD into the TiO_2 conduction band requires close QD-TiO_2 contact. Deposition of CdSe-QD layers on a conducting transparent substrate (usually fluorine doped tin oxide—FTO) covered with nano-crystalline(NC)-TiO_2 by using simple dipping has been reported to be applicable (direct absorption—DA) [19-23]. Anchoring QDs on TiO_2 by linker molecules containing both carboxylate and thiol functional groups is another approach to achieve close and stable contact (linker technique) [12,21,24]. Direct or in layer reaction of the cationic and anionic QD precursors in the TiO_2 pores is a further preparation method called chemical bath deposition (CBD) [25-27]. Electrodeposition of NC-semiconductor films by cathodic reduction [28-32], co-evaporation [29,33,34] and screen printing [35-37] are techniques for photo-sensitizing conducting electrodes with semiconductor QDs.

In this paper we introduce an alternative approach to achieve close QD-TiO_2 contact. We mix size-defined CdSe and CdS NCs as well as CdS_xSe_y nanocomposites with TiO_2 gel in defined volume ratios, deposit the mixture on FTO at defined thickness and anneal the sample at 450˚C. We call this technique "sintered mixture deposition" (SMD) and found it to be more simple, versatile and reproducible than DA or anchoring by linker molecules. In particular it is possible to optimise IPCE by independently adjusting the QD size and volume ratio of the mixture. This technique however requires the development of QDs which are thermally stable and do not react with the electrolyte. Much of this paper describes our development and control of suitable QDs for SMD. Using SMD we attained optimum IPCE of 15% - 60% over an extended wavelength range in the visible spectral region which was not achievable before with single junction QDSCs to the best of our knowledge.

2. Experiments

Two routes to synthesize CdSe nanoparticles suitable for SMD were followed. The size of the CdSe nanoparticles made by ***route* 1** was in the range of 2 - 50 nm [20]. As selenium source a 80 mM aqueous solution of sodium selenosulphate (Na_2SeSO_3) was prepared by stirring elemental selenium powder (63.2 mg, 0.8 mmol, Aldrich) with an excess of sodium sulfite (252 mg, 2 mmol, Aldrich) in 10 ml H_2O at 70˚C for several hours until the Se˚ powder was completely dissolved. The cadmium source was obtained by mixing 10 ml of a 80 mM aqueous cadmium sulphate solution ($CdSO_4{\cdot}8/3H_2O$, Fluka) with 120 mM nitrilotriacetic acid trisodium salt (Aldrich) dissolved in 10 ml H_2O. After combining the Se and Cd solution under rapid stirring, a light green colour appeared, which increased to an intense yellow colour within the next 10 minutes, then changed to orange within 48 hours and finally shifted to red after one week. The CdSe suspension was centrifuged for 5 minutes at 13,000 rmp, the aqueous phase was discarded and the residual solid was dried for 1 week in an exsiccator.

As first step of the synthesis ***route* 2** [38] we dissolved $Cd(ClO_4)_2{\cdot}6H_2O$ (985 mg, 2.35 mmol, Aldrich) in 125 mL water. Afterwards 5.7 mmol of the thiol-stabiliser 1-thioglycerol (TG, Aldrich), thioglycolic acid (TGA, Aldrich) or 3-mercaptopropionic acid (MPA, Aldrich)) were added during stirring and the pH of the solution was adjusted between 11.2 and 11.8 by dropwise addition of 1 M NaOH (Aldrich). Another three neck flask with Al_2Se_3 lumps (134 mg, 0.46 mmol, Aldrich) was connected and flushed with N_2 for about 30 minutes. The cadmium solution was deaerated in the same way. Subsequently 10 - 20 ml of 0.5 M H_2SO_4 (Aldrich) was added to the Al_2Se_3 lumps via a septum and the generated H_2Se gas was passed to the solution via a slow nitrogen flow for about 20 minutes. Depending on the stabiliser the colour of the solution changed due to the formation of CdSe particles (TG: yellow; TGA: orange; MPA: red). The CdSe precursors grew to CdSe nanocrystals under reflux (100˚C, 24 hours). For precipitation of the CdSe NCs isopropanol (Fluka) was added. The suspension was centrifuged for 5 minutes at 13,000 rmp, the aqueous phase was discarded and the residual solid was dried in an exsiccator for 1 week. This route gave particles with a size of 1 - 2 nm (TG), 2 - 4 nm (TGA) and 3 - 5 nm (MPA).

For comparison cadmium sulphide NCs were prepared with a size distribution of 2 - 4 nm [39]. 1-thioglycerol (TG, 1.87 ml, 22 mmol, Aldrich) was added to 30 ml of a 80 mM aqueous cadmium sulphate solution ($CdSO_4{\cdot}8/3H_2O$, Fluka). During constant stirring for 5 minutes ammonium sulphide (12.4 ml, 36 mmol, 20% aqueous solution, Al- drich) was added. An intense yellow solution with yellow precipitation was obtained thereafter, which was cen- trifuged and dried for 1 week.

Cadmium selenosulphide nanocomposites were prepared according to ***route* 3** [40]. In specific molar ratios 1:1 to 24:1 (S:Se) cadmium sulphide NCs and selenium powder were suspended in methanol. After 1 hour of stirring methanol was evaporated at 80˚C. The resulting CdS:Se solid was sintered at 550˚C for 15 minutes. Depending on the molar ratio the colour of the CdSxSey composites ranged from yellow (24:1) to dark violet (1:1).

TiO_2 nanopowder (5 g Aearosil TiO_2 P25, Evonik) was suspended in nitric acid (1 N, Fluka) to prepare a suitable TiO_2 paste. This suspension was heated at 80˚C for 24 hours. Afterwards the nitric acid was evaporated and the TiO_2 solid was dried for 3 days at 100°C. Finally the TiO_2 solid was treated with 25 ml water, acetylacetone (2.5 g, 24.97 mmol, Merck), Triton-X-100 (1.25 g, 1.92 mmol, Avocado) and polyethylene oxide (M.W. 100,000, Alfa Aesar).

A colourless and clear solution of TiO_2 nanoparticles was prepared as follows [41]. Titanium(IV)butoxide (8.5 ml, 0.025 mmol, Aldrich) was dissolved in ethanol (19 ml, Aldrich) and diethanolamine (2 ml, 20.8 mmol, Fluka). Subsequently 19 ml ethanol and 0.5 ml H_2O were added very slowly. The solution was stirred over night.

The SMD technique used for fabrication of QDSC will now be described in detail. For each type of CdSe-NP three different volume ratios of the solid QD-samples were mixed with the TiO_2-suspension (1:1, 1:3 and 1:5 QD:TiO_2). In the case of CdS_xSe_y 20 mg of the solid nanocomposite sample were mixed with 0.5 ml of the TiO_2 suspension. We used aluminoborosilicate glass coated with fluorine doped tin oxide (SnO_2:F) as conducting transparent substrate (resistivity ~10 ohm/cm^2; Solaronix SA, Switzerland). The active area was masked with scotch tape and was coated with CdSe/TiO_2, CdS/TiO_2 or CdS_xSe_y/TiO_2 suspension by use of a glass scraper and dried at 80°C for 10 min. After removal of the scotch tape we measured ~11 μm deposit thickness. This primed photoelectrode was sintered in a muffle furnace for ~45 min at 450°C. A thin platinium layer was spread on the FTO coating of the counter electrode (Platisol; Solaronix SA, Switzerland). Few drops of an aqueous electrolyte mixture of 0.1 M Na_2S (Alfa-Aesar) and 0.01 M Na_2SO_4 (Aldirch) were added to the photoelectrode for reduction of the oxidized QDs. In case of the I-V measurements the active redox couple iodine/iodide (0.1 M lithium iodide (Fluka), 0.05 M iodine (Aldrich), 0.6 M tetrabutylammonium iodide (Aldrich) and 0.5 M 4-tert-butylpyridine (Aldrich) in acetonitrile (Aldrich)) was used. Finally the counter electrode was clamped to the photoelectrode/electrolyte system.

Measurements of the wavelength dependence of the short circuit current (j_{sc}) were carried out in a light-proof box with a 70 W Xenon lamp (Oriel, Germany) and a grating monochromator (Zeiss, Germany) in the spectral range of 300 - 800 nm at 1 nm resolution. Data were transmitted to a computer and processed there. The light intensity incident on the electrode (I_{inc}) was measured with a power meter (Coherent, USA). The IPCE is defined by the following expression [42]:

$$IPCE(\%) = \frac{j_{sc}\left(\mathrm{A/cm^2}\right)\cdot 1240}{\lambda(\mathrm{nm})\cdot I_{inc}\left(\mathrm{W/cm^2}\right)}\cdot 100 \qquad (1)$$

The photocurrent-voltage (I-V) curves were measured using a 120 W Xenon lamp (Oriel, Germany) and a special filter, which was focused to give 100 mW/cm² (1 sun) at Air Mass (AM) 1.5 at the surface of the solar cell. The fill factors ff and overall efficiencies η were calculated according to Equations (2) and (3). Here MPP is the "maximum power point" and U_{oc} the open-circuit voltage [43].

$$ff = \frac{MPP\left(\mathrm{W/cm^2}\right)}{j_{sc}\left(\mathrm{A/cm^2}\right)\cdot U_{oc}(\mathrm{V})} \qquad (2)$$

$$\eta(\%) = \frac{j_{sc}\left(\mathrm{A/cm^2}\right)\cdot U_{oc}(\mathrm{V})\cdot ff}{I_{inc}\left(\mathrm{W/cm^2}\right)}\cdot 100 \qquad (3)$$

Optical absorption spectra of the synthesis products were recorded by using a Cary 300 UV-Vis spectrophotometer operated at a resolution of 1 nm. The absorption spectra of the electrodes were measured in a reflection arrangement. The particle size distribution was investigated by Transmission Electron Microscopy (HITACHI TEM 7500) at the Max-Planck institute for coal research (MPI Mülheim a.d. Ruhr).

3. Results and Discussion

CdSe NCs synthesized via ***route*** **1** exhibited nearly complete sulphide-selenide exchange with the electrolyte in QDSCs prepared with SMD, see Supplementary Material. We investigated CdSe NCs synthesized via ***route*** **2** for their stability against the electrolyte. Their TEM images are given in **Figure 2**. The size distribution dependent on the stabiliser is shown in the insets of **Figure 2**. Based on our electron microscope images we propose Ostwald ripening to be responsible for the formation of the nanoparticles.

Figure 3 shows the absorption spectra of CdSe NCs protected with different stabilisers in solution. As expected the absorption shifts to the red with increasing particle size [44]. The size of the CdSe-NPs influences the performance of the QDSCs significantly and thus the size distribution is a decisive factor to prepare QDSCs with high efficiencies.

For solar cells the absorption of the sintered mixture is more important than the absorption spectra of the QDs in aqueous solution. **Figure 4** shows the absorption spectra of the mixtures of TGA-stabilised CdSe NCs with TiO_2 on FTO coated glass as obtained by reflection spectroscopy. The absorption of pure TiO_2 nanoparticles is already subtracted here. Before annealing the characteristic absorption of CdSe/TGA NCs is in the range of 400 - 600 nm.

After annealing the orange coloured CdSe/TiO_2 sheets change its colour to purple. In contrast to the reflection spectrum of the CdSe NPs synthesized by ***route*** **1** (**Figure S3**) there is still considerable absorption in the range up to 700 nm. Obviously, after contact with the Na_2S/

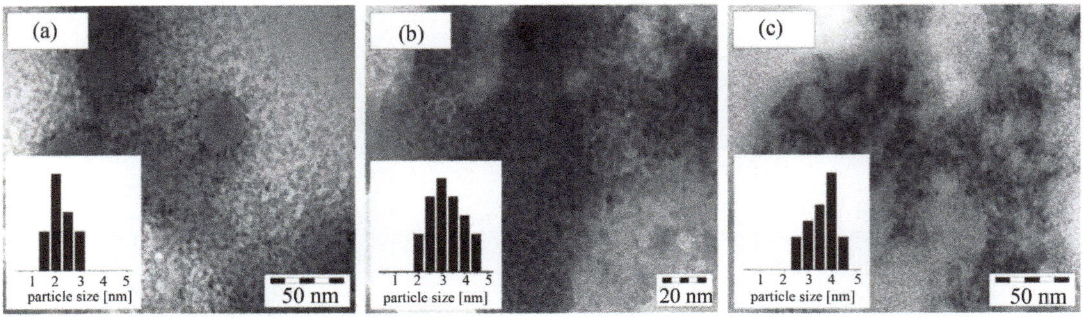

Figure 2. TEM images of CdSe NCs stabilised with three different reagents. Their synthesis is described in the experimental section (*route* 2). The insets indicate the size distribution as obtained from diameter measurements of 20-40 particles. Typical size ranges are 2 nm for the particles stabilised by TG, 2-4 nm by TGA and 4 nm by MPA. (a) TG; (b) TGA: (c) MPA.

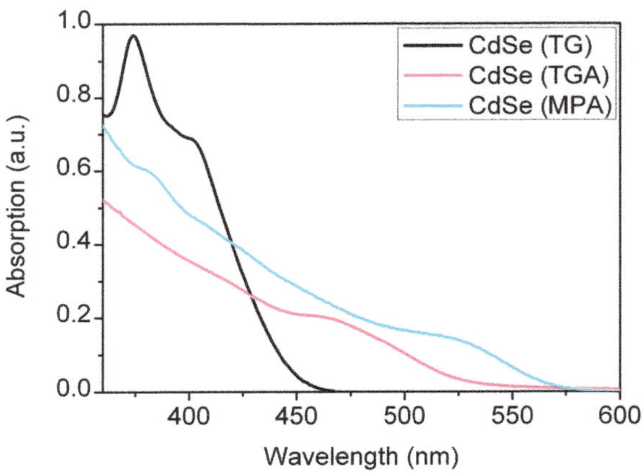

Figure 3. Absorption spectra of CdSe NCs (*route* 2) of different size measured in solution (after 24 hours heating).

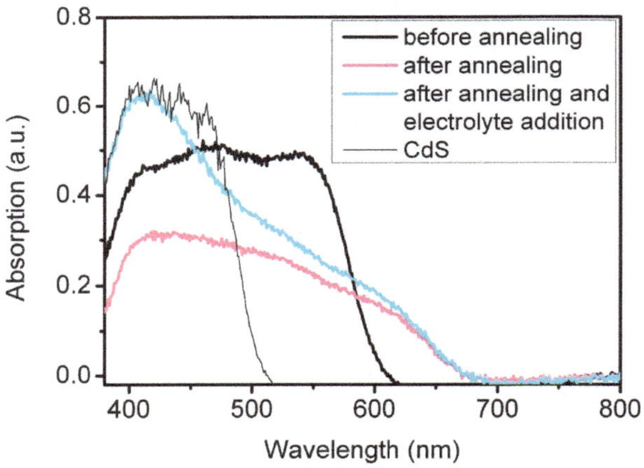

Figure 4. Absorption spectra of CdSe NCs (*route* 2) stabilised by TGA and CdS NCs measured in reflection before or after annealing at 450°C and electrolyte (Na_2S/Na_2SO_4) contact. After contact with the Na_2S/Na_2SO_4 electrolyte only very little selenide-sulphide exchange can be observed.

Na_2SO_4 electrolyte only very little selenide-sulphide exchange takes place. The main difference between ***route* 1** and **2** is the use of sulphur-containing stabilisers in ***route* 2**. Probably CdS layer is formed on the surface of ***route* 2** QDs during the sintering process at 450°C leading to core-shell particles. This CdS layer prevents a large selenide-sulphide exchange. The CdSe particles stabilised by TG and MPA show the same behaviour. Hence ***route* 2** QDs are promising candidates for SMD coated QDSCs.

Figure 5 displays IPCE curves of CdSe particles stabilised with TG, TGA and MPA in volume ratios of 1:1, 1:3 and 1:5 (CdSe:TiO_2) and Na_2S/Na_2SO_4 as redox couple. The smallest CdSe particles with a size of 2 nm (stabilised with TG) in a 1:1 volume ratio with TiO_2 exhibit the highest IPCE (**Figure 5(a)**).

In the range of 350 - 450 nm the IPCE reaches a value over 70%. But for the more interesting range of 500 - 700 nm a volume ratio of 1:3 shows the best result. Here an IPCE of about 20% is observed. The volume ratio 1:5 is only photoelectrically sensitive up to ~470 nm. For TGA as stabiliser the spectra in **Figure 5(b)** show an opposite behaviour in the range up to 500 nm. The IPCE increases with increasing volume ratio of TiO_2. A 1:5 volume ratio shows a maximum IPCE of 65% at ~370 nm and 10% - 20% in the range 450 - 600 nm. **Figure 5(c)** shows IPCEs when MPA stabiliser is used. Volume ratios of 1:3 and 1:5 have IPCE values of 45% at about 370 nm and 5% - 10% at 450 - 700 nm. In summary the smaller NCs obtained by use of TG stabiliser show the best IPCE. Probably they are more easily intercalated between the TiO_2 nanoparticles and therefore homogeneously mixed in large quantities with TiO_2. A possible however speculative explanation is that larger NCs do not fit into the TiO_2 lattice cavities. In larger quantities they might form NC layers which electrically isolate TiO_2. Whereas the smallest NCs show the best IPCE their absorption in the visible spectral range is limited due to their size (quantum confinement).

We therefore looked for small NCs which absorb further to the red and simultaneously show good temperature and electrolyte stability for effective sintered mixture deposition.

Cadmium selenosulphide CdS_xSe_y nanocomposites proved to be an excellent alternative. Because of their synthesis at 550°C in ***route* 3** they are stable during the

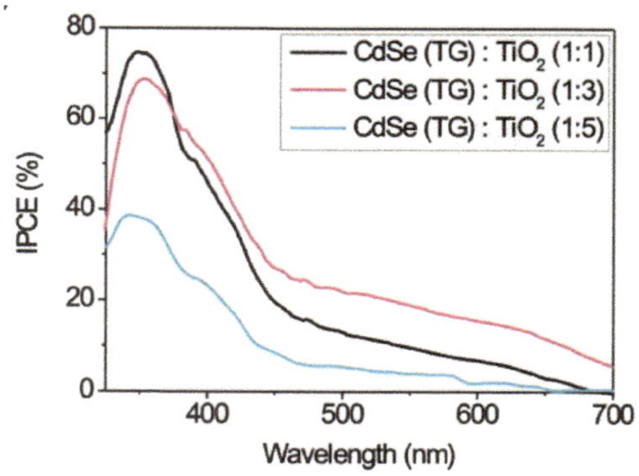

(a)

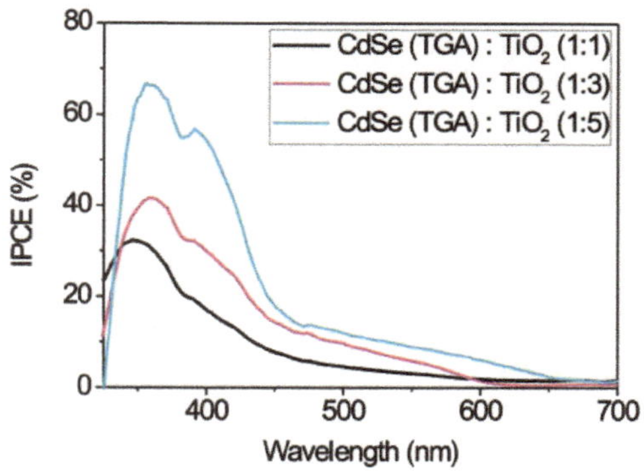

(b)

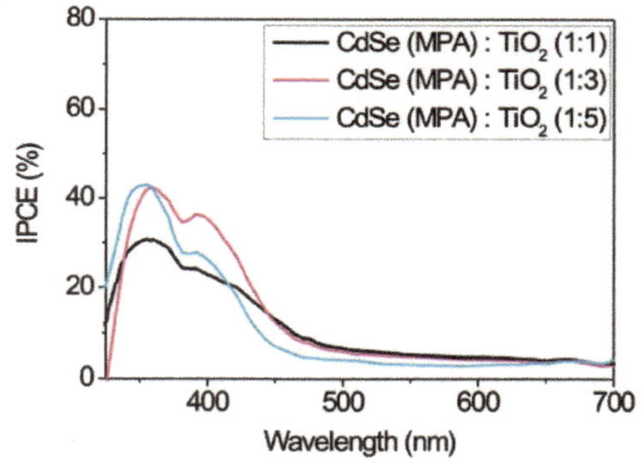

(c)

Figure 5. Comparison of the incident photon-to-current efficiencies (IPCE) of CdSe solar cells working with Na_2S/Na_2SO_4 redox couple. CdSe NCs of different size (depending on the stabiliser (a) TG; (b) TGA and (c) MPA (*route* 2)) and different NC/TiO_2 volume ratios of 1:1, 1:3 and 1:5 were used.

preparation of the CdS_xSe_y/TiO_2 solar cells with SMD. In contrast to CdSe NCs CdS_xSe_y nanocomposites are not soluble in standard solvents. This impedes the analysis of their size distribution and composition by TEM. Due to the high affinity of CdS_xSe_y to TiO_2 a part of them however binds to the surface of TiO_2 NPs in a colourless, clear ethanolic TiO_2 solution, see Experiments. Now it is possible to obtain meaningful TEM images. **Figure 6** displays images of the $CdS_{12}Se_1$ particles in contact with TiO_2. The composite particles exhibit a size of about 50 nm. Elemental analysis by local energy dispersive X-ray spectroscopy (EDX) indeed showed that the composite particles contain Ti, Cd, S and Se.

Figure 7 displays the absorption spectra of CdS_xSe_y particles with different S:Se ratios mixed with TiO_2 paste

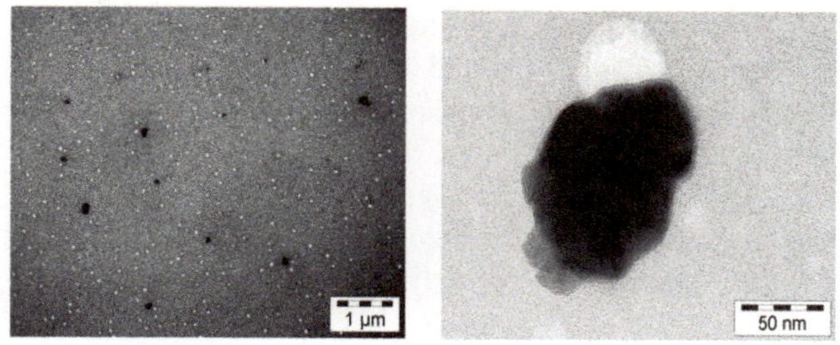

Figure 6. TEM images of $CdS_{12}Se_1/TiO_2$ nanocomposites obtained from dispersion of $CdS_{12}Se_1$ QDs in a colourless, clear TiO_2 solution (black spots in the left image; one of the spots is magnified in the right image).

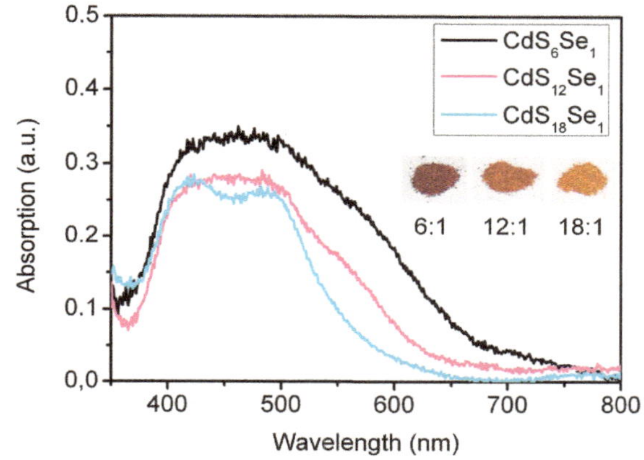

(a)

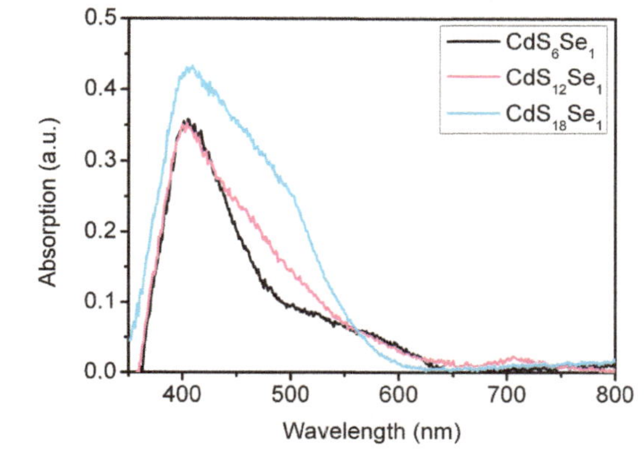

(b)

Figure 7. (a) Absorption spectra of CdS_xSe_y particles (*route* 3) of different molar ratios measured in reflection before annealing at 450°C. The inset shows the photograph of three samples with different S:Se molar ratios; (b) Absorption spectra of CdS_xSe_y NCs (*route* 3) of different molar ratios measured in reflection after annealing at 450°C and electrolyte (Na_2S/Na_2SO_4) contact.

on FTO coated glass as obtained by reflection spectroscopy. The absorption of pure TiO_2 nanoparticles was subtracted. Before sintering (**Figure 7(a)**) a broad absorption band in the range of 400 - 700 nm is observed. In the case of the 6:1 and 12:1 ratios (S:Se) absorption extends up to 800 nm. After annealing at 450°C and contact with the Na_2S/ Na_2SO_4 electrolyte the samples still show some absorption up to 800 nm (**Figure 7b)**). Only very little selenide-sulphide exchange could be observed. With increasing ratio of sulphur the absorption spectra are more similar to the spectrum of pure CdS-NCs (**Figure 4**). The CdS-NCs exhibit absorption only up to 500 nm.

Figure 8 compares the different TiO_2 sensitizers using SMD for cell preparation. Clearly the CdS_6Se_1 sensitizer shows the best results. At shorter wavelengths (400 - 600 nm) an IPCE of ~40% (520 nm) is observed. IPCE values in the visible range up to 800 nm are higher compared to the CdSe sensitizer.

Figure 8 shows that the sulphide/sulphate electrolyte leads to approximately two times higher IPCE values than the iodine/iodide electrolyte. When however using sulphide/sulphate as electrolyte the photocurrent decreases rapidly under irradiation at AM 1.5. A record of the I-V curves is not possible. During the irradiation the sulphate/sulphide electrolyte decomposes. The sheets show yellow coloured parts from elemental sulphur.

Figure 9 presents the I-V curves of CdS and CdS_xSe_y based QDSCs measured with an iodine/iodide redox system. The resulting fill factors and overall efficiencies are shown in **Table 1**.

Solar cells prepared with CdSe-NP's by using the SMD method show only the typical TiO_2 IPCE-spectrum. We have achieved a best efficiency η of 0.32% with a fill factor of 0.34 for QDSCs based on CdS_6Se_1-NCs and the iodine/iodide electrolyte. The order of the short-circuit photocurrent densities are in agreement with the trends of the IPCE-values (CdS_6Se_1 > $CdSe_{12}Se_1$ > $CdS_{18}Se_1$). The open-circuit photovoltages of these nanocomposites are similar. In comparison with QDSCs based on CdS the overall efficiences and the photocurrents of CdS_xSe_y sensitized QDSCs are substantially higher because CdS nanoparticles absorb only in a small range of the solar spectrum.

4. Conclusions

We synthesized thiol-stabilised cadmium selenide nanocrystals by passing selenium hydrogen gas in an aqueous solution of cadmium ions, precipitation and drying in order to get a solid CdSe nanopowder. In contrast to other preparation methods for QDSC we are able to control the ratio between titanium dioxide and CdSe more precisely in a large range by mixing them in defined quantities using SMD. In order to enlarge the absorption range of the solar cells we investigated nanoparticles of CdS_xSe_y type, which are quite promising. The best results were achieved with CdS_6Se_1/TiO_2 solar cells. These cells show reasonable IPCE values up to 800 nm. However, the overall efficiencies of our QDSCs are quite low in

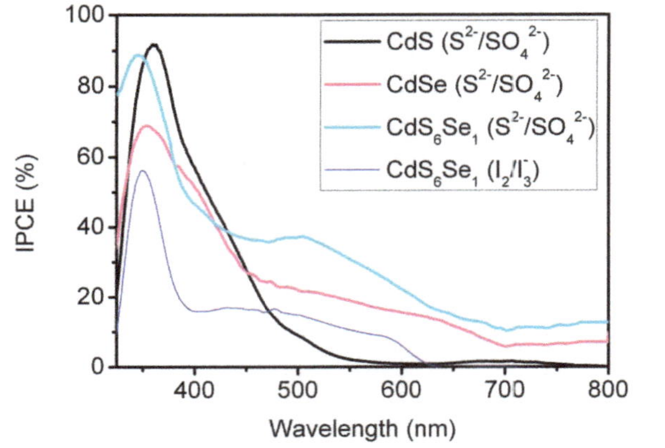

Figure 8. Comparison of the incident photon-to-current efficiencies (IPCE) of CdS_6Se_1, CdSe and CdS sensitized solar cells measured with Na_2S/Na_2SO_4. In the case of CdS_6Se_1 I_2/I_3^- were also used as electrolyte.

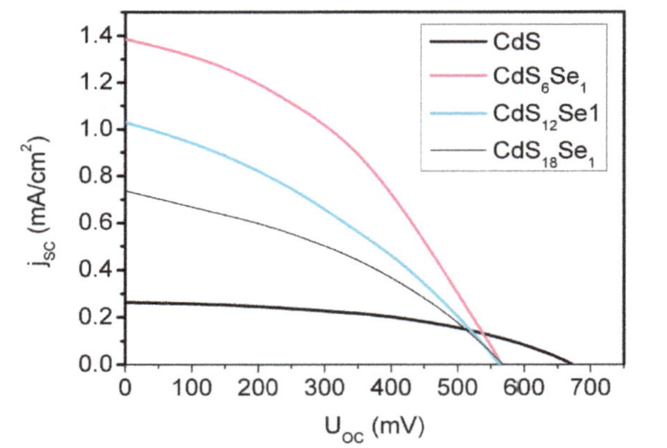

Figure 9. Photocurrent density vs voltage curves for QDSCs based on CdS- and CdS_xSe_y-NCs with iodine/iodide electrolyte under irradiation at AM 1.5 simulated solar light (100 mW/cm²).

Table 1. Photovoltaic performance of QDSCs based on CdS_xSe_y.

QDs	j_{sc}/mA·cm^{-2}	U_{oc}/mV	*ff*	η/%
CdS	0.26	672	0.47	0.08
CdS_6Se_1	1.38	565	0.34	0.32
$CdS_{12}Se_1$	1.03	563	0.31	0.20
$CdS_{18}Se_1$	0.73	567	0.38	0.16
TiO_2	0.05	53	0.32	0.001

comparison with cells, which were prepared by the linker method. Nevertheless our SMD method is a great alternative for unsolvable sensitizers, which are not suitable for the common techniques.

5. Acknowledgements

The authors gratefully acknowledge Dr. Christian W. Lehmann and his group for the implementation of the TEM measurements at the Max-Planck Institut für Kohlenforschung, Mülheim/Ruhr.

REFERENCES

[1] B. O'Regan and M. Grätzel, "A Low-Cost, High-Efficiency Solar Cell Based on Dye-Sensitized Colloidal TiO_2 Films," *Nature*, Vol. 535, No. 6346, 1991, pp. 737-740. doi:10.1038/353737a0

[2] M. Grätzel, "Photovoltaic and Photoelectrochemical Conversion of Solar Energy," *Philosophical Transactions of the Royal Society a-Mathematical Physical and Engineering Sciences*, Vol. 365, No. 1853, 2007, pp. 993-1005. doi:10.1098/rsta.2006.1963

[3] M. Grätzel, "Conversion of Sunlight to Electric Power by Nanocrystalline Dye-Sensitized Solar Cells," *Journal of Photochemistry and Photobiology A: Chemistry*, Vol. 164, No. 1-3, 2004, pp. 3-14. doi:10.1016/j.jphotochem.2004.02.023

[4] D. B. Kuang, C. Klein, Z. P. Zhang, S. Ito, J. E. Moser, S. M. Zakeeruddin and M. Gratzel, "Stable, High-Efficiency Ionic-Liquid-Based Mesoscopic Dye-Sensitized Solar Cells," *Small*, Vol. 3, No. 12, 2007, pp. 2094-2102. doi:10.1002/smll.200700211

[5] A. Zaban, O. I. Micic, B. A. Gregg and A. J. Nozik, "Photo-Sensitization of Nanoporous TiO_2 Electrodes with InP Quantum Dots," *Langmuir*, Vol. 14, No. 12, 1998, pp 3153-3156. doi:10.1021/la9713863

[6] R. Vogel, P. Hoyer and H. Weller, "Quantum-Sized PbS, CdS, Ag_2S, Sb_2S_3, and Bi_2S_3 Particles as Sensitizers for Various Nanoporous Wide-Bandgap Semiconductors," *Journal of Physical Chemistry B*, Vol. 98, No. 12, 1994, pp. 3183-3188. doi:10.1021/j100063a022

[7] R. Plass, S. Pelet, J. Krueger and M. Grätzel, "Quantum Dot Sensitization of Organic-Inorganic Hybrid Solar Cells," *Journal of Physical Chemistry B*, Vol. 106, No. 31, 2002, pp. 7578-7580. doi:10.1021/jp020453l

[8] L. M. Peter, K. G. U. Wijayantha and J. P. Waggett, "Band-Edge Tuning in Self-Assembled Layers of Bi_2S_3 Nanoparticles Used to Photosensitize Nanocrystalline TiO_2," *Journal of Physical Chemistry B*, Vol. 107, No. 33, 2003, pp. 8378-8381. doi:10.1021/jp030334l

[9] P. Yu, K. Zhu, A. G. Norman, S. Ferrere, A. J. Frank, A. J. Nozik, "Nanocrystalline TiO_2 Solar Cells Sensitized with InAs Quantum Dots," *Journal of Physical Chemistry B*, Vol. 110, No. 50, 2006, pp. 25451-25454. doi:10.1021/jp064817b

[10] L. M. Peter, D. J. Riley, E. J. Tull, K. G. U. Wijayantha, "Photosensitization of Nanocrystalline TiO_2 by Self-Assembled Layers of CdS Quantum Dots," *Chemical Communication*, Vol. 10, 2002, pp. 1030-1031. doi:10.1039/b201661c

[11] O. Niitsoo, S. K. Sarkar, C. Pejoux, S. Rühle, D. Cahen and G. Hodes, "Chemical Bath Deposited CdS/CdSe-Sensitized Porous TiO_2 Solar Cells," *Journal of Photochemistry and Photobiology a-Chemistry*, Vol. 181, No. 2-3, 2006, pp. 306-313. doi:10.1016/j.jphotochem.2005.12.012

[12] I. Robel, V. Subramanian, M. Kuno and P. V. Kamat, "Quantum Dot Solar Cells. Harvesting Light Energy with CdSe Nanocrystals Molecularly Linked to Mesoscopic TiO_2 Films," *Journal of the American Chemical Society*, Vol. 128, No. 7, 2006, pp. 2385-2393. doi:10.1021/ja056494n

[13] S. Rühle, M. Shalom and A. Zaban, "Quantum-Dot-Sensitized Solar Cells," *Journal of Chemical Physics and Physical Chemistry*, Vol. 11, 2010, pp. 2290-2304. doi:10.1002/cphc.201000069

[14] W. W. Yu, L. Qu, W. Guo and X. Peng, "Experimental Determination of the Extinction Coefficient of CdTe, CdSe, and CdS Nanocrystals," *Chemistry of Materials*, Vol. 15, No. 14, 2003, pp. 2854-2860. doi.:10.1021/cm034081k

[15] R. D. Schaller, V. M. Agranovich and V. I. Klimov, "High-Efficiency Carrier Multiplication through Direct Photogeneration of Multi-Excitons via Virtual Single-Exciton States," *Nature Physics*, Vol. 1, No. 3, 2005, pp. 189-194. doi:10.1038/nphys151

[16] M. C. Beard, K. P. Knutsen, P. R. Yu, J. M. Luther, Q. Song, W. K. Metzger, R. J. Ellingson and A. J. Nozik, "Multiple Exciton Generation in Colloidal Silicon Nanocrystals," *Nano Letters*, Vol. 7, No. 8, 2007, pp. 2506-2512. doi:10.1021/nl071486l

[17] A. J. Nozik, "Quantum Dot Solar Cells," *Physica E-Low-Dimensional Systems & Nanostructures*, Vol. 14, No. 1-2, 2002, pp.115-120. doi:10.1016/S1386-9477(02)00374-0

[18] V. I. Klimov, "Spectral and Dynamical Properties of Multiexcitons in Semiconductor Nanocrystals," *Annual Review of Physical Chemistry*, Vol. 58, No. 1, 2007, pp. 635-673. doi:10.1146/annurev.physchem.58.032806.104537

[19] K. C. Mandal and O. Savadogo, "Chemically Deposited n-CdSe Thin Film Photo-Electrochemical Cells: Effects of Zn^{2+}-Modification," *Journal of Materials Science*, Vol. 27, No. 16, 1992, pp. 4355-4360. doi:10.1007/BF00541566

[20] Q. Shen, T. Sato, M. Hashimoto, C. C. Chen and T. Toyoda, "Photoacoustic and Photoelectrochemical Characterization of CdSe-Sensitized TiO_2 Electrodes Composed of Nanotubes and Nanowires," *Thin Solid Films*, Vol. 499, No. 1-2, 2006, pp. 299-305. doi:10.1016/j.tsf.2005.07.019

[21] N. Fuke, L. B. Hoch, A. Y. Koposov, V. W. Manner, D. J. Werder, A. Fukui, N. Koide, H. Katayama and M. Sykora, "CdSe Quantum-Dot-Sensitized Solar Cell with ~100% Internal Quantum Efficiency," *ACS Nano*, Vol. 4, No. 11, 2010, pp. 6377-6386. doi:10.1021/nn101319x

[22] S. Gimenez, I. Mora-Sero, L. Macor, N. Guijarro, T. Lana-Villareal, R. Gomez, L. J. Diguna, Q. Shen, T. Toyoda and J. Bisquert, "Improving the Performance of Colloidal Quantum-Dot-Sensitized Solar Cells," *Nanotechnolgy*, Vol. 20, No. 29, 2009, Article ID: 295304. doi:10.1088/0957-4484/20/29/295204

[23] J. Chen, J. L. Song, X. W. Sun, W. Q. Deng, C. Y. Jiang, W. Lei, J. H. Huang and R. S. Liu, "An Oleic Acid-Capped CdSe Quantum-Dot Sensitized Solar Cell," *Applied Physics Letters*, Vol. 94, No. 15, 2009, Article ID: 153115. doi:10.1063/1.3117221

[24] H. J. Lee, D. Y. Kim, J. S. Yoo, J. Bang, S. Kim and S. M. Park, "Anchoring Cadmium Chalcogenide Quantum Dots (QDs) onto Stable Oxide Semiconductors for QD Sensitized Solar Cells," *Bulletin of the Korean Chemical Society*, Vol. 28, No. 6, 2007, pp. 953-958. doi:10.5012/bkcs.2007.28.6.953

[25] S. Gorer and G. Hodes, "Quantum Size Effects in the Study of Chemical Solution Deposition Mechanisms of Semiconductor Films," *The Journal of Physical Chemistry*, Vol. 98, No. 20, 1994, pp. 5338-5346. doi:10.1021/j100071a026

[26] Y. Lee and Y.-S. Lo, "Highly Efficient Quantum-Dot-Sensitized Solar Cell Based on Co-Sensitization of CdS/CdSe," *Advanced Functional Materials*, Vol. 19, No. 4, 2009, pp. 604-609. doi:10.1002/adfm.200800940

[27] T. Toyoda, K. Oshikane, D. Li, Y. Luo, Q. Meng and Q. Shen, "Photoacoustic and Photoelectrochemical Current Spectra of Combined CdS/CdSe Quantum Dots Adsorbed on Nanostructured TiO_2 Electrodes, Together with Photovoltaic Characteristics," *Journal of Applied Physics*, Vol. 108, 2010, Article ID: 114304. doi:10.1063/1.3517066

[28] T. Mahalingam, S. Thanikaikarasan, M. Raja, C. Sanjeeviraja, S. Lee, H. Moon, Y. D. Kim and P. J. Sebastian, "Studies on Electrochemically Grown Cd-Fe-Se Thin Films," *Journal of New Materials for Electrochemical Systems*, Vol. 10, 2007, pp. 33-37.

[29] R. Chandramohan, T. Mahalingam, J. P. Chu and P. J. Sebastian, "Preparation and Characterization of Semiconducting $Zn_{1-x}Cd_xSe$ Thin Films," *Solar Energy Materials and Solar Cells*, Vol. 81, 2004, pp. 371-378. doi:10.1016/j.solmat.2003.11.013

[30] X. Mathew, N. R. Mathews, P. J. Sebastian and C. O. Flores, "Deep Levels in the Band Gap of CdTe Films Electrodeposited from an Acidic Bath—PICTS Analysis," *Solar Energy Materials and Solar Cells*, Vol. 81, No. 3, 2004, pp. 397-405. doi:10.1016/j.solmat.2003.11.015

[31] X. Mathew, G. W. Thompson, V. P. Singh, J. C. McClure, S. Velumani, N. R. Mathews and P. J. Sebastian, "Development of CdTe Thin Films on Flexible Substrates—A Review," *Solar Energy Materials and Solar Cells*, Vol. 76, No. 3, 2003, pp. 293-303. doi:10.1016/S0927-0248(02)00281-7

[32] T. Mahalingam, C. Sanjeeviraja and S. R. Srikumar, "Studies on Electrochemically Deposited CdSe Thin Films," *Bulletin of Electrochemistry*, Vol. 12, 1996, pp. 148-150.

[33] P. Mahawela, G. Sivaraman, S. Jeedigunta, J. Gaduputi, M. Ramalingam, S. Subramanian, S. Vakkalanka, C. S. Ferekides and D. L. Morel, "II–VI Compounds as the Top Absorbers in Tandem Solar Cell Structures," *Materials Science and Engineering B-Solid State Materials for Advanced Technology*, Vol. 16, 2005, pp. 283-291. doi:10.1016/j.mseb.2004.05.054

[34] S. Vakkalanka, C. S. Ferekides, D. L. Morel, "Development of $ZnSe_xTe_{1-x}$ p-Type Contacts for High Efficiency Tandem Structures," *Thin Solid Films*, Vol. 515, No. 15, 2007, pp. 6132-6135. doi:10.1016/j.tsf.2006.12.160

[35] L. Y. Zeng, S. Y. Dai, W. W. Xu and K. J. Wang, "Dye-Sensitized Solar Cells Based on ZnO Films," *Plasma Science & Technology*, Vol. 8, No. 2, 2006, pp. 172-175. doi:org/10.1088/1009-0630/8/2/10

[36] T. Aramoto, F. Adurodija, Y. Nishiyama, T. Arita, A. Hanafusa, K. Omura and A. Morita, "A New Technique for Large-Area Thin Film CdS/CdTe Solar Cells," *Solar Energy Materials and Solar Cells*, Vol. 75, No. 1, 2003, pp. 211-217. doi:10.1016/FS0927-0248(02)00161-7

[37] H. Uda, H. Sonomura and S. Ikegami, "Screen Printed CdS/CdTe Cells for Visible-Light-Radiation Sensor," *Measurement Science & Technology*, Vol. 8, No. 6, 1997, pp. 86-91. doi:10.1088/0957-0233/8/1/012

[38] A. L. Rogach, A. Kornowski, M. Gao, A. Eychmüller and H. Weller, "Synthesis and Characterization of a Size Series of Extremely Small Thiol-Stabilized CdSe Nanocrystals," *The Journal of Physical Chemistry B*, Vol. 103, No. 16, 1999, pp. 3065-3069. doi:10.1021/jp984833b

[39] G. A. Martínez-Castañón, M. G. Sánchez-Loredo, J. R. Martínez-Mendoza, F. Ruiz, "Synthesis of CdS Nanoparticles: A Simple Method in Aqueous Media," *Azojomo*, Vol. 1, 2005, pp. 1-2. doi:10.2240/azojomo00170

[40] V. G. Kulkarni and P. D. Garn, "Study of the Formation of Cadmium Sulfoselenide," *Thermochimica Acta*, Vol. 99, 1986, pp. 33-36. doi:10.1016/0040-6031(86)85262-5

[41] J. Yu, X. Zhao, J. Du and W. Chen, "Preparation, Microstructure and Photocatalytic Activity of the Porous TiO_2 Anatase Coating by Sol-Gel Processing," *Journal of Sol-Gel Science and Technology*, Vol. 17, 2000, pp. 163-171. doi:10.1023/A:1008703719929

[42] S. Barazzouk and S. Hotchandani, "Enhanced Charge Separation in Chlorophyll a Solar Cell by Gold Nanoparticles," *Journal of Applied Physics*, Vol. 96, No. 12, 2004, pp. 7744-7746. doi:10.1063/1.1811775

[43] A. Hagfeldt, G. Boschloo, L. Sun, L. Klo, H. Pettersson, "Dye-Sensitized Solar Cells," *Chemical Reviews*, Vol. 101, No. 11, 2010, pp. 6595-6663. doi:10.1021/cr900356p

[44] E. Raevskaya, A. L. Stroyuk and S. Ya. Kuchmiy, "Preparation of Colloidal CdSe and CdS/CdSe Nanoparticles from Sodium Selenosulfate in Aqueous Polymers Solutions," *Journal of Colloid and Interface Science*, Vol. 302, No. 1, 2006, pp. 133-141. doi:10.1016/j.jcis.2006.06.018

Appendix

Figure S1 displays TEM images of the CdSe NCs synthesized as described in the experimental section (***route*** 1). The size distributions and flourescence colours are indicated in the insets of **Figure S1**.

Figure S2 shows the absorption spectra of the CdSe NCs of different size measured in an aqueous solution.

Samples were taken from the same reaction batch at different reaction times. The particles grew with time, so 1 - 2 nm sized particles are obtained after 1 minute, particles with a size of 30 nm after 2 days. The absorption increases and shifts with larger particles to the red.

Figures S3(a) and **(b)** show the absorption spectra of mixtures of CdSe NCs of different sizes with TiO_2 on FTO coated glass as obtained by reflection spectroscopy. The absorption of pure TiO_2 nanoparticles is already subtracted here. **Figure S3(a)** exhibits the absorption spectra of the samples before annealing at 500°C. The CdSe NCs up to a size of 10 nm show a fairly good absorption in the range from 350 to 650 nm. The CdS NCs which were prepared for comparison have a twice as large absorption in the range 350 - 500 nm compared to the CdSe samples. **Figure S3(b)** shows the situation after annealing at 500°C and contact with the Na_2S/Na_2SO_4 electrolyte. All reflection spectra up to a particle size of 10 nm are similar to the CdS spectrum. The emergence of the 400 - 500 nm band demonstrates that a selenide-sulphide exchange had taken place in the smaller CdSe particles. After sintering the TiO_2/CdSe sheets was colourless and they change to yellow upon addition of the electrolyte was visible to the naked eye. Presumably SeO_2 was formed and evaporated during the annealing at 500°C. The remaining cadmium ions reacted with the electrolyte to CdS. From these results we expect CdS to be the main TiO_2-photosensitizer at >400 nm.

Figure S4 displays IPCE curves with CdSe particles of different size (***route* 1**) in volume ratios of 1:1, 1:3 and 1:5 (CdSe:TiO_2) and Na_2S/Na_2SO_4 as redox couple. For smaller particle sizes (green and yellow samples) IPCE increases with increasing volume ratio of NCs while larger particles (orange and red samples) show an opposite behaviour. This can be explained by an easier intercalation of the smaller NCs between the TiO_2 nanoparticles. Therefore these particles can be mixed homogeneously in great quantities into the TiO_2 using the SMD procedure. Larger NCs do not fit into the free space between the TiO_2 particles. In larger quantities they form NC layers which electrically isolate TiO_2. The smallest CdSe particles of size 1 - 2 nm in a 1:1 volume ratio with TiO_2 exhibit the largest IPCE of about 40% with a photoelectrical sensitivity up to $\lambda \sim 500$ nm.

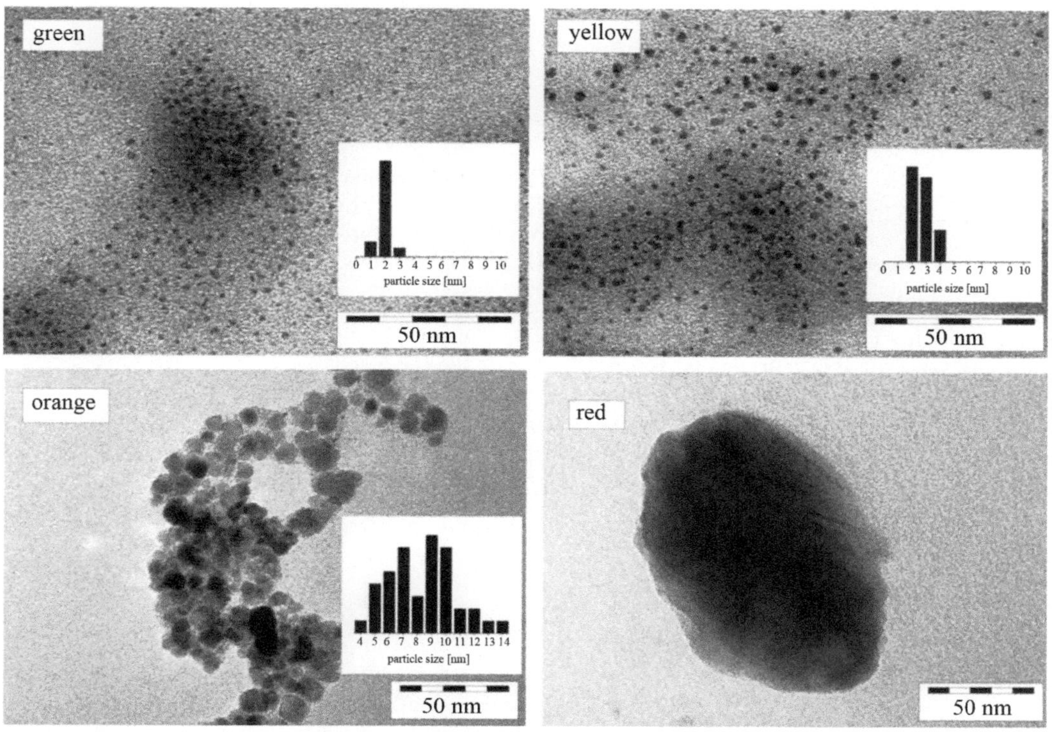

Figure S1. TEM images of CdSe NCs of different size and colour. Their synthesis is described in the experimental section (*route* 1). The insets indicate the size distribution as obtained from diameter measurements of 20 - 40 particles. Typical size ranges are 1 - 2 nm for the green fluorescing particles, 2 - 4 nm for the yellow, 6 - 12 nm for the orange and 30 - 50 nm for the red fluorescing particles.

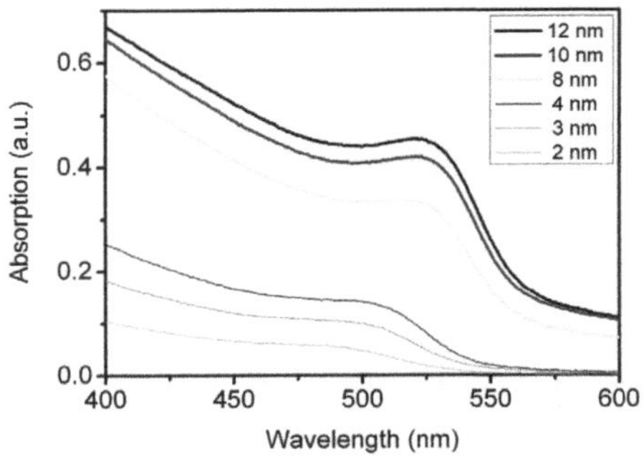

Figure S2. Absorption spectra of CdSe NCs (*route* 1) of different size measured in solution.

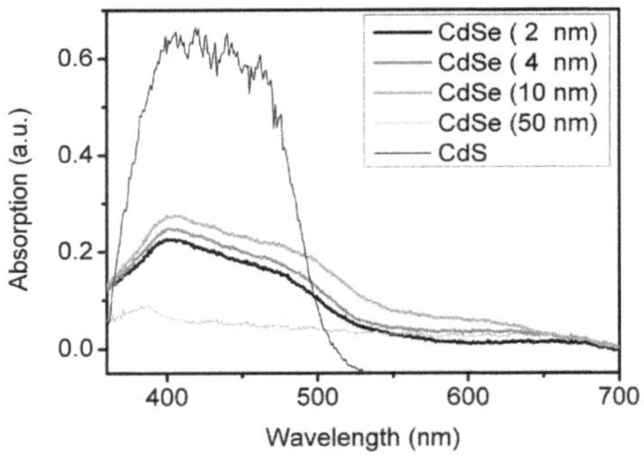

(a)

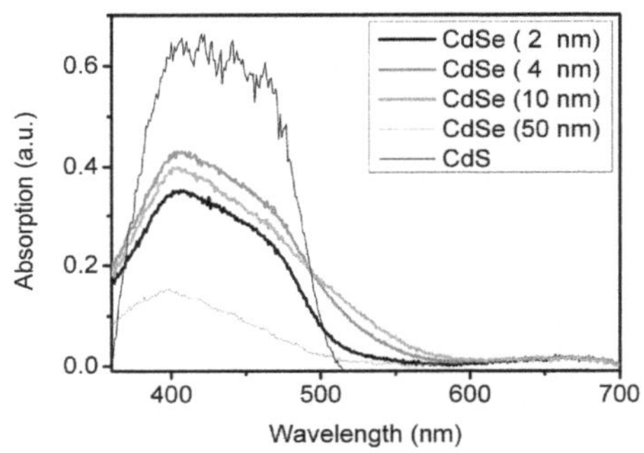

(b)

Figure S3. Absorption spectra of CdSe NCs (*route* 1) of different size and CdS NCs measured in reflection (a) before annealing at 500°C and (b) after annealing at 500°C and electrolyte (Na_2S/Na_2SO_4) contact. Due to $Se^{2-} \leftrightarrow S^{2-}$ exchange CdS is the main absorber in a CdSe/TiO_2 sintered mixture with electrolyte contact.

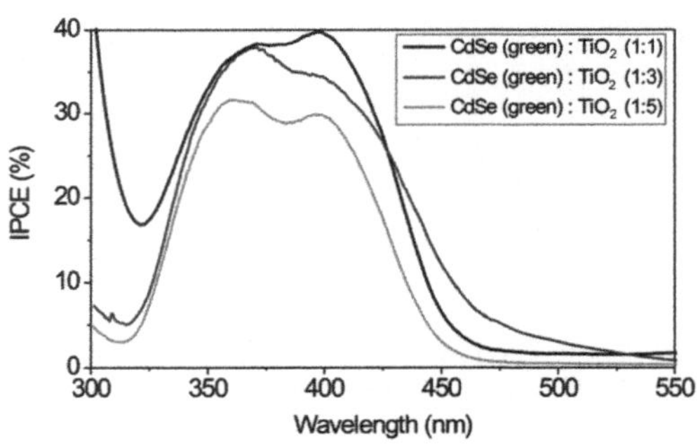

(a)

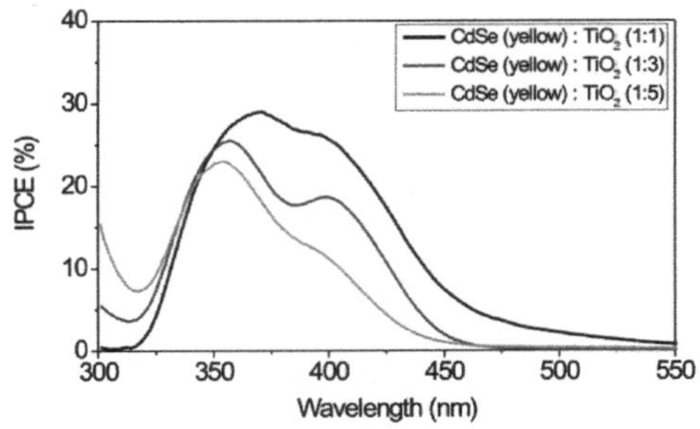

(b)

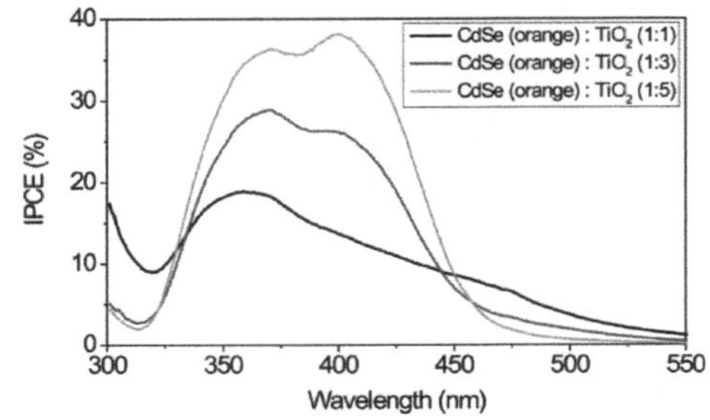

(c)

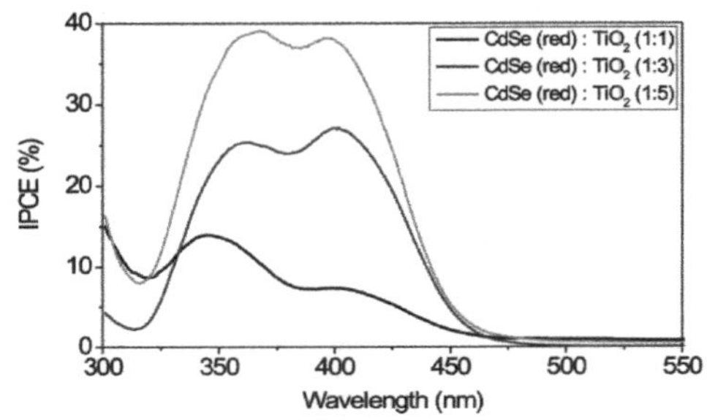

(d)

Figure S4. Comparison of the incident photon-to-current efficiencies (IPCE) of CdSe solar cells working with Na_2S/Na_2SO_4 redox couple and CdSe NCs of different size (*route* 1). NC/TiO_2 volume ratios of 1:1, 1:3 and 1:5 were used.

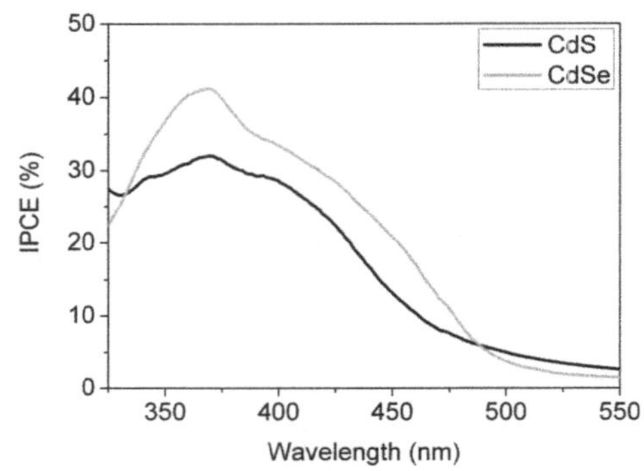

Figure S5. Comparison of the incident photon-to-current conversion efficiencies (IPCE) of CdSe and CdS solar cells working with a Na_2S/Na_2SO_4 redox couple, 1 - 2 nm NCs and 1:1 NC/TiO_2 volume ratio. Due to $Se^{2-} \leftrightarrow S^{2-}$ exchange CdS is the main sensitizer in a CdSe/TiO_2/Na_2S, Na_2SO_4 solar cell fabricated by sintered mixture coating of the photocathode.

In **Figure S5** the best IPCE spectrum of the TiO_2/CdSe samples is compared with a TiO_2/CdS solar cell. The IPCE spectral developing reflects the wavelength dependence of the CdS absorption in **Figure S3(a)** pointing again to CdS $Se^{2-} \leftrightarrow S^{2-}$ exchange and CdS as main sensitizer in a CdSe/TiO_2/Na_2S, Na_2SO_4 solar cell fabricated by sintered mixture coating. Synthesis of CdSe QDs via ***route* 1** is therefore not suitable for SMD and we used ***route* 2** synthesis for further work, see paper.

Theoretically Catalytic Synthesis of 5-Nitro-1,2,4-Triazol-3-One in Inert Gas Clustered System (X_6, X = He, Ne)

Min-Hsien Liu*, Ming-Yung Wu
Department of Chemical and Materials Engineering, Chung Cheng Institute of Technology National Defense University, Taiwan, China
Email: *minhsienliu@gmail.com

ABSTRACT

Inert gas-clustered systems (X_n, X = He, Ne, Ar and n = 2 - 20) were established in this study and their stability as a result of interparticulate interaction was examined. Ferric chloride and ferrous oxides were used as catalysts to promote reaction, and 5-nitro-1,2,4-triazol-3-one (NTO) was theoretically synthesized under an inert gas (X_6)-clustered environment in this study. The raw material, urea, initially underwent chlorination using chlorine as the reagent, followed by amination, formylation and nitration. Reaction routes closely related to the experimental processes were successfully constructed, and the corresponding energy barriers were estimated for each elementary reaction. The findings revealed that the average errors in the B3LYP/6-31G(d, p)-calculated geometry and vibrational frequency of NTO in an Ne_6 system relative to the observed values were 0.83% and 1.84%, respectively. The neon gas-clustered system achieved greater stabilization, which results from the difference in self-consistent field energy (E_{SCF}), than the corresponding stabilization acquired in a helium- or argon-based system. Ferric chloride serves as a good catalyst to reduce the energy barrier of the chlorination reaction, and ferrous oxide is suitable for catalyzing the amination, formylation and nitration reactions, although nitric acid is the better agent for nitration. The catalytic Ne_6-clustered reaction system is suggested to be a more feasible pathway for the synthesis of NTO.

Keywords: NTO; Inert Gas Clustered System; Metal Catalyst; Formylation Reaction

1. Introduction

In synthesis chemistry, specific materials are obtained efficiently under certain conditions through a well-designed chemical reaction. The design of an optimum reaction, including the use of suitable reagents and the setting of the reaction conditions, is an important issue and is usually considered prior to performing experiments. Generally, the whole scheme of kinetic details is confined to the fast rate of the general reaction, which cannot be easily understood through experimental techniques. The adoption of theoretical calculation methods [1-9] has been proposed to overcome the shortage of experiments in recent decades. Based upon our experience in this area, computational effort has been devoted to the development of new high-energy-density materials (HEDM), TNAD and TNAZ derivatives, [10,11] and to the exploration of the kinetic reaction pathways of FOX-7 and TNT synthesis [12,13] in our lab. There remains the question of the reliability of the computational results. Basic geometric and vibrational analyses are required to compare the calculated data with the observed values. Comparison of molecular geometries for a single molecule in the gaseous phase always shows apparent deviation from related X-ray analysis of molecules those with intermolecular interactions in condensed phase. Furthermore, interpretation and analysis of the accuracy of the theoretical frequencies should be performed in connection with gas-phase experimental results using a rare gas matrix [14-18]. The geometry and vibrational frequency apparently have a significant influence on the molecular thermodynamic properties and the reaction kinetics, and accordingly the phase environment has to be taken into consideration when performing calculations. It is said that short-lived, highly-reactive species such as radical ions and reaction intermediates may be observed using the matrix isolation technique and identified by spectroscopic means. [19,20] Molecular modeling in a rare gas-clustered system is regarded as being closely related to experiments using a gas matrix in the real gaseous world.

A promising material, 5-nitro-1,2,4-triazol-3-one (NTO), has emerged as a potential high-performance (e.g., cf. detonation velocity, detonation pressure) insensitive ex-

*Corresponding author.

plosive [21,22-26] that might endure a heavy mechanical and thermal impact, and this material is being explored as a substitute for RDX in munitions. NTO is easily prepared by the reaction of semicarbazide hydrochloride (SC) with formic acid followed by nitration with 70% nitric acid at 60°C [27]. There are many studies of NTO in the literature, including experiments that have been carried out to explore ways in which to improve the synthesis of the condensed phase of NTO [28-30], and some publications have discussed its thermal properties and decomposition kinetics [30-32]. Experimental and theoretical consideration of the structure of NTO and vibrational analysis [32-35] using quantum mechanical methods will provide more insight into the decomposition of NTO. In spite of theoretical aspects of chemical reactivity providing a broad overview of recent theoretical and computational advancements in the field of chemical reactivity [36-38], there exists little information regarding the application of theoretical techniques to simulate chemical reactions. This work is based on theory and refers to the related reaction mechanisms (**Scheme 1**) [27] in order to elucidate catalytic reaction routes for NTO synthesis in a special inert gas-clustered system. The reactivity of the reactants, which is generally inversely proportional to the molecular stability, indicates the rate at which a chemical substance tends to undergo a chemical reaction, was mainly concerned herein. Additionally, the integrated stabilization energy that arises from interparticlate interacttion in specific rare gas systems was examined and the related stabilization effect was inferred in order to identify the best gaseous reaction environment. In such stable systems, the proposed reaction profile included catalytic chlorination→amination→formylation→cyclocondensation→nitration in sequence using suitable reagents and catalysts. The comparable activation energies of each reaction stage were then used to identify favorable pathways to improve the synthesis of NTO.

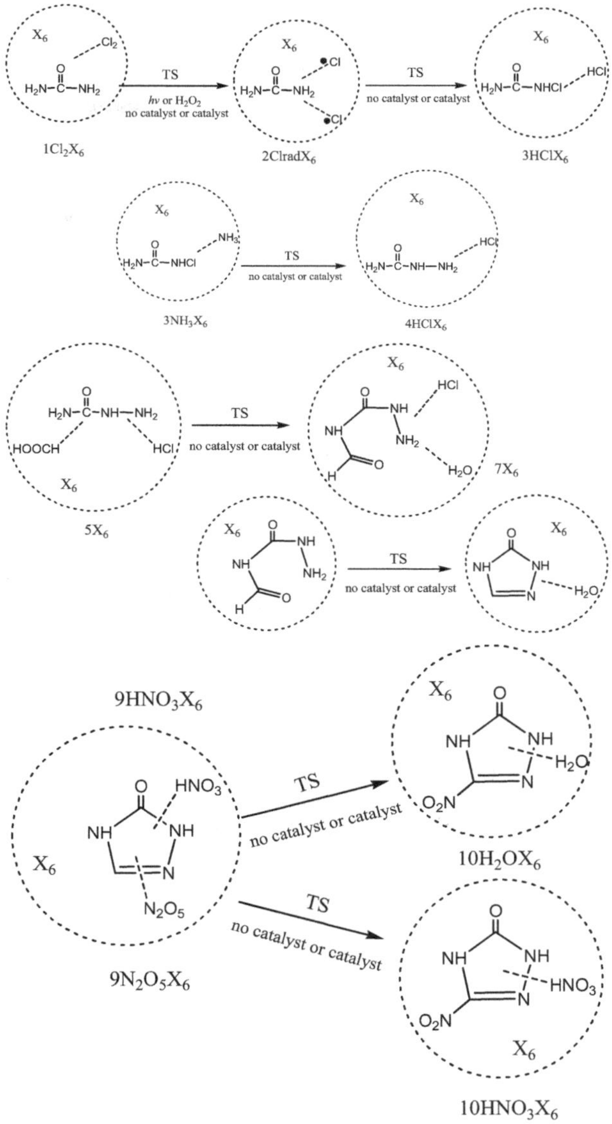

Scheme 1. Suggested reaction mechanisms of NTO synthesis in rare gas clustered system (X_6, X = He, Ne).

2. Computations

2.1. Geometrical Optimization and Molecular Thermodynamic Energy

Electronic structure calculations have been performed with the Gaussian 98 program [39]. Full geometry optimizations were made for all the stationary points using the B3LYP [40] hybrid density functionals and the 6-31G(d, p) basis set. Restricted calculations were used for closed shell systems and unrestricted ones for open shell systems. Frequency calculations were carried out for all the stationary points at the corresponding level of theory. Local minima and transition states were identified by the number of imaginary frequencies (0 or 1, respectively). The B3LYP/6-31G(d, p) method was used to calculate the molecular energy (E_{SCF}). Molecular thermal enthalpy (H) and Gibbs energy (G) were obtained by adding thermal correction to the molecular energy.

2.2. Transition-States Modeling

The transition-state species were modeled using the B3LYP/6-31G(d, p) calculation method and were then identified using the QST3-type optimization procedure in the program. [41,42] Any pairs of equally-sized molecular systems (with the same numbers of C, H, O and N atoms) from the optimized products of the reaction species were selected, and the created input file, with two sets of relative atomic Cartesian coordinates, was included in the QST3 calculation. The activation energy for each elementary reaction step was determined from the thermal enthalpy difference between a reactant and a transition complex.

3. Results and Discussion

3.1. Molecular Geometry and Energy

The B3LYP/6-31G(d, p)-modeled geometries of inert gas-clustered Xn (X = He, Ne, Ar, n = 2 - 20) indicated a shorter average interparticulae distance in the Ne system. (cf. 5.033 Å in Ne_6, 5.458 Å in He_6 and 5.701 Å in Ar_6; **Figure 1**) Additionally, the stabilization energy, which results from the difference in self-consistent field energy between $E_{SCF}(Xn)$ and $nE_{SCF}(X)$, indicated the relative stability of the Ne-clustered system (**Figure 2** and **Table 1**). The geometry of $NTONe_6$ (**Scheme 2**) was then compared with the experimental values [33,34,43,44] (**Table 2**), and the results showed a 0.83% average relative error in bond length and a 1.84% average relative error in vibrational frequency, which revealed that B3LYP/6-31G (d, p) was an accurate method for further computational use.

The related geometries of all species, including the optimized reactants and products (all with positive frequentcies) and transition complexes (one with an imaginary frequency), in a He_6 or Ne_6 gaseous environment indicated a weakly-bound molecular system (Figure in the supplementary material). The B3LYP/6-31G (d, p)-calculated molecular thermodynamic energies of all species are presented in **Tables 3-5**. The stabilization effect for all Ne_6-clustered reactants (with a stabilization energy of around 50 kJ/mol) is more obvious than the corresponding effect in the He6-clustered system (**Table 3**).

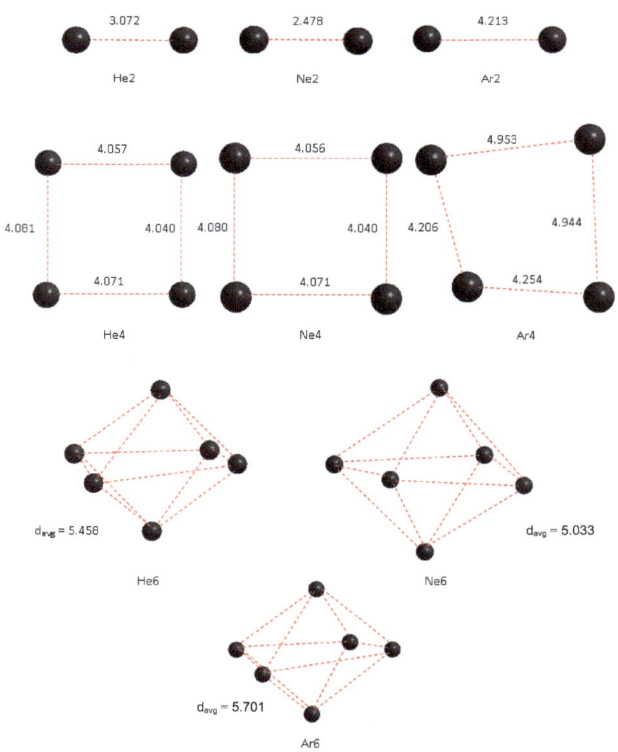

Figure 1. Optimized structure of inert gas clustered system (X_n, X = He, Ne, Ar, n = 2, 4, 6).

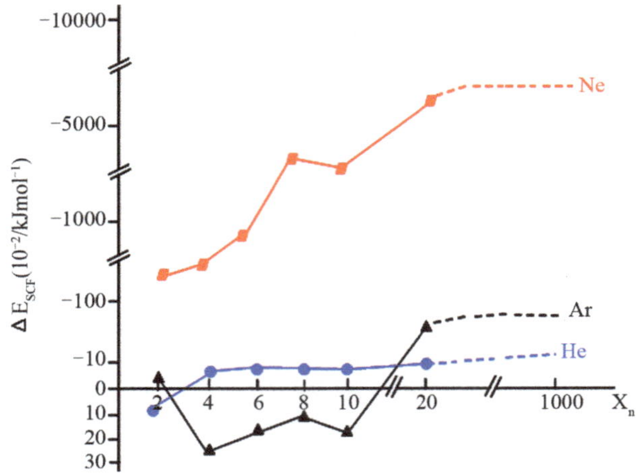

Figure 2. Stabilization on energy of inert gas clustered systems (Xn, X = He, Ne, Ar, n ≥ 2).

Table 1. Self-consistent field energy of weakly-bound inert gas systems.

Inert gas system	E_{SCF}	ΔE_{SCF} *	IR frequency#
He	–2.9070	0	-
He_2	–5.8141	0.0735	44.8
He_4	–11.6282	–0.0053	10.1 - 26.4
He_6	–17.4423	–0.0079	2.8 - 13.1
He_8	–23.2564	–0.0184	0.88 - 30.8
He_{10}	–29.0705	–0.0236	0.81 - 34.2
He_{20}	–58.1410	–0.0525	0.7 - 19.2
Ne	–128.8944	0	-
Ne_2	–257.7898	–2.7436	152.2
Ne_4	–515.5851	–6.1069	8.5 - 26.7
Ne_6	–773.3768	–9.8089	2.2 - 16.1
Ne_8	–1031.1721	–45.2899	44.0 - 206.3
Ne_{10}	–1288.9603	–43.8537	11.1 - 204.3
Ne_{20}	–2577.9206	–87.8020	1.0 - 208.4
Ar	–527.5171	0	-
Ar_2	–1055.0343	–0.0105	20.2
Ar_4	–2110.0685	0.2678	2.2 - 40.0
Ar_6	–3165.1028	0.0814	1.9 - 21.8
Ar_8	–4220.1371	0.0289	3.8 - 20.7
Ar_{10}	–5275.1714	0.1208	1.0 - 24.7
Ar_{20}	–10550.3431	–0.7509	1.2 - 21.3

* $\Delta E_{SCF} = [E_{SCF}(X_n) - nE_{SCF}(X)] \times 2625.5$, in kJ/mol unit; # in cm^{-1} unit

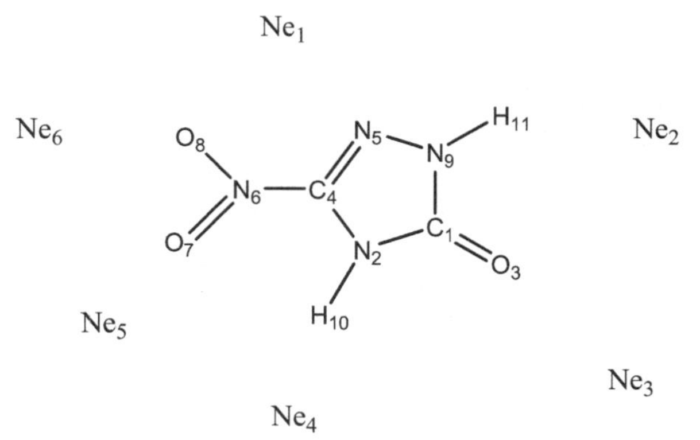

Scheme 2. Geometry of neon gas clustered-5-nitro-1,2,4-triazol-3-one ($NTONe_6$).

Table 2. Geometrical and vibrational analysis of NTO molecule.

Results* Bond type	Distance	Obs. a	b	b, c	Vib. Frequency (Scaled by 0.96)	Obs. b(1)	b(2)	d
C_4-N_6	1.442 (−0.35%)#	1.443	1.451	1.447	1157	-	1021	1018
C_1-N_2	1.393 (1.09%)	1.380	1.405	1.378	1055 (−2.77%)#	1085	1111	1109
C_1-N_9	1.388 (1.54%)	1.374	1.398	1.367	1224 (−2.63%)	1257	1282	1280
N_9-N_5	1.356 (−0.95%)	1.366	1.359	1.369	1375 (2.77%)	1338	1343	1340
C_4-N_2	1.368 (1.41%)	1.356	1.365	1.349	1422 (−2.80%)	1463	1477	1473
C_4=N_5	1.298 (0.62%)	1.299	1.290	1.290	1560	-	-	1541
N_6-O_7	1.224 (0.58%)	1.225	1.214	1.217	1573 (0.64%)	1563	1605	1546
C_1=O_3	1.225 (−0.08%)	1.234	1.203	1.226	1775 (−0.78%)	1789	1716	1712
N_9-H_{11}	1.010	1.009	1.008	0.916	3506 (0.49%)	3489	3200	3198
N_2-H_{10}	1.010	1.009	1.008	0.875	3507	-	-	3242
M. | R.E.	0.83%	-	-	-	1.84%	-	-	-

*B3LYP/6-31G (d, p) calculated; #Relative error (R.E.), refer to the experimental data in Ref. 34; a:see Ref. 43; b:see Ref. 34, b(1)/Ar matrix and b(2)/thin solid film; c:see Ref. 44; d:see Ref. 33.

Table 3. Self-consistent Field energy of helium and neon clustered reaction systems.

Molecular system	E_{SCF}	Molecular system	E_{SCF}	ΔE_{SCF}*
$1Cl_2He6$	−1163.078491	$1Cl_2$ + 6He	−1163.077688	−2.1083
2ClradHe6	−1163.035446	2Clrad + 6He	−1163.033733	−4.4975
3HClHe6	−1163.064750	3HCl + 6He	−1163.063498	−3.2871
$3NH_3He6$	−758.835059	$3NH_3$ + 6He	−758.833718	−3.5208
4HClHe6	−758.84490	4HCl + 6He	−758.843337	−4.1037
5He6	−948.635476	5 + 6He	−948.633465	−5.2799
6He6	−411.354934	6 + 6He	−411.353427	−3.9566
7He6	−948.596018	7 + 6He	−948.594356	−4.3636
$8H_2OHe6$	−411.373110	$8H_2O$ + 6He	−411.371673	−3.7728
$9HNO_3He6$	−615.830430	$9HNO_3$ + 6He	−615.828743	−4.4292
$9N_2O_5He6$	−820.267955	$9N_2O_5$ + 6He	−820.266964	−2.6019
$10H_2OHe6$	−615.866998	$10H_2O$ + 6He	−615.865742	−3.2976
$10HNO_3He6$	−820.326287	$10HNO_3$+6He	−820.325528	−1.9928
$1Cl_2Ne6$	−1919.018086	$1Cl_2Ne6$	−1919.001554	−43.4048
2ClradNe6	−1918.977538	2Clrad + 6Ne	−1918.957599	−52.3498
3HClNe6	−1919.006041	3HCl + 6Ne	−1918.987364	−49.0365
$3NH_3Ne6$	−1514.775193	$3NH_3$ + 6Ne	−1514.757584	−46.2324
4HClNe6	−1514.785991	4HCl + 6Ne	−1514.767203	−49.3279
5Ne6	−1704.575934	5 + 6Ne	−1704.557331	−48.8422
6Ne6	−1167.297423	6 + 6Ne	−1167.277293	−52.8513
7Ne6	−1704.542412	7 + 6Ne	−1704.518222	−63.5108
$8H_2ONe6$	−1167.315406	$8H_2O$ + 6Ne	−1167.295539	−52.1608
$9HNO_3Ne6$	−1371.773810	$9HNO_3$ + 6Ne	−1371.752609	−55.6640
$9N_2O_5Ne6$	−1576.208445	$9N_2O_5$ + 6Ne	−1576.190830	−46.2482
$10H_2ONe6$	−1371.809549	$10H_2O$ + 6Ne	−1371.789608	−52.3551
$10HNO_3Ne6$	−1576.270748	$10HNO_3$ + 6Ne	−1576.249394	−56.0649

* $\Delta E_{SCF} = \left[E_{SCF}(species \cdots X_6) - E_{SCF}(species + 6X)\right] \times 2625.5$, X = He and Ne, in kJ/mol unit.

3.2. Characteristics of Some Elementary Reactions

The synthesis of NTO from urea was performed in a six-membered rare gas-clustered system (He_6 and Ne_6), the reactions being in the sequence of chlorination, amination, formylation, and nitration, with the adoption of catalysts and corresponding reagents. In addition to the superior stabilization effect that was observed in the neon-clustered system (**Table 3**), comparative analysis of each reaction in which free radical or ionic-type substitution occurred was performed, as discussed below.

3.3. Chlorination of Urea

Chlorine gas was initially cleaved to produce chlorine radicals. Urea was then chlorinated in a neon-clustered and a helium-clustered reaction system, respectively, to prepare N-chlorine urea. The calculation results revealed that a total activation energy of 182.9 kJ/mol was required for chlorine cleavage ($Ea_{12radHe6}$ = 121.5 kJ/mol) followed by chlorine substitution of an amine hydrogen ($Ea_{23HClHe6}$ = 61.4 kJ/mol) in the He6 system, while a total of 173.5 kJ/mol was required for the same reaction to proceed in the Ne_6 system. The energy difference be-

Table 4. Thermodynamic energy of reaction species of NTO synthesis.

Energy* Species	B3LYP/6-31G(d,p)		
	E_{SCF}	H	G
1Cl$_2$He6	−1163.078491	−1162.986885	−1163.093681
2CIradHe6	−1163.035446	−1162.943360	−1163.049030
3HClHe6	−1163.064750	−1162.975878	−1163.078242
3NH$_3$He6	−758.835059	−758.717684	−758.813489
4HClHe6	−758.84490	−758.728555	−758.831724
5He6	−948.635476	−948.478824	−948.587206
6He6	−411.354934	−411.239798	−411.333159
7He6	−948.596018	−948.440155	−948.554078
8H$_2$OHe6	−411.373110	−411.257982	−411.354280
9HNO$_3$He6	−615.830430	−615.710020	−615.818177
9N$_2$O$_5$He6	−820.267955	−820.144442	−820.263175
10H$_2$OHe6	−615.866998	−615.747083	−615.850369
10HNO$_3$He6	−820.326287	−820.201318	−820.317571
1Cl$_2$Ne6	−1919.018086	−1918.925066	−1919.022028
2ClradNe6	−1918.977538	−1918.884995	−1918.976024
3HClNe6	−1919.006041	−1918.916523	−1919.007846
3NH$_3$Ne6	−1514.775193	−1514.656512	−1514.747725
4HClNe6	−1514.785991	−1514.669109	−1514.761627
5Ne6	−1704.575934	−1704.419063	−1704.515776
6Ne6	−1167.297423	−1167.180526	−1167.267036
7Ne6	−1704.542412	−1704.386050	−1704.485649
8H$_2$ONe6	−1167.315406	−1167.199561	−1167.288638
9HNO$_3$Ne6	−1371.773810	−1371.652991	−1371.745242
9N$_2$O$_5$Ne6	−1576.208445	−1576.084565	−1576.188654
10H$_2$ONe6	−1371.809549	−1371.689051	−1371.780684
10HNO$_3$Ne6	−1576.270748	−1576.145535	−1576.243347

*E_{SCF} , H (thermal enthalpy) and G (Gibbs free energy) in au.

Table 5. Thermodynamic energy of reaction species of NTO synthesis (continued).

Energy* Species	B3LYP/6–31G (d, p)		
	E_{SCF}	H	G
1Cl$_2$Ne6FeCl$_3$	-4563.256395	-4563.125531	-4563.190213
2ClradNe6FeCl$_3$	-4563.235088	-4563.104869	-4563.164119
3HClNe6FeCl$_3$	-4563.241093	-4563.112217	-4563.172849
3NH$_3$Ne6FeO	-2853.484392	-2853.333243	-2853.394204
4HClNe6FeO	-2853.481378	-2853.330736	-2853.391812
5Ne6FeO	-3043.210429	-3043.018312	-3043.088748
6Ne6FeO	-2505.932844	-2505.781961	-2505.836291
7Ne6FeO	-3043.263005	-3043.071021	-3043.144449
8H$_2$ONe6FeO	-2505.949772	-2505.797756	-2505.850440
9HNO$_3$Ne6FeO	-2710.512044	-2710.353101	-2710.407649
9N$_2$O$_5$Ne6FeO	-2914.846054	-2914.686007	-2914.737940
10H$_2$ONe6FeO	-2710.498635	-2710.340116	-2710.394519
10HNO$_3$Ne6FeO	-2914.911227	-2914.749782	-2914.801342

*E_{SCF} , H (thermal enthalpy) and G (Gibbs free energy) in au.

tween the above two reaction systems was attributed to the fact that the chlorine radical was separated by 2.284 - 2.435 Å and was distant from the amino H atom by 1.987 - 2.425 Å in the transition complexes ($TS_{12radNe6}$ and $TS_{23HClNe6}$, **Figure 3**) of the neon-clustered system, while the corresponding distances were shorter in the transition states of the helium-clustered system at 2.544 - 2.648 Å and ~2.436 Å, respectively ($TS_{12radHe6}$ and $TS_{23HClHe6}$). When a metal chloride catalyst was included in the reaction, a total of 131.8 kJ/mol ($Ea_{12radNe6FeCl3}$ + $Ea_{23HClNe6FeCl3}$) was required to complete the reaction. Therefore, ferric chloride exhibited a catalytic effect in this chlorination stage (**Figure 3**). Inspection of the molecular geometries of the transition state complexes showed that the two chlorine atoms were separated by only 2.109 Å, and they formed two strong hydrogen bonds (with H-bond distances of 1.843 Å and 1.847 Å) with the amine hydrogen in $TS_{12radNe6FeCl3}$, stabilizing the molecular system. This suggests that the use of a ferric chloride catalyst is advantageous for optimum progression of the chlorination stage in a neon gas environment.

3.4. Amination of N-Chloro Urea

In the next stage of the reaction, ammonia was used as the amination reagent for the amination of N-chloro urea in helium gas-clustered (3NH$_3$He6), neon gas-clustered (3NH$_3$Ne$_6$) and ferrous oxide-catalyzed neon gas-clustered (3NH$_3$Ne$_6$FeO) systems. The corresponding energy barrier for the introduction of an amino group into the main frame to form semicarbazide hydrochloride was lowest at 135.2 kJ/mol ($Ea_{34HClNe6FeO}$) when the reaction proceeded through the FeO catalytic route (3NH$_3$Ne$_6$FeO→ 4HClNe$_6$FeO); hence, ferrous oxide exhibited an obvious catalytic effect for the amination reaction (**Figure 4**). By surveying the molecular structure and atomic charge density, the following observations were made: 1) in 3NH$_3$Ne$_6$FeO, weak interaction between the Fe and N atoms led to the atomic charge of N being –0.6144, slightly higher than the related charge of –0.6231 in 3NH$_3$Ne$_6$, which elongated the N-Cl bond by 0.004 Å as compared with the length of 1.739 Å in 3NH$_3$He6 and 3NH$_3$Ne$_6$. The more positive N provides a more positive and reactive site for NH_3 to attack. 2) In $TS_{34HCLNe6FeO}$, the electron-withdrawing effect between FeO and the amino H caused N (charge density –0.6086) to favorably combine with NH_2^-, and a strong bonding interaction will form between the Cl^- ion and the ammonia H (Cl···H 1.578 Å); thus, the corresponding interparticulate force stabilized the transition state complex.

3.5. Nitration of 1,2,4-Triazol-3-One

Nitric acid (HNO_3) and dinitrogen pentoxide (N_2O_5)

were used as nitration reagents for the nitration of 1,2,4-triazol-3-one (TO) in non-catalytic rare gas-clustered systems (He_6 or Ne_6) or a catalytic neon gas-clustered system to obtain 5-nitro-1,2,4-triazol-3-one. The reactionusing HNO_3 as the reagent and ferrous oxide as the catalyst, which proceeded through electrophilic substitution via the route $9HNO_3Ne_6FeO \rightarrow 10H_2ONe_6FeO$, had an energy barrier of 81.6 kJ/mol ($Ea_{910H2One6FeO}$) that had to be overcome: this was lower than the energy barrier in the other reaction systems in which HNO_3 or N_2O_5 was used to nitrate 1,2,4-triazol-3-one in non-catalytic He_6 and non-catalytic Ne_6 environments (**Figure 5**). By inspect-

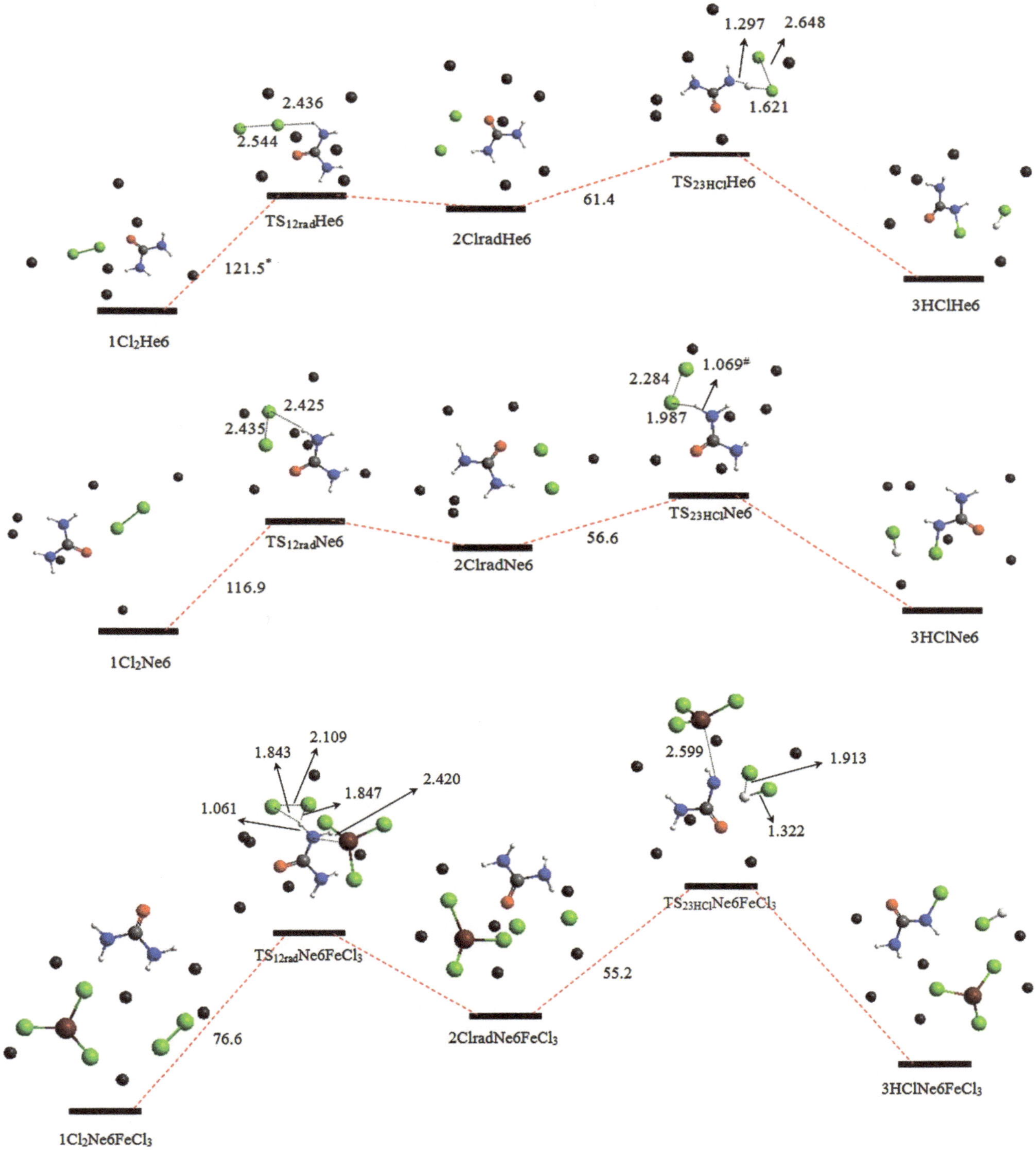

Figure 3. Comparison of chlorination of urea in inert gas clustered system (continued).

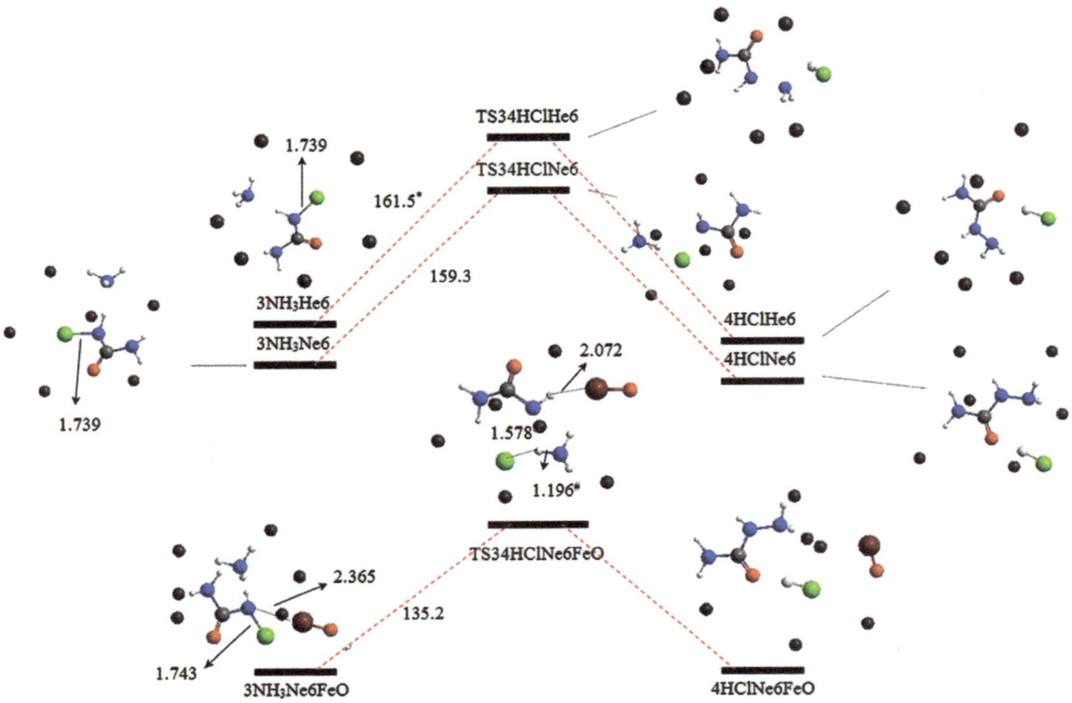

Figure 4. Comparison of amination of N-chloro urea in inert gas clustered system (*activation energy, in kJ/mole; #distance, in Å).

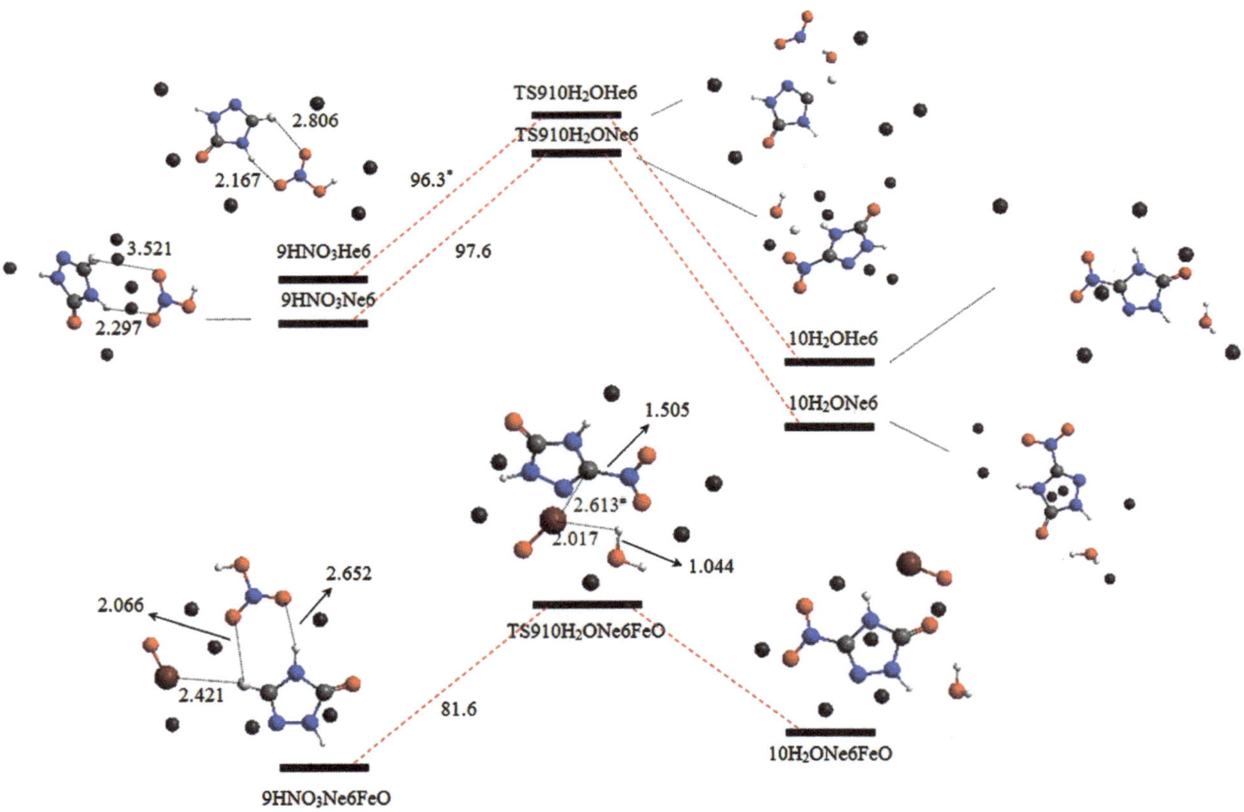

Figure 5. Comparison of amination of N-chloro urea in inert gas clustered system (*activation energy, in kJ/mole; #distance, in Å).

tion of the molecular structure and atomic charge density, the following characteristics were observed: 1) in the reactant system $9HNO_3Ne_6FeO$, the weak interacttion between FeO and H (distant by 2.421 Å) resulted in an electron-withdrawing effect of Fe, pulling the H atom away from the 1,2,4-triazol-3-one (TO) mainframe, resulting in the charge density of C (0.2725) being less positive than the related charge density of C in $9HNO_3He6$ and $9HNO_3Ne_6$ (0.3001 and 0.354, respectively), thus providing a more reactive site for further NO_2^+ substitution. 2) In $TS_{910H2One6FeO}$, the $^-$O-H anion in HNO_3 directly pulled away an amino H to form H_2O, which then interacted strongly with FeO, and the negatively-charged C atom could then easily combine with the NO_2^+ ion. The whole scheme transition complex scheme indicated that no separate active particles existed, thus resulting in the integrated stability of the intermediate system.

4. Conclusion

Rare gas-clustered systems (X_n, X = He, Ne, Ar, n = 2 - 20) were modeled using a quantum mechanical computation method. Neon-clustered systems were found to result in a greater stabilization energy (ΔE_{SCF}) than the respective helium- or argon-clustered systems owing to the interaction of individual gas atoms. B3LYP/6-31G(d, p)-calculated geometries and vibrational frequencies are close to the experimental values, and therefore this method can be implemented in cases in which experiments are unable to be performed to obtain the related data, especially for precious high-energy-density materials. Another important purpose of this research was the exploration of feasible pathways for the gaseous-phase synthesis of 5-nitro-1,2,4-triazol-3-one (NTO). Indeed, the adoption of theoretical techniques to model molecular reaction systems provides a new method of investigation of the synthesis of specific materials and could offer usefulinformation for reference prior to experiments being performed. The findings revealed the optimum conditions for a series of reactions, and showed for the studied reaction processes that chlorination is feasible in a ferric chloride-catalyzed Ne_6-clustered system. The results indicated that ferrous oxide is a good catalyst, effectively reducing the energy barrier to promote amination and nitration, and that nitric acid was a suitable nitration reagent. Together, the findings indicated an improved method of NTO synthesis.

5. Acknowledgements

The authors would like to thank the National Science Council of the Republic of China for financially supporting this research under Contract No. NSC 98-2113-M-606-001-MY2 and thank the National Center for High-performance Computing for the supporting of calculation facility.

REFERENCES

[1] Z. X. Yu, Q. Dang and Y. D. Wu, "Aromatic Dienophiles. 1. A Theoretical Study of an Inverse-Electron Demand Diels-Alder Reaction between 2-Aminopyrrole and 1,3,5-Triazine," *The Journal of Organic Chemistry*, Vol. 66, No. 18, 2001, pp. 6029-6036.

[2] C. S. Wannere, R. K. Bansal and P. V. R. Schleyer, "Diels-Alder Reaction of Phosphaethene with 1,3-Dienes: An AB Initio Study," *The Journal of Organic Chemistry*, Vol. 67, No. 26, 2002, pp. 9162-9174.

[3] X. Zhang, H. Du, Z. Wang, Y. D. Wu and K. Ding, "Experimental and Theoretical Studies on the Hydrogen-Bond-Promoted Enantioselective Hetero-Diels-Alder Reaction of Danishefsky's Diene with Benzaldehyde," *The Journal of Organic Chemistry*, Vol. 71, No. 7, 2006, pp. 2862-2869.

[4] A. G. Leach, K. N. Houk and J. L. Reymond, "Theoretical Investigation of the Origins of Catalysis of a Retro-Diels-Alder Reaction by Antibody 10F11," *The Journal of Organic Chemistry*, Vol. 69, No. 11, 2004, pp. 3683-3692.

[5] S. M. Bachrach and S. Jiang, "Theoretical Study of the Diels-Alder Reaction of Selenoaldehydes and Selenoketones," *The Journal of Organic Chemistry*, Vol. 64, No. 22, 1999, pp. 8248-8255.

[6] A. Barman, W. Taves and R. Prabhakar, "Insights into the Mechanism of Methionine Oxidation Catalyzed by Metal (Cu2+, Zn2+, and Fe3+)—Amyloid Beta (A*β*) Peptide Complexes: A Computational Study," *Journal of Computational Chemistry*, Vol. 30, No. 9, 2009, pp. 1405-1413. doi:10.1002/jcc.21167

[7] D. Duca, P. Baranyai and T. Vidóczy, "Monte-Carlo Model for the Hydrogenation of Alkenes on Metal Catalyst," *Journal of Computational Chemistry*, Vol. 19, No. 4, 1998, pp. 396-403. doi:10.1002/(SICI)1096-987X(199803)19:4<396::AID-JCC2>3.0CO;2-N

[8] M. Li, D. Tang, X. Luo and W. Shen, "Mechanism of Asymmetric Hydrogenation of Enamides with $[Rh(BisP^*)]^+$ Catalyst: Model DFT Study," *International Journal of Quantum Chemistry*, Vol. 102, No. 1, 2005, pp. 53-63. doi:10.1002/qua.20294

[9] S. Castillo, V. Bertin, E. S. Reyes, H. L. García, A. Cruz and E. Poulain, "Theoretical Studies on Hydrogen Activation by Iridium Dimers," *International Journal of Quantum Chemistry*, Vol. 70, No. 4-5, 1998, pp. 1029-1035. doi:10.1002/(SICI)1097-461X(1998)70:4/5<1029::AID-QUA50>3.0.CO;2-X

[10] M. H. Liu, C. Chen and Y. S. Hong, "Theoretical Study on the Detonation Properties of Energetic TNAD Molecular Derivatives," *Journal of Molecular Structure: THEOCHEM*, Vol. 710, No. 1-3, 2004, pp. 207-214. doi:10.1016/j.theochem.2004.09.014

[11] M. H. Liu, C. Chen and Y. S. Hong, "Theoretical Study of the Unimolecular Decomposition Mechanisms of Energetic TNAD and TNAZ Explosives," *International Journal of Quantum Chemistry*, Vol. 102, No. 4, 2005, pp. 398-408. doi:10.1002/qua.20284

[12] M. H. Liu, K. F. Cheng, C. Chen and Y. S. Hong, "Computational Study of FOX-7 Synthesis in a Solvated Reaction System," *International Journal of Quantum Chemistry*, Vol. 111, No. 9, 2011, pp. 1859-1869.

[13] M. H. Liu, K. F. Cheng, C. Chen and Y. S. Hong, "Computational Study of TNT Synthesis in Solvated Nitration Reaction Systems," *International Journal of Quantum Chemistry*, Vol. 109, No. 5, 2009, pp. 1009-1023.

[14] E. Whittle, D. A. Dows and G. C. Pimentel, "Matrix Isolation Method for the Experimental Study of Unstable Species," *The Journal of Chemical Physics*, Vol. 22, No. 11, 1954, p. 1943. doi:10.1063/1.1739957

[15] E. D. Becker and G. C. Pimentel, "Spectroscopic Studies of Reactive Molecules by the Matrix Isolation Method," *The Journal of Chemical Physics*, Vol. 25, No. 2, 1956, pp. 224-228. doi:10.1063/1.1742860

[16] R. A. Moss, Y. Ma, F. Zheng, R. R. Sauers, T. Bally, A. Maltsev, J. P. Toscano and B. M. Showalter, "Matrix Isolation, Time-Resolved IR, and Computational Study of the Photochemistry of Benzoyl Azide," *Physical Chemistry Chemical Physics*, Vol. 5, No. 6, 2003, pp. 1010-1018.

[17] G. Seybold, "Flash Thermolysis of Organic Compounds," *Angewandte Chemie International Edition in English*, Vol. 16, No. 6, 1977, pp. 365-373. doi:10.1002/anie.197703651

[18] E. Hedaya, "Techniques of Flash Vacuum Pyrolysis. Cyclopentadienyl Radical and Its Dimmer," *Accounts of Chemical Research*, Vol. 2, No. 12, 1969, pp. 367-373.

[19] Z. Zhu, T. Bally, L. L. Stracener and R. J. McMahon, "Reversible Interconversion between Singlet and Triplet 2-Naphthyl(Carbomethoxy)Carbine," *Journal of American Chemical Society*, Vol. 121, No. 12, 1999, pp. 2863-2874.

[20] T. Khasanova and R. S. Sheridan, "Photochemistry of 2-Benzofurylchlorocarbene: Characterization of a 2,3-Didehydro-2H-Pyran," *Journal of American Chemical Society*, Vol. 122, No. 35, 2000, pp. 8585-8586.

[21] K. Y. Lee and M. D. Coburn, "3-Nitro-1,2,4-Triazo-5-One, a Less Sensitive Explosive," *Report LA*-10302-*NM*, Los Alamos National Laboratory, Los Alamos, 1985.

[22] K. Y. Lee and M. D. Coburn, "3-Nitro-1,2,4-Triazol-5-One, a Less Sensitive Explosive," US Patent No. 4733610, 1988.

[23] X. Yi, H. Rongzu, Y. Chaoging, F. Guofu and Z. Tihua, "Studies on the Critical Temperature of Thermal Explosion for 3-Nitro-1,2,4-Triazol-5-One (NTO) and Its Salts," *Propellants, Explosives, Pyrotechnics*, Vol. 17, No. 6, 1992, pp. 298-302. doi:10.1002/prep.19920170608

[24] J. P. Agarwal, S. M. Walley and J. E. Field, "A High-Speed Photographic Study of the Impact Initiation of Hexanitro-Hexaaza-Isowurtzitane and Nitrotriazolone," *Combust and Flame*, Vol. 112, No. 1-2, 1998, pp. 62-72.

[25] G. Singh, I. P. S. Kapoor, S. M. Mannan and S. K. Tiwari, "Studies on Energetic Compounds: Part XI: Preparation and Thermolysis of Polynitro Organic Compounds," *Journal of Hazardous Materials*, Vol. 68, No. 3, 1999, pp. 155-178. doi:10.1016/S0304-3894(99)00056-4

[26] H. H. Licht, S. Braun, M. Schaefer, B. Wanders and H. Ritter, "Nitrotriazole: Chemische Struktur und explosive Eigenschaften Nitrotriazoles: Chemical Structure and Explosive Properties," *Proceedings of the 29th International Annual Conference of the ICT, Energetic Materials*, Karlsruhe, 30 June-3July 1998.

[27] G. Singh, I. P. S. Kapoor, S. K. Tiwari and P. S. Felix, "Studies on Energetic Compounds: Part 16. Chemistry and Decomposition Mechanisms of 5-Nitro-2,4-Dihydro-3H-1,2,4-Triazole-3-One (NTO)," *Journal of Hazardous Materials*, Vol. 81, No. 1-2, 2001, pp. 67-82. doi:10.1016/S0304-3894(00)00289-2

[28] K. Y. Lee and M. M. Stincipher, "Synthesis and Initial Characterization of Amine Salts of 3-Nitro-1,2,4-Triazol-5-One," *Propellants, Explosives, Pyrotechnics*, Vol. 14, No. 6, 1989, pp. 241-244. doi:10.1002/prep.19890140605

[29] L. Campion and J. Ouazzani, "Synthesis of 5-Amino-1,2,4-Triazole-3-One through the Nitroreduction of 5-Nitro-1,2,4-Triazole-3-One. Comparison between Chemical and Microbiologicalcatalysis," *Biocatalysis and Biotransformation*, Vol. 17, No. 1, 1999, pp. 37-44.

[30] P. B. Kulkarni, G. N. Purandare, J. K. Nair, M. Talawar, B. T. Mukundan and S. N. Asthana, "Synthesis, Characterization, Thermolysis and Performance Evaluation Studies on Alkali Metal Salts of TABA and NTO," *Journal of Hazardous Materials*, Vol. 119, No. 1-3, 2005, pp. 53-61.

[31] V. L. Zbarsky and N. V. Yudin, "Kinetics of the Synthesis of NTO in Nitric Acid," *Propellants, Explosives, Pyrotechnics*, Vol. 30, No. 4, 2005, pp. 298-302. doi:10.1002/prep.200500020

[32] G. Singh and S. P. Felix, "Studies on Energetic Compounds. Part 32: Crystal Structure, Thermolysis and Applications of NTO and Its Salts," *Journal of Molecular Structure*, Vol. 649, No. 1-2, 2003, pp. 71-83. doi:10.1016/S0022-2860(02)00717-2

[33] R. I. Hiyoshi, Y. Kohno and J. Nakamura, "Vibrational Assignment of Energetic Material 5-Nitro-2,4-Dihydro-1,2,4-Triazole-3-One (NTO) with Labeled Isomers," *The Journal of Physical Chemistry A*, Vol. 108, No. 27, 2004, pp. 5915-5920. doi:10.1021/jp049118i

[34] D. C. Sorescu, T. R. L. Sutton, D. L. Thompson, D. Beardall and C. A. Wight, "Theoretical and Experimental Studies of the Structure and Vibrational Spectra of NTO," *Journal of Molecular Structure*, Vol. 384, No. 2-3, 1996, pp. 87-99. doi:10.1016/S0022-2860(96)09343-X

[35] L. Turker and T. Atalar, "Quantum Chemical Study on 5-Nitro-2,4-Dihydro-3H-1,2,4-Triazol-3-One(NTO) and Some of Its Constitutional Isomers," *Journal of Hazardous Materials*, Vol. 137, No. 3, 2006, pp. 1333-1344. doi:10.1016/j.jhazmat.2006.05.015

[36] A. Toro-Labbe, "Characterization of Chemical Reactions from the Profiles of Energy, Chemical Potential, and Hardness," *The Journal of Physical Chemistry A*, Vol. 103, No. 22, 1999, pp. 4398-4403.

[37] A. Toro-Labbe, "Theoretical Aspects of Chemical Reactivity," Elsevier Science, Amsterdam, 2007.

[38] M. P. Sammes, "Chemistry of Heterocyclic Compounds: Pyrroles, Part One: The Synthesis and the Physical and Chemical Aspects of the Pyrrole Ring," John Wiley & Sons, New York, 2008.

[39] M. J. Frisch, *et al.*, "Gaussian 98, Revision A.7," Gaussian, Inc., Pittsburgh, 1998.

[40] C. Lee, W. Yang and R. G. Parr, "Development of the Colle-Salvetti Correlation-Energy Formula into a Functional of the Electron Density," *Physical Review B*, Vol. 37, No. 2, 1988, pp. 785-789.

[41] C. Peng and H. B. Schlegel, "Combining Synchronous Transit and Quasi-Newton Methods to Find Transition States," *Israel Journal of Chemistry*, Vol. 33, No. 2, 1993, pp. 449-454.

[42] C. Peng, P. Y. Ayala, H. B. Schlegel and M. J. Frisch, "Using Redundant Internal Coordinates to Optimize Equilibrium Geometries and Transition States," *Journal of Computational Chemistry*, Vol. 17, No. 1, 1996, pp. 49-56.

[43] E. A. Zhurova and A. A. Pinkerton, "Chemical Bonding in Energetic Materials: β-NTO," *Acta Crystallographica Section B*, Vol. 57, No. 3, 2001, pp. 359-365. doi:10.1107/S0108768100020048

[44] K. Lee, Y. R. Gilardi, R. W. Armstrong and J. J. Gilman, "Structure and Properties of Energetic Materials: Volume 296," Materials Research Society, Pittsburgh, 1993, p. 237.

20

Catalytic Reduction of Benzaldehyde under Hydrogen Flow over Nickel-Containing Mesoporous Silica Catalysts

Adel Saadi[1], Kahina Lanasri[1], Khaldoun Bachari[1,2], Djamila Halliche[1], Chérifa Rabia[1]
[1]Laboratoire de Chimie du Gaz Naturel, Faculty of Chemistry, University of Sciences and Technology Houari Boumediene, Algiers, Algeria
[2]Centre de Recherche Scientifique et Technique en Analyses Physico-Chimiques, Algiers, Algeria
Email: asaadi@usthb.dz

ABSTRACT

The hydrogenation of benzaldehyde over a series of nickel-containing mesoporous silicas with different nickel contents was studied at atmospheric pressure in the range temperature of 393 - 513 K under H_2 flow. These materials (noted Ni_n-HMS with n = Si/Ni = 50, 25, 15) have been prepared at room temperature using a route based on hydrogen bonding and self-assembly between neutral primary amine micelles (S^0) and neutral inorganic precursors (I^0). They were characterized by their chemical analysis, BET surface area, XRD, FT-IR, and SEM microscopy. The obtained products were benzylalcohol, toluene, benzene with yields depending on the nickel content (Si/Ni ratio) and reaction temperature. The products of benzaldehyde hydrogenation (benzylalcohol, and toluene) and hydrogenolysis (benzene) were preferentially formed at low/middle and high reaction temperature respectively. The mesoporous Ni-containing materials were very active hydrogenation catalysts with almost 90% selectivity to benzylalcohol product and showed excellent stability. A mechanism in which the reaction could be initiated by a benzaldehyde reduction over Ni_n-HMS materials under hydrogen flow with formation of reaction products is proposed.

Keywords: Nickel; HMS; Benzaldehyde; Reduction; Mesoporous Materials

1. Introduction

One class of reactions important to produce fine chemical compounds is the hydrogenation of aldehydes and ketones [1-5]. Benzaldehyde reduction is one of these reactions which can provide information about nature and properties of surface centers [6-11]. It was found that nickel, platinum and copper based catalysts preferentially hydrogenate the carbonyl group [12-15]. For the aromatic compounds, the benzene ring and the carbonyl group should be hydrogenated but hydrogenolysis should also compete [9,14,16]. The obtained selectivities mainly depended on the nature of the catalyst [3,4,15].

Heterogeneous catalytic studies of benzaldehyde hydrogenation have been largely devoted to the liquid phase medium [5,17,18] whereas few other studies have been reported for the gas phase medium [7,12,19]. The later system were studied over metal supported catalysts [6, 12,14] or metal oxides catalysts [7,13] but no works have been reported using metal-containing hexagonal mesoporous silica as catalysts. For the metal supported catalysts it was suggested that the observed performances were related to new active sites created in the metalsupport interfacial region [6,20] whereas for metal oxides catalysts it was found that they strongly depended on both the reducibility and surface acid-base properties [7,13]. We recently confirmed the role of metal and acid-base properties in the same reaction conducted over copper supported catalysts [10].

The substitution of supported catalysts or metal oxides catalysts in the case of the hydrogenation reactions is an important challenge that is in heterogeneous catalysis. Indeed, several solid catalysts have already been proposed which are efficient catalysts. The discovery of the new family of mesoporous silica molecular sieves with pore diameters in the 2.0 - 10 nm range, designated as M41S, is of considerable interest for heterogeneous catalysis and material science [21,22]. This family of materials is characterized by a regular array of pores with uniform diameter, high specific surface areas and pore volumes, which are advantageous for the adsorption and catalysis.

Depending on the synthesis conditions, and on the surfactant/SiO_2 ratio, different phases could be obtained, like the hexagonal phase MCM-41, the cubic one MCM-48, and the lamellar compound MCM-50 [21,22]. In 1994, a new pathway was proposed by Tanev *et al.* [23] to prepare mesoporous silicas at room temperature by a neutral templating route (S^0I^0). In this case, the organic surfactant is not quaternary ammonium cation but a primary amine, and the assembly involves hydrogen-bonding interactions between neutral primary amines and neutral inorganic precursors. These materials denoted HMS (hexagonal mesoporous silica) exhibit excellent catalytic ability for macromolecular reactions and offer new opportunities for transition metal incorporation into silica frameworks [24-27].

In the present paper, we report the synthesis and characterization of such materials incorporating nickel and their test as catalysts for the hydrogenation of benzaldehyde. The correlations between the catalytic activity and the surface property, the nature of active sites and the reaction mechanism have been investigated.

2. Experimental

2.1. Catalysts Preparation

The Ni-hexagonal mesoporous solid (HMS) catalysts were prepared as described previously [23]. Hexadecylamine (HDA), nickel nitrate ($Ni(NO_3)_3 \cdot 6H_2O$), and then tetraethylorthosilicate (TEOS) were added to water and ethanol (EtOH) solutions. The dissolution of each component was carried out until completion before the next one was added. After mixing, the resulting solutions with composition [29]:

$$SiO_2 - (1/2)_n\ NiO - 0.3HDA - 7EtOH - 35H_2O,$$

where n = Si/Ni = 50, 25, 15, were stirred at room temperature for 24 h.

The solids thus obtained were recovered by filtration, washed with distilled water, and air-dried at 393 K. Organic molecules occluded in the mesopores were removed by dispersing the solid (5 g/100ml) in a solution of ethanol containing NH_4Cl (1 g/100ml) and maintained under vigorous stirring for 2 h. The procedure was repeated twice before drying at 393 K and calcination at 823 K in air for 6 h. The samples are referred to as Ni_n-HMS, where n is the Si/Ni ratio in the solutions.

2.2. Characterization Methods

The chemical compositions of the samples were determined by atomic absorption using Perkin-Elmer 2380 spectrophotometer. The surface area was calculated based on the BET (Brunner-Emmett-Teller) method, while pore size distribution was computed using the BJH (Barrett-Joyner-Halenda) method. X-ray diffraction (XRD) patterns obtained using a Siemens D500 diffractometer and CuKα radiation, were recorded with 0.02° (2θ) steps and 1 s counting time per step between 1° and 10° (2θ) and between 10° and 80° (2θ). The FTIR spectroscopy analysis has been carried out on a Phillips 9800-FTIR spectroscopy type. The Ni_n-HMS catalysts have also been examined with a scanning electron microscope (SEM), a JEOL2000 electron microscope.

Catalytic Testing

The catalytic performances were carried out in a fixed-bed glass tubular reactor at atmospheric pressure with 30 mg samples and the total flow rate of 50 $cm^3 \cdot min^{-1}$. The reactant gas feed consisted of 4.8 torr of benzaldehyde (Aldrich, 99.98%) and 250 torr of H_2 diluted in N_2. Gaseous benzaldehyde was obtained by bubbling N_2 in liquid benzaldehyde maintained at constant temperature (323K) in a suitable saturator. Before testing, the catalysts were in-situ reduced for 2 h at 623 K in flowing H_2 (flow rate of 20 $cm^3 \cdot min^{-1}$).

The gaseous reactant and products were heated upstream and outstream and analyzed on line by a FID gas chromatograph (Delsi IGC 121 ML) equipped with a 10% CP-SIL 8 CB/Chromosorb W column. Each reaction temperature was maintained constant until the corresponding steady-state was reached as indicated by the gas chromatograph analysis of the exit gases samples. For each catalyst, the reaction temperature was changed in a crossing order: 473 K, 393 K, 513 K, 433 K. Using the Weisz modulus, varying sample weight and flow rate in the range of 10 $cm^3 \cdot min^{-1}$ to 80 $cm^3 \cdot min^{-1}$ results showed that no mass transfer limitations occurred when the sample weight was ≤ 30 mg.

3. Results and Discussion

3.1. Chemical Composition and BET Surface Area

The Si/Ni ratio of the compounds was varied over a very wide range (15 - 50) to characterize all of the nickel species that may be formed in this type of compound. The chemical composition of catalysts obtained by AAS shows that Si/Ni ratios are relatively closer to those fixed before the preparation and correspond well to the nominal values (see **Table 1**). This result confirms that

Table 1. Chemical composition and the characteristics of Ni_n-HMS catalysts.

Sample	Chemical analysis		S_{BET} ($m^2 \cdot g^{-1}$)	Pore volume ($cm^3 \cdot g^{-1}$)	Pore Diameter (Å)[a]	Wall thickness (Å)[a]
	Si/Ni (gel)	Si/Ni (AAS)				
HMS	-	-	1120	0.97	38	38.0
Ni_{50}-HMS	50	48.5	1143	0.90	28	27.7
Ni_{25}-HMS	25	23.6	1035	0.88	27	26.3
Ni_{15}-HMS	15	14.4	998	0.82	24	24.1

[a]BJH (Barrett, Joyner and Halenda); [b]Wall thickness = d-spacing (calcined sample)—pore diameter (BJH).

practically all the metal introduced in preparation gel is located in the catalyst.

The textural analysis of calcinated solids shows that the HMS structure exhibits a high specific surface (1120 $m^2 \cdot g^{-1}$) with pore diameter of about 38 Å. This result confirms the presence of wide pores due to the mesoporous structure of synthesized material (between 25 and 50 Å).The BET surface areas and pore volumes of the Ni_n-HMS samples are in the range of 990 - 1150 $m^2 \cdot g^{-1}$ and 0.8 - 0.9 $cm^3 \cdot g^{-1}$ respectively (**Table 1**).

In the present case, the surface areas and pore volumes of the Ni_n-HMS samples are lower than the value of HMS material. Indeed, after incorporation of Ni element in the HMS structure, the results show that BET surface area and, consequently the pore volume, decreases with decreasing of Si/Ni ratio.

3.2. FTIR and XRD Results

In order to investigate the binding interaction between Ni and the surface functional groups of HMS structure, the Ni_n-HMS samples were characterized by FT-IR spectra, and their IR spectra were shown in **Figure 1**.

Characteristic bands of the HMS structure have been observed at 3420 cm^{-1} for the silanol group Si-OH (not presented), between 800 and 700 cm^{-1} for Si-O-Si, and the band near 960 cm^{-1} corresponding to Si-O-Ni. Although there are still some disputes [20,21] to the assignment of the 960 cm^{-1} band, but more evidences, in which a similar band was previously observed in titanium siloxane polymers or in TiO_2-grafted on silica [25], have supported that the 960 cm^{-1} should be due to a modification of SiO_4 units directly related to the presence of a strong binding of the surface functional groups of HMS with Nickel. On the other hand, the IR spectra show that the intensity of Ni-O-Si band increases with increasing nickel content.

The X-ray diffraction is one of the most used techniques to identify HMS structure. It also gives information about the internal structure of solid, the nature and purity of crystallized material. The **Figure 2** shows that only one peak reflection (100) appeared in the range of $2\theta = 2° - 5°$, which is corresponding to the well-ordered hexagonal mesoporous structure [30,31]. The intensity of this peak decreases with increasing nickel content, demonstrating that adding nickel has probably a negative effect on crystallinity.

At the same time, small peaks corresponding to NiO phase appeared in the 10° - 80° (2θ) range and grew with the nickel content. **Figure 3** showed that, when the nickel content increased from n = Si/Ni = 50 to 15, clear diffraction peaks of NiO phase appeared (**Figure 3**). In fact, the Ni_{50}-HMS did not showed any XRD peaks of metal oxide. Indeed, the Ni_{15}-HMS showed more intense reflection than Ni_{25}-HMS. This result indicated that the increase of copper content favored the formation of larger NiO particles extra framework.

3.3. SEM Study

Morphology as well as structural ordering of calcinated nickel-containing HMS samples were analyzed by electron microscopy studies. SEM pictures of Ni_n-HMS samples show non-uniform aggregates of small, distinct particles. Indeed, all nickel-HMS catalysts show disordered spherical structures with the presence of agglomerated particles. An example of SEM pictures of Ni_{25}-HMS sample is shown in **Figure 4**.

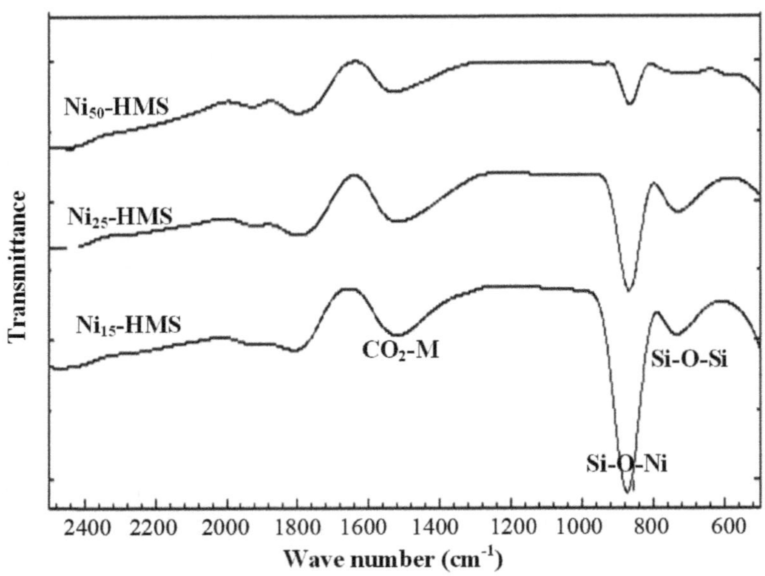

Figure 1. FT-IR spectra of Ni-HMS samples.

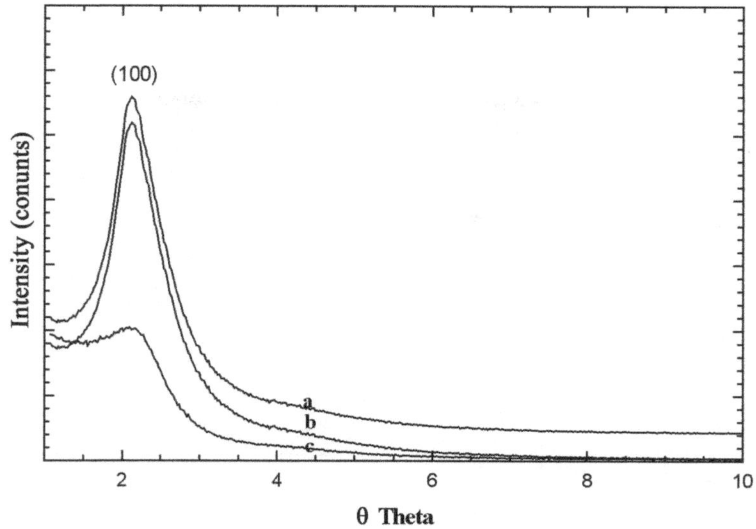

Figure 2. XRD patterns of, a: Ni_{50}-HMS; b: Ni_{25}-HMS; c: Ni_{15}-HMS in the range of 1° - 10° (2θ).

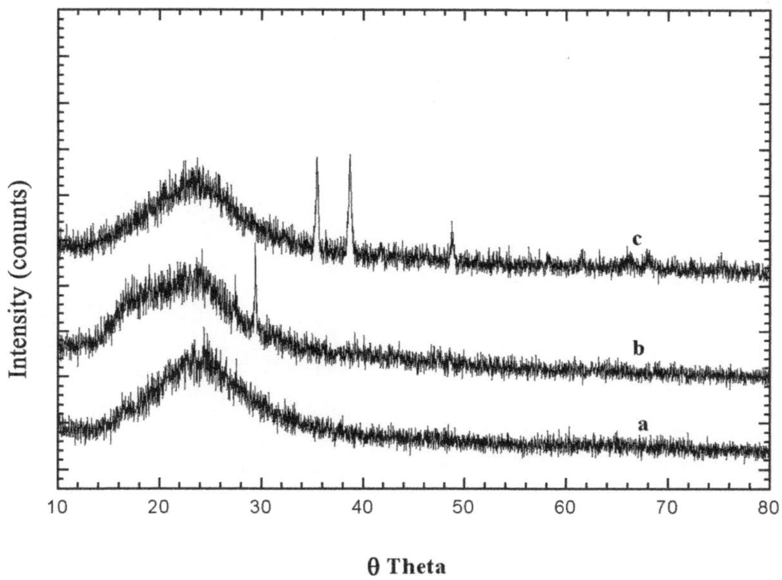

Figure 3. XRD patterns of, a: Ni_{50}-HMS; b: Ni_{25}-HMS; c: Ni_{15}-HMS in the range of 10° - 80° (2θ).

Figure 4. SEM images of calcinated Ni_{25}-HMS sample.

3.4. Benzaldehyde Reduction

The Ni_n-HMS catalysts exhibited initial deactivation before the steady-state was established and the degree (2% - 5%) and duration (1 - 3 h) of deactivation depended on the nickel content and reaction temperature. **Figure 5** reports activity results with time on stream in the whole range of reaction temperature. In any case, the deactivation was reversible since the activity was almost totally recovered after increasing the reaction temperature. The decrease of activity is attributable to the catalyst poisoning by strongly adsorbed oxygenate molecules [32].

3.4.1. Catalytic Activity

The steady-state conversion increased gradually with the reaction temperature for all catalysts (**Figure 6**). The level and order of conversions depended on both the reaction temperature and Si/Ni ratio. The Ni_{15}-HMS catalyst showed high activity between 373 K and 513 K (34% - 90% of conversion), whereas Ni_{50}-HMS catalyst was the less active and the order of activity was as follows: Ni_{15}-HMS > Ni_{25}-HMS > Ni_{50}-HMS.

Such an order seems to be probably related to the crystallite size of the Ni_n-HMS phase. Indeed, the weak activity of Ni_{50}-HMS is probably due to the presence of small crystallites of nickel-HMS and the degree of substitution of nickel in the HMS structure. Inversely, the high degree of substitution of Si by Ni in the HMS structure encouraged the NiO crystallites formation on the surface of HMS structure. The presence of interaction between NiO phase and HMS structure contributed to the increase of the catalytic activity.

3.4.2. Selectivity

The fact that the aromatic nucleus is unreactive while the carbonyl substituent undergoes hydrogenation and hydrogenolysis suggests that the aromatic aldehyde reacts principally with the catalyst via the carbonyl function. It should be assumed that in the adsorbed benzaldehyde both the aromatic ring and the carbonyl group lie parallel to the active surface [14]. In such a case, the adsorbed hydrogen preferably attack the carbonyl group due to energy barriers [14,15]. The reaction products were: benzyl alcohol, toluene and benzene. No aromatic ring hydrogenation was observed.

3.4.2.1. Benzyl Alcohol Selectivity

In the range of reaction temperature, benzyl alcohol was formed on all catalysts (**Figure 6**). The reduction of the carbonyl function was observed at rather low temperature with high selectivity, regardless of the activity level (**Table 2**). It was produced between 393 K and 473 K on Ni_{50}-HMS and Ni_{25}-HMS catalysts and between 393K and 433K on Ni_{15}-HMS catalyst. The elevation of reaction temperature is not favorable to benzyl alcohol formation.

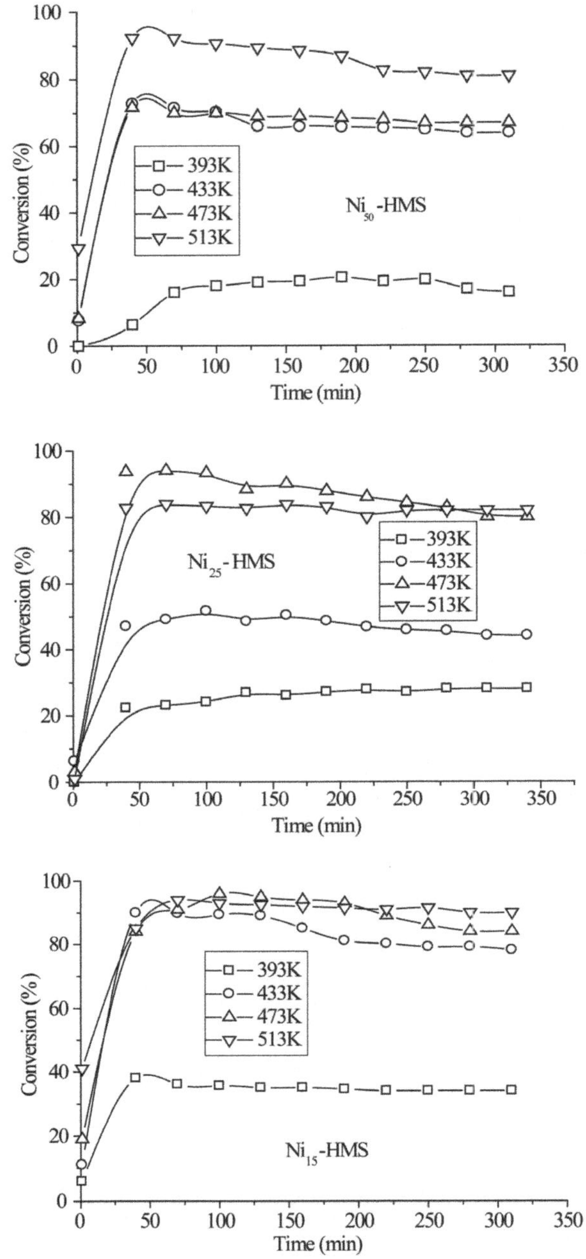

Figure 5. Conversion of benzaldehyde with time on stream in the range of reaction temperature 393 K - 513 K under H_2 flow for Ni_n-HMS catalysts. Pre-treatment: H_2/623 K.

Indeed, the **Figure 5** shows that this product decreased with increasing of reaction temperature and was totally disappeared after 473 K on Ni_{50}-HMS and Ni_{25}-HMS catalysts and after 433 K on Ni_{15}-HMS catalyst.

The analysis of results revealed that high selectivity to benzyl alcohol was obtained at 393 K on Ni_{15}-HMS (88% of selectivity) and it's decreased when the Ni content increased in order: Ni_{15}-HMS (88%) > Ni_{25}-HMS (80%) > Ni_{50}-HMS (75%). The order of selectivity evolution is

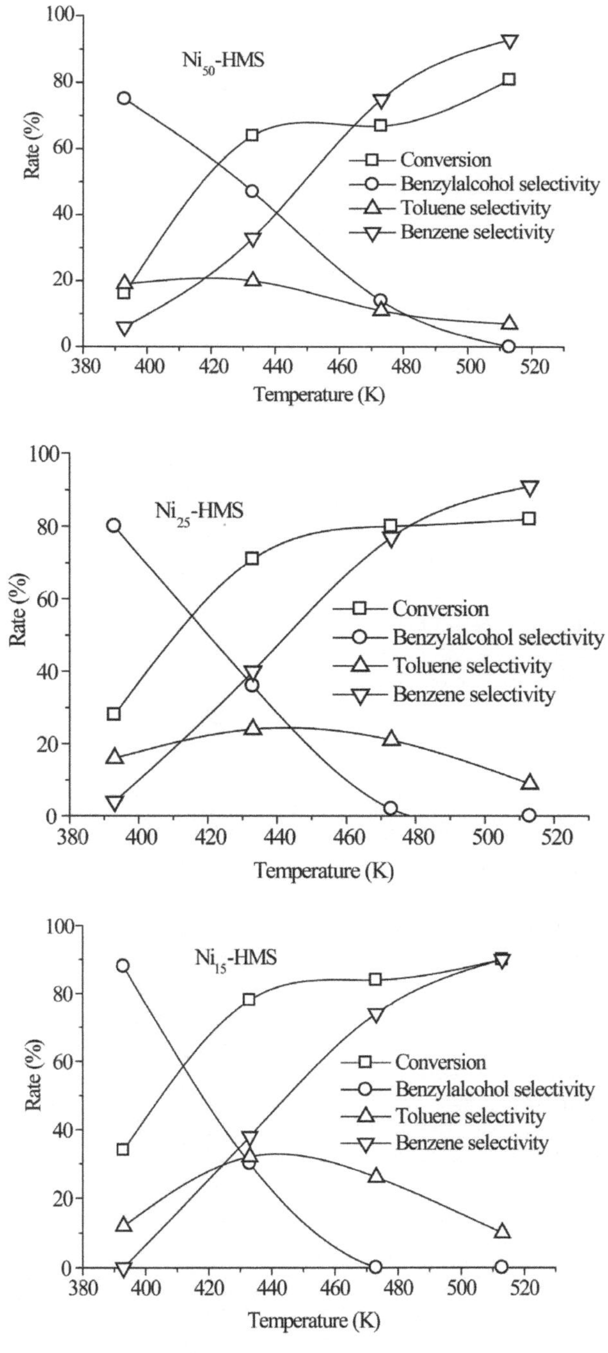

Figure 6. Steady-state conversion of benzaldehyde under H_2 flow as a function of the reaction temperature. Pre-treatment: H_2/623 K.

probably due to the degree of nickel incorporation into silica frameworks.

The chemical process of benzyl alcohol formation is believed to be a 1 - 2 nucleophilic addition, with a high polarization of the transition state, as in the case of the hydrogenation of conjugated carbonyl compounds [1,15, 20]:

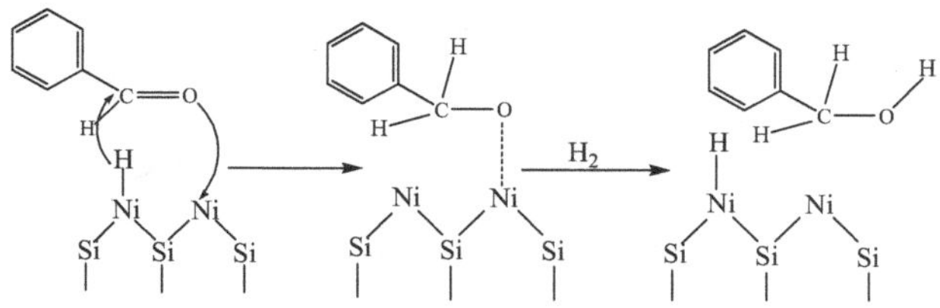

3.4.2.2. Benzyl Alcohol Selectivity

Low toluene selectivity (<35%) was observed for all catalysts in the whole range of reaction temperature. The maximum of selectivity appeared at 433 K for Ni_{15}-HMS (32%), Ni_{25}-HMS (24%) and Ni_{50}-HMS (20%) catalysts. On the other hand, the results show that, after 433 K, toluene selectivity decreased with elevation of temperature reaction for obtain low values at 513 K on all catalysts.

It is generally considered that toluene is the product of a consecutive reaction of hydrogenolysis of benzyl alcohol [1,7,32,33]. Indeed, flowing the nickel-HMS catalysts with the benzyl alcohol instead of benzaldehyde, selectively led to toluene production. In a parallel experiment, replacing benzaldehyde by benzyl alcohol in the reactant flow showed the formation of toluene with a selectivity of 100%. However, benzyl alcohol is a complex molecule possessing several active centers such as the acidic proton and basic oxygen of the O-H function and the reductive hydrogen of the C-H exocyclic function. All these centers may interact with the oxide surface or dihydrogen molecule [32].

In the case of nickel-HMS catalysts, the low activity in toluene should reflect the aptitude of the surface to both adsorb the alcohol and dissociate dihydrogen. The toluene formation probably resulted in the protonation of the alcohol function on a Brönsted acid site (H-O-Si-) followed by hydrogenolysis of the protonated species on a adjacent metal site (Ni-H):

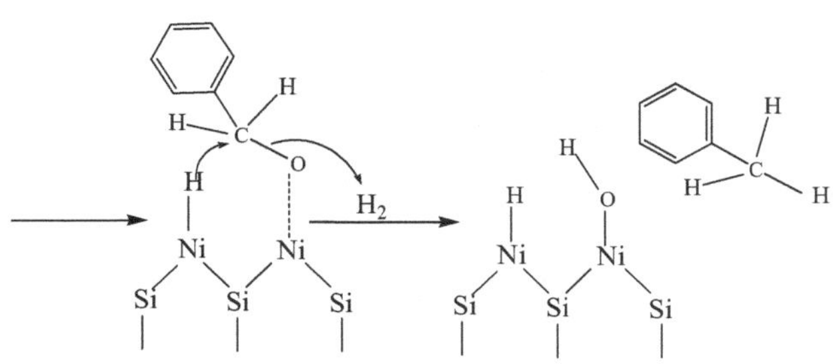

3.4.2.3. Benzene Selectivity

In the present case, benzene is formed from 313 K on all catalysts with very low selectivity (<7%). In contrast to benzyl alcohol and toluene, the benzene selectivity increased with the reaction temperature and the maximum of selectivity (90% - 93%) was obtained at a high reaction temperature (**Figure 6**). High selectivity to benzene was observed for all catalysts at 513 K (**Table 2**).

Table 2. Catalytic results for benzaldehyde hydrogenation over Ni_n-HMS catalysts.

Catalyst	$T_{reaction}$ K	Conv %	$Sel_{alcohol}$ %	$Sel_{toluene}$ %	$Sel_{benzene}$ %
Ni_{50}-HMS	393	16	75	19	6
	433	64	47	20	33
	473	67	14	11	75
	513	81	0	7	93
Ni_{25}-HMS	393	28	80	16	4
	433	71	36	24	40
	473	80	2	21	77
	513	82	0	9	91
Ni_{15}-HMS	393	34	88	12	0
	433	78	30	32	38
	473	84	0	26	74
	513	90	0	10	90

It is believed to be the product of the hydrogenolysis of the external C(aryl)-C bond of benzaldehyde [14,27]. Vannice and Poondi [6] were showed that benzene formation was produced directly from the reactant molecule and not from the alcohol or toluene intermediates. Carbon monoxide was identified as co-product of benzene [6]. On the other hand, the study of the activity as a function of time on stream showed that benzene also arose as a primary product (not presented). This result significantly shows that benzene was formed by an independent way from that of the products of benzaldehyde reduction:

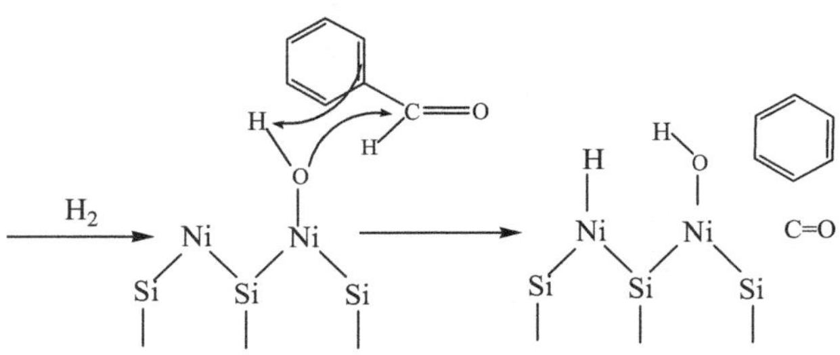

4. Conclusion

The hydrogenation of benzaldehyde over nickel-containing mesoporous silicas with different Ni contents at atmospheric pressure and 393 K - 513 K produced competitively benzyl alcohol, toluene and with yields depending on the Si/Ni ration and the reaction temperature. The order of activity was attributed to crystallite size and degree of nickel incorporation into silica frameworks. In connection, benzyl alcohol was selectively obtained at low reaction temperature. Toluene product arose from the hydrogenolysis of benzyl alcohol on probably bifunctional sites involving metal hydride species. As to benzene, it resulted from benzaldehyde hydrogenolysis; probably through a nickel oxygenate species.

REFERENCES

[1] M. A. Vannice and R. L. Garten, "CO Hydrogenation Reactions over Titania-Supported Nickel," *Journal of Catalysis*, Vol. 66, No. 1, 1980, pp. 242-247.

[2] M. Fiefelder, "Practical Catalytic Hydrogenation: Technics and Applications," Wiley-Interscience, New York, 1971.

[3] P. N. Rylander, "Catalytic Hydrogenation in Organic Synthesis," Academic Press, London, 1979.

[4] P. N. Rylander, "Hydrogenation Methods," Academic Press, London, 1985.

[5] K. Yoshida, C. Gonzalez-Arellano, R. Luque and P. L. Gai, "Efficient Hydrogenation of Carbonyl Compounds Using Low-Loaded Supported Copper Nanoparticles under Microwave Irradiation," *Applied Catalysis A: General*, Vol. 379, No. 1-2, 2010, pp. 38-44. doi:10.1016/j.apcata.2010.02.028

[6] R. Rao, A. Dandekar, R. T. K. Baker and M. A. Vannice, "Properties of Copper Chromite Catalysts in Hydrogenation Reactions," *Journal of Catalysis*, Vol. 171, No. 2, 1997, pp. 406-419. doi:10.1006/jcat.1997.1832

[7] D. Haffad, U. Kameswari, M. M. Bettahar, A. Chambellan and J. C. Lavalley, "Reduction of Benzaldehyde on Metal Oxides," *Journal of Catalysis*, Vol. 172, No. 1, 1997, pp. 85-92. doi:10.1006/jcat.1997.1854

[8] A. Saadi, Z. Rassoul, J. Barrault and M. M. Bettahar, "Hydrogénation Sélective de la Cinnamaldéhyde en Alcool Cinnamique sur de Catalyseurs Supportés à Base de Cuivre," *Journal of Algerian Chemical Society*, Vol. 9, No. 2, 1999, pp. 269-276.

[9] A. Aboulayt, A. Chambellan, M. Marzin, J. Saussey, F. Maugé, J. C. Lavalley, C. Mercier and R. Jacquot, "Study of the Hydrogenation of Methyl Benzoate to Benzaldehyde on Various Metal Oxides," *Studies in Surface Science and Catalysis*, Vol. 78, 1993, pp. 131-138. doi:10.1016/S0167-2991(08)63312-1

[10] A. Saadi, M. M. Bettahar and Z. Rassoul, "Reduction of Benzaldehyde on Copper Supported on SiO_2. Effect of

Method of Preparation," *Studies in Surface Science and Catalysis*, Vol. 130, 2000, pp. 2261-2266. doi:10.1016/S0167-2991(00)80805-8

[11] A. Saadi, R. Merabti, M. M. Bettahar and Z. Rassoul, "Catalyseurs Cu/MO (M = Mg, Ca, Sr). Caractérisation par des Réactions Tests," *Journal of Algerian Chemical Society*, Vol. 11, No. 2, 2001, pp. 231-240.

[12] R. Merabti, A. Saadi, Z. Rassoul and M. M. Bettahar, "Caractérisations Physico-Chimiques des Catalyseurs à Base de Nickel Supportés," *Journal of Algerian Chemical Society*, Vol. 16, No. 1, 2006, pp. 39-47.

[13] M. A. Keane, "Gas Phase Hydrogenation/Hydrogenolysis of Benzaldehyde and O-Tolualdehyde over Ni/SiO_2," *Journal of Molecular Catalysis A: Chemical*, Vol. 118, No. 2, 1997, pp. 261-269. doi:10.1016/S1381-1169(96)00386-X

[14] R. Hubaut, J. P. Bonnelle and M. Daage, "Selective Hydrogenation of Heavy Polyunsaturated Molecules on Copper-Chromium Catalysts," *Journal of Molecular Catalysis*, Vol. 55, No. 1, 1989, pp. 170-183. doi:10.1016/0304-5102(89)80251-2

[15] D. Poondi and M. A. Vannice, "The Influence of MSI (Metal-Support Interactions) on Phenylacetaldehyde Hydrogenation over Pt Catalysts," *Journal of Molecular Catalysis A: Chemical*, Vol. 124, No. 1, 1997, pp. 79-89. doi:10.1016/S1381-1169(97)00066-6

[16] M. W. De Lange, J. G. Van Ommen and L. Lefferts, "Deoxygenation of Benzoic Acid on Metal Oxides: 2. Formation of Byproducts," *Applied Catalysis A: General*, Vol. 231, No. 1-2, 2002, pp. 17-26. doi:10.1016/S0926-860X(01)00702-5

[17] S. Paganelli, U. Matteoli, A. Scrivanti and C. Botteghi, "Pt0-Complexes as Catalyst Precursors for Homogeneous Carbon=Carbon and Carbon=Oxygen Double Bond Hydrogenation," *Journal of Organometallic Chemistry*, Vol. 397, No. 3, 1990, pp. 375-381. doi:10.1016/0022-328X(90)85337-X

[18] J. A. Schreifels, P. C. Maybury and W. E. Swartz, "X-Ray Photoelectron Spectroscopy of Nickel Boride Catalysts: Correlation of Surface States with Reaction Products in the Hydrogenation of Acrylonitrile," *Journal of Catalysis*, Vol. 65, No. 1, 1980, pp. 195-206. doi:10.1016/0021-9517(80)90294-8

[19] A. Saadi, R. Merabti, Z. Rassoul and M. M. Bettahar, "Hydrogenation of Benzaldehyde over Supported Nickel Catalysts," *Journal of Molecular Catalysis A: Chemical*, Vol. 253, No. 1-2, 2006, pp. 79-85. doi:10.1016/j.molcata.2006.03.003

[20] M. A. Vannice, "Hydrogenation of co and Carbonyl Functional Groups," *Catalysis Today*, Vol. 12, No. 2-3, 1992, pp. 255-267.

[21] C. T. Kresge, M. E. Leonowicz, W. J. Roth and J. C. Beck, "Ordered Mesoporous Molecular Sieves Synthesized by a Liquid-Crystal Template Mechanism," *Nature*, Vol. 359, No. 6397, 1992, pp. 710-712. http://134.243.5.70/ASSETS/87CB6FBC90C84546BA13660DB519F346/200207j5.pdf

[22] J. S. Beck, J. C. Vartuli, W. J. Roth, M. E. Leonowicz, C. T. Kresge, K. D. Schmitt, C. T. W. Chu, D. H. Olson, E. W. Sheppard, S. B. McCullen, J. B. Higgins and J. L. Schlenker, "A New Family of Mesoporous Molecular Sieves Prepared with Liquid Crystal Templates," *Journal of American Chemical Society*, Vol. 114, No. 27, 1992, pp. 10834-10843. doi:10.1021/ja00053a020

[23] P. T. Tanev, M. Chibwe and T. J. Pinnavaia, "Titanium-Containing Mesoporous Molecular Sieves for Catalytic Oxidation of Aromatic Compounds," *Nature*, Vol. 368, 1994, pp. 321-323. doi:10.1038/368321a0

[24] A. Yin, J. Qu, X. Guo, W. L. Dai and K. Fan, "The Influence of B-Doping on the Catalytic Performance of Cu/HMS Catalyst for the Hydrogenation of Dimethyloxalate," *Applied Catalysis A: General*, Vol. 400 No. 1-2, 2011, pp. 39-47. doi:10.1016/j.apcata.2011.04.011

[25] A. Sayari, "Catalysis by Crystalline Mesoporous Molecular Sieves," *Chemistry of Materials*, Vol. 8, No. 8, 1996, pp. 1840-1852. doi:10.1021/cm950585+

[26] A. Yin, C. Wen, X. Guo, W. L. Dai and K. Fan, "Influence of Ni Species on the Structural Evolution of Cu/SiO_2 Catalyst for the Chemoselective Hydrogenation of Dimethyl Oxalate," *Journal of Catalysis*, Vol. 280, No. 1, 2011, pp. 77-88. doi:10.1016/j.jcat.2011.03.006

[27] C. Y. Ma, J. Cheng, H. L. Wang, Q. Hu, H. Tian, C. He and Z. P. Hao, "Characteristics of Au/HMS Catalysts for Selective Oxidation of Benzyl Alcohol to Benzaldehyde," *Catalysis Today*, Vol. 158, No. 3-4, 2010, pp. 246-251. doi:10.1016/j.cattod.2010.03.080

[28] E. S. Vasiliadou and A. A. Lemonidou, "Investigating the Performance and Deactivation Behaviour of Silica-Supported Copper Catalysts in Glycerol Hydrogenolysis," *Applied Catalysis A: General*, Vol. 396, No. 1-2, 2011, pp. 177-185. doi:10.1016/j.apcata.2011.02.014

[29] A. Tuel, I. Arcon and J. M. M. Millet, "Investigation of Structural Iron Species in Fe-Mesoporous Silicas by Spectroscopic Techniques," *Journal of the Chemical Society, Faraday Transactions*, Vol. 94, No. 23, 1998, pp. 3501-3510. doi:10.1039/A806912C

[30] X. Lu and Y. Yuan, "Copper-Containing Hexagonal Mesoporous Silicas for Styrene Epoxidation Using Tert-Butylhydroperoxide," *Applied Catalysis A: General*, Vol. 365, No. 2, 2009, pp. 180-186. doi:10.1016/j.apcata.2009.06.012

[31] K. Bachari, A. Touileb, N. Tahir, A. Saadi, D. Halliche and O. Cherifi, "Catalytic Propeties of Cr-HMS Materials in the Benzylation of Benzene with Benzylchloride," *Kinetics and Catalysis*, Vol. 52, No. 1, 2011, pp. 48-54. doi:10.1134/S0023-158411010022

[32] A. Saadi, Z. Rassoul and M. M. Bettahar, "Gas Phase Hydrogenation of Benzaldehyde over Supported Copper Catalysts," *Journal of Molecular Catalysis A: Chemical*, Vol. 164, No. 1-2, 2000, pp. 205-216. doi:10.1016/S1381-1169(00)00199-0

[33] R. Merabti, K. Bachari, D. Halliche, Z. Rassoul and A. Saadi, "Synthesis and Characterization of Activated Copper or Nickel and Their Catalytic Behavior towards Benzaldehyde Hydrogenation," *Reaction Kinetics, Mechanisms and Catalysis*, Vol. 101, 2010, pp. 195-208. doi:10.1007/S11144-0100215-x

KCl Single Crystals Growth with Mn, Ag and in Impurities by Czochralski Method and Study of Impurities Influence on Their Properties

Feridoun Samavat[1*], Ebrahim Haji Ali[2], Somayeh Solgi[1], P. Taravati Ahmad[1]
[1]Physics Department, Bu-Ali Sina University, Hamedan, Iran
[2]Physics Department, Imam Hossein University, Tehran, Iran
Email: *fsamavat@yahoo.com, ehajiali@yahoo.com, somayeh.solgi@yahoo.com, p_taravati@yahoo.com

ABSTRACT

In this research KCl, KCl:Mn, KCl:Ag and KCl:In single crystals have been grown by Czochralski method. X-ray diffraction confirms KCl single crystal formation. In this work also influence of mentioned impurities on the optical property of KCl single crystal has been studied by chemical etching. Then the hardness of these crystals have been measured by Vickers micro hardness. The result indicate the positive effect of the impurity on optical and mechanical properties.

Keywords: Single Crystal; KCl; Czochralski; X-Ray Diffraction; Chemical Etching; Vickers Micro Hardnes

1. Introduction

After one century from the start of the crystal growth, many efforts of researches have been devoted to it. This is because of vast application of crystal in the construction of physical devices such as detectors, solid-state lasers, scintillators and other optical pieces. In general three kinds of crystals are always considered. Scintillator crystal, crystals with the use of laser and optical crystals.

This is the three types of crystals have been grown with technology of crystal growth from melting by crystal growth Czochralski and Brigman methods [1]. Crystal with mentioned applications which often are alkali halides, after semiconductors such as silicon; arsenid gallium; and etc, have the most of world annual production of crystal [2]. One from the most important application of optical crystals is production of optical window of lasers. For construction of this component is needed to material with uniform structure, wide gap band and good mechanical strength[1]. Single crystal of KCl, with wide gap band (~8 eV), structure of uniform and simple cubic, is used in production of optical windows.

In a survey, windows made from KCl comparison with the windows from ZnS, Gar and etc have been had so much better efficiently in CO_2 laser windows [3]. Although this crystal has undesirable properties such as attractive moisture, poor thermal and mechanical properties. But with adding of heavy impurities it can replace for the crystals used in lasers, scintillators and telecommunications equipment optical fiber with low-cost and efficient [4]. In this study the crystal growth of KCl with heavy impurities and effect of increasing of impurity on the properties of these crystals are investigated.

*Corresponding author.

2. Crystal Growth

Crystal growth by Czochralski method is based on slow pull of seed from melt free surface. This action is associated with the exact regulation of the temperature of the furnace during slow pull. In this method, with changing of pulling speed of seed and also rate of increasing and the decreasing of the furnace temperature, can get a crystal with needed size and shape.

In this paper, we are describing the growth of the single crystal of KCl by Czochralski method with the impurities of Mn, In and Ag in approximate size diameter 50 mm, length 40 mm (single crystal with impurity In is more slender). For growth this crystals was used from Alumina (Al_2O_3) crucible. This crucible has good physical properties such as high mechanical strength and resistance to environmental factors such as moisturing. In addition, its melting temperature is higher than experiment material.

For growth of the single crystal of KCl with 0.02 impurity of Mn, 100 gr KCl powder with 0.03 gr $MnCl_2$ powder have been mixed in crucible. The crucible is placed within furnace of the growth system. This furnace has two elements, one for heating of the wall and the other for floor. The furnace has turn on by temperature control digital system (that temperature control is responsible for both elements) and is heated with rate of 20°C/min from room temperature to 930°C. For the confidence from complete melting and homogenization of the materials, the temperature increase to 930°C from the room temperature melting temperature of KCl powder is 771°C.

After complete melting of the material and getting a thermodynamic equilibrium, changing rate of furnace temperature is decreased from 3°C/min to 0.3°C/min is changed. Furnace temperature fluctuations are decreased and growth process is controlled easier. A single crystal of KCl with length of 30 mm is used as seed. The seed is contacted with melting free surface vertically by pulling the seed holder slowly and with rotation speed 30 rpm [5].

After peeling, again seed has been placed in temperature a little lower than the melting temperature and with pulling rate of 5 mm/h, so the initial nucleus is formed on seed.

After than, the neck part of crystal is being growth with speed of more pulling. The crystal neck is a narrow part with length of 10 mm to 15 mm. Finally, the crystal shoulder is forming with pulling speed of 6 mm/h and gradually reduce the temperature of the furnace. For forming of crystal body, the temperature is kept constant and pulling is done more quickly. **Figure 1** shows that the crystal of KCl:Mn in growing.

After ending of growth, annealing process is performed for 24 h in the temperature of 600°C. The single crystals KCl with impurities of In and Ag, also have been growth via mentioned method. **Figure 2** shows the single crystals KCl:Ag and KCl:In after annealing.

These crystals were analyzed then cut into small pieces for studying of structure and their quality.

3. Analysis of Grown Crystals

3.1. X-Ray Diffraction (XRD)

Radiation of X-beam to a sample is obtained information

Figure 1. The single crystal KCl:Mn in growing.

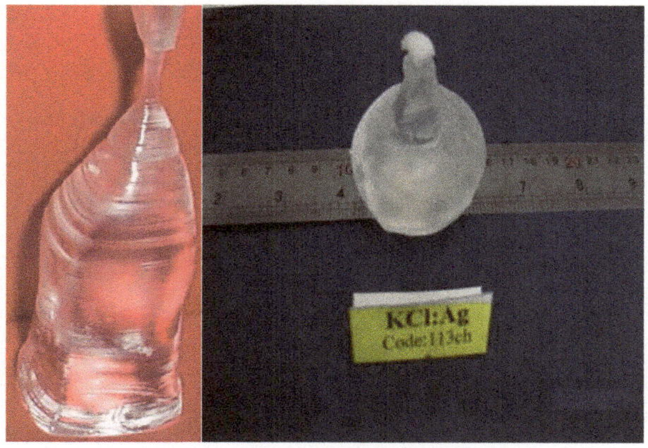

Figure 2. The single crystals KCl:Ag and KCl:In after exit of the furnace.

such as structure, material gender, elements amount in the sample. XRD of single crystal of KCl:In is obtained by apparatus of Ital Structure, that is set on 40 kV and 30 mA. Sample is scanned in range of 2θ =10° to 2θ = 20° with step of 2θ = 0.01°.

Figure 3 shows X-ray diffraction spectrum of single crystal of KCl:In.

Survey of XRD spectra of single crystals of KCl:Mn and KCl:Ag performed by Philips apparatus with diffract meter 1800 pw, which is set on 40 kV and 30 mA. Samples were scanned in range of 2θ = 10° to 2θ = 90° with step of 2θ = 0.02°. **Figures 4** and **5** show XRD spectra of these crystals.

3.2. Chemical Etching

Chemical etching is a method for determination crystal quality. In this method can be estimated crystal dislocation. Alkali halides are prepared often via growth from melting. Thus, they are in thermal stress. The thermal gradient import stress into the crystal and cause of combining of present dislocations in seed that continued to boundary of solid-melting [6].

If the crystal is placed in the condition that atoms is separated of surface, the atoms of around of dislocation will separate with higher speed from the crystal and holes are been formed in their site. That to they are been called the Etchpit. The Etchpit is criterion from further dislocation.

The used corrosion solution for crystals of KCl is solution of acetic acid with 4 gr from powder of $FeCl_3$ [7].

Fresh-cut pieces of crystals are contacted with the corrosive solution and this time is noted. Then crystal pieces are dried and their surfaces are observed with transmission microscope of model JNOEC J6-20 and with magnification of 400. The mentioned corrosion solution will not get clear Etchpit for single crystals of KCl:Mn and KCl:Ag, thus for they are used from corrosion solution of pure acetic acid.

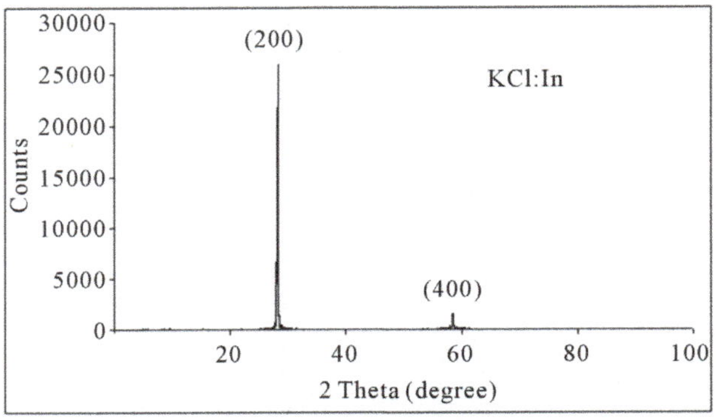

Figure 3. XRD spectrum of KCl:In.

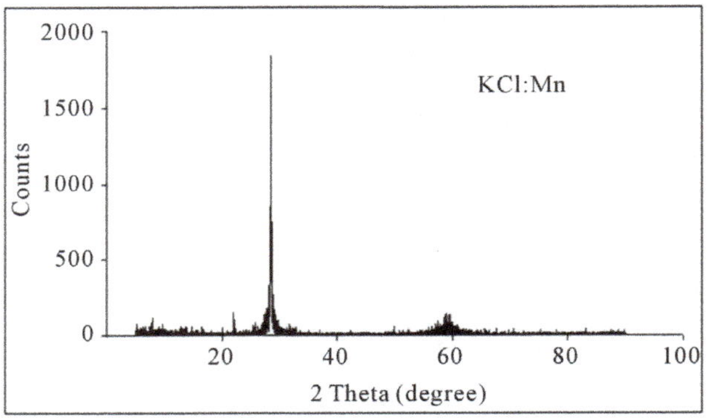

Figure 4. XRD spectrum of KCl:Mn.

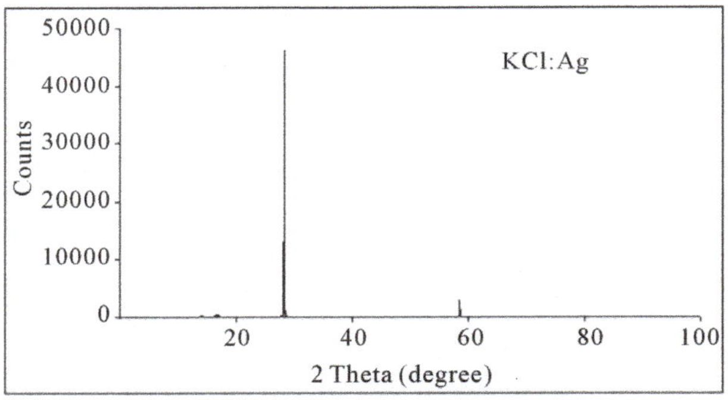

Figure 5. XRD spectrum of KCl:Ag.

Table 1 and **Figure 6** show the results of this experiment and picture of Etchpit of single crystal of KCl:Ag.

3.3. Hardness

Hardness is materials resistance to infiltration or to become punctured by more difficult material [8].

Hardness is divided to macroscopic and microscopic hardness. Vickers micro hardness is a kind from microscopic hardness which is used for measurement of hardness of thin metals and fragile. In this way, applied load to the sample is equal or less than 1000 gr [9].

In this experiment, the pieces are cut in same size from crystals. These pieces have been tested with mass of 10 gr at time of 15 s. The results of samples have been reported in **Table 2**.

Table 1. The results chemical etching.

Density of Etchpit (cm^{-2})	Time (S)	Corrosion solution	Crystal
3.4×10^5	80	$CH_3 + FeCl_3$	KCl:In
2.9×10^5	20	CH_3COOH	KCl:Ag
1.3×10^5	40	CH_3COOH	KCl:Mn
1×10^5	45	CH_3COOH	KCl

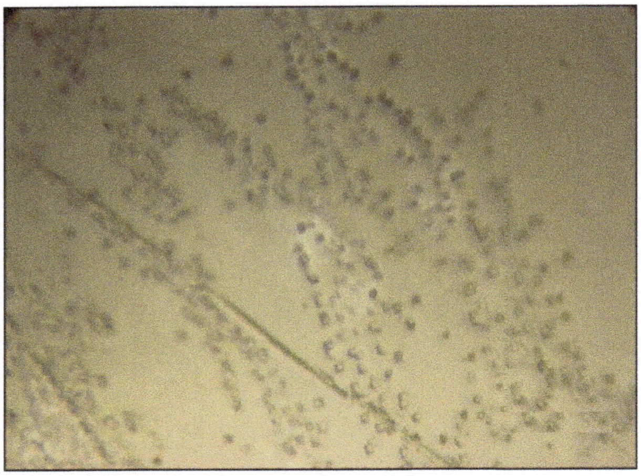

Figure 6. The picture of Etchpit from chemical etching in the surface of the crystal of KCl:Ag.

The above results show increasing of impurities in single crystal of KCl have positive effects in increasing of hardness of the single crystal. This result is clearer when impurity atom is heavier and bigger. The data of chemical etching **Table 1** confirm these results. **Figure 7** shows matching of results of two experiments. During growth process and particularly at the end of growth impurity atoms put in dislocation seat.

The existence of such great atoms is prevented from dislocation motion, because of large energy require for movement of the atoms.

4. Discussion and Conclusion

In XRD spectra of samples are observed high peak with low width ($2\theta = 0.06^\circ$) thus samples are single crystal. In all samples, the main peak is in $2\theta = 28.3^\circ$. This peak is corresponding to plane (200) than single crystal is grown in main direction of (200). Weaker peak is occurred in $2\theta = 58.6^\circ$. This peak is corresponding to diffraction from plane of (400) that is harmonic of plane (200). As chemical etching results show density of Etchpit of crystals that is in order 10^5 cm^{-2}. The density of crystals , that are used in optical purposes , is in order 10^4 to 10^6 cm^{-2} [10]. Thus the grown crystals have good optical quality

Table 2. Results of Vickers micro hardness.

Crystal	Vickers micro hardness	Mass number	r
KCl	12.9 HV_{10}	-	-
KCl:Mn	16.6 HV_{10}	Mn: 54	0.8 (2^+)
KCl:Ag	25.6 HV_{10}	Ag: 107	1.13 (1^+)
KCl:In	27 HV_{10}	In: 114	0.92 (3^+)

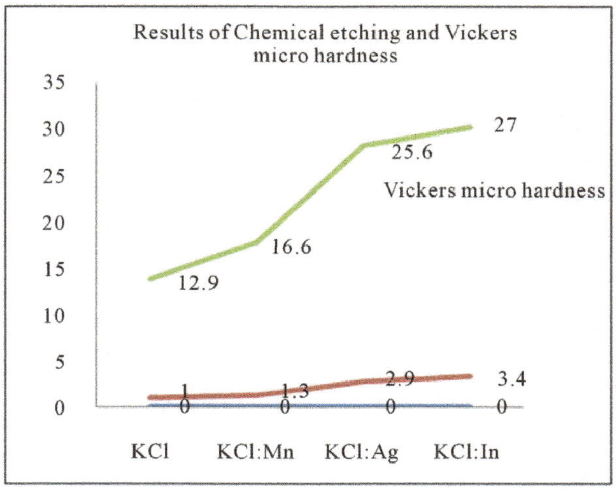

Figure 7. Comparison of results chemical etching and vickers micro hardness.

and are suitable for optical applications. As adding of impurity to pure crystal of KCl was increased hardness of this crystal, that is a positive step in improving of their mechanical properties.

REFERENCES

[1] S. G. Singh, M. Tyagi, D. G. Desai, A. K. Singh, B. Tiwari, S. SeN, A. K. Chauhan and S. C. Gadkari, "Development of Technologically Important Crystals and Devices," *BARC Newsletter*, No. 318, 2011, pp. 16-27.

[2] H. G. Scheel, "Estimated Shares of World Crystal Production," *Journal of Crystal Growth*, Vol. 211, No. 1-4, 2000, pp. 1-12. doi:10.1016/S0022-0248(99)00780-0

[3] G. Atanassov, D. Popov and P.-F. Gu, "Antireflection Coatings on KC1 Single-Crystal Windows with Low Optical Loss," *Thin Solid Films*, Vol. 261, 1995, pp. 256-262.

[4] S. Polosan, T. Tsuboi, E. Apostol and V. Topa, "Electrolytic Reduction of Tl+ Ions in KCl Crystals," *Optical Materials*, Vol. 30, No. 1, 2007, pp. 95-97. doi:10.1016/j.optmat.2006.11.006

[5] M. Tyagi, S. G. Singh, A. K. Singh and S. C. Gadkari, "Understanding Colorations in $PbMoO_4$ Crystals through Stoichiometric Variations and Annealing Studies," *Physica Status Solid A*, Vol. 207, No. 8, 2010, pp. 1802-1807. doi:10.1002/pssa.200925625

[6] F. Rosenberger and G. H. Westphal, "Low-Stress Physical VAPOR Growth," *Journal of Crystal Growth*, Vol. 43, No. 2, 1978, pp. 148-152. doi:10.1016/0022-0248(78)90162-8

[7] K. Sangwal, "Eching of Crystals," North Holland, 1987.

[8] www.leco.com/resources/met_tips/met_tip9.pdf

[9] D. B. Sirdeshmukh, L. Sirdeshmukh and K. G. Subhadra, "Micro and Macro Properties of Solids," Springer-Verlag Berlin Hidelberg, 2006.

[10] N. B. Singh, T. A. Gould and R. H. Hopkins, "Etching Studies on Crystals of Thalium Arsenic Selenide," *Journal of Crystal Growth*, Vol. 78, No. 1, 1996, pp. 43-50.

22

Conformational and Bonding Analysis of $C_2H_4^{2+}$

Gholam Hossein Shafiee
Department of Chemistry, Islamic Azad University of Abadeh, Abadeh, Iran
Email: shafiee@iaug.ac.ir

ABSTRACT

In this report, different models of bonding and structure such as Lewis, VSEPR, Ligand close packing (LCP), VB, qualitative MO and QTAIM have been applied to analyze the Bonds and structures of two equilibrium geometries (planar D_{2h} and perpendicular D_{2d}) of $C_2H_4^{2+}$. The geometries were optimized at near RHF and MP2 limit using cc-pVTZ basis set. While the above bonding models are successfully applied for predicting the low energy isomers of molecules, prior to solving the Schrödinger equation, it is shown that the cited models fail in predicting the existence of perpendicular, D_{2d} form of $C_2H_4^{2+}$. In this regard the interpretations of significant energetic stabilization of D_{2d} form over planar isomer has also been revisited. This is attributed to the hidden effect of the Pauli Exclusion principle.

Keywords: $C_2H_4^{2+}$; LCP; Hyperconjugation; QTAIM; Pauli Exclusion Principle

1. Introduction

The earliest work on the potential energy surface (PES) of $C_2H_4^{2+}$ has shown the existence of twisted (D_{2d}) $H_2CCH_2^{2+}$ minimum and planar (D_{2h}) transition state geometries [1]. The remarkably large rotational barrier energy, 28.1 kcal/mol has been described by the Hyperconjugation between two formally vacant orbitals and corresponding vicinal CH_2 group [1]. The substituted ethylene dications have also been extensively studied [2,3] and the results were explained using descriptive tools such as π donation/conjugation, orbital diagrams, lone pairs, steric repulsion and Hyperconjugation. The energy lowering in twisted geometry of ethylene radical cation [4,5] has also been described by the attractive interaction of a p orbital of C atom with the p-like group orbital of adjacent σ bonds [6,7]. We believe that the proposed explanations about the energy lowering in twisted geometry of $C_2H_4^{2+}$ are based on the mixed effects of overlapping, charge separation, Hyperconjugation models. It is the goal of this work to reinvestigate the effect of each of these models on the final geometry of $C_2H_4^{2+}$ individually. In this work the previous computations with small size Pople type basis sets [1,2] have been extended to larger correlation-consistent basis sets.

2. Computational Details

Ab initio computations were performed at both RHF and MP2 methods using cc-pVTZ basis set [8]. PC GAMESS 7.1.5 package [9] running on parallel mode on 5 PCs cluster was used for optimization and frequency calculations. The total electron densities (Rho functions) constructed from wave functions were analyzed, using AIM2000 and MORPHY99 softwares [10-13]. The useful utilities implemented in the ChemCraft program [14] were optimized for handling the outputs of PC GAMESS.

3. Results and Discussion

The resulted *ab initio* geometries and energies of C_2H_4 and $C_2H_4^{2+}$ are gathered in **Figure 1**. Some differences between the results of present computations at MP2 level and those of 6-31G* and 6-31G** [1,2] are discussed first. C-C distances which have been reported previously as 1.317 and 1.587 Å in C_2H_4 and $C_2H_4^{2+}$ (D_{2h}) respectively [1], are 1.332 and 1.603 Å, each are 0.015 Å longer at MP2 level. In contrast, the C-C distance in $C_2H_4^{2+}$ (D_{2d}) shrinks from 1.432 [1] to 1.393 Å, the significant shortening of 0.039 Å. The difference between C-C distances between two $C_2H_4^{2+}$ isomers is reported as 0.21 Å, 0.055 Å longer than previously computed value, *i.e.* 0.155 Å [1]. The energy difference between two isomers is reported as 30.9 kcal/mol at MP2 level, in agreement with previously reported value of 28.1 kcal/mol [1,2]. As depicted in **Figure 1**, two major geometrical changes occur from D_{2h} to D_{2d} form.

1) The perpendicular arrangement of Hydrogen atoms in D_{2d} form,

2) The significant shortening of C-C distances in D_{2d} C-H distances are almost equal between two $C_2H_4^{2+}$ isomers.

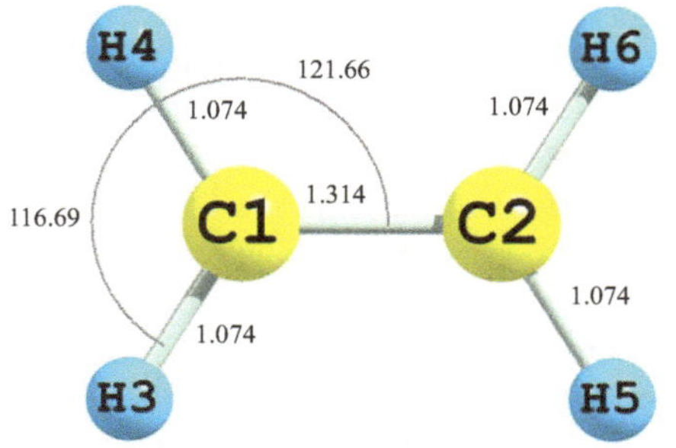

C_2H_4 (D_{2h})
E(RHF): –78.0647728914 *a.u*
H4-H6 = 2.442 Å

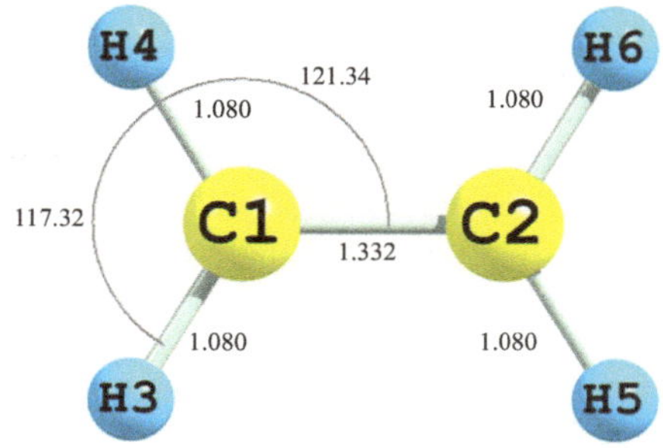

C_2H_4 (D_{2h})
E(MP2): –78.4014122454 *a.u*
H4-H6 = 2.456 Å, H3-H4 = 1.846 Å

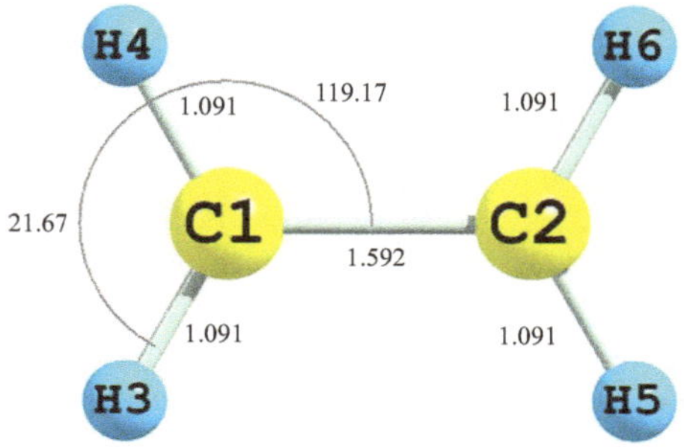

$C_2H_4^{2+}$, Min (D_{2d})
E(RHF): –77.0770048672 *a.u*
H4-H6 = 2.655 Å

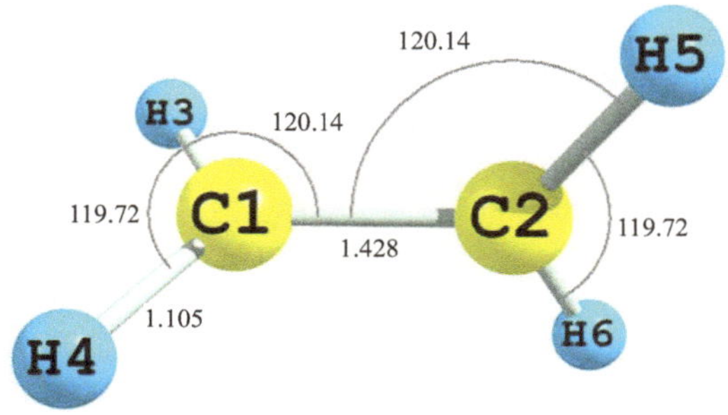

$C_2H_4^{2+}$, TS (D_{2d})
E(RHF): –77.1165234897 *a.u*
H4-H6 = 2.537 Å (In Planar Form)

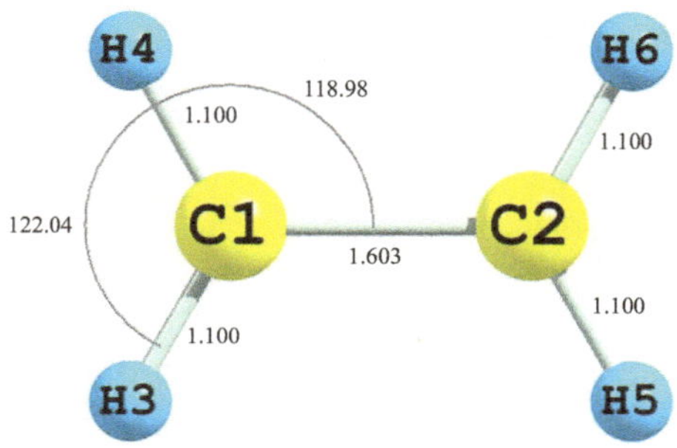

$C_2H_4^{2+}$, Min (D_{2d})
E(MP2): –77.3117689261 *a.u*
H4-H6 = 2.669 Å, H3-H4 = 1.924 Å, H3-H6 = 3.290 Å

$U(H6) = a^2(1/2.669 + 1/3.290 + 1/1.924)$
$U(H6) = 1.198\ a^2$, $V_{NN} = 30.11963$ *a.u*

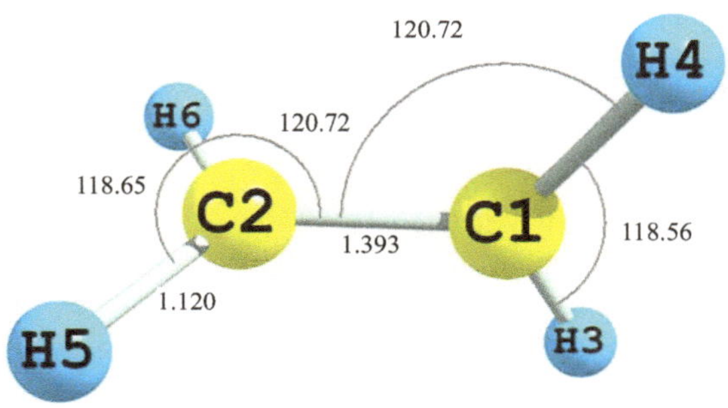

$C_2H_4^{2+}$, TS (D_{2d})
E(MP2): –77.3609539882 *a.u*
H4-H6 = 2.537 Å (In Planar Form), H3-H4 = 1.926 Å, H3-H6 = 2.879 Å

$U(H6) = b^2(1/2.879 + 1/2.879 + 1/1.924)$
$U(H6) = 1.210\ b^2$, $V_{NN} = 32.10244$ *a.u*

Figure 1. Geomertical parameters, total energie, germinal and vicinal HH distances in C_2H_4 and $C_2H_4^{2+}$ at RHF/cc-pVTZ and MP2(FC)/cc-pVTZ models. "TS" stands for transition state and "Min" stands for local minimum geometry on PES. U is the electrostatic repulsive potential energy. (a) and (b) are the charges of H atoms in two molecules.

How different chemical models predict or provide the explanation about these geometrical changes is discussed below. Here we start with basic Lewis model.

3.1. Lewis Model

In Lewis model, a planar structure composed of single bonds for $C_2H_4^{2+}$ can be written [15]. The formal charge of each C is +1 and that of H is zero. Simple electrostatic rule predicts that the C-C distances in planar Lewis structure should be longer than 1.5 Å, the normal C-C single bond in saturated hydrocarbons, because of the repulsion between two C atoms; each bears a positive formal charge. In this model the existence of more stable perpendicular form cannot be predicted or described.

3.2. VSEPR

The VSEPR model predicts that the three single bonding pairs around each C atom should adapt the trigonal arrangement and thus the molecular geometry could be planar [15,16]. However the model is only reliable for predicting the relative geometry around any one central atom and not the relative arrangements of the bonds around two adjacent atoms [16]. Usually the bond-pair bond-pair repulsion and Ligand-Ligand interaction (steric repulsion) effects are added to VSEPR to account for geometrical changes. The internuclear distances depicted in **Figure 1**, show that the vicinal CH bonds repulsion in planar isomer is less than twisted, since they are separated from each other by longer C-C distances. **Figure 1** indicates that in both $C_2H_4^{2+}$ isomers the vecinal HH distances are significantly longer than 2.40 Å (sum of the Van der Waals radia) so steric repulsion between them is negligible and therefore is not responsible for twisting the geometry to D_{2d} form. Therefore VSEPR together with steric repulsion model cannot predict or explain the existence of twisted $C_2H_4^{2+}$.

3.3. Ligand Close-Paking

Empirical Ligand close-packing (LCP) model which is the extension of VSEPR [16], treats ligands as hard sphere objects packed around central atom, *i.e.*, C atom. "Each ligand can be considered to be touching its neighbors and can be assigned a nonbonded radious, which is given by the half the ligand-ligand distance" [16]. The 1,3 nonbonded radius of H atom is 0.92 Å [16]. For example the MP2 geometry of ethylene shows a near perfect close-packing of two H atoms and one CH_2 group around each C atom, since C-C distance is 1.339 and germinal HH distance is 1.846 Å (twice that of their 1,3 nonbonded radius). No steric repulsion exists between two vicinal H atoms too. Close inspection of MP2 geometries of $C_2H_4^{2+}$, **Figure 1**, reveals that the internuclear distances in perpendicular $C_2H_4^{2+}$ is approximately consistent with the LCP model. H3-H4 distance is 1.926 Å and C-C is 1.393 Å. The observed lengthening of Ligand-Ligand distances, in comparison to ethylene, can be assigned to the electrostatic repulsion between positively charged ligands. Still the preferred D_{2d} isomer of $C_2H_4^{2+}$ obtained at MP2 level can not be predicted by LCP model.

3.4. VB, Hyperconjugation and Steric Repulsion

The bond angles about 120.0° around C nuclei, **Figure 1**, in both forms are consistent with the SP^2 hybridization in VB model. Assuming that each C atoms loses one electron, the valence atomic orbitals in C^+ with the electronic configuration as $1s^2\ 2s^2\ 2p^1$, could participate in hybridization process and produce three SP^2 orbitals to overlap with 1s orbitals of H atoms and SP^2 orbital of adjacent Carbon. The existence of twisted isomer is explained by adding hyperconjugation concept to VB. While VB tends to localize electrons between the nuclei, the hyperconjugation tends to delocalize them. The vacant p orbitals, one on each C atom, interact with corresponding vicinal σ bonds in CH_2 [1,6] and therefore lowers the molecular energy. According to our MP2 results this effect (hyperconjugation) should be abnormally strong if it solely responsible for significant lowering in energy (30 kcal/mol) of D_{2d} isomer and corresponding geometrical changes (0.21 Å C-C shortening). This is surprising since the hyperconjugation is a weak effect.

We examined the strength of steric effect which was used to account for the perpendicular arrangement of positively charged Hydrogen atoms [1]. Simple calculation of the electrostatic repulsion energy (U) for each H atom (assuming them as point charges) has been done on MP2 geometries, **Figure 1.** The results demonstrate that if each H atom bears more positive charges on D_{2d} isomer, then each H atom will suffer greater repulsion in comparison to D_{2h} form. If the H atoms bear same charge in two isomers, then D_{2d} form will again suffer from greater repulsion between hydrogen atoms, or at least the same as D_{2h}. In agreement with our simple analysis the V_{NN} (nuclear-nuclear potential energy) value is also greater in D_{2d} form at MP2 level, **Figure 1**.

It is concluded that the energetic and geometric changes cannot be explained based on the weak hyperconjugation or even reduction of steric repulsion between H atoms in D_{2d} isomers.

3.5. Qualitative MO

Analyzing the orbital density [7,17] is other method for tracking the geometrical changes. The densities, of HOMO and HOMO-1 for each isomer of $C_2H_4^{2+}$ are depicted in **Figure 2**. The shape of MOs in D_{2h} form support the VB picture of bonding. The HOMO and HOMO-1 of D_{2d} are degenerate and they are more delocalized than HOMO of D_{2h}. Non of the MOs in D_{2d} show the orbital density between C nuclei, the density which is necessary for accounting the significant shortening of C-C distances in this isomer. More inspection revealed that The HOMO-2 of the D_{2d} shows orbital density between C nuclei. HOMO-2 shape of D_{2d} is the same in shape to HOMO-1 in D_{2h}.

Again we did not find any reasonable explanations for the two major geometrical changes which we mentioned earlier in this work. We suggest that more sophisticated orbital analysis such as Walsh diagrams [18] can be applied to predict the lowest isomer of $C_2H_4^{2+}$.

3.6. The QTAIM

In the framework of QTAIM theory, the total molecular electron density, Rho(ρ) [19], is the source of chemical information about the molecule. The integration of Rho over all space is exactly equal to the total number of electrons in the molecule [7,19].

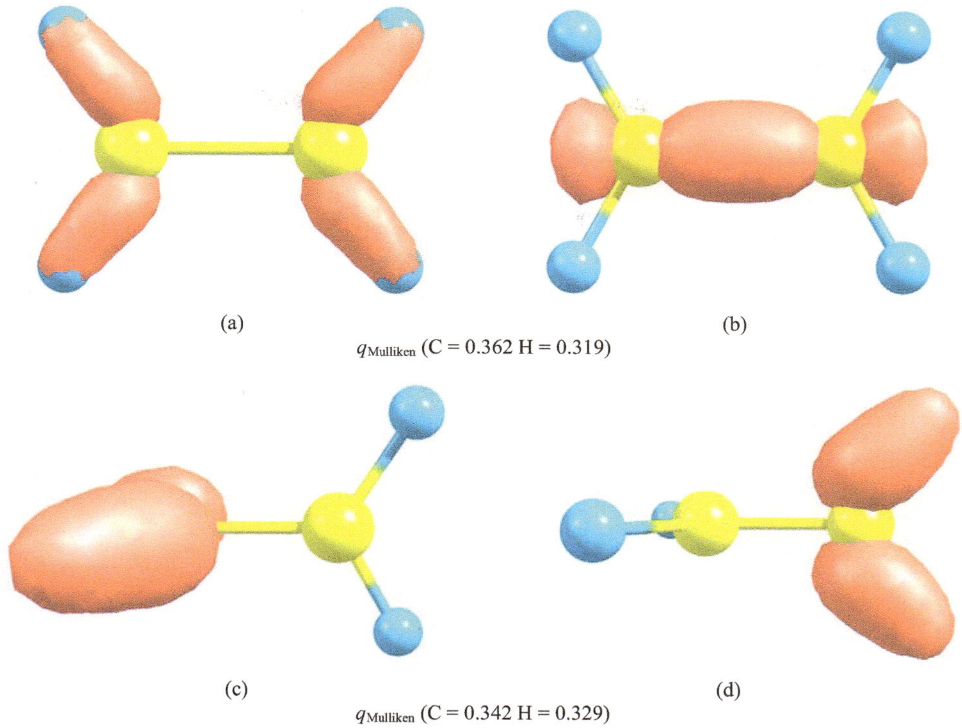

Figure 2. The RHF orbital density (Ψ^2_{MO}) of the HOMO (a, c), HOMO-1 (b, d) and Mulliken atomic charges in planar (a, b) and perpendicular (c, d) forms of $C_2H_4^{2+}$.

We skip the normal topological analysis of electron density at this stage and continue the discussion using the Rho(ρ) values at the geometric midpoint between C-C and C-H in two cited isomers of $C_2H_4^{2+}$.

At MP2 level the ρ values at the chosen points in D_{2h} are 0.232, 0.345 a.u., in D_{2d} are 0.343, 0.316 a.u. respectively. These numbers nicely represent the correct picture of distribution of electronic charge from C-H regions to C-C region when the geometry is changed from D_{2h} to D_{2d}.The accumulation of charge density between C nuclei in twisted form, result in the decrease of positive charge on C atoms and the marked decrease of C-C length [20]. Because of this transferring of negative charge, the H atoms bear more positive charge in D_{2d}. While the ρ between C-C changes by 0.111 a.u. from D_{2h} to D_{2d}, the corresponding value in each CH region is only 0.029 a.u. This explains why C-H distances remain almost the unchanged between two isomers.

Does the ρ provide the explanation for why the D_{2d} isomer is the lowest energy isomer? The quantum theory of atoms in molecules has provided the physical basis for the LCP and VSEPR models based on the arrangement of maxima in the Lapalcian map of ρ around the central atom [21]. Instead of electron pairs, the concept of electron domains is used when using LCP or VSEPR in their new forms [15]. It is expected that at least three maxima be found around each C atom. The maps of Laplacian of electron density for C_2H_4 and $C_2H_4^{2+}$ geometries are depicted in **Figure 3**. The number and the orientation of the maxima around C atoms are the same, 3 maxima in trigonal orientation in all cases, **Figures 3(a1)-(c1)**. The number of domains of charge concentration are also the same, **Figures 3(a3)-(c3)**. The comparison of contour maps of Laplacian among C_2H_4 and two isomers of $C_2H_4^{2+}$, **Figures 3(a)-(c)** show the existence of distinct regions of charge accumulation between C nuclei in C_2H_4. In both $C_2H_4^{2+}$ isomers this region disappears. This is due to the removal of two electrons from the parent molecule (C_2H_4). The Laplacian maps of the two $C_2H_4^{2+}$ isomers are essentially the same. It is not clear yet why $C_2H_4^{2+}$ preferred geometry is the D_{2d} isomer.

Finally we present the full QTAIM analysis [15,19-20, 22], Atomic basin charge and energy values in **Table 1**. The trends of basin charges are the same as what has been predicted by the ρ values at the midpoints of CH and C-Cs. The positive charge of each C basin in D_{2d} is half of the value in D_{2h}. The comparison of atomic charges (**Table 1**) indicates that 0.05 unit of charge is transferred from each H basin to C basins when the geometry is changed from D_{2h} to D_{2d}. The volumes of C basins increase and those of H basins decrease by this charge transfer. Energy analysis demonstrates that each C basin is stabilized by 68.15 kcal/mol but each H basin is destabilized by 26.36 kcal/mol, from D_{2h} to D_{2d} form.

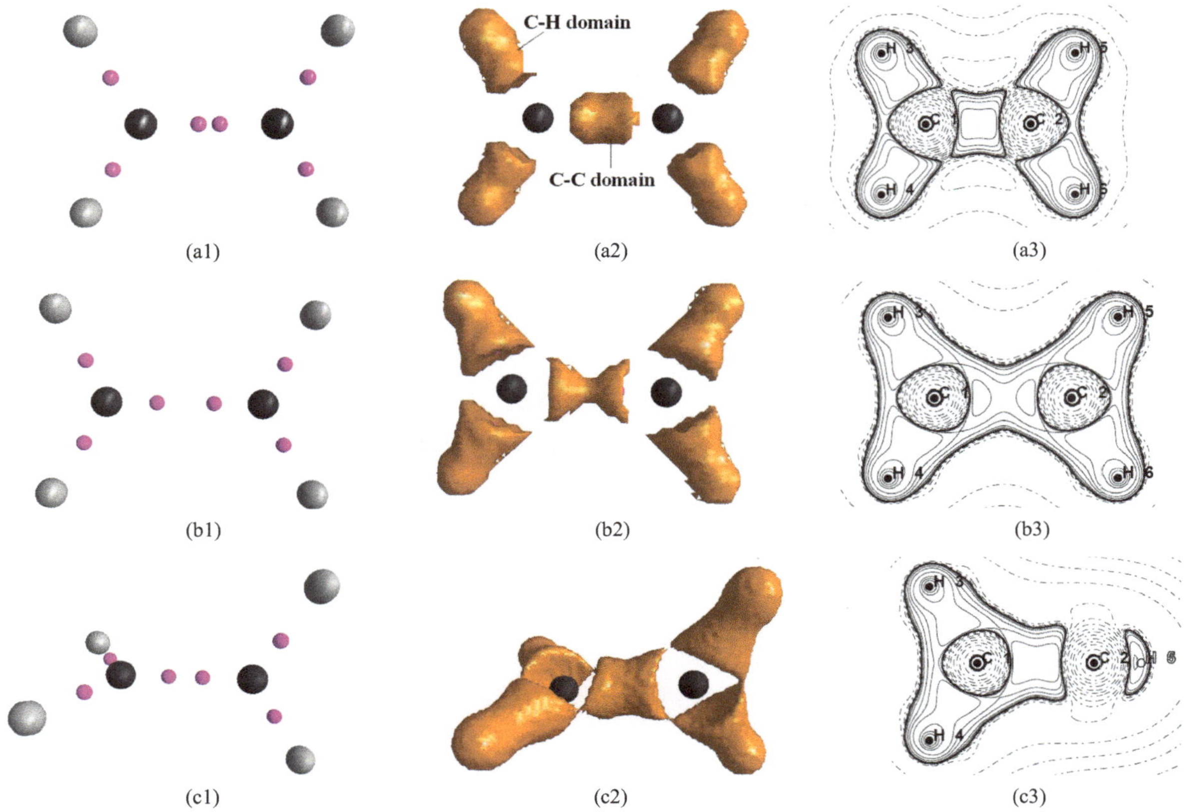

Figure 3. Laplacian maps of C_2H_4 (a), $C_2H_4^{2+}$ planar (b), $C_2H_4^{2+}$ perpendicular (c) at MP2/cc-pVTZ level.Maximum points (small circles) on Laplacian of Rho, (a1)-(c1). 3D pictures of isosurfaces), –0.1 a.u, of charge concentration *i.e.* negative value of Laplacian, (a2)-(c2). Contour plots of Laplacian maps (dash lines = charge depletion, solid line = charge concentration), (a3)-(c3).

Table 1. Atomic properties (in atomic units) from QTAIM analysis at MP2/cc-pVTZ.

	Q‡	E(Ω)**	V(Ω)*
$C_2H_4(D_{2h})$			
C1	–0.0650	-37.9776	98.69
C2	–0.0650	-37.9776	
H3	0.0327	–0.6116	48.75
H4	0.0327	–0.6116	
H5	0.0327	–0.6116	
H6	0.0327	–0.6116	
Δ|(*AIM-Ab initio*)|	0.0008	0.2 kcal/mol	
$C_2H_4^{2+}$ (D_{2d})			
C1	0.0954	–37.8837	89.97
C2	0.0954	–37.8837	
H3	0.4523	–0.3984	27.52
H4	0.4523	–0.3984	
H5	0.4523	–0.3984	
H6	0.4523	–0.3984	
Δ|(*AIM-Ab initio*)|	0.0000	0.002 kcal/mol	
$C_2H_4^{2+}$ (D_{2h})			
C1	0.2012	–37.7751	88.13
C2	0.2012	–37.7751	
H3	0.3994	–0.4404	29.01
H4	0.3994	–0.4404	
H5	0.3994	–0.4404	
H6	0.3994	–0.4404	
Δ|(*AIM-Ab initio*)|	0.0002	0.1 kcal/mol	

‡Total atomic charge (a.u); **Total atomic energy (a.u) which is calculated by integrating over atomic basin (Ω); *Atomic volume bounded by rho contour surface of 0.001 a.u.

4. Conclusion

In 1954 Lennard-Jones published the paper, the distribution functions of neutral fermionic particles with same spin moving in the closed ring and Surface of sphere [23]. The orientation of maxima of probability of finding the particles on the ring or sphere, were demonstrated mathematically as 180° for two particles, 120° for three particles and tetrahedral arrangement for four particles moving on the surface of sphere. The paper has been written to show the net effect of PEP in distributing the electrons in confined region of space. Adding the presented discussion (Failure of different models) it seems that the molecular geometry is controlled by hidden effect of Pauli Exclusion Principle, PEP [21] which is imposed to the wave function.

REFERENCES

[1] K. Lammertsma, M. Barazaghi, G. A. Olah, J. A. Pople, A. J. Kos and P. V. R. Schleyer, "Carbodications. 6. The Ethylene Dication: A Theoretical Study of the Ethylene Dication ($C_2H_4^{2+}$) Potential-Energy Surface," *Journal of the American Chemistry Society*, Vol. 105, No. 16, 1983, pp. 5252-5257. doi:10.1021/ja00354a012

[2] G. Frenking, "Substituted Ethylene Dications: Planar or Perpendicular?" *Journal of the American Chemistry Society*, Vol. 113, No. 7, 1991, pp. 2476-2481. doi:10.1021/ja00007a020

[3] T. Ohwada and K. Shudo, "Substituted Ethylene Dications. Structures of Phenylmethyl Cations Substituted with an Electron-Withdrawing Group on the Cation Center," *Journal of the American Chemistry Society*, Vol. 111, No. 1, 1989, pp. 34-40. doi:10.1021/ja00183a006

[4] L. A. Eriksson, S. Lunell and R. J. Boyd, "Electronic Structure Calculations of Hydrocarbon Radical Cations: A Density Functional Study," *Journal of the American Chemistry Society*, Vol. 115, No. 15, 1993, pp. 6896-6900. doi:10.1021/ja00068a055

[5] M. L. Abrams, E. F. Valeev, C. D. Sherrill and T. D. Crawford, "The Equilibrium Geometry, Harmonic Vibrational Frequencies, and Estimated *ab Initio* Limit for the Barrier to Planarity of the Ethylene Radical Cation," *Journal of Physical Chemistry A*, Vol. 106, No. 11, 2002, pp. 2671-2675. doi:10.1021/jp0134143

[6] R. V. I. Rauk, "Orbital Interaction Theory of Organic Chemistry," Wiley Inter-Science, New York, 2001.

[7] C. J. Cramer, "Essentials of Computational Chemistry Theories and Models," Wiley, New York, 2008.

[8] T. H. Dunning Jr., "Gaussian Basis Sets for Use in Correlated Molecular Calculations. I. The Atoms Boron through Neon and Hydrogen," *Journal of Chemical Physics*, Vol. 90, No. 2, 1989, pp. 1007-1023. doi:10.1063/1.456153

[9] M. W. Schmidt, K. K. Baldridge, J. A. Boatz, S. T. Elbert, M. S. Gordon, J. H. Jensen, S. Koseki, N. Matsunaga, K. A. Nguyen, S. J. Su, T. L. Windus, M. Dupuis and J. A. Montgomery, "General Atomic and Molecular Electronic Structure System," *Journal of Computational Chemistry*, Vol. 14, 1993, pp. 1347-1363.

[10] F. Biegler-König, "Calculation of Atomic Integration Data," *Journal of Computational Chemistry*, Vol. 21, No. 12, 2000, pp. 1040-1048. doi:10.1002/1096-987X(200009)21:12<1040::AID-JCC2>3.0.CO;2-8

[11] F. Biegler-König, J. Schönbohm and D. Bayles, "AIM 2000—A Program to Analyze and Visualize Atoms in Molecules," *Journal of Computational Chemistry*, Vol. 22, 2001, pp. 545-559.

[12] P. L. A. Popelier and R. G. A. Bone, "MORPHY99," A Topological Analysis Program, UMIST, Engl, EU.

[13] P. L. A. Popelier, "MORPHY, a Program for an Automated 'Atoms in Molecules' Analysis," *Computer Physics Communications*, Vol. 93, No. 2-3, 1996, pp. 212-240. doi:10.1016/0010-4655(95)00113-1

[14] G. A. Zhurko and D. A. Zhurko, "ChemCraft, Tool for Treatment of the Chemical Data." http://www.chemcraftprog.com

[15] P. L. A. popelier and R. J. Gillespie, "Chemical Bonding and Molecular Geomerty from Lewis to Electron Dnsities," Oxford University Press, Oxford, 2001.

[16] R. J. Gillespie and I. Hargittai, "The VSEPR Model of Geometry," Allyn & Bacon, Boston, 1991.

[17] D. A. McQuarrie, "Quantum Chemistry," Oxford University Press, Oxford, 1983.

[18] B. Gimarc, "Application of Qualitative Molecular Orbital Theory," *Accounts of Chemical Research*, Vol. 7, No. 11, 1974, pp. 384-392. doi:10.1021/ar50083a004

[19] R. F. W. Bader, "Atoms in Molecules: A Quantum Theory," Clarendon Press, Oxford, 1990.

[20] R. F. W. Bader, "Chem 106 Seminar #2," Case Western Reserve University, 2008.

[21] R. F. W. Bader, R. J. Gillespie and P. J. MacDougall, "A Physical Basis for the VSEPR Model of Molecular Geometry," *Journal of the American Chemistry Society*, Vol. 110, No. 22, 1988, pp. 7329-7336. doi:10.1021/ja00230a009

[22] P. L. A. Poplier, "Atoms in Molecules: An Introductory," Prentice Hall, Upper Saddle River, 2000.

[23] S. J. Lennard-Jones, "New Ideas in Chemistry," *Advanced Science*, Vol. 11, No. 54, 1954, p. 136.

23

In Silico Study on Sulfated and Non-Sulfated Carbohydrate Chains from Proteoglycans in *Cnidaria* and Their Interaction with Collagen

Thomas Eckert[1,2], Sabine Stötzel[1], Monika Burg-Roderfeld[1], Judith Sewing[3,4], Thomas Lütteke[1], Nikolay E. Nifantiev[5], Johannes F. G. Vliegenthart[6], Hans-Christian Siebert[1,7]

[1]Institut für Biochemie und Endokrinologie, Veterinärmedizinische Fakultät, Justus-Liebig-Universität, Gießen, Germany
[2]Klinik für Geburtshilfe, Gynäkologie und Andrologie, Veterinärmedizinische Fakultät, Justus-Liebig-Universität, Gießen, Germany
[3]Coastal Research & Management, Kiel-Holtenau, Germany
[4]Institut für Virologie und Zellbiologie, Lübeck Universität, Lübeck, Germany
[5]N. D. Zelinsky Institute of Organic Chemistry, Russian Academy of Sciences, Moscow, Russian Federation
[6]Department of Bio-Organic Chemistry, Bijvoet Center for Biomolecular Research, Utrecht University, Utrecht, The Netherlands
[7]RI-B-NT - Research Institute of Bioinformatics and Nanotechnology, Kiel, Germany
Email: info@ri-b-nt.de

ABSTRACT

Proteoglycans and collagen molecules are interacting with each other thereby forming various connective tissues. The sulfation pattern of proteoglycans differs depending on the kind of tissue and/or the degree of maturation. Tissues from *Cnidaria* are suitable examples for exploration of the effects in relation to the presence and the absence of sulfate groups, when studying characteristic fragments of the long proteoglycan carbohydrate chains *in silico*. It has been described that a non-sulfated chondroitin appears as a scaffold in early morphogenesis of all nematocyst types in *Hydra*. On the other hand, sulfated glucosaminoglycans play an important role in various developmental processes of *Cnidaria*. In order to understand this biological phenomenon on a sub-molecular level we have analysed the structures of sulfated and non-sulfated proteoglycan carbohydrate chains as well as the structure of diverse collagen molecules with computational methods including quantum chemical calculations. The strong interactions between the sulfate groups of the carbohydrates moieties in proteoglycans and positively charged regions of collagen are essential in stabilizing various *Cnidaria* tissues but could hinder the nematocyst formation and its proper function. The results of our quantum chemical calculations show that the sulfation pattern has a significant effect on the conformation of chondroitin structures under study.

Keywords: Quantum Chemical Calculations; Proteoglycans; Collagen; Atom Force Microscopy; *Cnidaria*

1. Introduction

Proteoglycans represent a special class of glycoproteins, which are heavily glycosylated. These bio-macromolecules consist of a core protein with one or more covalently attached glycosaminoglycan (GAG) chain(s). The glycosaminoglycan chains are long, linear carbohydrate polymers that are negatively charged under physiological conditions, due to the occurrence of sulfate and uronic acid groups. Chondroitin sulfate is the most prevalent GAG. Its linkage geometry between predominant disaccharide units is GlcAβ1-3GalNAcβ1. The monomeric residues are GlcA or GlcA(2*S*) and GalNAc or GalNAc(4*S*) or GalNAc(6*S*) or GalNAc(4*S*,6*S*) [1]. Beside chondroitin, hyaluronan is another GAG, which consists of the same disaccharide units as chondroitin sulfate but is the only GAG that is exclusively non-sulfated. Since the sulfation pattern in *Cnidaria* differs depending on its locus, tissue or organelle type [2] we used various tissue probes from these organisms and analysed them with AFM methods as well as with optic microscopy techniques. This strategy allows an optimal preparation of collagen-proteoglycan samples and enables to estimate the influence of sulfation on tissue differentiation. *Cnidaria* are *Anthozoa* (corals, sea anemones, sea fanes), *Hydrozoa* (hydra), *Cubozoa* (box jellyfishes) and *Scyphozoa* (jellyfishes). Like in tissues of other animals sulfated and non-sulfated proteoglycan carbohydrate chains have been found in *Cnidaria* [2,3]. The proteoglycan carbohydrate chains consist of numerous saccharide moieties. In order to design reliable models of fragments of these large carbohydrate chains we have generated its diverse building blocks with the highest possible degree of accuracy in the calculation process. First, five different hexasaccharides have been constructed, which can be considered as the characteristic fragments of the carbohydrate proteoglycan chains. The impact of the presence and the absence of sulfate groups on the hydrogen bond network in these carbohydrate chains was studied with quantum chemical calculations.

The intra-molecular interaction mechanisms could be described by molecular modelling calculations on various levels of theory when performing quantum chemical calculations. We have analysed five different hexasaccharides with various sulfation profiles, which share the common carbohydrate structure L-4-en-4-deoxy-*thr* Hexp-A(αl-3) GalNAc(β1-4)GlcA(β1-3)Gal(β1-3)Gal(βl-4) Xylp-β (see **Figure 1**). Residue L-4-en-4-deoxy-thr-HexpA-unit A on **Figure 1** an unsaturated derivative of D-glucuronic acid) is formed within the degradation of polysaccharide chains. Parent compound **I** does not contain any sulfate group. Three compounds are mono-sulfated, two of them have an *O*-sulfate group at either C6 (**II**) or C4 (**III**) of the GalNAc residue. The third mono-sulfated compound **(IV)** has an *O*-sulfate group at C4 of the Gal residue preceding the GlcA residue, whereas the GalNAc is not sulfated. The disulfated compound **(V)** has sulfate groups at C4 of both the Gal residue preceding GlcA and the GalNAc residue. The structural heterogeneity of these hexasaccharides reflects the polydiversity in the linkage region of the chondroitin sulfate chains: L-4-en-4-deoxy-*thr*HexpA(αl-3)GalNAc(4-*O*-or6-*O*-sulfate)(β1-4)GlcA and L-4-en-4-deoxy-*thr*HexpA (αl-3) GalNAc(4-*O*-or6-*O*-sulfate) (β1-4)GlcA(β1-3)Gal-NAc (4-*O*-sulfate), respectively. We have used the quantum chemical approach for a precise determination of energies and conformational angles [4-8].

2. Materials and Methods

Methods for theoretical calculation and modeling

At first, we have generated a conformation analysis of the trisaccharide building blocks using the force field CHARMM27 and refined the conformations with the semi-empirical method AM1 with the Hyperchem 8.0 prof. packet [9]. The results are a prerequisite for our density functional theory (DFT) calculations. The models were used as starting structures for DFT calculations at a deeper level of theory with the B3LYP/6-31G* approach implemented in Gaussian 03 [10]. We have also performed calculations with B3LYP/6-31+G* and could show that the extension of time by a factor of 5 to 10 was the only significant result. Considerable alterations in the calculated structures did not occur. The constructed fragments were grouped together to five different hexasaccharides **(I)-(V)** and calculated again with DFT using B3LYP/ 6-31G*. The docking calculations between the glycans and colla- gen were performed with the Molegro trial version [11]. The following geometric and energy optimizations were carried out again with the force field CHARMM27 in- cluded in the Hyperchem 8.0 prof. packet [9].

Preparation of exumbrella tissue samples

Exumbrella of salted eatable jellyfish *Rhopilema esculentum* was washed several times in distilled water and then equilibrated for at least one hour in PBS. Small pieces were prepared for cryo-cutting by embedding in Tissue Tek® (Sakura) and freezed at –20°C. Cryo-cutting was carried out with a Leica CM 3050S machine. For DAPI-staining (Roche diagnostics) a 1 µg/ml solution in PBS was dropped on the slices, incubated at room temperature for half an hour in darkness and then rinsed with PBS. The optic microscopy study was performed using fluorescence dyes. The AFM measurements have been performed as described in literature [12]. For the AFM analysis of the exumbrella tissue of salted eatable jellyfish.

Rhopilema esculentum

We have dissolved 2.8 mg in 1 ml PBS buffer and diluted with pure water to a concentration corresponding to 10 ng/ml. From this probe we laid 50 µl on the mica plates and dried it about 20 min under nitrogen.

3. Results and Discussion

Calculations

The quantum-chemical calculations were started with core glycan structures and consisting of trisaccharide building blocks. These building blocks were used for the construction of five different Glycan structures **Figure 1 I-V**, which differ by their sulfation pattern as described above. It was our aim to calculate precisely defined fragments of the huge proteoglycan carbohydrate chains in order to generate the building blocks for an extensive *in silico* interaction study with diverse collagen molecules occurring in different tissues of *Cnidaria*. The calculated energy minimum structures are shown in **Figures 2 (I)-(V)**.

In a first step, the energies and conformations of trisaccharides with and without sulfate groups were calculated with Gaussian 03/Hyperchem 8.0, (Preliminary calculations: AM1, Gnorm = 0.00001). The quantum chemical

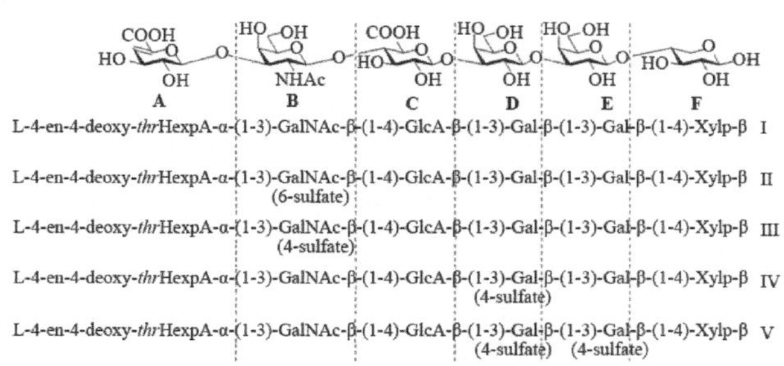

Figure 1. Schematic presentation of the hexasaccharides I-V.

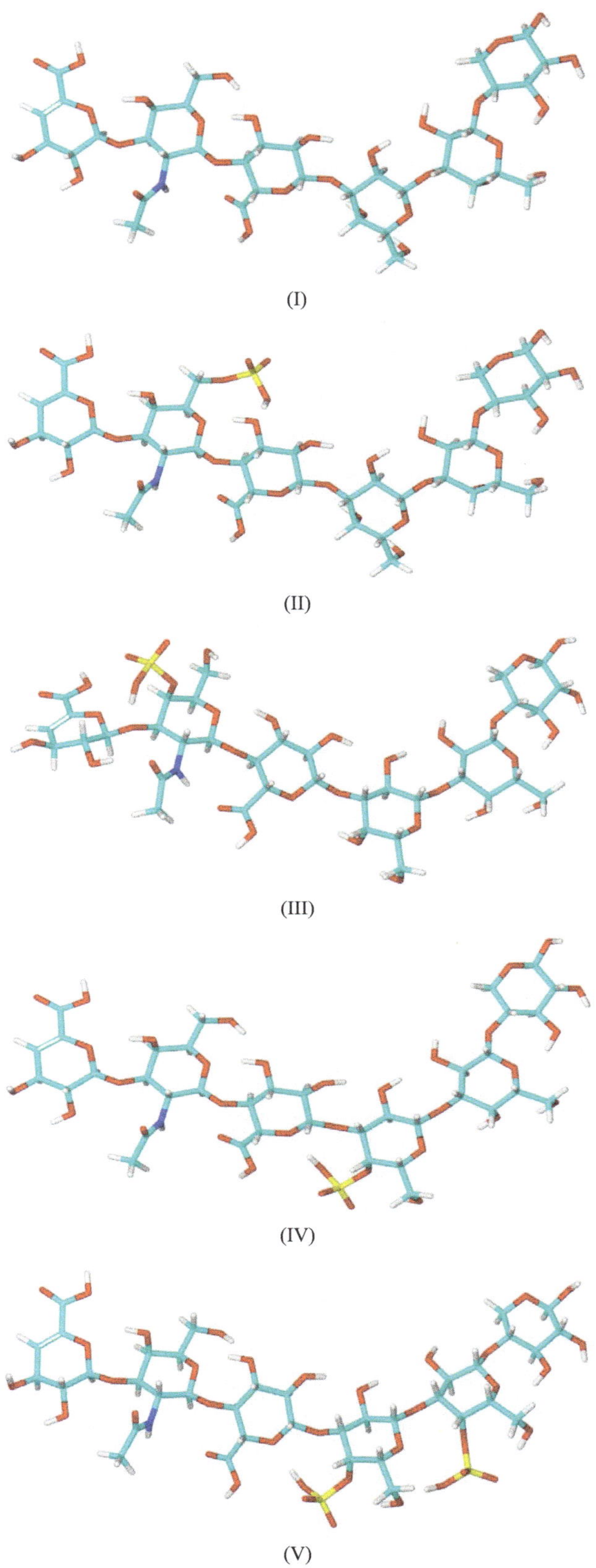

Figure 2. Minimal energy conformations of hexasaccharides (I); (II); (III); (IV) and (V).

Under physiological pH conditions the COO^- and SO_4^- groups are de-protonated. However, in our DFT calculations using B3LYP/6-31G* we have chosen the non-ionized state. It turned out that the protonated state corresponds to the conditions in the water environment much better than the de-protonated state in the vacuum. Certain energy minima are overestimated others are underestimated when quantum chemical calculations are carried out under vacuum conditions. Independent of the protonation state certain energy-minima are adopted under classical conditions when running MD simulations in a water environment. For the *in silico* molecular docking studies the de-protonated Glycan forms have been used since the electrostatic interactions between the Glycans and the proteins are of major importance.

The calculated glycosidic linkages are shown in **Table 1**. Corresponding trisaccharides have been constructed and energetically minimized with semi-empirical methods (AM1). The models were used as starting structures for DFT (Density Functional Theory) calculations in a deeper level of theory with the RB3LYP/6-31G* approach.

The trimeric saccharides with the lowest energies were combined to the larger hexameric saccharide chains of Glycans **I-V**. The energies and conformational angles Φ and Ψ were again calculated with the DFT approach at the RB3LYP/6-31G* level of theory.

The conformation of Glycan **I** (**Figure 2**) listed in **Table 2** is mainly stabilized by hydrogen bonds, which are displayed for all five glycan stuctures in **Table 3**. In the case of Glycan **II** (**Figure 2**), the sulfate group on Gal-NAc in position 6 has no significant effect on the Φ and Ψ angles due to the high mobility of the sulfate group. Therefore, Φ and Ψ angles similar to those calculated for Glycan **(I)** occur in **Table 2**. The sulfate group in 4 position of GalNAc, which is present in Glycan **III** (**Figure 2**) suppresses hydrogen bond formation. Therefore, the Φ1- and Ψ1-angle values differ significantly from those of Glycan **I**. The sulfate group also influences the CH_2OH group of the GalNAc, which has an impact on the Φ2 and Ψ2 angle values. When calculating the glycosidic angles of Glycan **IV** (**Figure 2**) it is obvious that the presence of a sulfate group of the Gal-residue influences the glycosidic angles Φ3 and Ψ3 between GlcA and Gal. In the case of the disulfated Glycan **V** (**Figure 2**) the glycosidic angles Φ3 und Ψ3 are similar to those of Glycan **IV**, as expected. However, the second sulfate group of the second galactose (unit E) causes an alteration in the Φ4 and Ψ4 glycosidic angles due to the hydrogen bonding between this sulfate group and the CH_2OH group of the other galactose (unit D). As has already been shown for fucobiosides and fucoidans [13], it is possible to determine conformational differences in dependence of the sulfation profile. In contrast to the polysaccharides from *Cnidaria*, which are discussed here, fucoidans are a group of highly sulfated polysaccharides

Table 1. Energies and conformational characteristics (angles in degrees) of trisaccharides corresponding to the fragments of hexasaccharides I-V.

Conformation	Energy (kcal/mol)	Φ1	Ψ1	Φ2	Ψ2
L-4-en-4-deoxy-*thr*HexpA(*α*1-3)GalNAc(*β*1-4)GlcA*β*	**(units A, B, C of Glycan I)**				
1	−1,325,674.17	37.73	4.10	51.78	5.33
2	−1,325,677.04	34.42	−9.16	46.63	−3.38
3	−1,325,678.05	29.70	−53.27	46.90	−3.20
4	−1,325,678.07	29.50	−51.34	44.27	−9.67
5	−1,325,674.35	163.34	1.33	39.59	−8.76
6	−1,325,673.47	30.04	−53.37	30.00	176.72
7	−1,325,677.76	−1.49	155.07	43.33	−9.47
8	−1,325,671.99	148.30	14.94	43.73	−8.65
GalNAc(*β*1-4)GlcA(*β*1-3)Gal*β*	**(units B, C, D of Glycan I)**				
1	−1,327,155.39	64.68	−9.04	5.43	35.74
2	−1,327,149.67	64.13	−8.23	165.63	−39.00
3	−1,327,155.04	55.48	−8.53	5.77	36.01
4	−1,327,156.34	8.00	−50.19	8.18	31.94
5	−1,327,160.99	35.72	−35.97	9.99	−12.69
GlcA(*β*1-3)Gal(*β*1-3) Gal*β*	**(units C, D, E of Glycan I)**				
1	−1,243,852.36	66.88	76.94	16.30	−11.38
2	−1,243,834.10	34.07	−49.54	132.12	−55.56
3	−1,243,850.13	6.66	31.60	19.66	156.20
Gal(*β*1-3) Gal*β*(1-4) Xylp*β*	**(units D, E, F of Glycan I)**				
1	−1,125,529.81	33.67	−42.24	44.24	0.50
2	−1,125,534.58	8.08	32.38	46.17	−19.12
3	−1,125,530.20	14.72	18.41	176.06	−0.18
4	−1,125,530.33	−168.52	54.77	46.47	−20.40
5	−1,125,526.82	34.82	−51.59	11.92	179.52
6	−1,125,534.29	6.44	−13.92	49.49	−21.52
L-4-en-4-deoxy-*thr*HexpA(*α*1-3)GalNAc(6-sulfate)(*β*1-4)GlcA*β*	**(units A, B, C of Glycan II)**				
1	−1,717,116.15	30.93	−53.65	48.59	−2.95
2	−1,717,101.57	39.96	−0.60	38.53	14.57
3	−1,717,099.50	40.80	−20.39	27.84	−16.45
4	−1,717,105.26	28.60	−51.08	44.73	15.60
5	−1,717,105.88	38.48	1.03	21.30	−47.82
6	−1,717,104.44	40.00	3.89	43.36	14.87
7	−1,717,104.45	172.81	−1.35	35.24	10.29
8	−1,717,104.73	29.40	−41.92	53.41	−151.40
9	−1,717,110.20	35.29	−53.65	133.92	14.62
10	−1,717,107.64	3.24	164.65	16.72	−56.74
GalNAc(6-sulfate)(*β*1-4)GlcA(*β*1-3)Gal*β*	**(units B, C, D of Glycan II)**				
1	−1,718,589.49	52.84	3.35	6.88	−11.88
2	−1,718,584.90	21.76	−48.54	3.41	40.02
3	−1,718,588.12	29.97	−37.19	48.28	44.32
4	−1,718,581.39	155.35	−0.87	10.61	32.35
5	−1,718,584.19	115.39	19.50	45.16	44.61
6	−1,718,586.30	148.49	−5.48	52.28	39.44
7	−1,718,596.38	35.18	−27.02	21.29	153.27
8	−1,718,586.77	30.34	−36.82	14.87	160.41
9	−1,718,585.38	12.66	11.85	34.68	−53.65
10	−1,718,581.34	157.41	53.79	−2.33	52.07
11	−1,718,597.12	37.26	−22.60	176.98	6.15
12	−1,718,585.70	31.15	−35.63	56.50	80.85
13	−1,718,586.92	20.48	−45.85	8.84	−12.35
14	−1,718,587.44	29.82	−36.92	10.77	30.83
15	−1,718,597.12	37.26	−22.60	176.98	6.15
16	−1,718,583.79	26.99	−164.09	178.90	5.63
17	−1,718,585.35	137.61	0.54	15.69	160.68

Continued

GlcA(β1-3)Gal(β1-3)Galβ	**(units C, D, E of Glycan II)**				
1	−1,243,852.36	66.88	76.94	16.30	−11.38
2	−1,243,834.10	34.07	−49.54	132.12	−55.56
3	−1,243,850.13	6.66	31.60	19.66	156.20
Gal(β1-3)Gal(β1-4)Xylpβ	**(units D, E, F of Glycan II)**				
1	−1,125,529.81	33.67	−42.24	44.24	0.50
2	−1,125,534.58	8.08	32.38	46.17	−19.12
3	−1,125,530.20	14.72	18.41	176.06	−0.18
4	−1,125,530.33	−168.52	54.77	46.47	−20.40
5	−1,125,526.82	34.82	−51.59	11.92	179.52
6	−1,125,534.29	6.44	−13.92	49.49	−21.52
L-4-en-deoxy-*thr*HexpA(α1-3)GalNAc(4-sulfate)(β1-4)GlcAβ		**(units A, B, C of Glycan III)**			
1	−1,717,104.54	39.84	0.78	30.07	−23.65
2	−1,717,109.00	45.82	−11.80	33.87	−19.89
3	−1,717,104.09	−15.48	−28.56	37.21	−11.59
4	−1,717,108.57	12.33	34.24	−3.30	173.06
5	−1,717,104.93	168.99	−0.92	35.73	−15.71
6	−1,717,097.74	42.38	−3.74	34.32	171.61
7	−1,717,105.52	47.15	−17.11	162.52	−3.91
GalNAc(4-sulfate)(β1-4)GlcA(β1-3)Galβ	**(units B, C, D of Glycan III)**				
1	−1,7185,88.10	44.49	−13.74	13.43	28.55
2	−1,718,585.83	45.40	−12.44	−177.26	1.80
3	−1,718,587.51	32.99	−35.88	11.02	31.10
4	−1,718,584.85	21.60	−48.98	3.47	−9.62
GlcA(β1-3)Gal(β1-3)Galβ	**(units C, D, E of Glycan III)**				
1	−1,243,852.36	66.88	76.94	16.30	−11.38
2	−1,243,834.10	34.07	−49.54	132.12	−55.56
3	−1,243,850.13	6.66	31.60	19.66	156.20
Gal(β1-3)Gal(β1-4)Xylpβ	**(units D,E,F of glycan III)**				
1	−1,125,529.81	33.67	−42.24	44.24	0.50
2	−1,125,534.58	8.08	32.38	46.17	−19.12
3	−1,125,530.20	14.72	18.41	176.06	−0.18
4	−1,125,530.33	−168.52	54.77	46.47	−20.40
5	−1,125,526.82	34.82	−51.59	11.92	179.52
6	−1,125,534.29	6.44	−13.92	49.49	−21.52
L-4-en-deoxy-*thr*HexpA(α1-3)GalNAc(β1-4)GlcAβ		**(units A, B, C of Glycan IV)**			
1	−1,325,674.17	37.73	4.10	51.78	5.33
2	−1,325,677.04	34.42	−9.16	46.63	−3.38
3	−1,325,678.05	29.70	−53.27	46.90	−3.20
4	−1,325,678.07	29.50	−51.34	44.27	−9.67
5	−1,325,674.35	163.34	1.33	39.59	−8.76
6	−1,325,673.47	30.04	−53.37	30.00	176.72
7	−1,325,677.76	−1.49	155.07	43.33	−9.47
8	−1,325,671.99	148.30	14.94	43.73	−8.65
GalNAc(β1-4)GlcA(β1-3)Gal(4-sulfate)β	**(units B,C,D of Glycan IV)**				
1	−1,718,582.52	29.07	−47.06	56.34	48.57
2	−1,718,592.73	39.27	−13.78	46.36	−21.85
3	−1,718,576.04	30.36	−32.25	−172.26	−12.42
4	−1,718,590.74	36.66	−14.39	165.01	5.37
5	−1,718,591.20	135.47	19.91	57.97	55.41
6	−1,718,586.04	37.75	−13.98	36.05	−4.63
GlcA(β1-3)Gal(4-sulfate)(β1-3)Galβ	**(units C, D, E of Glycan IV)**				
1	−1,635,282.35	42.76	−25.88	14.63	−7.15
2	−1,635,285.69	27.59	−27.19	4.64	−5.96
3	−1,635,281.14	−13.74	−46.86	47.75	37.06
4	−1,635,286.47	24.20	−24.86	12.07	29.60
5	−1,635,285.70	27.68	−27.20	4.61	−5.99
6	−1,635,284.44	29.05	−31.40	177.58	6.48
7	−1,635,276.23	5.93	−6.16	7.40	−9.21
8	−1,635,284.60	−10.95	−13.13	7.27	−8.74

Continued

Gal(4-sulfate)(*β*1-3)Gal(*β*1-4)Xylp*β*		**(units D, E, F of Glycan IV)**			
1	−1,516,955.89	12.30	−28.74	80.32	−8.80
2	−1,516,950.45	51.44	31.15	28.80	−47.75
3	−1,516,955.61	11.43	29.19	28.83	−44.02
4	−1,516,959.46	6.64	−11.57	49.09	−22.07
L-4-en-deoxy-*thr*HexpA(*α*1-3)GalNAc(*β*1-4)GlcA*β*		**(units A, B, C of Glycan V)**			
1	−1,325,674.17	37.73	4.10	51.78	5.33
2	−1,325,677.04	34.42	−9.16	46.63	−3.38
3	−1,325,678.05	29.70	−53.27	46.90	−3.20
4	−1,325,678.07	29.50	−51.34	44.27	−9.67
5	−1,325,674.35	163.34	1.33	39.59	−8.76
6	−1,325,673.47	30.04	−53.37	30.00	176.72
7	−1,325,677.76	−1.49	155.07	43.33	−9.47
8	−1,325,671.99	148.30	14.94	43.73	−8.65
GalNAc(*β*1-4)GlcA(*β*1-3)Gal(4-sulfate)*β*		**(units B, C, D of Glycan V)**			
1	−1,718,582.52	29.07	−47.06	56.34	48.57
2	−1,718,592.73	39.27	−13.78	46.36	−21.85
3	−1,718,576.04	30.36	−32.25	−172.26	−12.42
4	−1,718,590.74	36.66	−14.39	165.01	5.37
5	−1,718,591.20	135.47	19.91	57.97	55.41
6	−1,718,586.04	37.75	−13.98	36.05	−4.63
GlcA(*β*1-3)Gal(4-sulfate)(*β*1-3)Gal(4-sulfate)*β*		**(units C, D, E of Glycan V)**			
1	−2,026,722.96	43.15	−30.52	43.36	−31.17
2	−2,026,723.02	13.17	27.27	41.71	−30.23
3	−2,026,711.65	9.11	29.55	55.98	37.49
4	−2,026,720.45	−168.38	12.31	41.43	−31.07
5	−2,026,720.03	44.08	−28.74	25.44	−45.21
6	−2,026,722.95	43.15	−30.51	43.39	−31.22
7	−2,026,719.90	48.84	29.31	175.46	−0.47
8	−2,026,720.60	−157.78	39.25	26.77	−40.07
9	−2,026,719.34	11.23	32.53	176.52	2.70
10	−2,026,715.53	171.39	3.09	177.87	6.53
11	−2,026,714.42	49.39	23.29	44.50	100.00
12	−2,026,716.69	48.68	25.46	43.83	75.77
13	−2,026,712.26	−26.94	−16.68	27.89	−45.95
Gal(4-sulfate)(*β*1-3)Gal(4-sulfate)(*β*1-4)Xylp*β*		**(units D, E, F of Glycan V)**			
1	−1,908,378.29	8.37	28.49	27.66	−46.69
2	−1,908,381.53	8.33	29.82	31.43	−29.16
3	−1,908,391.38	−29.89	30.79	29.11	10.42
4	−1,908,387.17	−33.03	28.34	−4.11	175.65
5	−1,908,389.08	29.54	37.09	28.50	−32.32
6	−1,908,391.01	13.08	7.50	177.11	−2.12
7	−1,908,386.78	10.43	25.95	15.07	−178.49
8	−1,908,388.69	170.04	5.79	176.06	−2.58
9	−1,908,396.97	49.58	62.71	38.08	−28.43

Table 2. Calculated conformational angles Φ and Ψ for inter-unit linkages of Glycans I-V (in degrees).

Glycan	Φ1	Ψ1	Φ2	Ψ2	Φ3	Ψ3	Φ4	Ψ4	Φ5	Ψ5
I	29.86	−52.21	44.27	−9.80	1.81	−13.79	6.82	−13.47	49.83	−22.01
II	30.58	−52.99	47.13	−2.60	1.41	−9.96	6.71	−10.68	49.22	−21.58
III	46.25	−12.17	32.59	−22.67	0.62	−12.12	8.07	−11.78	50.46	−22.02
IV	29.63	−50.78	45.71	−8.28	−29.49	−16.70	9.15	−10.76	50.78	−21.91
V	30.35	−52.64	47.38	−8.55	−30.22	−14.58	30.62	−40.40	49.96	−23.85

Table 3. Hydrogen-bonds (in nm) in the structures of the Glycan I - V.

Hydrogen-bonds	Glycan **I**	Glycan **II**	Glycan **III**	Glycan **IV**	Glycan **V**
Ring-O unit A with OH-4 unit B	0.220	0.219	-	0.223	0.219
O-1 unit A with OH-4 unit B	0.258	0.261	-	0.256	0.259
Sulfate-H unit B with O-1 unit A	-	-	0.186	-	-
OH-6 unit B with O-5 unit B	-	-	2.43	-	-
OH-6 unit B with O-3 unit C	0.242	-	-	0.245	0.244
Ring-O unit B with OH-3 unit C	0.196	0.208	0.193	0.195	0.194
O-1 unit B with OH-3 unit C	0.252	0.244	0.266	0.252	0.253
Sulfate-H unit B with O-3 unit C	-	0.164	-	-	-
NH unit B with carboxy-O unit C	-	-	0.205	-	-
OH-2 unit C with O-2 unit D	0.199	0.194	0.197	0.191	0.193
O-1 unit C with OH-4 unit D	0.219	0.219	0.219	-	-
Sulfate-H unit D with ring-O unit C	-	-	-	0.182	0.180
OH-6 unit D with O-4 unit D	0.196	0.196	0.196	0.217	0.209
OH-2 unit D with O-2 unit E	0.199	0.195	0.198	0.197	0.277
O-1 unit D with OH-4 unit E	0.222	0.221	0.223	0.224	-
Sulfate-H ring E with O-6 ring D	-	-	-	-	0.165
OH-6 unit E with sulfate-O unit E	-	-	-	-	0.200
OH-6 unit E with O-4 unit E	0.199	0.200	0.200	0.200	0.267
Ring-O unit E with OH-3 unit F	0.196	0.196	0.196	0.196	0.200
OH-2 unit F with O-3 unit F	0.223	0.222	0.223	0.223	0.221

of brown seaweeds, which have received increasing interest as readily available biopolymers having many promising biological activities. Another source of sulfated carbohydrates are marine fishes. The structural characterization, the anti-inflammatory and anticoagulant activity of chondroitin sulfates from cartilage of Atlantic salmon (*Salmo salar*), Greenland shark (*Somniosus microcephalus*), Blackmouth catshark (*Galeus melastomus*), Birdbeak dogfish (*Deania calcea*) and Arctic skate (*Amblyraja hyperborea*) have recently been described [14].

The absence or the presence of a sulfate group is not only important for the energetic minima of the Glycans, the interactions of the Glycans with collagen molecules also depend on the position of the sulfate groups. Therefore, we have highlighted these functional parts of the proteoglycan molecules in their homo-orbital presentations (**Figure 3**).

In relation to our docking studies it is remarkable that the sulfate groups with their significant HOMO orbitals are essential contact groups for various collagen molecules (**Figures 4** and **5**).

Molecular interaction partners of proteoglycans in *Cnidaria* tissues are mini-collagens with charged ends (**Figure 4**) [15] as well as triple-helical collagen fragments [12,16]. The positively charged contact points consist of Arg and Lys residues [12,15,16]. Also in the case of triple-helical collagen structures Arg residues are the most suited contact points (**Figure 5**).

We have performed *in silico* interaction studies of the five designed Glycan structures (Glycan **I-V**) and mini-collagen (**Figure 4**). In the same way we have studied the interactions between Glycans **I-V** with triple-helical collagen structures (**Figure 5**). It turned out that the glycosidic angles Φ and Ψ, which correspond to the low-energy conformation of the ligand-free state differ slightly from those of the components interacting with the collagen molecules.

Complexes between mini-collagen fragments and characteristic Glycans (**Figure 4**) have been performed by a feasible molecular docking program (the Molegro trial version [11]). The same software was used for the docking studies, in which triple helical collagen fragments are interacting with the Glycan hexamers under study (**Figure 5**). The triple helical structure was taken from an X-ray model of an integrin-collagen complex (1dzi. pdb) [16]. The Glycans and the Arg residues, which are preferentially interacting with the proteoglycan sulfate groups are highlighted in their van-der-Waals representation.

Mini-collagen molecules and small collagen fragments [12,15,17-19] as well as longer collagen fragments [12,17] are interacting in multiple manners with the Glycan structures under study. These results differ significantly from our observations concerning collagen integrin interactions [12]. Sophisticated microscopic techniques are needed to clarify how the absence and presence of sulfate groups in proteoglycans from *Cnidaria* trigger its morphogenesis.

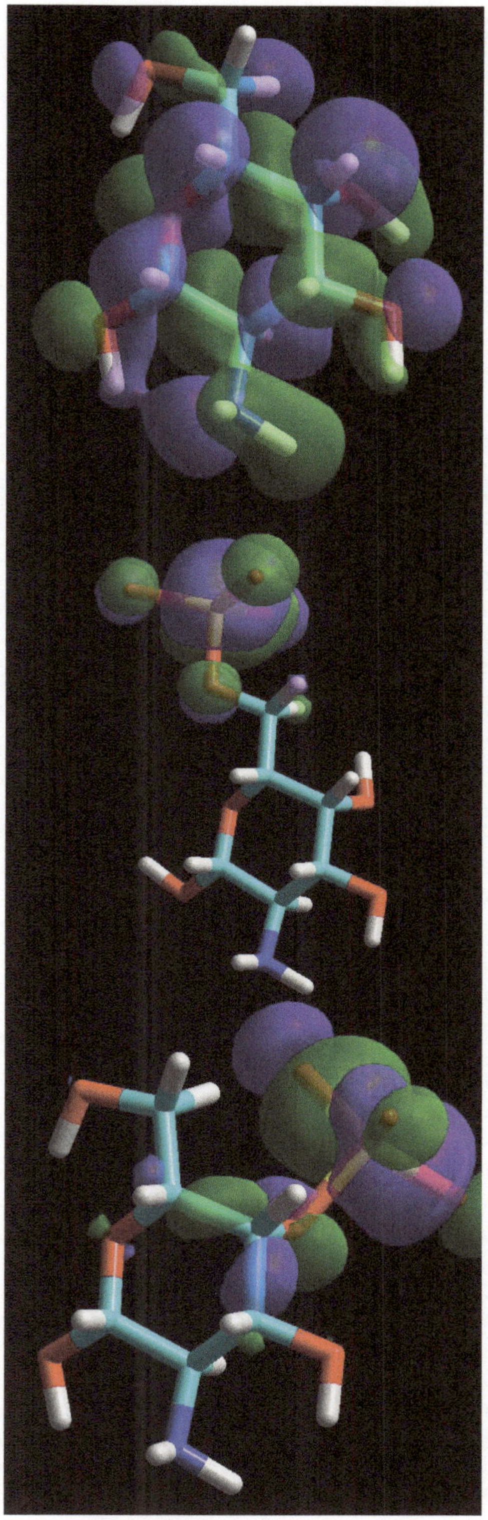

Figure 3. Homo-orbital presentation of GalNAc, non-sulfated (top) and sulfated at position 6 (middle) as well as at position 4 (bottom). The sulfate groups are charged. Orbitals have been calculated for the whole molecules. When a sulfate group is present the orbitals are located at the corresponding position.

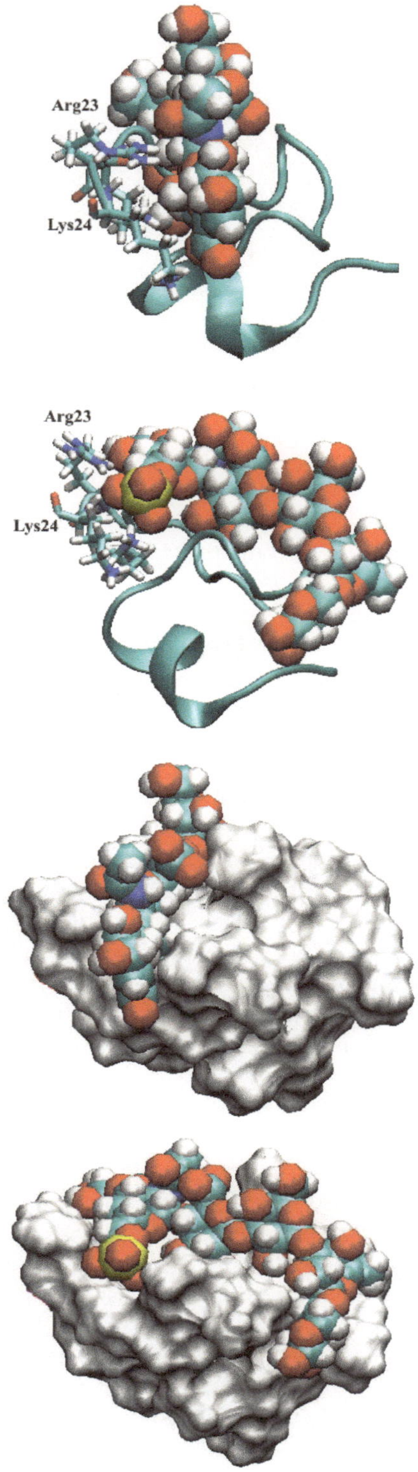

Figure 4. Mini-collagen fragment (1sop.pdb) [15, 17] in complex with non-sulfated Glycan I (top) and sulfated Glycan II (below the top) in their backbone presentations with highlighted Arg23 and Lys24 residues. The surface presentations are given in the same orientation: Gycan I (above bottom) and sulfated Glycan II (bottom).

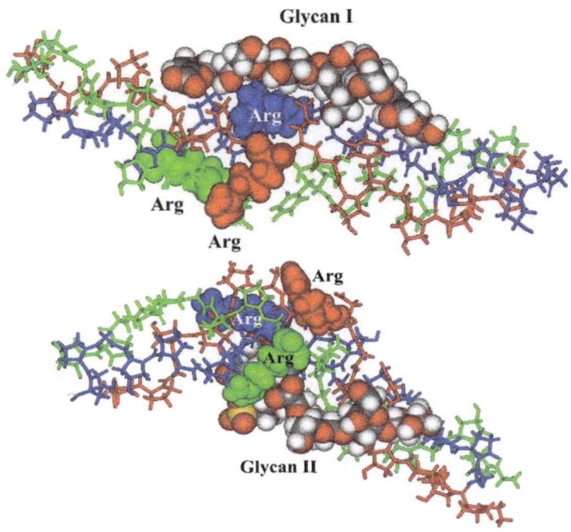

Figure 5. Triple-helical collagen structure in complex with Glycan I (top) and Glycan II (bottom).

Furthermore, the nematocyst discharge processes can be explained in a better way when the interaction mechanisms between non-sulfated proteoglycans and mini-collagen molecules can be described on a sub-molecular level. In a nematocyst poly-gamma-glutamate rich mini-collagens are synthesized during the formation of nematocyst capsules in Hydra [20]. Together with the available information about the differentiation process [21-25] a first step is made to simulate nematocyst discharge processes (coulomb explosions) as well as proteoglycan collagen interactions in cartilage tissues with computational methods. As outlined in two recent publications a solid knowledge about the structural properties of Glycan sulfate groups is a prerequisite for a detailed understanding in respect to their biological function [26,27].

4. Results and Discussion

Experimental Part

For the AFM analysis probes of the exumbrella tissue of salted eatable *Rhopilema* jellyfish were put on mica plates **Figure 6** (top). We compared the AFM probes with probes from the same exumbrella tissue, which were prepared for light microscopic analysis by Dapi staining - **Figure 6** (bottom). After treatment with an aprotic solvent we recognized that the addition of DMSO can disturb and destroy the collagen-proteoglycan network so that only small tissue pieces remain (**Figure 7**).

In our present study the quantum chemical calculations and the computational docking analysis of the five Glycans are flanked by an AFM analysis of various jellyfish tissues. It is described in the literature that hydrolyzed collagen can induce chondrogenic differentiation of equine adipose tissue-derived stromal cells [28]. Although the impact of proteoglycan fragments in such differentiation

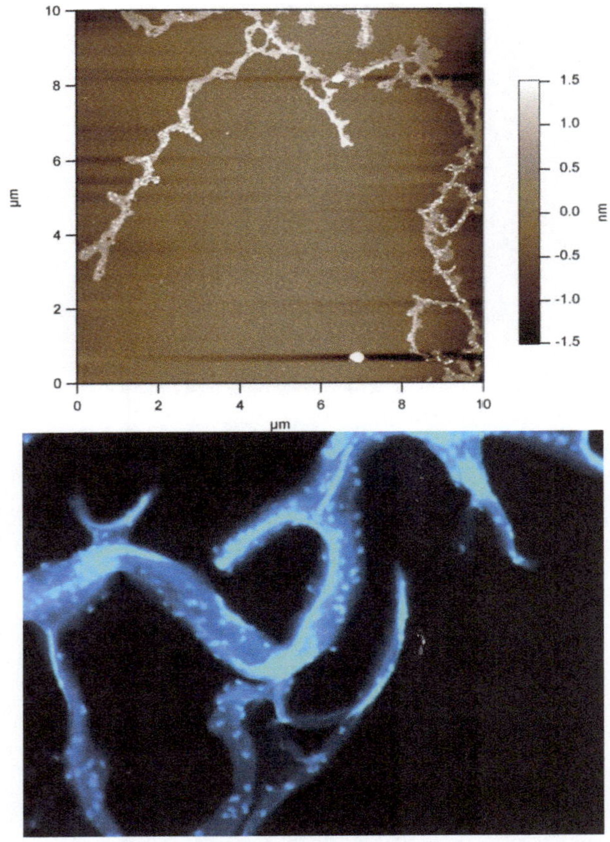

Figure 6. AFM and fluorescent pictures of the exumbrella tissue of salted eatable *Rhopilema* jellyfish.

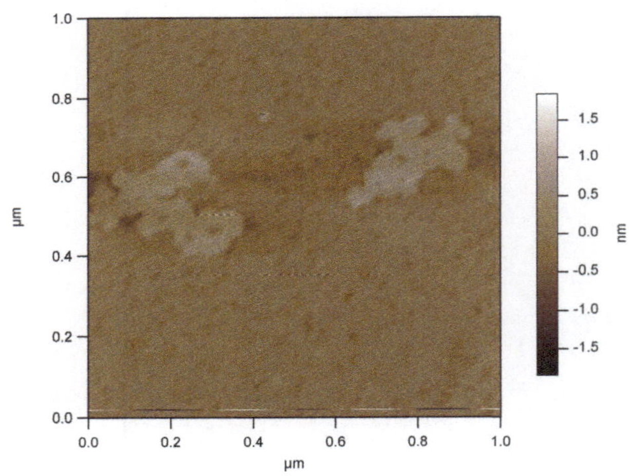

Figure 7. Impact of DMSO on the exumbrella tissue of the jellyfish *Rhopilema esculentum*. Only small pieces of the jellyfish tissue remain in the aprotic solvent. The sample was dried under nitrogen.

processes is unknown it is now possible to combine the theoretically derived data about the collagen-proteoglycan interactions with results obtained by AFM and light microscopic techniques (**Figure 8**). After this study the theoretical background, which allows us a better defined

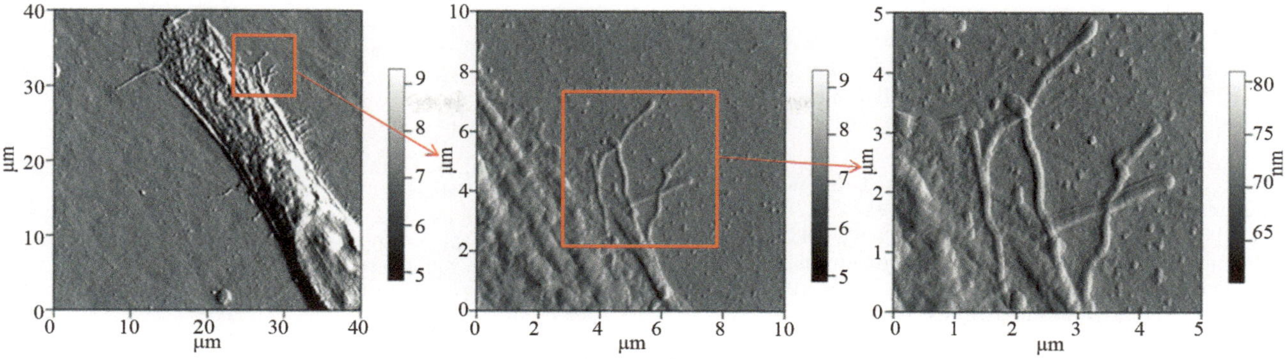

Figure 8. AFM pictures of an equine mesenchymal stem cell (cultured on a collagen surface). The middle and right picture shows a detailed presentation of their cilia.

discussion of the influence of collagen fragments and proteoglycans from various species on a sub-molecular level. Equine mesenchymal stem cells can be cultured on various media consisting of collagen (**Figure 8**) or of collagen-proteoglycan mixtures. Thereby, considering studies of stem cells in hydra [29,30]. We recognized that the differentiation processes are strongly depending on the proteoglycan-collagen ratio of the growth media.

Beside the unspecific proteoglycan collagen interacttion our theoretical and experimental results argue in favor that for the stabilization of the nematocyst tissue also a specific carbohydrate binding protein, *i.e.* nematogalectin [31] must be present. However, this kind of galectin is only specific for non-sulfated carbohydrates as it is the case for galectins in general [32]. To summarize our studies we have opened new routes to investigate proteoglycan-collagen interactions at an atomic size-level, as described for collagen hydrolysates and collagen fragments from various species [33].

5. Acknowledgements

We thank Mr. Robert Drews at the Institut für Virologie und Zellbiologie at the Universität Lübeck for his kind assistance when generating fluorescent pictures of ex-umbrella tissues. Elements of the project are financed by the European Commission's Framework Program 7 (BIO-NMR-00007)-Bio-NMR grant: Jellyfish protein NMR (BIO-NMR-00007)-IEP.

REFERENCES

[1] T. De Beer, A. Inui, H. Tsuda, K. Sugahara and J. F. G. Vliegenthart, "Polydispersity in Sulfation Profile of Oligosaccharide Alditols Isolated from the Protein-Linkage Region and the Repeating Disaccharide Region of Chondroitin 4-Sulfate of Bovine Nasal Septal Cartilage," *European Journal of Biochemistry*, Vol. 240, No. 3, 1996, pp. 789-797. doi:10.1111/j.1432-1033.1996.0789h.x

[2] P. Adamczyk, C. Zenkert, P. G. Balasubramanian, S. Yamada, S. Murakoshi, K. Sugahara, J. S. Hwang, T. Gojobori, T. W. Holstein and S. Özbek, "A Non-Sulfated Chondroitin Stabilizes Membrane Tubulation in Cnidarian Organelles," *The Journal of Biological Chemistry*, Vol. 285, 2010, pp. 25613-25623. doi:10.1074/jbc.M110.107904

[3] S. Yamada, H. Morimoto, T. Fujisawa and K. Sugahara, "Glycosaminglycans in *Hydra Magnipapillata* (Hydrozoa, Cnidaria): Demonstration of Chondroitin in the Developing Nematocyst, the Sting Organelle, and Structural Characterizayion of Glycosaminoglycans," *Glycobiology*, Vol. 17, No. 8, 2007, pp. 886-894. doi:10.1093/glycob/cwm051

[4] H.-C. Siebert, E. Tajkhorshid and J. Dabrowski, "Barrier to Rotation around the C_{sp}^2-C_{sp}^2 Bond of the Ketoaldehyde Enol Ether MeC(O)==CH-OEt as Determined by ^{13}C NMR and AB Initio Calculation," *The Journal of Physical Chemistry A*, Vol. 105, No. 3, 2001, pp. 8488-8494. doi:10.1021/jp004476g

[5] J. H. van Lenthe, D. H. W. de Boer, R. W. A. Havenith, R. Schauer and H.-C. Siebert, "AB Initio Calculations on Various Sialic Acids Provide Valuable Information about Sialic Acid-Specific Enzymes," *Journal of Molecular Structure: THEOCHEM*, Vol. 667, No. 19-3, 2004, pp. 29-37. doi:10.1016/j.theochem.2004.01.013

[6] M. Remko and M. Hricovíni, "Theoretical Study of Structure and Properties of Hexuronic Acid and D-Glucosamine Structural Units of Glycosaminoglycans," *Structural Chemistry*, Vol. 18, No. 5, 2007, pp. 537-547. doi:10.1007/s11224-007-9181-y

[7] M. Hricovíni, E. Scholtzová and F. Bízik, "B3LYP/6-311++G** Study of Structure and Spin-Spin Coupling Constant in Heparin Disaccharide," *Carbohydrate Research*, Vol. 342, No. 10, 2007, pp. 1350-1356. doi:10.1016/j.carres.2007.03.020

[8] E. Scholtzová, P. Mach and M. Hricovíni, "Structure of Sulfated Monosaccharides Studied by Quantum Chemical Methods," *Molecules*, Vol. 8, No. 11, 2003, pp. 770-779. doi:10.3390/81100770

[9] "Hyperchem Professional 8.0," Hypercube, Inc., Gainesville, 2007.

[10] M. J. Frisch, *et al.*, "Gaussian 03, Revision B.03," Gaussian, Inc., Pittsburgh, 2003.

[11] "Molegro-Trial Version 3.0.0," 2008.

[12] H.-C. Siebert, M. Burg-Roderfeld, T. Eckert, S. Stötzel, U. Kirch, T. Diercks, M. J. Humphries, M. Frank, R. Wechselberger, E. Tajkhorshid and S. Oesser, "Interaction of tne α2A Domain of Integrin with Small Collagen Fragments," *Protein & Cell*, Vol. 1, No. 4, 2010, pp. 393-405.

[13] A. G. Gerbst, N. E. Ustuzhanina, A. A. Grachev, E. A. Khatuntseva, D. E. Tsvetkov, A. S. Shashkov, A. I. Usov, M. E. Preobrazhenskaya, N. A. Ushakova and N. E. Nifantiev, "Synthesis, NMR and Conformational Studies of Fucoidan Fragments. V.[1] Linear 4,4',4''-Tri-O-Sulfated and Parent Non-Sulfated (1→3)-Fucotrioside Fragments," *Journal of Carbohydrate Chemistry*, Vol. 22, No. 2, 2003, pp. 109-122. doi:10.1081/CAR-120020481

[14] V. B. Krylov, A. A. Grachev, N. E. Ustyuzhanina, N. A. Ushakova, M. E. Preobrazhenskaya, N. I. Kozlova, M. N. Portsel, I. N. Konovalova, V. U. Novikov, H.-C. Siebert, A. S. Shashkov and N. E. Nifantiev, "Preliminary Structural Characterization, Anti-Inflammatory and Anticoagulant Activities of Chondroitin Sulfates from Marine Fish Cartilage," *Russian Chemical Bulletin*, Vol. 60, 2011, pp. 746-753.

[15] E. Pokidysheva, A. G. Milbradt, S. Meier, C. Renner, D. Haussinger, H. P. Bachinger, L. Moroder, S. Grzesiek, T. W. Holstein, S. Ozbek and J. Engel, "The Structure of the Cys-Rich Terminal Domain of *Hydra* Minicollagen, Which Is Involved in Disulfide Networks of the Nematocyst Wall," *The Journal of Biological Chemistry*, Vol. 279, 2004, pp. 30395-30401. doi:10.1074/jbc.M403734200

[16] J. Emsley, G. Knight, R. Farndale, M. Barnes and R. Liddington, "Structural Basis of Collagen Recognition by Integrin $\alpha 2\beta 1$," *Cell*, Vol. 101, No. 1, 2000, pp. 47-56. doi:10.1016/S0092-8674(00)80622-4

[17] A. G. Milbradt, C. Boulègue, L. Moroder and C. Renner, "The Two Cysteine-rich Head Domains of Minicollagen from *Hydra* Nematocysts Differ in their Cystine Framework and Overall Fold Despite an Identical Cysteine Sequence Pattern," *Journal of Molecular Biology*, Vol. 354, No. 3, 2005, pp. 591-600. doi:10.1016/j.jmb.2005.09.080

[18] S. Meier, D. Haussinger, E. Pokidysheva, H. P. Bachinger and S. Grzesiek, "Determination of a High-Precision NMR Structure of the Minicollagen Cysteine Rich Domain from Hydra and Characterization of Its Disulfide Bond Formation," *FEBS Letters*, Vol. 569, No. 1, 2004, pp. 112-116. doi:10.1016/j.febslet.2004.05.034

[19] J.-O. Lee, P. Rieu, M. A. Arnaout and R. Liddington, "Crystal Structure of the A Domain from the a SubUnit of Integrin CR3 (CD11 b/CD18)," *Cell*, Vol. 80, No. 4, 1995, pp. 631-638. doi:10.1016/0092-8674(95)90517-0

[20] S. Szczepanek, M. Cikala and C. N. David, "Poly-Gamma-Glutamate Synthesis during Formation of Nematocyst Capsules in Hydra," *Journal of Cell Science*, Vol. 115, 2002, pp. 745-751.

[21] S. Berking and K. Herrmann, "Formation and Discharge of Nematocysts Is Controlled by a Proton Gradient across the Cyst Membrane," *Helgoland Marine Research*, Vol. 60, No. 3, 2006, pp. 180-188. doi:10.1007/s10152-005-0019-y

[22] S. Özbek, P. G. Balasubramanian and T. W. Holstein, "Cnidocyst Structure and the Biomechanics of Discharge," *Toxicon*, Vol. 54, No. 8, 2009, pp. 1038-1045. doi:10.1016/j.toxicon.2009.03.006

[23] T. W. Holstein, M. Benoit, G. V. Herder, G. Wanner, C. N. David and H. E. Gaub, "Fibrous Mini-Collagens in *Hydra* Nematocysts," *Science*, Vol. 265, No. 5170, 1994, pp. 402-404. doi:10.1126/science.265.5170.402

[24] G. Tettamanti, A. Grimaldi, T. Congiu, G. Perletti, M. Raspanti, R. Valvassori and M. de Eguileor, "Collagen Reorganization in Leech Wound Healing," *Biology of the Cell*, Vol. 97, No. 7, 2005, pp. 557-568. doi:10.1042/BC20040085

[25] N. Ohta, M. Sato, K. Ushida, M. Kokubo, T. Baba, K. Taniguchi, M. Urai, K. Kihira and J. Mochida, "Jellyfish Mucin May Have Potential Disease-Modifying Effects on Osteoarthritis," *BMC Biotechnology*, Vol. 9, No. 98, 2009, pp. 1-11. doi:10.1186/1472-6750-9-98

[26] N. Sapay, E. Cabannes, M. Petitou and A. Imberty, "Molecular Modeling of the Interaction between Heparan Sulfate and Cellular Growth Factors: Bringing Pieces Together," *Glycobiology*, Vol. 21, No. 9, 2011, pp. 1181-1193. doi:10.1093/glycob/cwr052

[27] D. O. Croci, *et al.*, "Fucans, but Not Fucomannoglucuronans, Determine the Biological Activities of Sulfated Polysaccharides from *Laminaria Saccharina* Brown Seaweed," *PLoS ONE*, Vol. 6, 2011, pp. 1-10. doi:10.1371/journal.pone.0017283

[28] O. Raabe, C. Reich, S. Wenisch, A. Hild, M. Burg-Roderfeld, H.-C. Siebert and S. Arnhold, "Hydrolyzed Fish Collagen Induced Chondrogenic Differentiation of Equine Adipose Tissue-Derived Stromal Cells," *Histochemistry & Cell Biology*, Vol. 134, No. 6, 2010, pp. 545-554. doi:10.1007/s00418-010-0760-4

[29] U. Technau and R. E. Steele, "Evolutionary Crossroads in Developmental Biology: Cnidaria," *Development*, Vol. 138, 2011, pp. 1447-1458. doi:10.1242/dev.048959

[30] T. C. G. Bosch, "Hydra and the Evolution of Stem Cells," *BioEssays*, Vol. 31, No. 4, 2009, pp. 478-486. doi:10.1002/bies.200800183

[31] J. S. Hwang, Y. Takaku, T. Momose, P. Adamczyk, S. Özbek, K. Ikeo, K. Khalturin, G. Hemmrich, T. C. G. Bosch, T. W. Holstein, C. N. David and T. Gojobori, "Nematogalectin, a Nematocyst Protein with GlyXY and Galectin Domains, Demonstrates Nematocyte-Specific Alternative Splicing in *Hydra*," *PNAS*, Vol. 107, No. 43, 2010, pp. 18539-18544. doi:10.1073/pnas.1003256107

[32] A. M. Wu, T. Singh, J.-H. Liu, S. André, M. Lensch, H.-C. Siebert, M. Krzeminski, A. M. J. J. Bonvin, H. Kaltner, J. H. Wu and H.-J. Gabius, "Adhesion/Growth-Regulatory Galectins: Insights into Their Ligand Selectivity Using Natural Glycoproteins and Glycotopes, " *Advances in Experimental Medicine and Biology*, Vol. 705, 2011, pp. 117-141. doi:10.1007/978-1-4419-7877-6_7

[33] S. Stötzel, M. Schurink, H. Wienk, U. Siebler, M. Burg-Roderfeld, T. Eckert, B. Kulik, R. Wechselberger, J. Sewing, J. Steinmeyer, S. Oesser, R. Boelens and H.-C. Siebert, "Molecular Organization of Different Collagen Hydrolysates and Collagen Fragments as Revealed by a Combination of Atomic Force Microscopy (AFM) and Diffusion Ordered NMR Spectroscopy (DOSY)," *ChemPhysChem*, 2012.

24

Sensitivity of the C-N Vibration to Solvation in Dicyanobenzenes: A DFT Study

Arianna Gladney, Changyong Qin*, Helene Tamboue
Department of Biology, Chemistry and Environmental Health Science Benedict College, Columbia, USA
Email: *qinc@benedict.edu

ABSTRACT

Using DFT calculations, we have studied the C-N vibration of dicyanobenzenes in different solvents. The effects of solvent polarity and hydrogen bonding were evaluated. The frequency is decreased by 4 cm^{-1} when the solvent polarity increases. In the current study, the red shift due to the solvent polarity was underestimated by PCM. For hydrogen bonding, the frequency is increased by 10 cm^{-1} because of the charge transfer from the lone pair of nitrogen in C≡N to water, which slightly increases the bond strength of C-N. The C-N vibration of dicyanobenzenes is calculated to be sensitive to solvation.

Keywords: Dicyanobenzene; Vibration; Solvation; Infrared Spectrum; Density Functional Theory

1. Introduction

The vibration of the cyano group (C≡N) in organic and biological molecules has been studied as infrared (IR) probe of local environment in biological systems [1-11]. Previous studies show that the C-N vibration is highly localized, causes only negligible perturbation to the local three-dimensional structure, and the frequency is also very sensitive to the environmental change [12]. All of those make it an ideal probe for the local biological environment such as electrostatic potential, dielectric constant, ionic strength, and hydrogen bonding.

To predict the local environment of a targeted domain, one or more cyano groups have to be inserted into the biomolecule. The IR spectrum will then be taken and analyzed. Based on the frequency shift, peak width and intensity change of the C-N stretching, the local environment can be predicted. Therefore, a molecular level understanding of the frequency shift, peak width and intensity change caused by various local environmental conditions is of increasing importance and will significantly advance the predictive capabilities. Nowadays, theoreticcal modeling has advanced to a point where vibrations inside of a molecule can be examined at a level of quantum theory [13-17]. Furthermore, the solvation effects can be evaluated by using the Polarizable Continuum Model (PCM) [18] and also by including explicit solvent molecules in the quantum model.

Previous studies have shown that a single cyano group can provide predictive information about the local environment in biological systems, but there are still uncertainties in predictions since the inner cancellation between different environmental factors is not considered. For example, the C-N stretching frequency decreases with the dielectric constant of the solvent, but increases very quickly if hydrogen bonding occurs, which makes the prediction very difficult [1,2,19]. Recently, Krummel and Zanni have shown that the coupling between two cyano groups is detectable within a distance of even more than 4.5 Å and is very promising in monitoring DNA structures with a Fourier Transform Infared (FTIR) [20]. This has opened a door to development of new dicyano probes with higher accuracy since the coupling between the two cyano groups will possibly provide additional information about the local environment and will in turn improve the prediction. In the present study, the sensitivity of the C-N vibration to solvation in three isomers of dicyanobenzene was examined using Density Functional Theory (DFT) methods.

2. Theory and Method

The ground-state geometries of 1,2-dicyanobenzene (o-DCB), 1,3-dicyanobenzene (m-DCB) and 1,4-dicyanobenzene (p-DCB) were optimized using the Generalized Gradient Approximation (GGA) hybrid functional of B3LYP [21,22] in combination with the basis set of 6-31G(d, p) [23-25]. B3LYP has been proven to be very reliable in treating electronic exchange and correlation in small organic molecules. All calculations were performed by Gaussian09 [26] suite of quantum programs. The sol-

*Corresponding author.

vation effects were basically evaluated using the Polarizable Continuum Model (IEFPCM) [18], while explicit solvent molecules were included in the corresponding quantum model where hydrogen bonding occurs. The vibrational frequencies were calculated from analytic second derivatives using a harmonic oscillator model. In addition, the infrared (IR) intensities were calculated from the transition matrix elements of the dipole moment between the initial and final vibrational states.

3. Results and Discussion

3.1. Geometries, Relative Energies, Dipole Moments, Vibrational Frequencies and IR Intensities in the Gas Phase

The structures of dicyanobenzenes are optimized in the gas phase at the B3LYP/6-31G (d, p) level and displayed in **Figure 1**. Selected bond distances, dipole moments and relative energies are listed in **Table 1**. It is found that p-DCB is more stable than m-DCB and o-DCB with a lower energy of 0.38 kcal/mol and 2.01 kcal/mol, respecttively. Obviously, larger separation between two cyano groups leads to less steric repulsion and lower the total energy from o-DCB, m-DCB to p-DCB. Correspondingly, the dipole moment of each molecule decreases in the same order, which is largely determined by the relative position of the two cyano groups. In p-DCB, the two groups are aligned linearly, but in opposite directions, which leads to a zero polarity. However, they are nearly parallel and in the same direction in o-DCB, which gives the largest dipole moment of 7.08 Debye. Between p-DCB and o-DCB is m-DCB with a dipole moment of 4.27 Debye.

For the selected bond distances in **Table 1**, the calculated values are slightly larger than those form x-ray diffraction, but within 0.04 Å and have shown a fairly good agreement. Since all calculations here are based on single molecules in the gas phase only, no intermolecular interactions are considered, which is different from the x-ray experiment [27,28]. It is notable that the C-C bonds on the benzene ring close to the cyano groups are elongated compared to those further away. For example, C1-C2 and C2-C3 are longer than C3-C4 and C4-C5 in o-DCB. Electron population analysis gives the Mulliken charge of C1, C3, C7 and N8 in o-DCB as +0.140e, –0.053e, +0.236e and –0.462e, respectively. Significant negative charge of 0.226e is detected on C≡N, compared to a calculated value of 0.128e for acetonitrile. The high electronegativity of nitrogen has led to a charge transfer from benzene to C≡N, which should destablize the ring structure and make the neighboring bonds slightly longer. The electron density is finally injected into the lowest unoccupied molecular orbital (LUMO) of C≡N. Since the LUMO is of antibonding character, this electron trans fer makes the C-N bond slightly longer than in acetonitrile, which has a calculated C-N bond distance of 1.160 Å at the same level of theory. In addition, the π-conjugation in benzene and C≡N is merged and extended to the whole molecule, which brings partial π-bonding character to C1-C7 and makes the bond (~1.43 Å) significantly shorter than a C-C single bond (~1.50 Å). Very similar results were observed for m-DCB and p-DCB as well.

Since the C-N vibrations are highly localized and will only be used for the probing in biological systems, only the two C-N vibrational modes, symmetric and asymmetric, will be discussed in this report. In the symmetric

Table 1. Calculated bond distances (Å), relative energies (kcal/mol), dipole moments (Debye), vibrational frequencies (cm^{-1}) and IR intensity (km/mol) for dicyanobenzenes.

Structure	o-DCB		m-DCB		p-DCB	
	Curr.	Exp.[a]	Curr.	Exp.[a]	Curr.	Exp.[a]
d_{C1-C2}	1.416	1.401	1.401	1.372	1.405	1.378
d_{C2-C3}	1.402	1.378	1.401	1.379	1.389	1.374
d_{C3-C4}	1.393	1.354	1.406	1.393	1.405	1.378
d_{C4-C5}	1.396	1.397	1.393	1.373	1.405	1.378
d_{C1-C7}	1.433	1.430	1.435	1.432	1.434	1.441
d_{C7-N7}	1.163	1.149	1.163	1.144	1.163	1.133
E_{rel}	2.01	-	0.38	-	0.00	-
μ	7.08	-	4.27	-	0.00	-
Vibration	Freq.	IR Int.	Freq.	IR Int.	Freq.	IR Int.
Symm.	2352.5	9.09	2355.4	5.33	2349.9	0.00
Asym.	2355.1	5.20	2354.4	23.26	2352.4	30.02

Figure 1. Optimized structures of dicyanobenezenes.

mode, both C and N in the two cyano groups move towards or away from each other at the same pace. However, in the asymmetric mode, the two atoms move towards each other in one cyano group, while they move away from each other in the other cyano group. Calculated vibrational frequencies and infrared (IR) intensities in the gas phase are also tabulated in **Table 1**. For p-DCB, the symmetric mode is not IR active because two C-N vibrations cancel each other and gives no change to the net dipole moment. However, the asymmetric mode has the highest intensity of 30.02 km/mol. For o-DCB, two modes are separated by 2.7 cm^{-1}, indicating a possible detectable long-range vibrational coupling. Also both modes are sensitive to IR. For m-DCB, the two modes are very close with the asymmetric mode of only 1.0 cm^{-1} lower in energy, but it has a much higher IR intensity. Although the gas phase calculations cannot represent the real vibrational environment, they truly can provide an easy model to understand the C-N vibration in the selected dicyanobenzenes. Furthermore, the treatment of solvation is needed to provide accurate and reliable predictions.

3.2. Vibrations in Different Solvents

In this section, the solvation is treated using PCM, in which the solvents are electric dipoles and hydrogen bonding is not fully taken into account. The molecular structures are same as those optimized in the gas phase, while vibrational frequencies and IR intensities are calculated in different solvents, including tetrachloromethane (CCl_4, ε_r = 2.2), chloroform ($CHCl_3$, ε_r = 4.8), dichloromethane (CH_2Cl_2, ε_r = 8.9), acetone (CH_3COCH_3, ε_r = 20.7), acetonitrile (CH_3CN, ε_r = 37.5) and water (H_2O, ε_r = 78.3). In **Figure 2**, the calculated solvation energy by PCM is plotted as a function of the solvent dielectric constant. It is found that the solvation energy increases with the polarity of solvent. From the gas phase ε_r = 0.0) to dichloromethane, it is increased by about 6.5 kcal/mol. However, the change is almost negligible from acetone to water, less than 1.0 kcal/mol. From p-DCB, m-DCB to o-DCB, the solvation energy increases slightly with the molecular polarity. In addition, the calculated vibrational frequencies and IR intensities are plotted as a function of the dielectric constant, and are displayed in **Figure 3**.

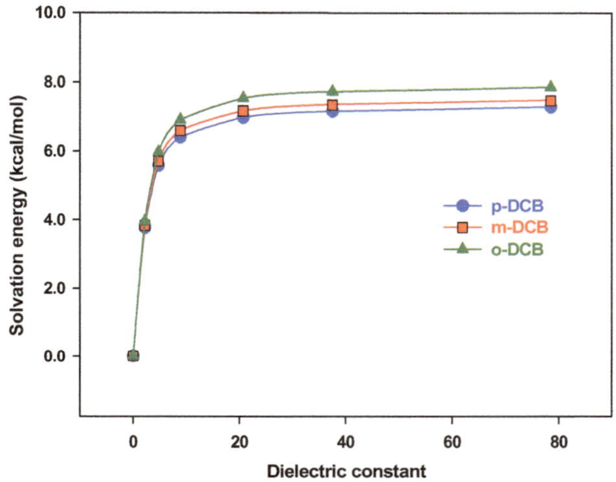

Figure 2. Solvation energy vs dielectric constant.

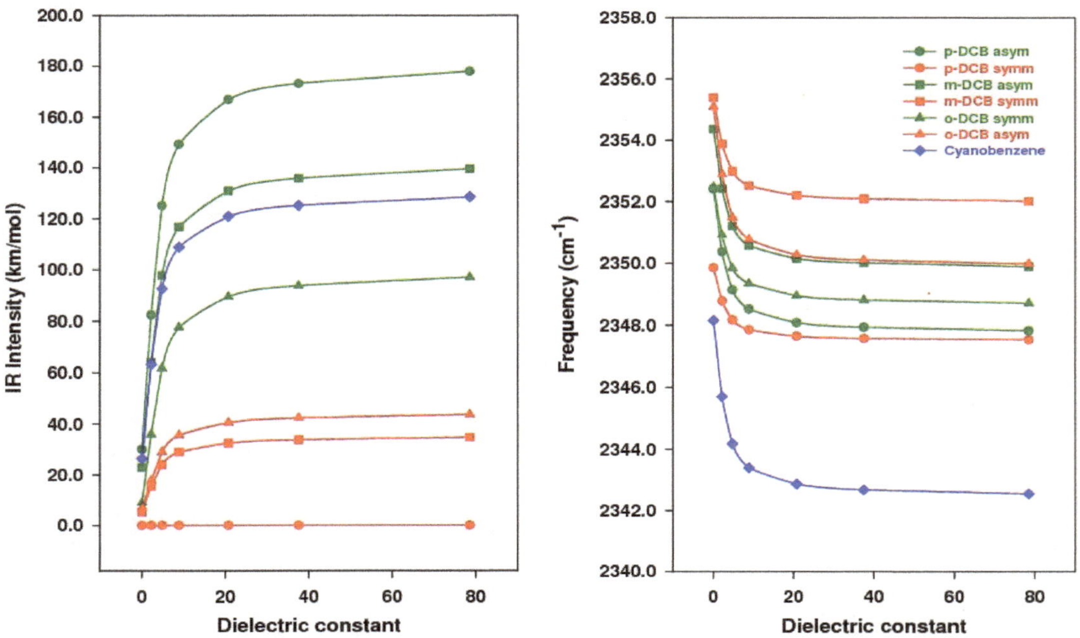

Figure 3. Vibrational frequency (bottom) and IR intensity (top) vs dielectric constant.

In **Figure 3**, the vibrational frequency of either symmetric or asymmetric mode decreases with the dielectric constant, while the IR intensity increases. From the gas phase to water, the frequency is decreased by 4.0 cm^{-1} for all except the symmetric mode of p-DCB, which is 2.0 cm^{-1}. To verify the reliability of PCM in predicting shifts of the C-N vibrational frequency in different solvents, the C-N vibrational frequency of acetonitrile is calculated to be 2379.1 cm^{-1}, 2376.6 cm^{-1}, 2375.1 cm^{-1}, and 2374.1 cm^{-1} at the same level of theory in the gas phase, tetrachloromethane, chloroform, and dichloromethane, respectively. From gas phase to tetrachloromethane, the red shift is 2.5 cm^{-1}, 4.0 cm^{-1} to chloroform, and 5.0 cm^{-1} to dichloromethane. The experimental value is 11.5 cm^{-1}, 11.0 cm^{-1}, 12.0 cm^{-1} in tetrachloromethane, chloroform and methylene chloride, respectively. Obviously, the shift is underestimated by PCM, indicating the importance of including intermolecular interactions in the theoretical model. However, PCM is simple and fast, which provides easiest model to predict the C-N frequency shift. In addition, the C-N vibrational frequency in benzonitrile (BN) is calculated at the same level of theory and shows a decrease of 6.0 cm^{-1} from gas phase to water, which is larger than the shift calculated for dinitriles in this study.

In an IR probing, the intensity change is also taken into account to provide additional information about the local environment. **Figure 3** (right graph) shows the calculated IR intensity as a function of the solvent dielectric constant. With the increase of the solvent polarity, the IR intensity increases dramatically at the beginning. For the asymmetric mode of p-DCB in the gas phase and tetrachloromethane, the intensity changes from 20.0 km/mol to 150.0 km/mol, indicating the highest sensitivity to the solvation. However, the change for benzonitrile is only about 75.0 km/mol from gas phase to tetrachloromethane. In terms of the IR intensity, the theoretical calculations show that p-DCB is more sensitive to the solvent polarity than m-DCB, and o-DCB. Therefore, it is possible to use the IR frequency and intensity change to determine the local bonding structure of the cyano group in biological systems.

3.3. Effects of Hydrogen Bonding

In the PCM model, majorly electrostatic effect is considered. To include the effect of hydrogen bonding, two water molecules were added to each structure and the DCB-water complexes in the gas phase were then optimized at the B3LYP/6-31G (d, p) level. For the geometry of benzene ring, no significant changes have been found, while the local H-bonding structure is displayed in **Figure 4** with calculated Mulliken charges. In addition, there is no notable difference in the three isomers in terms of hydrogen bonding, and here only the local structure of p-DCB-water is shown in **Figure 4** and discussed in details.

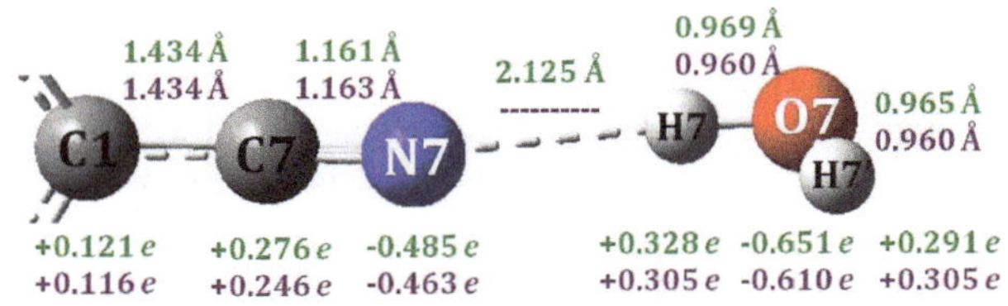

Figure 4. H-Bonding structure of p-DCB-water (Green: p-DCB-water; Purple: p-DCB).

In **Figure 4**, bond distances and Mulliken charges of p-DCB-Water are displayed in green, while purples for single water and p-DCB. The binding energy is calculated as $E_{DCB\text{-}Water}$-E_{DCB}-E_{Water} and to be 4.5 kcal/mol for all DCB-Water complexes, indicating significant hydrogen bonding between DCB and water. It is found that the hydrogen bonding shortens the C7-N7 bond by 0.002 Å, indicating a slightly stronger C-N bond and higher C-N vibrational frequency. A charge transfer of 0.032e is detected from p-DCB to water. The charge is mainly from the molecular orbital of nitrogen lone pairs. Since the lone pair orbital is of anti-bonding character between C7 and N7, losing electron density should lead to a stronger C-N bond and higher C-N vibrational frequency. The calculated C-N vibrational frequency is 2360.1 cm^{-1} and 2361.9 cm^{-1} for the symmetric and asymmetric mode, respectively. The hydrogen bonding of water to p-DCB has increased the C-N vibrational frequency by 10 cm^{-1}. The C-N frequency shift of acetonitrile in dimethyl sulfoxide (DMSO) and water was measured to be 17.5 cm^{-1} and 6.7 cm^{-1}, respectively [1]. Approximately, we can assume that the effect of the solvent polarity on the C-N vibration is same in the two solvents, and the difference in the C-N vibration is then mainly caused by hydrogen bonding. This effect is calculated to be +10.8 cm^{-1} and is in an excellent agreement with the value from the current study. In addition, the C-N vibrational frequency of p-DCB-Water is calculated to be 4.0 cm^{-1} lower in PCM water, which is exactly same as we found for p-DCB in water.

4. Conclusion

We have reported a DFT study of the C-N vibration in dicyanobenzenes (DCB) relevance to the IR probing of local environment in biological systems. The effect of solvent polarity (dielectric constant) and hydrogen bonding were evaluated using PCM and by including two water molecules in the quantum model. The C-N vibrational frequency is decreased by up to 4 cm^{-1} from the PCM calculations when the solvent polarity increases. However, a large red shift of 17.5 cm^{-1} was reported for acetonitrile in DMSO [1], which implies that the red shift due to the solvent polarity was underestimated by PCM in the current study. For the hydrogen bonding, the frequency is increased by 10 cm^{-1} because of the charge transfer from the lone pair of nitrogen to water, which slightly increases the C-N bond strength. The C-N vibration of dicyanobenzenes is calculated to be sensitive to both solvent polarity and hydrogen bonding.

5. Acknowledgements

The authors gratefully acknowledge both the Department of Energy and Department of Defense for financial support of this work (Grant Number: DE-FG52-08NA28800 and W911NF-10-R-006).

REFERENCES

[1] J. R. Reimers and L. E. Hall, "The Solvation of Acetonitrile," *Journal of the American Chemical Society*, Vol. 121, No. 15, 1999, pp. 3730-3744. doi:10.1021/ja983878n

[2] S. S. Andrews and S. G. Boxer, "Vibrational Stark Effects of Nitriles I. Methods and Experimental Results," *The Journal of Physical Chemistry A*, Vol. 104, No. 51, 2000, pp. 11853-11863. doi:10.1021/jp002242r

[3] I. T. Suydam, C. D. Snow, V. S. Pande and S. G. Box, "Electric Fields at the Active Site of an Enzyme: Direct Comparison of Experiment with Theory," *Science*, Vol. 313, No. 5784, 2006, pp. 200-204. doi:10.1126/science.1127159

[4] Z. Getahun, C. Huang, T. Wang, B. D. Leon, W. F. DeGrado and F. Gai, "Using Nitrile-Derivatized Amino Acids as Infrared Probes of Local Environment," *Journal of the American Chemical Society*, Vol. 125, No. 2, 2003, pp. 405-411. doi:10.1021/ja0285262

[5] M. J. Tucker, Z. Getahun, V. Nanda, W. F. DeGrado and F. Gai, "A New Method for Determining the Local Environment and Orientation of Individual Side Chains of Membrane-Binding Peptides," *Journal of the American Chemical Society*, Vol. 126, No. 16, 2004, pp. 5078-5079. doi:10.1021/ja032015d

[6] M. J. Tucker, R. Oyola and F. Gai, "A Novel Fluorescent Probe for Protein Binding and Folding Studies: *p*-Cyano-Phenylalanine," *Biopolymers*, Vol. 83, No. 6, 2006, pp. 571-576. doi:10.1002/bip.20587

[7] M. M. Waegele, M. J. Tucker and F. Gai, "5-Cyanotryptophan as an Infrared Probe of Local Hydration Status of Proteins," *Chemical Physics Letters*, Vol. 478, No. 4-6, 2009, pp. 249-253. doi:10.1016/j.cplett.2009.07.058

[8] K. C. Schultz, L. Supekova, Y. Ryu, J. Xie, R. Perera and P. G. Schultz, "A Genetically Encoded Infrared Probe," *Journal of the American Chemical Society*, Vol. 128, No. 43, 2006, pp. 13984-13985. doi:10.1021/ja0636690

[9] L. N. Silverman, M. E. Pitzer, P. O. Ankomah, S. G. Boxer and E. E. Fenlon, "Vibrational Stark Effect Probes for Nucleic Acids," *The Journal of Physical Chemistry B*, Vol. 111, No. 40, 2007, pp. 11611-11613. doi:10.1021/jp0750912

[10] M. D. Watson, X. S. Gai, A. T. Gillies, S. H. Brewer and E. E. Fenlon, "A Vibrational Probe for Local Nucleic Acid Environments: 5-Cyano-2′-Deoxyuridine," *The Journal of Physical Chemistry B*, Vol. 112, No. 42, 2008, pp. 13188-13192. doi:10.1021/jp8067238

[11] B. A. Lindquist, D. E. Furse and S. A. Corcelli, "Nitrile Groups as Vibrational Probes of Biomolecular Structure and Dynamics: An Overview," *Physical Chemistry Chemical Physics*, Vol. 11, No. 37, 2009, pp. 8119-8132. doi:10.1039/b908588b

[12] M. G. Maienschein-Cline and C. H. Londergan, "The CN Stretching Band of Aliphatic Thiocyanate Is Sensitive to Solvent Dynamics and Specific Solvation," *The Journal of Physical Chemistry A*, Vol. 111, No. 40, 2007, pp. 10020-10025. doi:10.1021/jp0761158

[13] S. D. Dalosto, J. M. Vanderkooi and K. A. Sharp, "Vibrational Stark Effects on Carbonyl, Nitrile, and Nitrosyl Compounds Including Heme Ligands, CO, CN, and NO, Studied with Density Functional Theory," *The Journal of Physical Chemistry B*, Vol. 108, No. 20, 2004, pp. 6450-6457. doi:10.1021/jp0310697

[14] B. A. Lindquist and S. A. Corcelli, "Nitrile Groups as Vibrational Probes: Calculations of the CN Infrared Absorption Line Shape of Acetonitrile in Water and Tetrahydrofuran," *The Journal of Physical Chemistry B*, Vol. 112, No. 20, 2008, pp. 6301-6303. doi:10.1021/jp802039e

[15] B. A. Lindquist, R. T. Haws and S. A. Corcelli, "Optimized Quantum Mechanics/Molecular Mechanics Strategies for Nitrile Vibrational Probes: Acetonitrile and Para-Tolunitrile in Water and Tetrahydrofuran," *The Journal of Physical Chemistry B*, Vol. 112, No. 44, 2008, pp. 13991-14001. doi:10.1021/jp804900u

[16] J. Choi, K. Oh, H. Lee, C. Lee and M. Cho, "Nitrile and thiocyanate IR Probes: Quantum Chemistry Calculation Studies and Multivariate Least-Square Fitting Analysis," *The Journal of Chemical Physics*, 128 (2008) 134506. doi:10.1063/1.2844787

[17] K.-I. Oh, J.-H. Choi, J. Lee, J. Han, H. Lee and M. Cho, "Nitrile and Thiocyanate IR Probes: Molecular Dynamics Simulation Studies," *The Journal of Chemical Physics*, Vol. 128, No. 13, 2008, pp. 154504-154511. doi:10.1063/1.2904558

[18] S. Miertu and J. Tomasi, "Approximate Evaluations of the Electrostatic Free Energy and Internal Energy Changes in Solution Processes," *Chemical Physics*, Vol. 65, No. 2, 1982, pp. 239-245. doi:10.1016/0301-0104(82)85072-6

[19] S. Kaur, E. S. Eberhardt, A. Doucette, A. Chase and C. Dalby, "Solvent Effects on Barrier to Rotation of Enaminonitriles Using Inversion Transfer 1H NMR Spectroscopy and FTIR Spectroscopy," *The Journal of Organic Chemistry*, Vol. 67, No. 11, 2002, pp. 3937-3940. doi:10.1021/jo025516a

[20] A. T. Krummel and M. T. Zanni, "Evidence for Coupling between Nitrile Groups Using DNA Templates: A Promising New Method for Monitoring Structures with Infrared Spectroscopy," *The Journal of Physical Chemistry B*, Vol. 112, No. 5, 2008, pp. 1336-1338. doi:10.1021/jp711558a

[21] C. Lee, W. Yang and R. G. Parr, "Development of the

Colle-Salvetti Correlation-Energy Formula into a Functional of the Electron Density," *Physical Review B*, Vol. 37, 1988, pp. 785-789. doi:10.1103/PhysRevB.37.785

[22] A. D. Becke, "Density-Functional Exchange-Energy Approximation with Correct Asymptotic Behavior," *Physical Review A*, Vol. 38, 1988, pp. 3098-3100. doi:10.1103/PhysRevA.38.3098

[23] V. A. Rassalov, J. A. Pople, M. Ratner and T. L. Windus, "6-31G* Basis Set for Atoms K through Zn," *The Journal of Chemical Physics*, Vol. 109, 1998, pp. 1223-1229. doi:10.1063/1.476673

[24] P. C. Harihan and J. A. Pople, "The Influence of Polarization Functions on Molecular Orbital Hydrogenation Energies," *Theoretical Chemistry Accounts*, Vol. 28, No. 3, 1973, pp. 213-222. doi:10.1007/bf00533485

[25] M. M. Francl, W. J. Pietro, W. J. Hehre, J. S. Binkley, M. S. Gordon, D. J. DeFree and J. A. Pople, "Self-Consistent Molecular Orbital Methods. XXIII. A Polarization-Type Basis Set for Second-Row Elements," *The Journal of Chemical Physics*, Vol. 77, No. 7, 1982, pp. 3654-3665. doi:10.1063/1.444267

[26] M. J. Frisch *et al.*, "Gaussian 09, Revision B.01," Gaussian Inc., Pittsburgh, 2009.

[27] J. T. López Navarrete, J. J. Quirante, M. A. G. Aranda, V. Hernández and F. J. Ramirez, "Structure, Polarized Micro-Raman and FT-IR Spectra, and AB Initio Calculations of 1,2-Dicyanobenzene," *The Journal of Physical Chemistry*, Vol. 97, No. 41, 1993, pp. 10561-10569. doi:10.1021/j100143a009

[28] J. Janczak and R. Kubiak, "Molecular Structure and Ring Distortions of 1,3-Dicyanobenzene in the Gas Phase and in the Crystal," *Journal of Molecular Structure*, Vol. 553, No. 1-3, 2000, pp. 157-166. doi:10.1016/S0022-2860(00)00578-0

Evaluation of TiO_2, ZnO, CuO and Ga_2O_3 on the Photocatalytic Degradation of Phenol Using an Annular-Flow Photocatalytic Reactor

Flavia Cristina Sertori Paschoalino[1], Matheus Paes Paschoalino[2], Elizabete Jordão[1], Wilson de Figueiredo Jardim[2]
[1]Department of Chemical Systems Engineering, Laboratory of Catalytic Processes Development (LDPC), Faculty of Chemical Engeneering, UNICAMP-University of Campinas, Campinas, Brazil
[2]Department of Analytical Chemistry Institute, UNICAMP-University of Campinas, Campinas, Brazil
Email: flaviasertori@feq.unicamp.br

ABSTRACT

Even with rigorous environmental regulations, phenol still is a major contaminant. One possible solution is the use of heterogeneous photocatalysis due to low chemical addition, feasibility and reliability to be implanted on cost-effective industrial process. TiO_2 is the most employed photocatalyst because of its favorable (photo) chemical properties and ZnO is considered one of the best alternative for that. Other oxides were tested in lesser proportions, like CuO and Ga_2O_3. When the photocatalyst is dispersed as slurry, higher degradation rates are achieved due to high solid to liquid contact area when compared with supported form. The aim of this work was to develop a batch recirculating photocatalytic reactor and evaluate its efficiency when assisted by the photocatalysts TiO_2 P25, ZnO, CuO and β-Ga_2O_3. TiO_2 achieved 95% mineralization after 200 min reaction in an average degradation rate of 0.68 $mg{\cdot}L^{-1}{\cdot}min^{-1}$ and ZnO was less efficient (0.41 $mg{\cdot}L^{-1}{\cdot}min^{-1}$). Ga_2O_3 and CuO presented poor performance, mainly due to low surface area for the CuO syntesized and the absorption of the UV radiation by the reactor walls, decreasing Ga_2O_3 activity. Degradation intermediates were detected in diverse concentrations and at different operational times for each oxide tested, which indicate different degradation mechanisms.

Keywords: Phenol; TiO_2; ZnO; CuO; Ga_2O_3; Heterogeneous Photocatalysis

1. Introduction

Despite the increase of rigor about environmental regulations, phenols persist as one of the major contaminants to aquatic life [1,2]. These compounds are highly carcinogenic and toxic to all forms of life and can be detected in higher concentrations on the industrial wastewater from petrochemical, oil-refineries, paper-making, coking and iron-smelting processes [3]. Due to its importance and recalcitrance to traditional degradation processes, it is a common model compound adopted in advanced water studies, mainly those envolving Advanced Oxidation Processes (AOPs).

The AOPs are process that generate high reactive radicals, mainly hydroxyl (•OH), which could mineralize organic pollutants completely if all chemical and engineerng conditions are well stabilished. Between them, ozonization, Fenton process and heterogeneous photocatalysis (HP) are the most promising technologies for final wastewater treatment stage, but HP is becoming the future of water treatment due to low chemical addition, easibility to be implanted on cost-effective industrial process [3,4]. This technique combines the use of a radiation with adequate energy to activate a semiconductor resulting on the generation of oxidative and reuductive sites on the photocatalyst surface.

TiO_2 P25 (70% anatase, 30% rutile) is the most employed photocatalytic semiconductor because of its chemical inertness, photostability, low cost, and atoxicity [5]. Anatase form has been successfully used for photocatalytic treatment of contaminants due to its faster electron transfer when exposed to UV radiation. As an alternative to TiO_2, ZnO has also been reported as an effective photocatalyst, due to its wide band-gap energy (3.37 eV) and large exciton binding energy (60 meV) [6]. A few studies have supported the assertion that ZnO is a better photocatalyst than TiO_2, especially for chlorinated compounds using hydroxyl radicals [7]. ZnO was also reported as being more efficient than TiO_2 in visible light photocatalytic degradation of some organic compounds in aqueous solution [5,8,9], but certainly it is not stable as titanium dioxide [4].

Other oxides were tested in lesser proportions, like CuO and Ga_2O_3. Copper oxide (II) has low cost and toxicity, in addition to high availability. This oxide is among the few with the band gap energy near to visible light, on the degradation of methylene blue [10]. Gallium oxides exist in a lot of polymorphs, and the thermodynamic stable form is β-Ga_2O_3. This material has much wider bandgap (4.8 eV) than TiO_2 doing that photogenerated electrons in the conductive band have much higher reductive capability. β-Ga_2O_3 is also an environmental friendly material according to Worksafe Australia criteria and some authors reported that it appears to be selective on the photocatalytic degradation of aromatics compounds, like benzene [11-13].

The photocatalyst could be used in slurry or supported forms. When the photocatalyst is dispersed as slurry inside the reactor, higher degradation rates are achieved due to high solid to liquid contact area and high axial flow rates are necessary to prevent the catalyst from settling. If the conversion per pass is low, recycling of the process fluid becomes necessary [14]. Some inconvenients of slurries are that the powders are not easy to precipitate and recover from water, preventing their regeneration and reuse, but several engineering/chemical solutions are being investigated, from incorporating titania on the reactor walls and the use of slurry reactors, to immobilization techniques on different supports [15-18].

In this context, the aim of this work was to develop a batch recirculating photocatalytic reactor and evaluate its efficiency when assisted by the photocatalysts TiO_2 P25, ZnO, CuO and βGa_2O_3.

2. Experimental

2.1. Photocatalytic Reactor Assembly

All the experiment were carried out in a batch recirculation mode. It was built a reactor with recirculation fluid composed by three main elements: 1) Photoreactor: consisting of Pyrex^{TM} cylindrical glass with hollow center formed by three concentric 3.5 mm layers isolating two chambers, the first with a 290 mL volume for the test solution flow and the second for a cold water stream to exchange heat between fluids and maintaining the system temperature at 30°C. In the reactor central annulus it was coupled a mercury vapor lamp (250 W, Osram) without bulbe involucres. The reactor was covered with aluminum to prevent radiation loss, and thus allowing a greater use of the emitted photons. The diagram of photoreactor with some measures is showed on **Figure 1**; 2) Recirculation pump (60 Hz) to promote the test solution flow

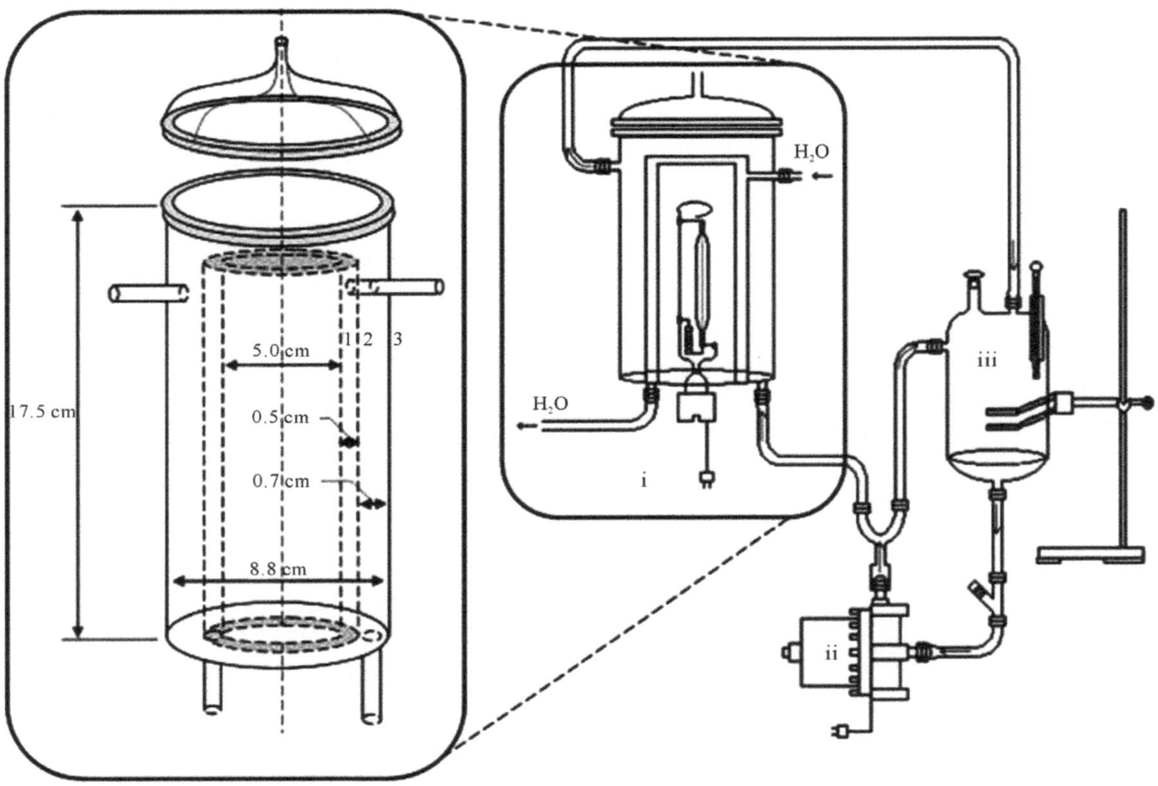

Figure 1. Scheme of the photocatalytic reactor assembled for the trials: (i) photocatalytic reactor and its detailed chambers; (ii) recirculation pump; (iii) recycling tank.

throughout the system, and to provide an adequate turbulence to keep catalyst in suspension; 3) Recycling tank: a 550 mL cylindrical glass container where the test solution passed through during the recirculation process in order to create enlightened and non-illuminated regimes during the process.

2.2. Photocatalytic Trials

Before each experiment, a catalyst (TiO_2 P25, ZnO, CuO or β-Ga_2O_3) amount of 0.3 g was added to distilled water and the resulting slurry was sonicated for 10 min to ensure an uniform catalyst dispersion and the recirculation flow was setted to 80 $L \cdot h^{-1}$, verified (checked) with a online rotameter. 1.2 mL of a phenol stok solution (50 $g \cdot L^{-1}$) was added to this slurry and mixed well to get the initial phenol concentration (100 $mg \cdot L^{-1}$). P25 and ZnO were purchased from Degussa and Sigma-Aldrich, respectively. CuO and Ga_2O_3 samples were synthesized in accord [17,18], respectively. The photocatalytic reaction was initiated once the Hg-lamp was turned on. Under these conditions, illuminated working volume was 290 mL, and trials time was standardized at 135 min after phenol addition. Preliminary tests were performed in order to recycle, where the solution was continuously through the reactor, and samples of 3 mL were withdrawn periodically and immediately filtered through 0.45 μm syringe filter (PTFE) for further analysis.

During the experiments, three different processes were evaluated and two of them used as controls the photocatalytic experiment: 1) photolysis: test solution treatment without using catalyst, but with the light on; 2) photocatalysis: It involved the treatment of test solution with the use of a catalyst, and the light on; 3) adsorption: consisted of test solution with the same amount of catalyst of photocatalysis but with the lamp off.

2.3. Analytical Procedures

The concentration of phenol and the intermediates (hydroquinone, benzoquinone and catechol) formed during the processes were evaluated by high performance liquid chromatography (HPLC) on a Shimadzu (SCL 10AVP) with C18 column and photodiode array detector (DAD, SPD-M10VP). Methanol-water at gradient elution of 40% - 100% in 10 min was used as the solvent. The injection volume was 20 μL. **Figure 2** shows a typical chromatogram for those analytes separation. Total organic carbon (TOC) also was measured before and after the trials using a Shimadzu TOC-V CPN Total Organic Carbon Analyser.

2.4. Radiation Source Characterization

UV intensity was measured using a radiometer Cole-Parmer Instruments Company, at a wavelength of 254 nm (model 9811-56), 312 nm (model 9811-54) and 365 nm (model 9811-50). Measurements were performed 15 minutes after the lamp accionament and also for the radiation transmitted through the glassy walls (three 3.5 mm layers and 0.5 cm of flowing water). The integral radiant flux of the incident light was measured with a power-meter (Newport 1830-C USA).

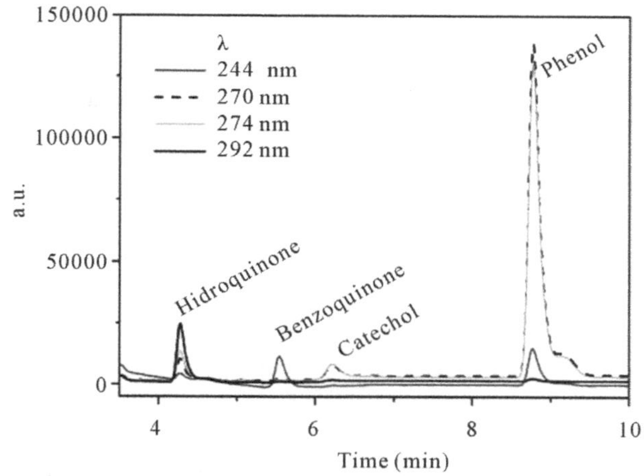

Figure 2. Typical HPLC chromatogram for separation of phenol and the degradation intermediates (hydroquinone, benzoquinone and catechol). HPLC configuration: Shimadzu (SCL 10AVP), C18 column; photodiode array detector (DAD, SPD-M10VP). Methanol-water at gradient ellution of 40% - 100% in 10 min; injection volume was 20 μL.

3. Results and Discussion

The reactor was operated in batch mode to prevent catalyst settling and simultaneously provide atmospheric oxygen absorption to guarantee the minimal leves necessary to generate hydroxyl radicals [14]. Despite the pH influence, no pH alterations were made to simulate realistic interactions between the produced species and the photocatalysts [1].

Control experiments using the photocatalysts without radiation exposition (dark) showed adsorption values varying from 0.8% to 3.2% (not showed here).

Despite the use of lamp with an irradiation spectrum varying from UV to visible regions, it is observed on **Figure 3** that only the photocatalysts which are activated mainly by UV-A radiations (TiO_2 and ZnO) were capable to produce significant results on the mineralization of phenol.

Analyzing **Figures 3** and **4** simultaneously it is possible to relationate mineralization with phenol degradation, and also adquire information about mechanisms differences. Photolysis graphs indicate 15% TOC remotion and 20% degradation, respectively. Considering high volatility of phenol and the fact that reactor used operated

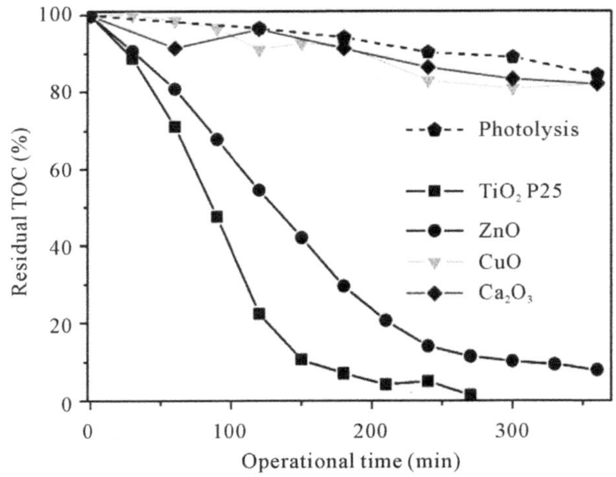

Figure 3. Mineralization results for the photocatalytic and photolysis processes, Residual TOC in accord with operational time.

not sealed, it is adequate consider this value as a combination of photolysis plus volatilization process. The indicative of phenol partial photolysis is the higher degradation than mineralization and the detection of hidroquinone intermediate in trace levels (1 to 2.5 mg·L^{-1}).

TiO_2 achieved more than 95% mineralization after 200 min reaction in an average degradation rate of 0.68 mg·L^{-1}·min^{-1}. During the first hour, after achieved 30% mineralization and 60% degradation it is observed the intermediates generation peak (catechol 2 mg·L^{-1}; benzoquinone 4 mg·L^{-1}; hifroquinone 14 mg·L^{-1}). The intermediates profile for TiO_2 is unique because all intermediates are formed at the same time, suggesting that the generation of radicals is different from the other catalysts and the intermediates considered here are the main formed compounds after phenol degradation, since mass balance between mineralization, degradation and intermediates generated are perfectly accurate. These results are similar to other studies using different reactor, as [16] that identified benzoquinone and hidroquinone as the predominant intermediates for P25, mainly during the first hour of irradiation [14].

Degradation results with ZnO were similar to TiO_2, but mineralization was less efficient (0.41 mg·L^{-1}·min^{-1}). This fact reflected on intermediates profile, starting with catechol (1.5 mg·L^{-1}) after 30 min, followed by hidroquinone (2 mg·L^{-1}) at 120 min and benzoquinone (0.9 mg·L^{-1}). Considering mass balance it is possible to infer that other intermediates might be present than the substances considered here. Organic acids such as the formic acid, acetic acid, oxalic acid and succinic acid are substances already reported in other works using different process as final intermediates before CO_2 formation [3]. ZnO is considered a promising photocatalyst as an alternative to TiO_2, showing effective oxidation of phenol even under low-powered UV radiation [7].

Gallium oxide is expected to be stimulated by UV-C and copper oxide is activated by visible radiation, but despite the radiation source be adequate to activate these photocatalysts, its performances were similar to mineralization behaviors for photolysis process. The poor performance of Ga_2O_3 could be due to glass absorption of the UV-C radiation from 30 to 10 mW·cm^{-2}, even with

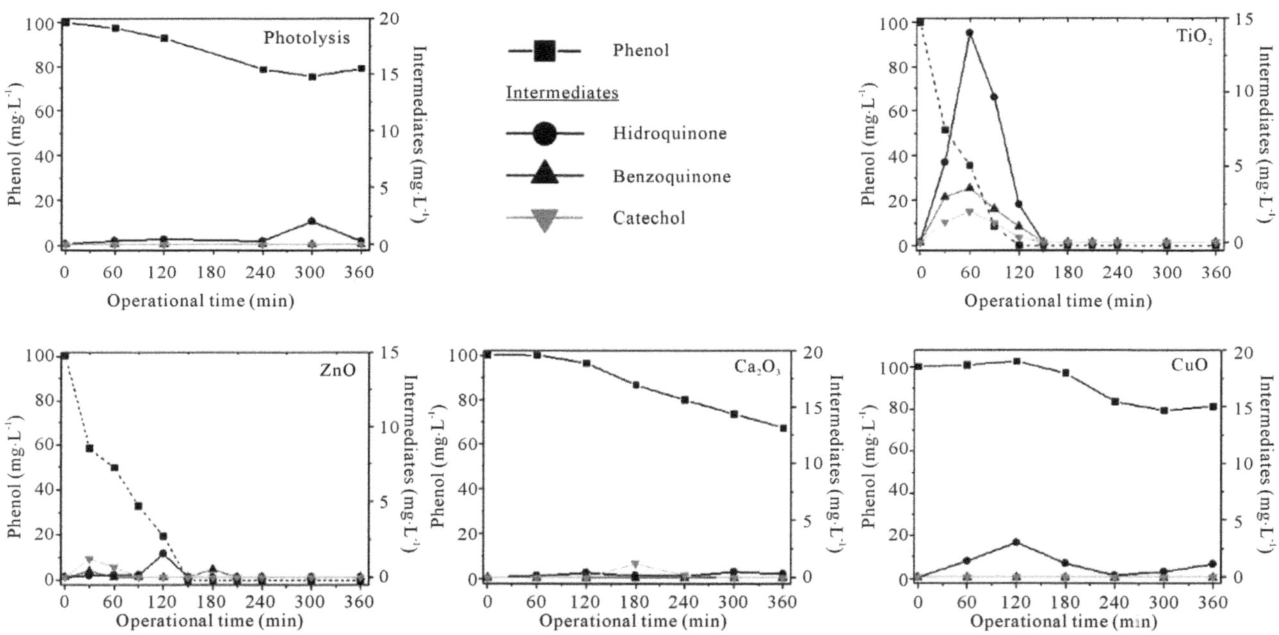

Figure 4. Degradation results for the photocatalytic and photolysis processes, Phenol and degradation intermediates (benzoquinone, hidroquinone, catechol) concentration in accord with operation time.

the use of PYREX glass. Zhang *et al.* [12] also reported a low catalytic activity for gallium oxide, being necessary 8 h irradiation using a 300 W lamp to obtain degradation of similar compounds. Hou *et al.* [19] revealed that β-Ga_2O_3 sample prepared with ethylene glycol showed the highest photocatalytic activity and this could be due to high surface area, abundant hydroxyl groups, and wide band gap in this case.

The difference between CuO and Ga_2O_3 degradation mechanisms is evident on **Figure 4**. Surface area and microstructure could be also influenced the reaction performance. CuO is recognized as active when surface área is higher than 40 $m^2 \cdot g^{-1}$, a big difference when compared with the CuO tested (0.5 $m^2 \cdot g^{-1}$). Karunakaran *et al.* [20] tested TiO_2, Fe_2O_3, CuO, ZnO, ZnS, CdO and Nb_2O_5 on the phenol degradation under UV-A and observed that the use of two semiconductors together in suspension could enhance photocatalytic activity due to interparticle electron-transfer.

4. Conclusion

The process described here is ecofriendly, cost-effective, and reusable for the treatment of phenol-contaminated water. Results showed that when feeding the reactor with TiO_2 the efficiency is higher than the other oxides but very similar to ZnO. Other interesting information about degradation mechanisms is that the degradation intermediates concentration varies differently for each photocatalyst tested, inferring different mechanism even when reaction rates were similar.

5. Acknowledgements

This project was financially supported by National Council for Scientific and Technological Development (CNPq) and National Council for the Improvement of Higher Education (CAPES).

REFERENCES

[1] K. M. Parida and A. C. Pradhan, "Fe/meso-Al_2O_3: An Efficient Photo-Fenton Catalyst for the Adsorptive-Degradation of Phenol," *Industrial & Engineering Chemistry Research*, Vol. 49, No. 18, 2010, pp. 8310-8318. doi:10.1021/ie902049s

[2] H. G. Oliveira, D. C. Nery and C. Longo, "Effect of Applied Potential on Photocatalytic Phenol Degradation Using Nanocrystalline TiO_2 Electrodes," *Applied Catalysis B: Environmental*, Vol. 93, No. 3-4, 2010, pp. 205-211. doi:10.1016/j.apcatb.2009.09.030

[3] Y. H. Huang, Y. J. Huang, H. C. Tsai and H. T. Chen, "Degradation of Phenol Using Low Concentration of Ferric Ions by the Photo-Fenton Process," *Journal of the Taiwan Institute of Chemical Engineers*, Vol. 41, No. 6, 2010, pp. 699-704. doi:10.1016/j.jtice.2010.01.012

[4] U. I. Gaya, A. H. Abdullah, M. Z. Hussein and Z. Zainal, "Photocatalytic Removal of 2,4,6-Trichlorophenol from Water Exploiting Commercial ZnO Powder," *Desalization*, Vol. 263, No. 1-3, 2010, pp. 176-182. doi:10.1016/j.desal.2010.06.055

[5] L. C. Chen, C. M. Huang, M. C. Hsiao and F. R. Tsai, "Mixture Design Optimization of the Composition of S, C, SnO_2-Codoped TiO_2 for Degradation of Phenol under Visible Light," *Chemical Engineering Journal*, Vol. 165, No. 2, 2010, pp. 482-489. doi:10.1016/j.cej.2010.09.044

[6] J. Rodríguez, F. Paraguay-Delgado, A. López, J. Alarcón and W. Estrada, "Syntesis and Characterization of ZnO Nanorod Films for Photocatalytic Disinfection of Contaminated Water," *Thin Solid Films*, Vol. 519, No. 2, 2010, pp. 729-735. doi:10.1016/j.tsf.2010.08.139

[7] P. Shukla, I. Fatimah, S. Wang, H. M. Ang and M. O. Tadé, "Photocatalytic Generation of Sulphate and Hydroxyl Radicals Using Zinc Oxide under Low-Power UV to Oxidise Phenolic Contaminants in Wastewater," *Catalysis Today*, Vol. 157, No. 1-4, 2010, pp. 410-414. doi:10.1016/j.cattod.2010.04.015

[8] B. Krishnakumar and M. Swaminathan, "Solar Photocatalytic Degradation of Acid Black 1 with ZnO," *Indian Journal of Chemistry*, Vol. 49a, 2010, pp. 1035-1040.

[9] E. S. Elmolla and M. Chaudhuri, "Effect of Fenton Operating Conditions on the Performance of the Combined Fenton-SBR Process for Antibiotic Wastewater Treatment," *Journal of Hazardous Materials*, Vol. 173, No. 1-3, 2010, pp. 445-449. doi:10.1016/j.jhazmat.2009.08.104

[10] A. P. L. Batista, H. W. P. Carvalho, G. H. P. Luz, P. F. Q. Martins, M. Gonçalves and L. C. A. Oliveira, "Preparation of CuO/SiO_2 and Photocatalytic Activity by Degradation of Methylene Blue," *Environmental Chemistry Letters*, Vol. 8, No. 1, 2010, pp. 63-67. doi:10.1007/s10311-008-0192-8

[11] Y. Hou, X. Wang, L. Wu, Z. Ding and X. Fu, "Efficient Decomposition of Benzene over a β-Ga_2O_3 Photocatalyst under Ambient Conditions," *Environmental Science & Technology*, Vol. 40, No. 18, 2006, pp. 5799-5803. doi:10.1021/es061004s

[12] H. Zhang, Y. Chen, L. Fu and J. Ma, "Syntesis, Thermal Stability, and Photocatalytic Activity of Nanocrystalline Gallium Nitrite via the Reaction of Ga_2O_3 and NH_4Cl at Low Temperature," *Journal of Alloys and Compounds*, Vol. 499, No. 2, 2010, pp. 269-272. doi:10.1016/j.jallcom.2010.03.185

[13] B. Zhao and P. Zhang, "Photocatalytic Decomposition of Perfluorooctanoic Acid with β-Ga_2O_3 Wide Bandgap Photocatalyst," *Catalysis Communications*, Vol. 10, No. 8, 2009, pp. 1184-1187. doi:10.1016/j.catcom.2009.01.017

[14] M. Subramanian and A. Kannan, "Photocatalytic Degradation of Phenol in a Rotating Annular Reactor," *Chemical Engineering Science*, Vol. 65, No. 9, 2010, pp. 2727-2740. doi:10.1016/j.ces.2010.01.004

[15] A. Akyol and M. Bayramoglu, "Photocatalytic Performance of ZnO Coated Tubular Reactor," *Journal of Hazardous Materials*, Vol. 180, No. 1-3, 2010, pp. 466-473. doi:10.1016/j.jhazmat.2010.04.053

[16] L. F. Velasco, B. Tsyntsarski, B. Petrova, T. Budinova and N. Petrov, "Carbon Foams as Catalyst Supports for Phenol Photodegradation," *Journal of Hazardous Materials*, Vol. 184, No. 1-3, 2010, pp. 843-848. doi:10.1016/j.jhazmat.2010.08.118

[17] M. P. Paschoalino, J. Kiwi and W. F. Jardim, "Gas-Phase Photocatalytic Decontamination Using Polymer Supported TiO_2," *Applied Catalysis B: Environmental*, Vol. 68, No. 1-2, 2006, pp. 68-73. doi:10.1016/j.apcatb.2006.08.001

[18] V. P. Silva, M. P. Paschoalino, M. C. Gonçalves, M. I. Felisberti, W. F. Jardim and I. V. P. Yoshida, "Silicone Rubbers Filled with TiO_2: Characterization and Photocatalytic Activity," *Materials Chemistry and Physics*, Vol. 113, No. 1, 2009, pp. 395-400. doi:10.1016/j.matchemphys.2008.07.104

[19] Y. Hou, J. Zhang, Z. Ding and L. Wu, "Syntesis, Characterization and Photocatalytic Activity of β-Ga_2O_3 Nanostructures," *Powder Technology*, Vol. 203, No. 3, 2010, pp. 440-446. doi:10.1016/j.powtec.2010.06.004

[20] C. Karunakaran, R. Dhanalakshmi, P. Gomathisankar and G. Manikandan, "Enhanced Phenol-Photodegradation by Particulate Semicondustor Mixtures: Interparticle Electron-Jump," *Journal of Hazardous Materials*, Vol. 176, No. 1-3, 2010, pp. 799-806. doi:10.1016/j.jhazmat.2009.11.105

The Influence of γ-Irradiation on Structuration in Solutions of BNR and Properties of Films

Shiraz M. Mammadov[1], Adil A. Garibov[1], Oktay A. Akparov[2], Sehrana A. Rzayeva, Akif K. Salehov[1], Sahile S. Ahmadova[1]
[1]Institute of Radiation Problems of ANAS, Baku, Azerbaijan
[2]Baku State University, Baku, Azerbaijan
Email: siraz_mm55@mail.ru

ABSTRACT

Studied influence of γ-irradiation on stucturation in butadiene nitrile rubber solution and films properties studied also. Structuration in solutions has defined by rheological method with rotation viscometer. Shown influence of solvent nature (methyl ethyl ketene: toluene) to strength properties and structures of films obtained from this solvents. It has been found that during irradiation of solvent, films and butadiene nitrile rubber solution characteristic viscosity ή increased, increasing of Huggins constant (K') observed as well. Changing of spatial mesh of polymers occurs at the cost of changed size of macromolecule balls. Shown, that in the issue of crossing the properties including thermo dynamical features of films and butadiene nitrile rubber solution has changed.

Keywords: Cross-Linking; Solvent; Rheology; Relaxation; Spectrometer; Radiation; Viscosity; Irradiation

1. Introduction

Physical-mechanical properties of polymeric materials along with the chemical nature of macromolecules in large extent are determined by the structure which is formed under the various factors accompanying the process of recycling. It has been shown in several investigations that structural-mechanical properties of solutions [1-5], and generated films from them [6] can be vary over by combining solvents and precipitants in the wide range. Maximum physical-mechanical properties meet the formation of ordered structures characterized by the highest perfection of relaxation processes [7,8]. Consequently, the structure that determined the mechanical properties of the material originates in the solution and in the certain degree captured in the transition to the solid state.

The research of the influence of the various factors (γ-irradiation) in the structuration process defining the complex of the mechanical properties of the materials is the actual in the light. It was done on base of the example of BNR solutions at current research.

2. Method and Object of Research

The object of study was BNR, (butadiene-nitrile-rubber-BNR-40). It means by the joint coagulation of butadiene-nitrile latex (with a cohesive acrylonitrile 37% - 40%) as a result of studies by the IR-spectroscopy method in the part of butadiene under temperature 300 K the polymerization of the isomeric compounds of double bonds in the investigated polymer is 1,2-isomer 14.3%, 1,4-isomer 14.8% and trans 1,4-isomers 70.9%.

The structure formation in solutions was researched by rheological method in rotary viscometer range 4.3×10^{-3} - 480 c^{-1}.

The process of the structuration was judged by the magnitude of the maximum $\acute{\eta}_0$ and minimum $\acute{\eta}_\infty$ of constant viscosity, the nature of the dependence of viscosity on the shear tension $\acute{\eta} = f(\mathrm{p})$, the limiting magnitude of the shear tension P_K. The structure and properties of the films was identified by the kinetics of the isothermal increase of strength after fracture.

The quality of solvent was evaluated by the capillary viscometer as viscometer type Ubbelohde with capillary diameter 0.60 mm.

The optical properties of system were investigated with spectrometer. For irradiation of BNR solutions and films with shown compounds were placed in vials (ampoules) made of glass like "Ray". Irradiation of the sealed ampoule was performed γ-rays under power Co^{60} at room temperature 4.9 D/c. The absorbed dose was 50 kGy.

The mechanical properties of the films was determined by a standard method at the strain-rate 20 mm/min on tensile testing machine "Instron".

The result of measurement of mechanical properties:

tensile strength σ and comparative elongation in the time of tear of film ε are presented in **Table 1**.

3. Results and Discussion

It is obvious that, the dependence of σ from the solvent compound has extreme nature: maximum strength for films, formed from system MEK:Toluene = 1:0.5. Also the films have enough high strength, obtained from a mixture 1:1. The further increase of toluene content leads to a decrease of strength of the films, however, in that case *σ* is still much higher than *σ* films which formed from MEK.

In **Table 1** presented the result of researches of dilute solutions by the capillary viscometry method. It is obvious that, with the increasing concentration of toluene up to 30% characteristic viscosity [*ή*] reduced slightly, but viscometric constant Huggins K' increases (*i.e.* the quality of solvent decreases), herewith the size of the macromolecule coils do not change practically. It is possible only during confirmatory rebuilding of BNR macromolecule consisting of units of butadiene and acrylonitrile.

The further increase of toluene proportion in the mixture leads to low decrease of [*ή*] and increase of K' *i.e.* the quality of solvent degrades the size of the macromolecule coils decrease. In the compound range, corresponding to the high content of toluene a sharp decrease of BNR solvability and the transition of the system to two-phase state is observed.

The spectrophotometric study of BNR solutions in MEK shows that even at 1% concentration of polymer in solution is observed deviation of the optical density, *i.e.* the macromolecules present in solution as associates, which is caused, apparently, the ability to form strong carbon bonds, as well as the tendency of the polymer to crystallization. The deterioration of thermodynamic quality of the solvent shifts the threshold of aggregation in low concentration, as in MEK solution: Toluene compound 1:3 and 1:4 is the critical concentration of 0.5 and 0.25% respectively. Thus, the observed extreme changes of mechanical properties of BNR films are the result of deterioration of thermodynamic quality of solvent.

Table 1. Influence of the solvent compounds on properties of films and BNR solutions.

Solvent compounds MEK:Toluene	[*ή*]	K'	*σ*, Pa	ε, %
1:0	0.665	0.47	0.10	260
1:0.5	0.66	0.50	2.60	780
1:1	0.69	0.52	2.10	800
1:2	0.55	0.60	1.40	740
1:3	0.59	0.30	-	-
1:4	0.53	0.28	-	-

The results of radiation chemical researches of structure of films and BNR solutions formed from the thermodynamic quality solution are shown in **Table 2**.

It is obvious that during irradiation of the same solvent compounds, films and BNR solutions characteristic viscosity [*ή*] increases, the growth of Huggins constant K' is observed, the parameter of the spatial grid of the polymer changes due to alteration in the size of the coils of the macromolecules, the three-dimensional structure of the transmission to the cross-linking forms, and the physical-mechanical properties of films and BNR solutions including the thermodynamic quality solvent changes.

In concentrated solutions area the influence of thermodynamic quality of the solvent on the structural state of the system observes more sharply (**Figure 1**).

The increase of the toluene leads not only the increase of the absolute values of viscosity, but also the increase

Table 2. Influence of *γ*-irradiation on the solvent compounds, properties of film and solutions BNR (D = 50 kGy).

Solvent compounds MEK:Toluene	[*ή*]	K'	*σ*, Pa	ε, %
1:0	0.87	0.58	1.14	350
1:0.5	0.91	0.62	3.60	670
1:1	0.86	0.70	3.30	690
1:2	1.0	1.1	2.30	650
1:3	1.12	0.87	1.45	800
1:4	0.61	1.30	1.30	850

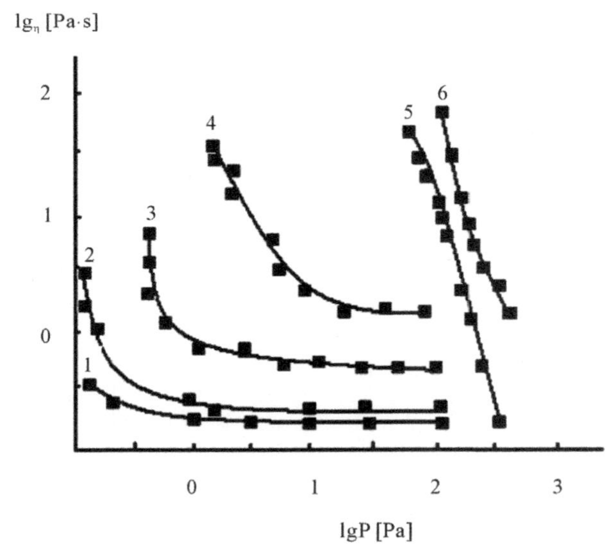

Figure 1. Influence of thermodynamic quality of the solvent on the rheological properties of 10% BNR solutions. MEK:T = 1:0 (1), 1:0.5 (2), 1:1.45 (3), 1:1.8 (4), 1:2.21 (5), and 1:4.28 (6).

of viscosity anomaly, *i.e.* the increase of structuring of the system. At sufficient high substance of toluene (curves 5, 6) along with the sharp increase of viscosity changes the dependency character $\acute{\eta} = f(\mathrm{p})$: curve is mixed in the zone of high shear tension, but the transition from $\acute{\eta}_0$ into $\acute{\eta}_\infty$ occurs rapidly in the narrow range of high shear tension.

Draws attention the fact that in this case the value $\acute{\eta}_\infty$ is in lower value obtained for the previous systems (curves 1 - 4). It is due to phase change system.

The phase diagrams of the ternary BNR-MEK-Toluene was constructed by the method of isothermal titration BNR in MEK precipitator (toluene). Unfortunately, to establish the compound of the precipitating phase failed.

Thus, deterioration of thermodynamic quality of solvents alters the character of the structure of macromolecules, *i.e.* the resulting spatial patterns: in case of a solution this spatial grid is formed by the fluctuating nature, but in case of dispersion clutch assemblies the new phase of macromolecules BNR.

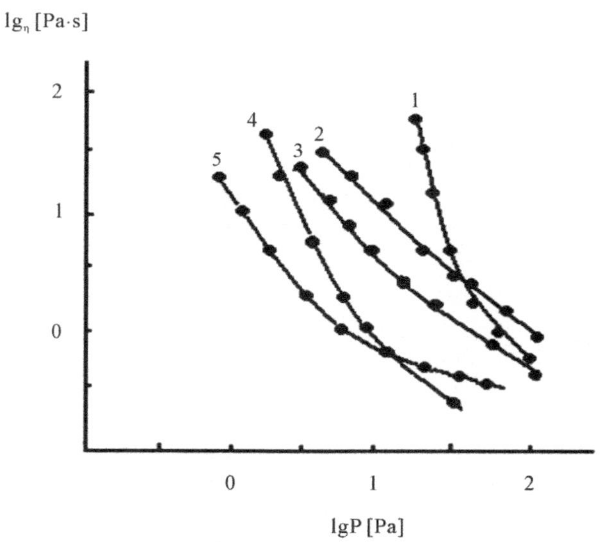

Figure 2. Influence of radiation on the rheological properties of 15% BNR solutions in the solvent of different thermodynamic quality.

Due to the fact that the deterioration of thermodynamic quality of the solvent leads to increased interaction of polymers, the strength of inter-structural contacts increases, the total destruction of the structure occurs at higher shear stresses.

In **Figure 2** the data were shown characterizing the influence of irradiation dose on the strength of inter-structural contacts. It is obvious that in case of the dose increasing of the solution leads to a decrease $\acute{\eta}_\infty$, but structuration is enhanced in the area of the low loads with increasing doses. In the two-phases systems the dose increasing leads to a decrease of strength of inter-structural contacts (**Figure 2**, curves 1, 2), so that the structural viscosity decreases, but $\acute{\eta}_\infty$ increases.

If the system is in a state close to the bundle, in this case the temperature rise reduces as a structural and effective viscosity of the solution. Curves (**Figure 2**) corresponding to the system at fifth represents the mixed case, when dose increasing of irradiation leads to a transition from a two-phase in a single-phase region.

REFERENCES

[1] S. P. Pankov, "Ravnovesiya faz v Sisteme Polimer—Rastvoritel," *M.: Chemistry*, 1981, p. 234.

[2] S. P. Pankov, "Studneobraznoe Sostoyanie Polimerov," *M.: Chemistry*, 1984, p. 237.

[3] V. G. Kulichikhin, A. Y. Malkin and S. P. Pankov, *Visokomol. Soed. A*, Vol. 24A, No. 11, 1982, p. 2291.

[4] Mandelkern, "Kristalizasiya Polimerov," Frenkel S.Y. L.: Chemistry, 1976, p. 296.

[5] M. E. Kurbanaliyev, I. K. Dustov and A. Y. Malkin, *Visokomolek. Soed. A*, Vol. 24, No. 11, 1984, p. 2291.

[6] P. I. Zubov, A. I. Zemcev and L. L. Sukhareva, *Kolloid. Zhurn.*, Vol. 38, No. 4, 1986, p. 66.

[7] S. A. Tashmuhamedov, S. A. Azizov, B. U. Sagdullayev and Z. G. Sagdiyeva, *Visokomolek. Soed. A*, Vol. 20, No. 1, 1988, p. 1180.

[8] I. Prigozhin and R. Dephey, Khimicheskaya Termodinamika. M., Khimiya, 1978, p. 306.

Effect of Annealing Temperature on the Optical Properties of Palladium Thin Film

Feridoun Samavat*, Farzane Mahmoodi, Parisa Taravati Ahmad, Mohammad Faraz Samavat, Mohammad Hossein Tavakoli, Sanaz Hadidchi
Department of Physics, Bu-Ali Sina University, Hamedan, Iran
Email: *Fsamavat@yahoo.com

ABSTRACT

In this paper, the effect of annealing temperature on the optical properties of palladium thin films has been investigated. The Pd thin films with thickness of about 123 Å were deposited on glass substrates by electron beam coating method. Then the palladium thin films were annealed in air at 350˚C, 450˚C, 550˚C and 650˚C for 1 h and then cold slowly. All the initial Pd films were found to have amorphous structure. Their optical properties were studied as function of annealing temperature using Ultraviolet-Visible (UV-Vis) spectroscopy. The direct band gaps in Pd films have been determined. Decreasing the values of the direct band gap shows also increasing the Pd crystallite nanostructure with increasing annealing temperature.

Keywords: Pd Properties; Coating; XRD; Band Gap; Annealing; Crystallite Nanostructure

1. Introduction

The structure and microstructure of systems containing nanoparticles have recently been extensively investigated. This is due to a rich number of fascinating properties observed in materials containing small submicrometer clusters, which imply promising applications in various areas of nanotechnology [1]. Thin films of metal nanoparticles have been receiving increasing attention because of the optical, electronic, and catalytic properties of such materials [2].

Thin films palladium have been the focus of many studies because of their ability to separate hydrogen from other gases [3]. Furthermore Pd has remained a focus of the material science community due to its important band gap, conversion efficiency, high absorption coefficient, stability and there exists a vast literature covering theoretical and experimental studies of electronic band structure [4-6]. Pd has a wide and direct band gap, n-type semiconducting material and found in crystalline form fcc phase [2]. A number of deposition techniques are being used for the fabrication of Pd thin films including thermal evaporation, chemical bath deposition, vacuum evaporation, chemical vapour deposition, spray pyrolysis, metal organic vapour-phase epitaxy, closed space sublimation, photochemical deposition, radio frequency sputtering, vapour transport deposition electro deposition, screen printing , pulsed laser deposition and couting [7].

*Corresponding author.

2. Experimental

In this research program, palladium metal (99/95% pure) was purchased from goodfellow Co. Thin films of palladium were prepared by electron beam coating method on glass substrates using graphite crucible. Substrates were cleaned for about 5 min by ultrasonic cleaner. The system was pumped to a base pressure of less than 3×10^{-5} mbar before deposition. Other coating parameter were cathode voltage 8.5 kV, substrate temperature 32˚C, coating rate 12 $Å \cdot s^{-1}$. Film thickness was estimate to be about 123 Å. These films were then annealed in air at various temperatures ranging from 350˚C to 650˚C for a fixed time of 1 h.

The structure of these films was studied by X-ray diffraction (XRD) using Cu $K\alpha$ (1.5418 Å) radiation with operating voltage/current of 40 kV. The optical property including optical band gap were calculated from the transmission spectra between 300 - 900 nm recorded by UV-Vis spectrometer.

In this research, Pd thin films have been coated by electron beam coating and their properties surveyed by UV-Vis spectrometer.

3. Results and Discussion

3.1. Structure Analysis

The structures of the Pd thin films were studied using X-ray diffraction (XRD) technique. The diffraction spectra were measured at 2θ scanning ranging from 0 to 90 degrees diffraction angle. The XRD spectra of Pd thin films with same thicknesses on glass substrates and annealed different temperature are shown in **Figure 1**.

These patters show that with increasing of temperature, intensity of obtained peaks increases. Intensities of reflections corresponding to the planes indifferent samples are different. However the strong reflections in all the samples indicates crystallites corresponding to (004) plane (2θ = 35°) and (111) plane (2θ = 41°). The calculated crystallite sizes of samples were found in the rang of 3 - 19 nm by applying the Scherrer formula. Generally the crystallite size depends on the annealing temperature [8].

3.2. Optical Analysis

The optical adsorption spectra annealed palladium films were recorded using a UV-visible spectrometer in the wave-length rang 300 - 900 nm. These spectra displayed in **Figure 2**.

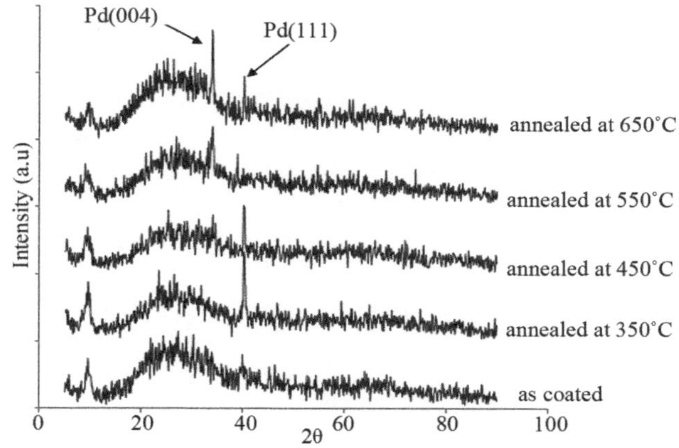

Figure 1. XRD spectra of Pd thin films.

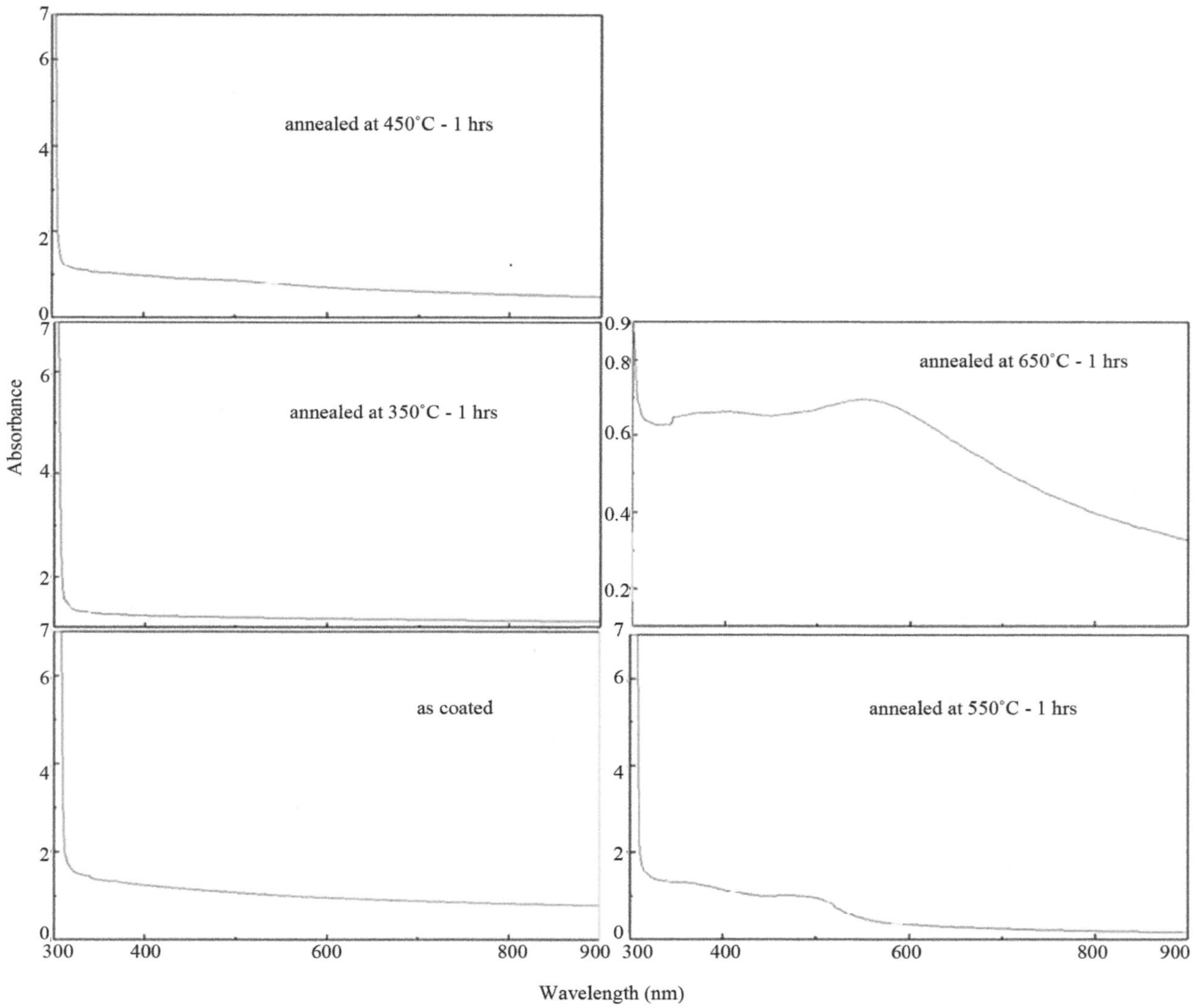

Figure 2. UV-visible spectra.

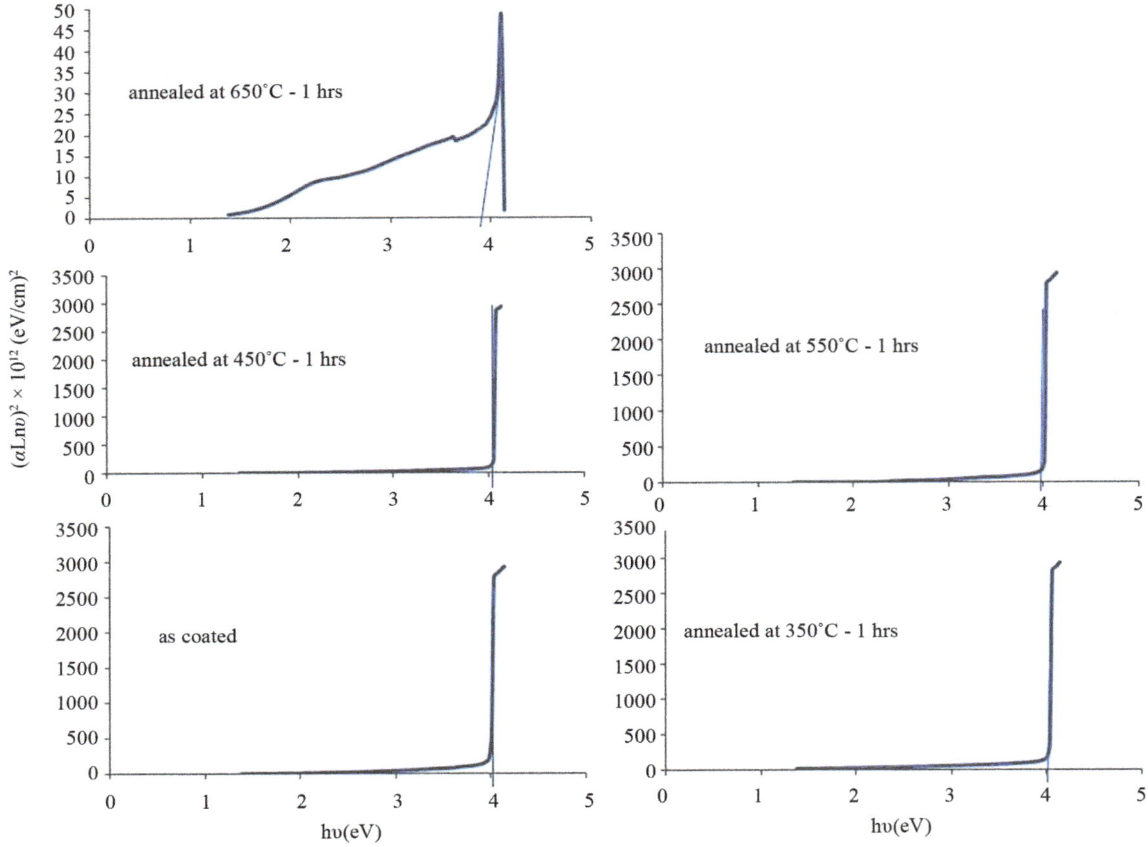

Figure 3. Plot of absorption coefficient, *α*, vs. photon energy (*hν*) of Pd thin films.

The optical absorption coefficient (α) was calculated from the absorbance (A). after correction for reflection losses α can be obtained using the following equation [9]:

$$\alpha(\nu) = 2.303\frac{A}{\mathrm{d}}$$

where d is the film thickness in cm and defined by $A = \mathrm{Log}(I/I_0)$ where I and I_0 are the intensities of the incident and transmitted beams, respectively.

The optical band gaps (E_g) of the films can be obtained according to its depence on α and energy hv of the incident photon, as expressed by following equation (Davis and Mott; 1970) [10,11]:

$$\alpha h\nu = B\left(h\nu - E_g\right)^n$$

where $h\nu$ is the energy of incidence photon, E_g is the value of the optical energy gap between the valence band and the the condition band, B is a constant that depends on the electronic transition probability and the exponent is a parameter which depends on the type of electronic transition responsible for absorption. Values of n = 2 and n = 1/2 correspond, respectively, to allowed indirect and allowed direct optical transitions [8,12,13].

The usual method for the determination of the value of E_g involves plotting $(\alpha h\nu)^{1/n}$ against ($h\nu$). In this study the most satisfactory results were obtained by plotting $(\alpha h\nu)^2$ as a function of the photon energy ($h\nu$). Such plots have been shown in **Figure 3**.

E_g, at various temperature were estimated by extrapolating the linear portion of $(\alpha h\nu)^2$. The optical band gap was shifted toward the lower energy from 4.1 to 3.87 eV as the annealing temperature rose from room temperature to 650°C. These show that with increasing temperature, values of obtained band gap decreased.

REFERENCES

[1] J. B. PelÇka, W. Paszkowicz, P. Dluzewski, M. Brust, C. J. Kiely, M. Knapp and E. Czerwosz, "Characterisation of Thin Films Containing Au and Pd Nanoparticles by Grazing-Incidence X-Ray Diffraction and Related Methods," *Journal of Alloys and Compounds*, Vol. 328, No. 1-2, 2001, pp. 248-252. doi:10.1016/S0925-8388(01)01303-2

[2] K. R. Patil, Y. K. Hwang, M.-J. Kim, J.-S. Chang and S.-E. Park, "Perparation of Thin Films Comprising Palladium Nanoparticles by a Solid-Liquid Interface Reaction Technique," *Journal of Colloid and Interface Science*, Vol. 276, No. 2, 2004, pp. 333-338. doi:10.1016/j.jcis.2004.04.001

[3] K. Zhang, H. Y. Gao, Z. B. Rui, Y. S. Lin and Y. D. Li, "Preparation of Thin Palladium Composite Membranes and Application to Hydrogen/Nitrogen Separation," *Chi-*

nese Journal of Chemical Engineering, Vol. 15, No. 5, 2007, pp. 643-647. doi:10.1016/S1004-9541(07)60139-6

[4] E. Kowalska, E. Czerwosz and J. Radomska, "The Methods of Synthesis of Nanopores Carbon-Palladium Materials," *Electronics*, Vol. 1, 2009, pp. 32-35.

[5] C. Benvenuti, P. Chiggiato, F. Cicoira, Y. L'Aminot and V. Ruzinov, "Vacuum Properties of Palladium Thin Film Coatings," *Vacuum*, Vol. 73, No. 2, 2004, pp. 139-144. doi:10.1016/j.vacuum.2003.12.022

[6] X. Wei, T. Wei, H. Xiao and Y. S. Lin, "Nano-Structured Pd-Long Period Fiber Gratings Integrated Optical Sensor for Hydrogen Detection," *Sensors and Actuators B: Chemical*, Vol. 134, No. 2, 2008, pp. 687-693. doi:10.1016/j.snb.2008.06.018

[7] N. A. Shah, A. Nazir, W. Mahmood, W. A. A. Syed, S. Butt, Z. Ali and A. Maqsood, "Physical Properties and Characterization of Ag Doped CdS Thin Films," *Journal of Alloys and Compounds*, Vol. 512, No. 1, 2012, pp. 27-32. doi:10.1016/j.jallcom.2011.08.081

[8] A. F. Khan, M. mehmood, A. M. Rana and T. Muhammad, "Effect of Annealing on Structural, Optical and Electrical Properties of Nanostructured Ge Thin Films," *Applied Surface Science*, Vol. 256, No. 7, 2010, pp. 2031-2037. doi:10.1016/j.apsusc.2009.09.043

[9] M. F. Zaki, "Gamma-Induced Modification on Optical Band Gap of CR-39 SSNTD," *Brazilian Journal of Physics*, Vol. 38, No. 4, 2008, pp. 17504-17508.

[10] T. Dhannia, S. Jayalekshmi, M. C. SanthoshKumar, T. PrasadaRao and A. ChandraBose, "Effect of Aluminium Doping and Annealing on Structural and Optical Properties of Cerium Oxide Nanocrystals," *Journal of Physics and Chemistry of Solids*, Vol. 70, No. 11, 2009, pp. 1443-1447. doi:10.1016/j.jpcs.2009.09.001

[11] S. Varghese, M. Iype, E. J. Mathew and C. S. Menon, "Determination of the Energy Band Gap of Thin Films of Cadmium Sulphide, Copper Phthalocyanine and Hybrid Cadmium Sulphide/Copper Phthalocyanine from Its Optical Studies," *Materials Letters*, Vol. 56, No. 6, 2002, pp. 1078-1083. doi:10.1016/S0167-577X(02)00682-1

[12] J.-W. Jeon, D.-W. Jeon, T. Sahoo, M. Kim, J.-H. Baek, J. L. Hoffman, N. S. Kim and I.-H. Lee, "Effect of Annealing Temperature on Optical Band-Gap of Amorphous Indium Zinc Oxide Film," *Journal of Alloys and Compounds*, Vol. 509, No. 41, 2011, pp. 10062-10065. doi:10.1016/j.jallcom.2011.08.033

[13] T. P. Kumar, S. Saravanakumar and K. Sankaranayanan, "Effect of Annealing on the Surface and Band Gap Alignment of CdZnS Thin Films," *Applied Surface Science*, Vol. 257, No. 6, 2011, pp. 1923-1927. doi:10.1016/j.apsusc.2010.09.027

Influence of Single Use and Combination of Reductants on the Size, Morphology and Growth Steps of Gold Nanoparticles in Colloidal Mixture

Ahmad Alshammari[1*], Angela Köckritz[2], Venkata Narayana Kalevaru[2], Abdulaziz Bagabas[1], Andreas Martin[2]
[1]National Nanotechnology Research Center, King Abdulaziz City for Science and Technology, Riyadh, Saudi Arabia
[2]Leibniz-Institute for Catalysis at University of Rostock, Rostock, Germany
Email: [*]aalshammari@kacst.edu.sa

ABSTRACT

A comprehensive investigation on the formation mechanism of gold nanoparticles (AuNPs) in colloidal mixture obtained from the reduction of chloroauric acid ($HAuCl_4$) solution using a single reducing agent (sodium citrate; process-I), (tannic acid; process-II), and a combination of two reducing agents (sodium citrate plus tannic acid; process-III) is reported. The growth steps at different time intervals during synthesis of colloidal AuNPs were monitored in situ and ex situ using various methods for all the three processes. The measurement of changes in the surface plasmon band position of colloidal AuNPs, along with dynamic light scattering results gave important information for the first assessing of particle size, shape and distribution. Besides, the size and morphological changes at different stages during different processes were also analyzed by transmission electron microscopy. The final Au particles of processes-I & II exhibited different shapes (spherical and nanowires) with particle size and nano wire diameter of 12 nm and 17 nm, respectively. Nevertheless, combination of two reductants (process-III) surprisingly leads to drastically reduced size (ca. 3 nm) with spherical morphology compared to their parent solutions with either of single reducing agent. This result clearly indicates that the combination of reductants has a significant influence on the particle size, morphology and formation mechanism.

Keywords: Colloidal Gold Nanoparticles; Reducing Agents; Size Distribution; Morphology

1. Introduction

Metal nanoparticles (NPs) have been the subject of extensive research in recent times due to their fascinating size-dependent electrical, optical, physical and chemical properties [1-3]. In particular, colloidal gold nanoparticles (AuNPs) have attracted increasing research attention in the field of nanocatalysis [4-8]. Since the properties of AuNPs strongly depend on their size, size distribution and shape [9,10]. Many studies on colloidal AuNPs have been focused on the control of their morphology, which also directly correlates to the catalytic activity [11-12].

Several approaches have been reported for the preparation of colloidal AuNPs such as conventional chemical reduction [13], radiation chemical reduction [14], and photochemical reduction [15]. The most simple and widely used method to synthesize colloidal AuNPs is the reduction of chloroauric acid (HAuCl4) with sodium citrate, a method pioneered by Turkevich *et al.* [16]. In general, such type of preparation method involves the formation of NPs mainly in three different stages such as 1) an initial step of reduction process of ionic Au^{3+} species, 2) a second step of formation of nuclei, and 3) a third step of particle growth. The first and second steps are somewhat difficult to identify due to the involvement of ionic species during synthesis while the third step can easily be detected due to the growth process of the particles. La Mer firstly reported the formation process of mono-dispersed particles [17]. This model showed that formation of seed occurred, which then proceeds by its successive growth of the particles. In addition, Chow and Zukoski [18] investigated the formation mechanism of particles at different conditions and proposed that aggregation of nanocrystals occurred at low reaction temperatures. A mechanism proposed by Watzky and Finke [18] showed that the nucleation process was rather slow, which resulted in nearly monodisperse size distribution. Among several proposals, the most accepted mechanism suggests a two-step process, *i.e.* nucleation followed by

[*]Corresponding author.

successive growth of particles. It was also reported elsewhere [19,20] that different morphologies of AuNPs such as rod-like particles or polyhedrons could be obtained during the further growth of gold seed particles. In addition, the formation mechanism of colloidal particles in presence of stabilizers (*i.e.* probucol/polyvinylpyrrolidone (PVP)) ensuring high dispersion was also investigated [21]. A recent work presented results on the photo-reduction of gold ionic species in aqueous surfactant solutions of dodecyltrimethylammonium chloride and polyethylene glycol lauryl ether [22]. Furthermore, some theoretical studies on the growth mechanism of colloidal AuNPs were also reported [23,24]. Even though some articles appear sporadically, systematic investigations approved with different spectroscopic and microscopic techniques, which monitor the formation growth mechanism of colloidal AuNPs, are still missing in the literature. In addition, studies on the application of two reducing agents on the size of AuNPs are very rare in the literature [25]. It is assumed that the second reductant has an additional reducing and protecting effect during colloidal gold formation as claimed elsewhere [26]. Therefore, there is a need to explore and understand the growth process of AuNPs in suspension, which certainly allows a tailor-made synthesis regarding catalytic applications. Furthermore, most of these studies listed above have used single reducing agents for AuNPs formation. To the best of our knowledge, there is not even single source of publication on exploring the growth steps of AuNPs formation at different stages of synthesis using the combination of two or more reducing agents.

Against this background, the present study is aimed at investigating the influence of combination of two reducing agents on the size, distribution and morphology of AuNPs during the formation process in colloidal mixture and compares such results with that of those obtained from the usage of single reducing agent. We further make attempts to understand the formation growth mechanism of such colloidal AuNPs in more detail. The objective is also to characterize these materials by various techniques such as UV-vis, TEM, zeta potential, and DLS for better understanding of growth process and material properties.

2. Experimental Section

2.1. Materials

Tetrachloroauric acid ($HAuCl_4 \cdot 3H_2O$) and tri-sodium citrate dihydrate ($C_6H_5Na_3O_7 \cdot 2H_2O$; SC) were purchased from Fluka. Tannic acid ($C_{76}H_{52}O_{46}$; TA) was obtained from Sigma-Aldrich. All chemicals were used as received. All solutions were prepared with distilled water.

2.2. Experimental Procedure

Process-I (single reducing agent-SC): 0.186 mmol of 1.0 mM $HAuCl_4$ solution was heated in a 400 ml beaker to 60°C. In a second vessel, 0.75 mmol of 15.0 mM SC solution was also heated to 60°C. In a typical experiment, the solution of SC was added at once and quickly to the solution of $HAuCl_4$ under stirring (1000 rpm). 20 seconds after mixing, a 50 ml aliquot of the reaction mixture was assayed for characterization. The sampling procedure was repeated and aliquots were taken from the reaction mixture at different reaction intervals in the range of 0 - 60 min.

Process-II (single reducing agent-TA): The same procedure and similar reaction conditions/concentrations as described above were applied except the type of reducing agent. The samples were assayed at three different time intervals such as 2, 5 and 9 sec, respectively due to fast reduction by TA as strong reducing agent.

Process-III (combination of two reducing agents-SC & TA): The same procedure and similar reaction conditions/ concentration as described above were applied. However, instead of using a single reductant, a combination of two reducing agents such as SC (7.5 mM) plus TA (7.5 mM) was used. Samples were collected at different time intervals (*i.e.* 2, 6 and 10 sec).

2.3. Analytical Methods

Ultraviolet-visible Spectroscopy (*UV-vis*): The optical properties (absorbance) of colloidal solutions were acquired with a UV-vis spectrometer (Avantes-2048, light source combined deuterium-halogen). The spectra were collected over a range of 200 - 1100 nm (with an optical path length of 0.4 cm) as a function of reaction time by directly dipping the optical probe in the reaction vessel under constant stirring.

Dynamic Light Scattering (DLS): Size, size distribution and zeta potential of colloidal AuNPs were determined using DLS with Malvern Instruments Zetasizer (ZS ZEN 4003) at room temperature. Samples were assayed at defined times from the reaction and transferred to cuvette for size and size distribution analysis.

Transmission Electron Microscopy (TEM): Size analysis of colloidal AuNPs was carried out using HRTEM JEM-2100F at a voltage of 200 kV. The samples were prepared by mounting a drop of the aqueous suspension containing the AuNPs on a carbon grid, which then was placed on filter paper to absorb excess solvent. The average particle diameter and size distribution were calculated using Java image tool software (Image J), based on the data of an average of 100 - 150 particles.

3. Results and Discussion

3.1. Formation of Colloidal Aunps with SC (Process-I)

The simultaneous mixing of $HAuCl_4$ and SC solutions

used in process-I led to the formation of colloidal AuNPs by the appearance of the typical ruby-red color of the final slurry. During the course of AuNPs growth, a succession of color changes in the reaction solution was observed passing from pale yellow to colorless, grey, light blue, purple, ruby and finally to red (**Figure 1**). No further change in the color of the gold suspension was observed after 60 min of reduction indicating completion of reaction. This applied preparation method is similar to those reported previously [15,26] but the aim of the present investigations was focused on comparative *in-situ* and *ex-situ* studies to understand the growth steps involved at different stages of synthesis.

First, UV-vis spectroscopy was applied as an effective method to illustrate the evolution of metallic species during the formation of colloidal metal clusters [27]. Absorbance or wavelength of Surface Plasmon (SP) band provides a measure of particles size, shape, morphology, concentration, and dielelectric medium properties [28,29]. The *in-situ* measurement of SP band during the origination of AuNPs certainly provides useful information of assessing the reaction kinetics. Colloidal AuNPs are known to exhibit a strong absorption in the visible range (380 - 780 nm) due to Localized Surface Plasmon Resonance (LSPR), which is caused by collective oscillations of the conduction electrons upon excitation with visible light [30]. Mie first described the theoretical basis of this behavior [31]. Besides, changes in the wavelength and bandwidth (Full Width at Half Maximum—FWHM) of this SP band depend significantly on various factors such as particle size [32,33], shape [34], dielectric environment [35], and the sign and size of the core charge [36,37]. The formation kinetics was followed by measuring changes in the absorption bands of resulting AuNPs at the maximum absorption wavelength (λ_{max}). As reported elsewhere [29,38,39], the position of the absorption band of metal species strongly depends on the size and length of the nanoclusters. The time-resolved UV-vis spectra obtained during the reduction of $HAuCl_4$ with SC are shown in **Figure 2**. The evolution of the band position revealed the formation history of the AuNPs. The base spectrum of pure $HAuCl_4$ solution (light yellow in color, *i.e.*, before mixing with reducing agent) showed no distinct absorption in the range of 400 - 1000 nm. However, upon the addition of SC solution to $HAuCl_4$ solution, the initial light yellow solution became colorless, indicating

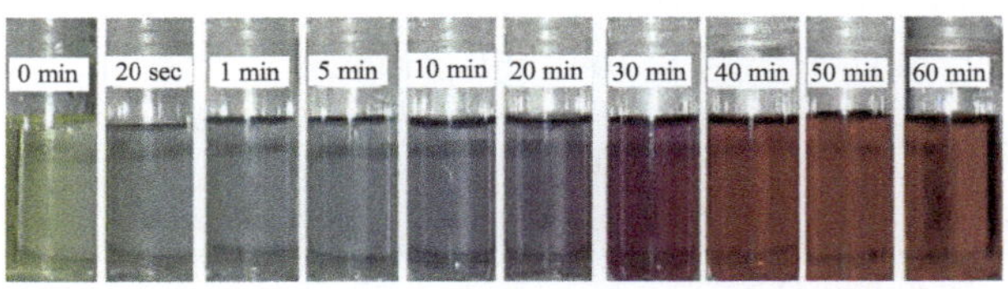

Figure 1. Color changes of the reaction solution during the course of colloidal AuNPs formation with SC as reductant.

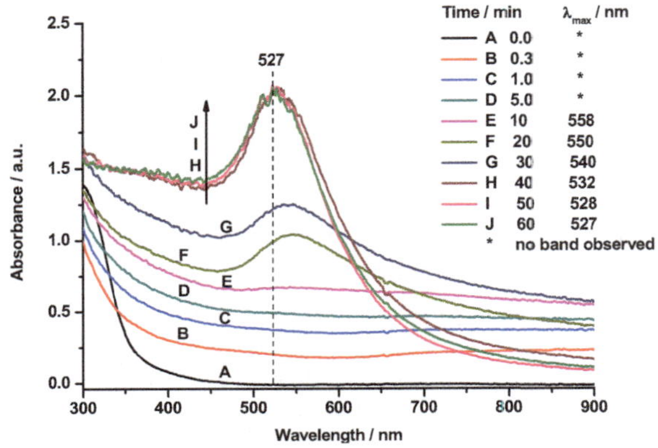

Figure 2. Time-resolved UV-vis spectra during the reduction of 1.0 mM $HAuCl_4$ solution with 15 mM SC solution at 60°C and an $[Au^{3+}]/[SC]$ ratio of 1:4.

the beginning of the reduction process. After 0.3, 1 and 5 min of reaction time, a considerable change in the color of colloidal solution could be observed by both visibly and analytically by recording the UV-vis absorption spectra, particularly in the range of 500 - 550 nm. Even though, no clear SP bands were found in UV-vis spectra, the base spectrum was found to be shifted upwards. Hence, one can assume that larger flower-like Au nanocrystals were probably formed during the initial stages of the reduction, which could not exhibit any SP band in the visible range between 400 - 1000 nm. Therefore, this band might be red-shifted, however, this was not detectable with our UV-vis instrument [40]. As a matter of support to this assumption, some previous reports [41,42] also claimed that very large nanoparticles absorb and scatter light in the near-infrared and mid-infrared region of the spectrum. On the other hand, TEM could provide further hints on the formation of expected flower-like nanocrystals. To confirm such proposition, TEM analysis at different stages of reaction was also carried out and discussed below in a separate section. However, after 10 min of reaction time, a flat absorbance band with a broadening of the curve profile begins to appear. As the reaction progressed further (20 - 30 min), symmetrical SP bands distinctly appeared at 550 and 540 nm, respectively. Finally, when the solution turned to ruby-red after 50 - 60 min, the SP band continued to exhibit a blue-shift with a peak maximum at 528 nm, and at the same time the peak was narrowed, which is a clear indication for the formation of colloidal AuNPs. There-after, this ruby-red color remained constant over two months, indicating good stability of such gold colloids. Additionally, in good agreement with the literature, the present results revealed strong evidence that the position of the absorption band strongly depends on the size and shape of the nano-clusters [43].

As mentioned above, various steps involved during the course of AuNPs formation were also tracked by TEM, and the results are shown in **Figure 3**. Here four main stages of AuNPs development were observed: 1) in the very initial stage (**Figure 3(a)**, after 1 min, colorless solution), the formation of large nanocrystals having flower-like morphology was observed. Individual flower-like or flake-like nanoclusters reached sizes between 80 - 100 nm with irregular shapes. Such type of AuNPs shape has also been reported elsewhere [44]. This result undeniably supports the view observed from UV-vis spectra. At initial stages of reduction, large nanocrystals were formed which might exhibit SP bands at higher wavelengths (in the region of near IR) that is beyond the scope of the detection limit (400 - 1000 nm) and hence no clear bands could be detected in UV-vis spectra under this set of conditions (see **Figure 2**). 2) Subsequently, wire-like networks with a smaller diameter ranging from 10 - 20 nm were observed when the reaction time was prolonged to 10 min (**Figure 3(b)**, light blue solution). These wire-like networks exhibited a Face-Centered Cubic (FCC) lattice with distinct facets that mostly showed Au(200) lattice planes. However, such wire-like structures could not be verified again with UV-vis spectra (see **Figure 2**: 10 and 20 min) but some indirect hints from the shift in the band position could be realized. Wang and co-workers [45] reported that the corresponding absorption band of gold nano-wires gave an extremely broadened band in the range of 650 - 1000 nm due to the longitudinal mode of the SP absorption and depending on the aspect ratio of the nanochains. In addition, it is reported elsewhere [46] that the colloidal solution having a nanowires structure gives a blue suspension, which is in good agreement with our results. Link and El-Sayed [37,47] also observed a strong longitudinal SP absorption band in the range of 700 - 1000 nm in the case of AuNPs with cylindrical shape and various aspect ratios. Nevertheless, none of such phenomena was observed in the present experiments. 3) When the reaction solution started becoming purple after 30 min, these nano-wires began to breakdown to form NPs with a diameter ranging from 12 to 16 nm (**Figure 3(c)**). 4) The final stage was observed after a period of 60 min where TEM images showed highly uniform NPs with an average diameter of 12 nm and a polyhedral, sphere-like morphology with some spherical particles exhibiting (111) and (200) lattice planes (**Figure 3(d)**). Interestingly, no further change in the morphology of the particles was observed after 60 min. The large aggregates of Au flower-like nanocrystals observed in the first stage (see **Figure 3(a)**) may be later separated due to the absorption of citrate ions on the surface of the gold particles, which may affect a strong repelling layer preventing the nanoparticles from coming into close contact [48].

Furthermore, supporting investigations regarding the formation of such colloidal AuNPs were also conducted using other technique such as DLS, which provide a more direct measurement of colloidal AuNPs in solution. DLS measures the hydrodynamic diameter of particles (solvation sphere) including agglomerates. **Figure 4** portrays time-resolved size distribution curves recorded by DLS during the reduction with SC over a period of 60 min. Initially, the average diameter of the gold cluster amounted to 220 nm immediately after the addition of SC to $HAuCl_4$ solution. When the reaction progressed, the average particle diameter rapidly decreased, indicating disaggregation of the Au nano-clusters flakes to a size of about 30 - 25 nm after 10 - 20 min. Finally, the smallest AuNPs with an average size of ca.13 nm appeared after 60 min. Interestingly, these small nanoparticles did not change further even after 2 hours of reaction time. This evidence well supports the observations made above by TEM analysis.

3.2. Formation of Colloidal AuNPs with TA (Process-II)

Additionally, colloidal AuNPs were also synthesized using other type of single reducing agent such as Tannic Acid (TA). **Figure 4** shows a typical example of the variation of suspension color when the reductant to the gold chloride solution is added. Upon the addition of TA, the initial yellow solution (at 0 sec) became light blue

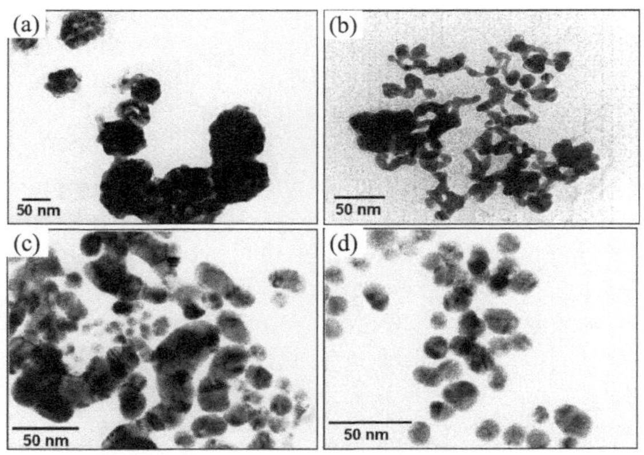

Figure 3. TEM images taken during the formation of colloidal AuNPs obtained by reduction of 1.0 mM $HAuCl_4$ solution with 15 mM SC solution at 60˚C and an $[Au^{3+}]/[SC]$ ratio of 1:4 at reaction time of (a) 1 min; (b) 10 min; (c) 30 min; and (d) 60 min.

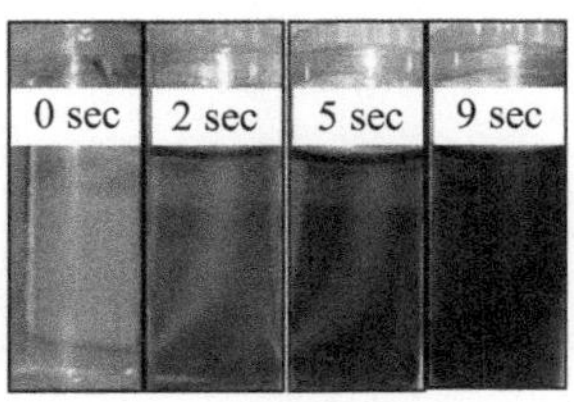

Figure 4. Color changes of the reaction solution during the course of colloidal AuNPs formation with TA as reducing agent.

within 2 sec indicating that the first stage of the reduction took place. Subsequently, the solution turned from light blue to blue (5 sec), which finally transferred to dark blue within 9 sec. No further change was observed afterwards, which indicates the completion of the reaction. In addition to the physical observation of color changes at different time intervals, the corresponding changes in the morphology and size distribution was also characterized using different methods, which are reported below one after the other.

Figure 5 illustrates UV-vis absorption spectrum colloidal AuNPs recorded from the suspension during the reduction process that corresponding to different reaction times (*i.e.* 2, 5 and 9 sec). The appearance of broad absorption band in the range from 500 to 900 nm observed after 2 sec provides indirect hints on the formation of nanowires, which is however confirmed from TEM. No further change in the plasmon band position and shape was observed even after 9 sec. However, it is interesting to note that clear changes in the color of colloidal solution were watched at different times (from yellow to dark blue). As we have described earlier on the usage of SC alone, Au nanowires showed a flat absorption pattern with a broad peak observed from 500 to 900 nm (see **Figure 3(b)**). Similar such observation of broadening plasmon band when nanowires are formed is also reported elsewhere [47]. From this result, it can be speculated that the reduction of gold ions by TA is producing gold metal having nanowires morphology. Such phenomena of nanowires structure can also be found in the literature as supporting evidence [46].

The corresponding TEM images of colloidal AuNPs obtained at different reaction times are shown in **Figure 6**. TEM images confirmed that the gold particles are in nanowires shape. At the early stage (**Figure 6(a)**, after 2 sec, light blue solution), Au nanowires extended from a big nanostructure. After 5 sec (blue solution), a clear Au nanowires shape was observed with diameter of approximately 17 nm as shown in **Figure 6(b)**. At final stage (**Figure 6(c)**, 9 sec, dark blue), the large aggregates

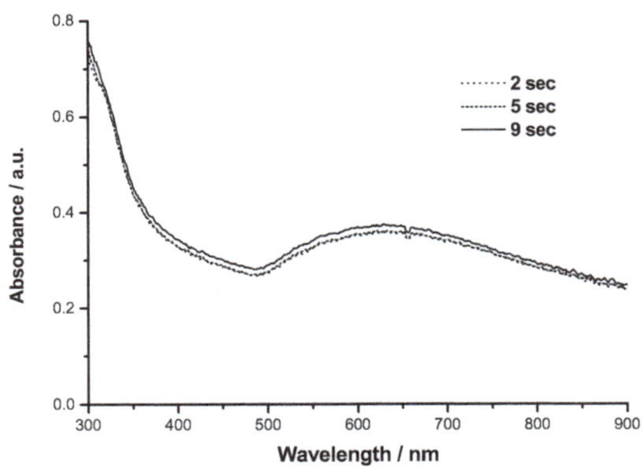

Figure 5. Time-resolved UV-vis spectra during the reduction of 1.0 mM $HAuCl_4$ solution with 15 mM TA at 60°C.

of Au nanowires was observed due to instability of AuNPs prepared using tannic acid, which is also confirmed using zeta potential instrument. The nanowires network remained without changes after 5 min of reaction, which corresponded to the broad absorption spectrum of gold colloids. Based on this observation, such nanowires structure was also reported in the very early stage when the gold ions reduced by SC (**Figure 3(b)**).

Furthermore, the histogram obtained from DLS (not shown here) depicts the evolution of gold colloids of average sizes between 16 and 45 nm vs. reaction time. Another interesting observation is that the formation process of colloidal AuNPs using TA was extremely fast, which was completed within 9 sec. Such quick reduction is expected from the stronger nature of reducing agent. At first stage, a broad size distribution with an average particle size of 35 nm was recorded in the suspension. After that, the average size was slightly decreased from 35 to 31 nm. The final stage showed a better distribution with average size of 27 nm.

3.3. Formation of Colloidal AuNPs by Combination of SC and TA (Process-III)

Further studies were focused on the application of combination of two reducing agents (process-III) and check their influence on the size and growth steps. The application of a mixture of SC and TA surprisingly caused an extremely faster reduction compared to SC while comparable time with that of TA when used as single reducing agent. Some changes in the color of the solution were also observed during the formation process of AuNPs using a combination of SC and TA. Such changes are shown in **Figure 7**.

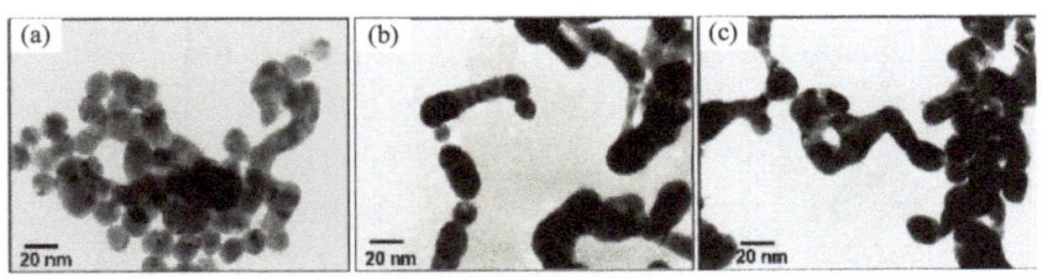

Figure 6. TEM images taken during the formation of colloidal AuNPs obtained by reduction of 1.0 mM $HAuCl_4$ solution with 15 mM TA solution at 60°C at reaction time of (a) 2 sec; (b) 5 sec; and (c) 9 sec.

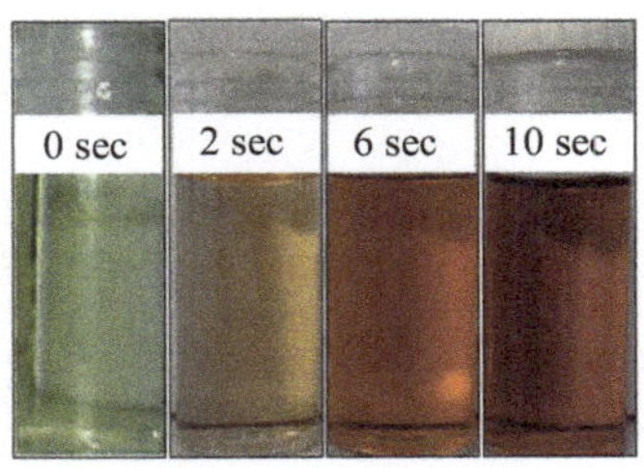

Figure. 7. Color changes of the reaction solution during the course of colloidal AuNPs formation with a combination of SC and TA as reducing agent.

The light yellowish solution of starting materials turned to yellow-orange immediately after the addition of the mixture of SC and TA (after 2 sec), which then turned to orange-red (6 sec), and the final colloidal solution exhibited a ruby-red color within 10 sec. The stirring of the solution was continued for another 5 min, but no further change in color was observed. Hence, it can be considered that the completion of reduction is achieved within 10 sec.

In addition, the samples were also collected at different stages of reduction and characterized by UV-Vis, TEM and DLS techniques with a view to follow the growth steps. **Figure 8** shows the UV-Vis spectra monitoring the formation of AuNPs at different reaction times. Amazingly, the maximum of plasmon band position was recorded in the region of around 520 nm (even after 2 sec), which is a characteristic of the formation of AuNPs. This result points to the fact that the formation of AuNPs with spherical morphology is occurring within the very first stage (2 sec). This observation completely differs from the usage of any of single reducing agents. In the subsequent stages (6 and 10 sec), no considerable shift in the band position could be noticed, which is again an indicative of single step reduction when combination is used. To get further confirmation of UV-Vis results on the fast and one-step reduction, TEM analysis was also carried out. Interestingly, a good complementary evidence for such single step reduction could be obtained from TEM, which is described below.

TEM studies were conducted to provide the better evidence for the evolution of structure, morphology, and size of AuNPs in solution at different reaction stages. Appropriate samples were prepared by dipping a carbon grid in the colloidal mixture collected at different time intervals. **Figure 9** portrays a time succession of TEM images of colloidal AuNPs at different reaction times. The corresponding particle size distributions of the respective steps were also included in this Figure. After 2 sec, monodispersed AuNPs were formed mostly with spherical morphology; however a few percentage of particles with irregular shapes were also present (**Figure 9(a)**). The particle size distributions were more or less narrowed with an average size of 6.3 nm. With increasing reaction time (6 sec), the average particle size slightly decreased from 6.3 nm to 4.9 nm (**Figure 9(b)**), and finally to 3.2 nm after 10 sec (**Figure 9(c)**). The size distribution obviously narrowed again at the final stage of the formation of colloidal AuNPs. No further change either in the morphology or in the size of AuNPs was noticed by TEM even after 5 min, which further supports the results of UV-Vis that the reaction was completed in short time (10 sec). Achieving such the smallest nanoparticles of 3.2 nm using the combination of two reductants is indeed the remarkable outcome of this study. One common observation between the usages of single or combination of two reducing agents is that always bigger Au particles are formed irrespective of type of reducing agent used especially during the initial stages of reduction process, and then the size of AuNPs decreased to a minimum where the reduction is complete. Interestingly, this observation is often noticed in all investigations and is also supported by different characterization techniques. The probable reason for obtaining different morphology with smaller particles of less than 4 nm using combination of reductants (against 12 nm with SC-alone and 17 nm TA alone) might be due to the hydrophobicity of ligand, charge and binding affinity to colloidal

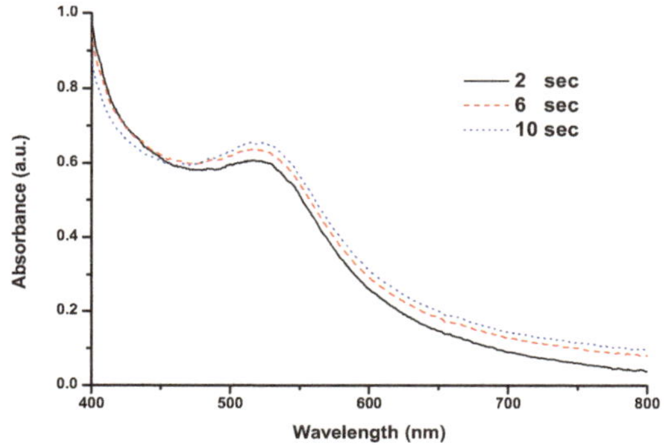

Figure 8. Time-resolved UV-vis spectra during the reduction of 1.0 mM $HAuCl_4$ solution with 15 mM SC solution in the presence of TA at 60°C.

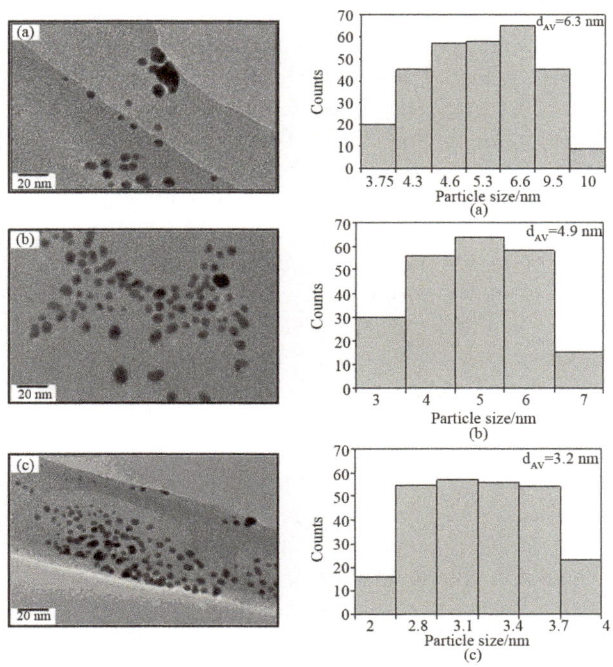

Figure 9. TEM images taken during the formation of colloidal AuNPs obtained by reduction of 1.0 mM $HAuCl_4$ solution with 15 mM SC solution in the presence of TA at 60°C at reaction time of (a) 2 sec; (b) 6 sec; (c) 10 sec.

mixture. Possibly, this assumption is also displayed by the changes in zeta potential (ζ) values with changed reductant, which are discussed after DLS results in more detail.

The formation of colloidal AuNPs produced by the combination of two reductants can also be conveniently followed by DLS measurements. The histogram (**Figure 9**) shows the evolution of gold colloids of average sizes between 3.5 and 15 nm vs. reaction time. It can also be seen that the formation process of colloidal AuNPs in process-III was completed within 10 sec, which is much faster than SC (60 min) but comparable with TA (9 sec). Initially (*i.e.* 2 sec), after the addition of SC + TA mixture to Au^{3+} solution, a broad size distribution with an average particle size of 11 nm was characterized in the suspension. After 6 sec, the average particle size slightly decreased to 9.5 nm with a more narrow size distribution, indicating that some changes in this process occurred. At final stage (after 10 sec), the mean average size of AuNPs reached 8.5 nm with a very narrow size distribution. In addition, another sample was also tested after 1 min but no further changes occurred. Overall, the difference in the average size of colloidal AuNPs formed at various reaction times is relatively small indicating that these results are in reasonable agreement with the UV-vis results where no significant change was found in the maximum wavelength of the SP bands.

In addition to UV-Vis, TEM and DLS investigation, the samples (final stage) were also characterized by zeta potential to get further insights on the stability of colloiddal AuNPs obtained from processes-I to III. The magnitude of zeta potential gives an indication of potential stability of colloid. In general, the particles with zeta potential values more positive than +30 mV or more negative than –30 mV are considered stable. On the other hand, the colloids are least stable at isoelectric point, where the zeta potential is zero. In other words, the colloids with high zeta potentials (positive or negative) are electrically stabilized while the ones with low zeta potentials tend to coagulate, *i.e.* a dividing line between stable and unstable colloids is usually taken at either +30 mV or –30 mV. In the present study, the ζ values are varied in the range from –17 mV to –39 mV, depending upon the type of reducing agent/process applied. Thus, the changes in the size of AuNPs are also reflected by the changes in ζ values when we move from process-I to -III. Our results showed that the process-II (TA as reductant) displayed the lowest zeta potential (–17 mV) and hence there is no force to prevent the particles coming together and floculate, which is called dispersion instability. Therefore, process-II gives the least stable AuNPs compared to others. Alternatively, the highest stability of AuNPs is obtained from process-I (*i.e.* SC reductant) with a ζ value of –39.3 mV. Fascinatingly, the combination (process-III) also gave not only the smallest particles of 3.2 nm but also stable particles with a ζ value of –32.4 mV. Based on these results, it can be stated that the combination of two reducing agents has a clear influence on the size and stability of AuNPs. However, the possibility of other parameters that are responsible for the formation of smaller particles during process-III cannot be ruled out. In general, various reasons can be taken into account to understand the formation of different morphologies of AuNPs such as 1) repulsive electrostatic forces, which stabilize the particles, 2) attractive electrostatic forces between non-equivalent metal particles due to charge redistribution, 3) attractive size-dependent dispersive forces between identical particles, 4) attractive dipole forces between in-plane magnetic or electric dipoles etc. However, further studies are necessary to clarify these aspects and identify the precise reasons for the formation of the smallest particles, when combination of two reductants is applied.

Comparing the effects of single reducing agent with that of combination, a few interesting observations can be made. The effects induced by the combination of reducing agents are: 1) clear changes in the colour of suspensions, 2) formation of the smallest of AuNPs could be produced, which is a beneficial effect, 3) the reduction process was fast probably due to the presence of strong reducing agent (TA). This might also be useful with respect to saving time, and 4) finally the growth steps and size distribution are different when we compare the combination effects with that of individual (single) reducing agents.

To confirm these results, the synthesis and the measurements were repeated three times under identical conditions and similar results were obtained. This fact shows that the formation of smaller particles and growth steps involved at different stages are quite reproducible.

Based on the above results, the different steps involved in the formation processes of colloidal AuNPs, using a single reducing agent (process-I & II), and a combination of two reducing agents (process-III), can be schematically depicted in **Figure 10**. This imagination is mainly based on the results of TEM analysis. In comparison, processes I, II and III, which differ considerably in particle morphology as well as growth steps involved during AuNPs formation at different stages. Process-I shows three distinct particle shapes during the reduction steps, namely: large Au flower/flake-like nanocrystals (after 1 min), wire-like networks (after 10 min), and spherical Au particles (almost polyhedrons) after 60 min. Process-II gives different steps which includes some nanocrystals (after 2 sec), nanowires structures after 5 sec which then aggregated at the end of the reaction. In contrast, such considerable differences in particle morphology could not be found in case of process-III where the reduction

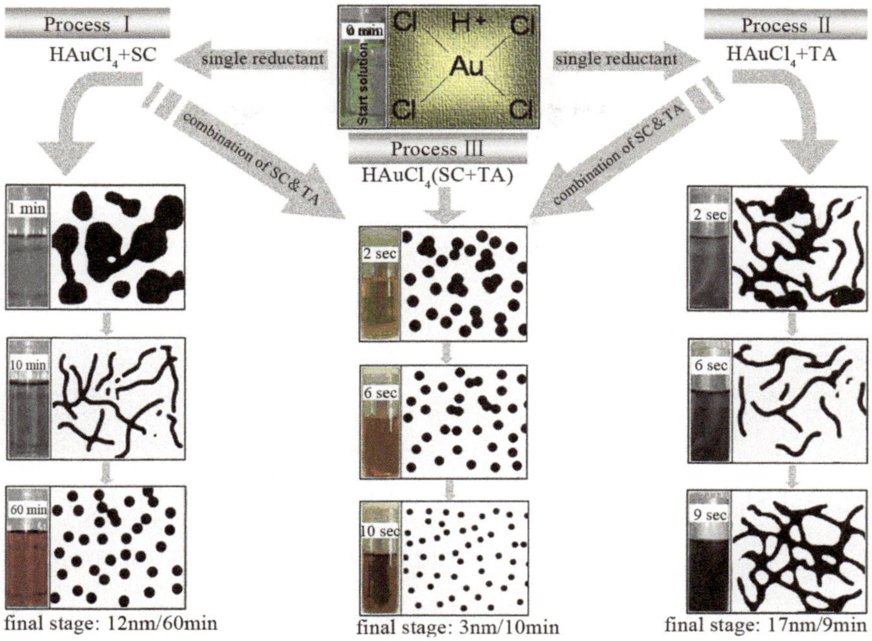

Figure 10. Scheme of the formation mechanism of colloidal AuNPs in solution. Process I shows the different steps of AuNPs formation using single reductant. Process II represents the steps of AuNPs formation using a combination of two reductants.

was achieved quickly and in a single step. In addition, the colour of colloidal suspensions was also found to depend upon the nature of reductant used.

4. Conclusion

The present study has shown a first overview of the effect of the combination of two reducing agents on the formation mechanism steps and size of colloidal AuNPs in solution. At first, the mechanism of the formation of AuNPs using an individual reductant like sodium citrate and tannic acid was elucidated. The results of SC showed clear evidence of a multi-stage formation process, which includes varying morphologies of AuNPs starting from flower/flake-like nanocrystals, wire-like networks to spherical Au particles at different stages of reduction. On the other hand, application of TA also showed some similarities with that of SC particularly during the first two steps of reduction and gave more or less similar morphology. However, TA differed from SC in the final stage in three ways such as 1) exhibiting nanowire morphology, 2) agglomeration of such nanowires and 3) low stability. Interestingly, the combination of two reductants (SC plus TA, process-III) led to a significant change in the formation mechanism. The data analysis of combination revealed no indication of multiple stages of particle formation during reduction instead the formation of AuNPs occurred quickly and in a single step. The most striking feature is that the smallest gold nanoparticles (3.2 nm) could be successfully achieved during the combination of reducing agents. Conversely, application of single reducing agents gave bigger AuNPs (ca. 12 nm in process-I, 17 nm in process-II). Moreover, it is worth noting that the results obtained from all the described techniques at various reaction time intervals showed comparable tendencies and complementary evidences. In summary, it can be said that the nature of reducing agent (either single or combination) has shown considerable influence on the size, morphology, distribution, growth steps and the stability of the formed AuNPs.

REFERENCES

[1] A. P. Alivisatos, "Semiconductor Clusters, Nanocrystals, and Quantum Dots," *Science*, Vol. 271, No. 5251, 1996, pp. 933-937. doi:10.1126/science.271.5251.933

[2] M.-C. Daniel and D. Astruc, "Gold Nanoparticles: Assembly, Supramolecular Chemistry, Quantum-Size-Related Properties, and Applications toward Biology, Catalysis, and Nanotechnology," *Chemical Reviews*, Vol. 104, No. 1, 2004, pp. 293-346. doi:10.1021/cr030698+

[3] K. E. Geckeler and E. Rosenberg, "Functional Nanomaterials," American Scientific Publicationn, Valencia, 2006.

[4] A. S. K. Hashmi and G. J. Hutchings, "Gold Catalysis," *Angewandte Chemie International Edition*, Vol. 45, No. 118, 2006, pp. 7896-7936. doi:10.1002/ange.200602454

[5] R. Narayana and M. A. El-Sayed, "Shape-Dependent Catalytic Activity of Platinum Nanoparticles in Colloidal Solution," *Nano Letters*, Vol. 4, No. 7, 2004, pp. 1343-1348.

[6] M. C. Daniel and D. Astruc, "Gold Nanoparticles: Assembly, Supramolecular Chemistry, Quantum-Size-Related Properties, and Applications toward Biology, Cata-

lysis, and Nanotechnology," *Chemical Reviews*, Vol. 104, No. 1, 2004, pp. 293-346. doi:10.1021/cr030698+

[7] B. L. V. Prasad, S. I. Stoeva, C. M. Sorensen, V. Zaikovski and K. J. Klabunde, "Gold Nanoparticles as Catalysts for Polymerization of Alkylsilanes to Siloxane Nanowires, Filaments, and Tubes," *Journal of the American Chemical Society*, Vol. 125, No. 35, 2003, pp. 10488-10489. doi:10.1021/ja035046h

[8] M. Haruta, "Gold Rush," *Nature*, Vol. 437, No. 7062, 2005, pp. 1098-1099. doi:10.1038/4371098a

[9] D. V. Leff, P. C. Ohara, J. R. Heath and W. M. Gelbart, "Thermodynamic Control of Gold Nanocrystal Size: Experiment and Theory," *The Journal of Physical Chemistry*, Vol. 99, No. 18, 1995, pp. 7036-7041. doi:10.1021/j100018a041

[10] R. L. Whetten and W. M. Gelbart, "Nanocrystal Microemulsions: Surfactant-Stabilized Size and Shape," *The Journal of Physical Chemistry*, Vol. 98, No. 13, 1994, pp. 3544-3549. doi:10.1021/j100064a042

[11] N. R. Jana, "Shape Effect in Nanoparticle Self-Assembly," *Angewandte Chemie International Edition*, Vol. 43, No. 12, 2004, pp. 1536-1540.

[12] A. Gole and C. Murphy, "Seed-Mediated Synthesis of Gold Nanorods: Role of the Size and Nature of the Seed," *Journal of Materials Chemistry*, Vol. 16, No. 19, 2004, pp. 36633-36640. doi:10.1021/cm0492336

[13] G. Schmid, "Large Clusters and Colloids. Metals in the Embryonic State," *Chemical Reviews*, Vol. 92, No. 8, 1992, pp. 1709-1727. doi:10.1021/cr00016a002

[14] K. Esumi, A. Suzuki, N. Aihara, K. Usui and K. Torigoe, "Preparation of Gold Colloids with UV Irradiation Using Dendrimers as Stabilizer," *Langmuir*, Vol. 14, No. 12, 1998, pp. 3157-3159. doi:10.1021/la980162x

[15] S. Kapoor, "Preparation, Characterization, and Surface Modification of Silver Particles," *Langmuir*, Vol. 14, No. 5, 1998, pp. 1021-1025. doi:10.1021/la9705827

[16] J. Kimling, M. Maier, B. Okenve, V. Kotaidis, H. Ballot and A. Plech, "Turkevich Method for Gold Nanoparticle Synthesis Revisited," *The Journal of Physical Chemistry B*, Vol. 110, No. 32, 2006, pp. 15700-15707. doi:10.1021/jp061667w

[17] V. K. LaMer and R. H. Dinegar, "Theory, Production and Mechanism of Formation of Monodispersed Hydrosols," *Journal of the American Chemical Society*, Vol. 72, No. 11, 1950, pp. 4847-4854. doi:10.1021/ja01167a001

[18] M. K. Chow and C. F. Zukoski, "Gold Sol Formation Mechanisms: Role of Colloidal Stability," *Journal of Colloid and Interface Science*, Vol. 165, No. 1, 1994, pp. 97-109. doi:10.1006/jcis.1994.1210

[19] M. A. Watzky and R. G. Finke, "Transition Metal Nanocluster Formation Kinetic and Mechanistic Studies. A New Mechanism When Hydrogen Is the Reductant: Slow, Continuous Nucleation and Fast Autocatalytic Surface Growth," *Journal of the American Chemical Society*, Vol. 119, No. 43, 1997, pp. 10382-10400. doi:10.1021/ja9705102

[20] K. R. Brown and M. Natan, "Hydroxylamine Seeding of Colloidal Au Nanoparticles in Solution and on Surfaces," *Langmuir*, Vol. 14, No. 4, 1998, pp. 726-728. doi:10.1021/la970982u

[21] L. Cao, T. Zhu and Z. Liu, "Formation Mechanism of Nonspherical Gold Nanoparticles during Seeding Growth: Roles of Anion Adsorption and Reduction Rate," *Journal of Colloid and Interface Science*, Vol. 293, No. 1, 2006, pp. 69-76. doi:10.1016/j.jcis.2005.06.012

[22] A. Pongpeerapat, C. Wanawongthai, K. Moribe and K. Yamamoto, "Formation Mechanism of Colloidal Nanoparticles Obtained from Probucol/PVP/SDS Ternary Ground Mixture," *International Journal of Pharmaceutics*, Vol. 352, No. 1-2, 2008, pp. 309-316. doi:10.1016/j.ijpharm.2007.10.052

[23] M. Harada, K. Saijo and N. Sakamoto, "Characterization of Metal Nanoparticles Prepared by Photoreduction in Aqueous Solutions of Various Surfactants Using UV-Vis, EXAFS and SAXS," *Colloids and Surfaces A*, Vol. 349, No. 1-3, 2009, pp. 176-188. doi:10.1016/j.colsurfa.2009.08.015

[24] A. Cacciuto, S. Auer and D. Frenkel, "Onset of Heterogeneous Crystal Nucleation in Colloidal Suspensions," *Nature*, Vol. 428, No. 6981, 2004, pp. 404-406. doi:10.1038/nature02397

[25] H. Mühlpfordt, "The Preparation of Colloidal Gold Particles Using Tannic Acid as an Additional Reducing Agent," *Expirentia*, Vol. 38, No. 9, 1982, pp. 1127-1128. doi:10.1007/BF01955405

[26] G. Frens, "Controlled Nucleation for the Regulation of the Particle Size in Monodisperse Gold Suspensions," *Nature Physical Science*, Vol. 241, No. 105, 1973, pp. 20-22.

[27] A. I. Kirkland, P. P. Edward, D. A. Jefferson and D. G. Duff, "The Structure, Characterization, and Evolution of Colloidal Metals," *Annual Reports Section C (Physical Chemistry)*, Vol. 87, 1990, pp. 247-304. doi:10.1039/pc9908700247

[28] S. Link, C. Burda, Z. L. Wang and M. A. El-Sayed, "Electron Dynamics in Gold and Gold-Silver Alloy Nanoparticles: The Influence of a Nonequilibrium Electron Distribution and the Size Dependence of the Electron-Phonon Relaxation," *Journal of Chemical Physics*, Vol. 111, No. 3, 1999, pp. 1255-1264. doi:10.1063/1.479310

[29] J. L. Yao, G. P. Pan, K. H. Xue, D. Y. Wu, B. Ren, D. M. Sun, J. Tang, X. Xu and Z. Q. Tian, "A Complementary Study of Surface-Enhanced Raman Scattering and Metal Nanorod Arrays," *Pure and Applied Chemistry*, Vol. 72, No. 1-2, 2000, pp. 221-228. doi:10.1351/pac200072010221

[30] S. Link and M. A. El-Sayed, "Optical Properties and Ultrafast Dynamics in Metallic Nanocrystals," *Annual Review of Physical Chemistry*, Vol. 54, 2003, pp. 331-366. doi:10.1146/annurev.physchem.54.011002.103759

[31] G. Mie, "Simulation of the Colour Effects Connected with Colloidal Gold Particles," *Annals of Physics*, Vol. 25, No. 3, 1908, pp. 377-445. doi:10.1119/1.1621030

[32] C. L. Haynes and R. P. Van Duyne, "Nanosphere Lithography: A Versatile Nanofabrication Tool for Studies of Size-Dependent Nanoparticle Optics," *The Journal of Physical Chemistry B*, Vol. 105, No. 24, 2001, pp. 5599-

5611. doi:10.1021/jp010657m

[33] S. Link, M. B. Mohamed and M. A. El-Sayed, "Simulation of the Optical Absorption Spectra of Gold Nanorods as a Function of Their Aspect Ratio and the Effect of the Medium Dielectric Constant," *The Journal of Physical Chemistry B*, Vol. 103, No. 16, pp. 3073-3077. doi:10.1021/jp990183f

[34] R. Jin, Y. Cao, C. A. Mirkin, K. L. Kelly, G. C. Schatz and J. G. Zheng, "Photoinduced Conversion of Silver Nanospheres to Nanoprisms," *Science*, Vol. 294, No. 5548, 2001, pp. 1901-1903. doi:10.1126/science.1066541

[35] M. D. Malinsky, K. L. Kelly, G. Schatz and R. P. Van Duyne, "Chain Length Dependence and Sensing Capabilities of the Localized Surface Plasmon Resonance of Silver Nanoparticles Chemically Modified with Al Kanethiol Self-Assembled Monolayers," *Journal of the American Chemical Society*, Vol. 123, No. 7, 2001, pp. 1471-1482. doi:10.1021/ja003312a

[36] P. Mulvaney, "Surface Plasmon Spectroscopy of Nanosized Metal Particles," *Langmuir*, Vol. 12, No. 3, 1996, pp. 788-800. doi:10.1021/la9502711

[37] I. O. Sosa, C. Noguez and R. G. Barrera, "Optical Properties of Metal Nanoparticles with Arbitrary Shapes," *The Journal of Physical Chemistry B*, Vol. 107, No. 26, 2003, pp. 6269-6275. doi:10.1021/jp0274076

[38] M. M. Alvarez, J. T. Khoury, T. J. Schaaff, M. N. Shafigullin, I. Vezmar and R. L. Whetten, "Optical Absorption Spectra of Nanocrystal Gold Molecules," *Journal of Physical Chemistry*, Vol. 101, No. 12, 1997, pp. 3706-3719. doi:10.1021/jp962922n

[39] U. Kreibig and L. Genzel, "Optical Absorption of Small Metallic Particles," *Surface Science*, Vol. 156, No. 2-3, 1985, pp. 678-700. doi:10.1016/0039-6028(85)90239-0

[40] M. Tréguer-Delapierre, J. Majimel, S. Mornet, E. Duguet and S. Ravaine, "Synthesis of Non-Spherical Gold Nanoparticles," *Gold Bulls*, Vol. 41, No. 2, 2008, pp. 195-207. doi:10.1007/s13404-011-0037-2

[41] T. R. Jensen, R. P. van Duyne, S. A. Johnson and V. A. Maroni, "Surface-Enhanced Infrared Spectroscopy: A Comparison of Metal Island Films with Discrete and Non-discrete Surface Plasmons," *Applied Spectroscopy*, Vol. 54, No. 3, 2000, pp. 371-377.

[42] T. R. Jensen, M. D. Malinsky, C. L. Haynes and R. P. van Duyne, "Nanosphere Lithography: Tunable Localized Surface Plasmon Resonance Spectra of Silver Nanoparticles," *Journal of Physical Chemistry B*, Vol. 104, No. 45, 2000, pp. 10549-10565. doi:10.1021/jp002435e

[43] E. Hutter and J. H. Fendler, "Exploitation of Localized Surface Plasmon Resonance," *Advanced Materials*, Vol. 16, No. 19, 2004, pp. 1685-1709. doi:10.1002/adma.200400271

[44] B. K. Jena and C. R. Raj, "Synthesis of Flower-Like Gold Nanoparticles and Their Electrocatalytic Activity towards the Oxidation of Methanol and the Reduction of Oxygen," *Langmuir*, Vol. 23, No. 7, 2007, pp. 4064-4070. doi:10.1021/la063243z

[45] C. D. Chen, Y. T. Yeh and C. R. Wang, "The Fabrication and Photo-Induced Melting of Networked Gold Nanostructures and Twisted Gold Nanorods," *Journal of Physics and Chemistry of Solids*, Vol. 62, No. 9-10, 2001, pp. 1587-1597. doi:10.1016/S0022-3697(01)00098-1

[46] L. Pei, K. Mori and M. Adachi, "Formation Process of Two-Dimensional Networked Gold Nanowires by Citrate Reduction of $AuCl_4^-$ and the Shape Stabilization," *Langmuir*, Vol. 20, No. 18, 2004, pp. 7837-7843. doi:10.1021/la049262v

[47] S. Link and M. A. El-Sayed, "Spectral Properties and Relaxation Dynamics of Surface Plasmon Electronic Oscillations in Gold and Silver Nanodots and Nanorods," *Journal of Physical Chemistry B*, Vol. 103, No. 40, 1999, pp. 8410-8426. doi:10.1021/jp9917648

[48] B. K. Pong, H. I. Elim, J. X. Chong, B. L. Trout and J. Y. Lee, "New Insights on the Nanoparticle Growth Mechanism in the Citrate Reduction of Gold (III) Salt: Formation of the Au Nanowire Intermediate and Its Nonlinear Optical Properties," *Journal of Physical Chemistry C*, Vol. 111, No. 17, 2007, pp. 6281-6287. doi:10.1021/jp068666o

Inhibition Mechanism of Pitting Corrosion of Nickel in Aqueous Medium by Some Macrocyclic Compounds

Fatma Mohamed Mahgoub[1,2], Ahmed Mohamed Hefnawy[2]
[1]Department of Chemistry, Faculty of Science, King Abdaziz University, Jeddah, Saudi Arabia
[2]Department of Materials Science, Institute of Graduate Studies and Research, Alexandria University, Alexandria, Egypt
Email: ftm_mahgoub@yahoo.com

ABSTRACT

Anodic polarization of nickel was studied by potentiostatic technique in neutral media in presence of two macrocyclic ligands. Pit initiation was detected by measuring pitting potential, E_p and the charge transfer, Q during the anodic polarization. Initiation of pitting and Q were found to be dependent on the structure and concentrations of inhibitors. Under steady state conditions, the inhibition efficiency was in the order 1, 4, 8, 11 tetraazacyclotetradecane (*cyclam*) > N, N'-bis (2-aminoethyl) 1, 3-propandiamine (2,3,2-*tet*) due to macrocyclic effect. These findings were confirmed by fitting the data to different adsorption isotherms. Adsorption on nickel surface was obeyed kinetic-thermodynamic model for the two inhibitors. Influence of [Cl^-] on corrosion behavior of nickel in sulphate solution containing inhibitors was investigated. A mechanism involving three competitive equilibria of dissolution of nickel in presence and absence of the inhibitors was proposed. Empirical and theoretical kinetic equations were compared and discussed. There is a good agreement between the calculated and observed rate constants.

Keywords: Organic Inhibitors; Polarization; Pitting

1. Introduction

The importance of nickel as basic material for corrosion resistant alloys and its catalytic properties with respect to many organic reactions is well recognized [1]. Although nickel is known as corrosion resistant metal in many aggressive media, it corrodes at a significant rate in the presence of mineral acids or in neutral medium.

Increased resistance of passive metals to localized corrosion in aqueous-chloride containing solutions often can be obtained by addition of certain compounds at concentrations much lower than that of chloride ion [2]. Jha *et al.* studied the effect of dicyandiamide, biguanide, and guanylurea on corrosion of nickel in neutral medium [3-5]. They found that the presence of the inhibitors promote considerable inhibition at higher concentration as well as with increasing the temperature. The inhibition characteristics and quantum chemical data show that corrosion current depends mainly upon the energy of the Highest Occupied Molecular Orbital (HOMO) while Lowest Unoccupied Molecular Orbital (LUMO) gives an idea about electron affinity of the molecule. Polarization characteristics of Ni were measured in 0.5 M K_2SO_4 in the presence and absence of dimethylacrylamide (DMA) [6]. The changes in the composition of the surface of the electrode were examined by XPS which indicated a change in the chemical nature of the surface; the hydroxide layer is covered with an additive film (NiOOH/DMA). The high efficiency is due to the presence of two double bonds and a tertiary nitrogen atom. Effect of a number of benzimidazole derivatives in inhibiting the anodic dissolution of nickel in aqueous solutions of lithium perchlorate were made [7]. The effectiveness of inhibition depended on the electrode potential and the nature of the substituent in the organic molecule. The effect of the concentration and chemical nature of anions-activators, the pH and solution temperature on the inhibition process of pitting of nickel has been studied [8]. The relationship between the potential of pitting (E_P) and chemical nature of the anion was described quantitatively using the hydrophobic and electronic characteristics of anion. The existence of the limited value of (E_p), which does not depend on concentration but only on its chemical nature, was shown.

The mechanisms of inhibition of localized corrosion are quite difficult to establish in view of the complexity of the localized corrosion process coupled with all of the possible interactions of the inhibitor with surface or inter phase chemistry [9-12]. Literature survey showed little work appears to have been made on the use of chelating ligands as corrosion inhibitors for nickel, particularly in neutral medium. Multidentate ligands [13] making available two or more donor atoms for binding and hence are thought to have superior inhibition efficiency to monodentate ligands. The cyclic nature of these ligands is thought to enhance ligation [14-17]. Just as a chelating n-dentate ligand gives a more stable complex than unidentate ligand of a similar type, a phenomenon known as the chelate effect. So an n-dentate macrocyclic ligand gives even more stable complexes than the most similar n-dentate open chain ligand. This so called macrocyclic effect results from a favorable entropy change assisted, usually by favorable enthalpy change as well [18,19].

The purpose of the present study is to investigate the corrosion inhibition characteristics of the qundridenate macrocyclic ligand 1, 4, 8, 11, tetraaza-cyclotetradecane (*cyclam*) and its open chain analogous N, N'-bis (2-amino ethyl) 1, 3-propandiamine (2,3,2-*tet*) on nickel dissolution in neutral media. In addition, to study the kinetics and mechanism of pitting corrosion of nickel in presence of inhibitors.

2. Experimental

Nickel dissolution rate was determined using DC cyclic anodic polarization. Electrochemical polarization measurements were achieved by connecting the cell to Wenking potentiostat. The working electrode was nickel wire (BDH, 99.99%) with 1.25 mm in diameter. Before the measurements, the specimens being used were polished with a series of emery papers with different grades (320 - 600 - 1200 grit), starting with a coarse one and proceeding in steps to fine grades. The rods were washed thoroughly with distilled water and dried with ethanol. All the experiments were thermostated at room temperature. The sweeping rate used in this study was 1 mV/sec. Electrode potentials were measured vs. reference Saturated Calomel Electrode (SCE).

The macrocyclic inhibitors studied were N,N'-bis (2-aminoethyl)-1,3-propane diamine (2,3,2-*tet*), and 1,4,8, 11-tetra-azacyclo-tetradecane, (*cyclam*) were purchased from BDH and Aldrich chemical companies respectively and were used without further purification.

The inhibition efficiency of the compounds used in retarding the corrosion of nickel could be measured by evaluating the percentage inhibition, %*P*, from the following relationship:

$$\%P = 100\times\left(1-\frac{q_i}{q}\right) \quad (1)$$

where q and q_i are the charge transfer per unit area in the absence and the presence of inhibitor respectively.

$$q = \frac{Q}{A\cdot F}\ \text{electron - equivalent/cm}^2 \quad (2)$$

where A is the electrode area, F is Faraday constant and Q is the number of coulombs of electric charge consumed in the electrode process. Q is evaluated by integration of i-V $f(t)$ diagrams of the cyclic polarization according to the following equation:

$$Q = \int i.dt\ coulombs \quad (3)$$

3. Results and Discussion

Inhibition effects of macrocyclic ligands on pitting corrosion of nickel

The potentiodynamic scan began at –0.1 V vs SCE and extended to 1.0 V (vertex point), followed by a reverse scan back to a final potential of –0.1 V. **Figure 1** represents the cyclic polarization i-V *f*(t) curves of nickel in deareated 0.1 M KCl solution with and without different concentrations of the 2,3,2-*tet*. The effect of *cyclam* and 2,3,2-*tet* on the electrochemical parameters is manifested in **Table 1**. The degree of hysteresis indicates the tendency toward localized corrosion or pit formation. From these curves, it can be observed that the presence of 2,3,2-*tet* in the medium leads to improve the inhibition of pitting corrosion by shifting the critical pitting potential of initiation (E_p) towards more positive values. Besides, the hysteresis area became less and accompanied with a reduction of the integral charge transfer of the anodic dissolution process (q). Diminished area of hysteresis means that the protected nickel suffers fewer tendencies towards pitting corrosion and its corrosion resistance increases. Findings reveal also that Ep shifts to more positive values and enhancing the concentration of *cyclam* and 2,3,2-*tet* in the medium leads the quantity of charge required for passivation gradually decreased suggesting that the adsorption of the inhibitors molecules on nickel surface inhibits the dissolution of nickel.

Figure 2 shows the effect of 2 × 10^{-3} M of 2,3,2-*tet* and *cyclam* on anodic polarization cycle in chloride solution. **Figure 3** shows the variation of the percentage inhibition (%*P*) as a function of the logarithm of molar concentration of the two inhibitors. The role of the inhibitor on passivating metals is to repair the breakdowns in the oxide film which are assumed to occur mainly at

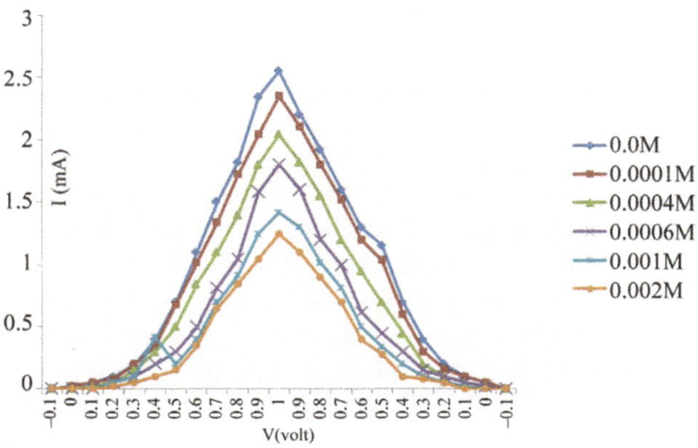

Figure 1. Effect of increasing the concentration of 2,3,2-*tet* on polarization cycle in 0.1 M KCl.

Table 1. Electrochemical parameters obtained from D.C. polarization measurements of nickel in 0.1 M KCl.

Inhibitor	Cyclam			2,3,2-*tet*		
Conc. M	$q \times 10^{-5}$ electron equiv./cm^2	E_p V	%P	$q \times 10^{-5}$ electron equiv./cm^2	E_p V	%P
0.0	1.72	0.00	-	1.72	0.00	-
1×10^{-4} M	1.48	0.09	13.7	1.59	0.05	7.47
4×10^{-4} M	1.12	0.14	34.8	1.31	0.07	23.65
6×10^{-4} M	0.82	0.20	52.2	1.01	0.10	41.20
1×10^{-3} M	0.65	0.31	62.5	0.81	0.18	52.95
2×10^{-3} M	0.55	0.40	68.2	0.68	0.21	60.30

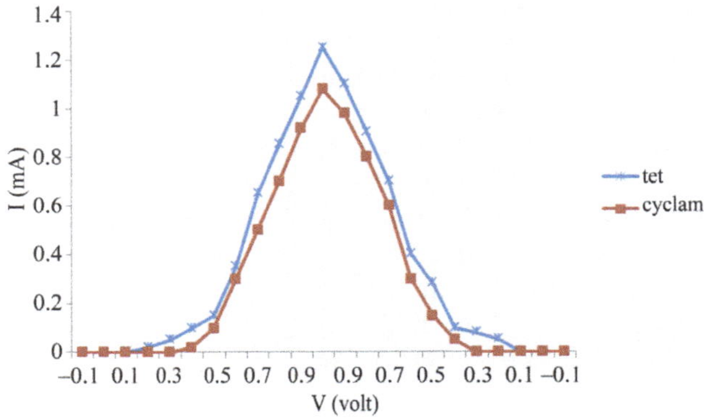

Figure 2. Effect of 0.002 M inhibitors on anodic polarization cycle in 0.1 M KCl.

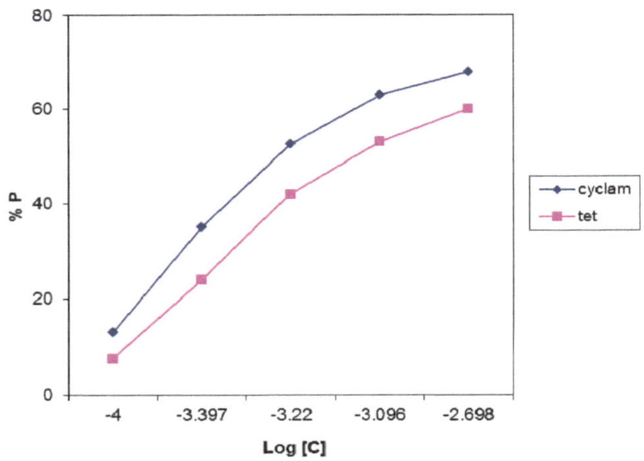

Figure 3. Relationship between percentage inhibition and log concentration of inhibitors.

the inclusion sites or grain boundaries [20] or to block the oxide surface. **Figures 2** and **3** show that the inhibition of *cyclam* is more efficient than 2,3,2-*tet*. These could be attributed to the macrocyclic effect where *cyclam* is a cyclic compound while 2,3,2-*tet* is an open chain one. The cyclic nature of these ligands is thought to enhance ligation. Besides, macrocyclic ligands are known to form stable complexes with metals in their zero oxidation state [19] or in this study as nickel hydroxide/inhibitor complex.

Kinetic-thermodynamic model and adsorption isotherms analyses

The data obtained from DC measurements have been fitted to the kinetic thermodynamic model in terms of active site occupancy parameter, y, and binding constant of the inhibitor with the metallic surface, K. The obtained results are compared with fits executed from the application of Langmuir, Frumkin and Flory-Huggins adsorption isotherms.

The curve fitting for 2,3,2-*tet* and *cyclam* data to the Langmuir isotherm (Equation (4)), Frumkin isotherm (Equation (5)) and Flory-Huggins isotherm (Equation (6)) are given in **Figures 4(a)-(c)**, respectively. The results indicate that the data does not fit any of the above isotherms.

$$\frac{\theta}{1-\theta} = KC \quad (4)$$

$$\left[\frac{\theta}{1-\theta}\right]\exp\left[-2a\theta\right] = KC \quad (5)$$

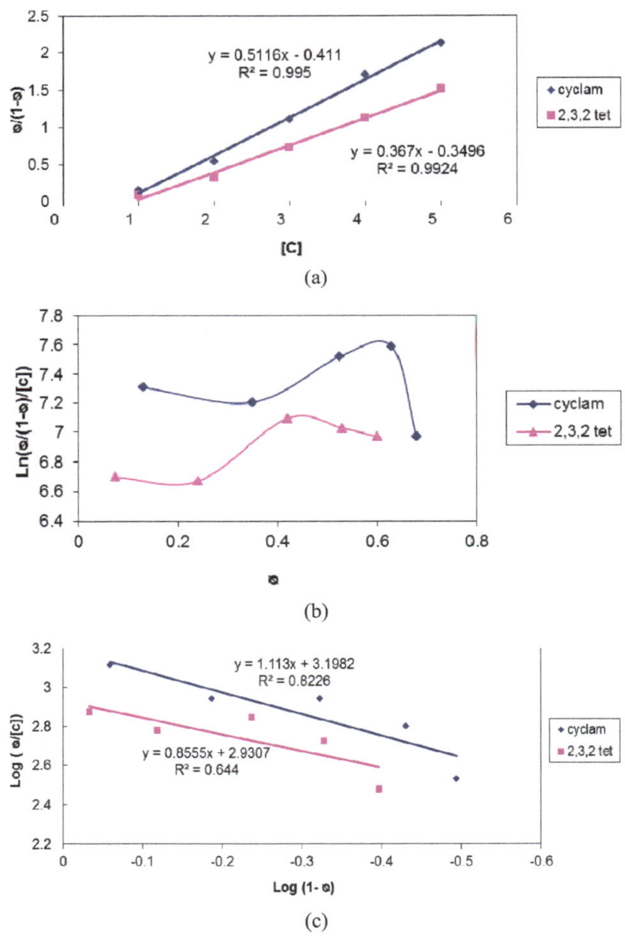

Figure 4. (a) Test of Langmuir adsorption isotherm; (b) Test of Frumkin adsorption isotherm; (c) Test of Flory-Huggins adsorption isotherm.

$$\frac{\theta}{x\,(1-\theta)^{x}} = KC \quad (6)$$

where θ is surface coverage, C is the inhibitor concentration, a is a molecular interaction parameter depending on the molecular interactions in the adsorption layer and on the degree of heterogeneity of the surface, and x is the size ratio and is simply the number of water molecules replaced by one molecule of organic adsorbate.

The curve fitting of the obtained data using *cyclam* and 2,3,2-*tet* to kinetic-thermodynamic model (Equation (7)) is given in **Figure 5**.

$$\log\frac{\theta}{1-\theta} = \log K' + y\log C \quad (7)$$

$$K = K'^{(1/y)} \quad (8)$$

As can be seen, good straight lines fit are given for the compounds used. The slope of the lines is representing y, the number of inhibitor molecules occupying a single active site. The intercept is $\log K'$. $(1/y)$ gives the number of active sites occupied by a single inhibitor molecule. **Table 2** gives the adsorption parameters obtained from Kinetic-thermodynamic model using *cyclam* and 2,3,2-*tet*. The table displays that the value of K increases from 931 to 1442 for 2,3,2-*tet* and *cyclam* respectively.

This clearly indicates that the strength of electrical interactions by adsorbing molecules and surface of the metal increases in the same order noticed previously and hence the percentage inhibition. The results show also that $(1/y)$ values are approximately equal to the unity indicating that the adsorption process takes place by the occupation of one active site per a single inhibitor molecule. ΔG^{o}_{ads} was –ve values mean spontaneous adsorption of inhibitor on the metal surface.

Kinetics study of pitting corrosion of nickel in Na_2SO_4 media containing Cl^- ions

Figure 6 shows the anodic potentiodynamic polarization curves of nickel in 1.0 M Na_2SO_4 containing different concentrations of KCl. Inspection of this Figure shows that pitting potential was shift to the less positive potential direction with increasing $[Cl^-]$. MacDougall [21] discussed this behavior on the basis that E_p becomes progressively easier to initiate pitting at higher $[Cl^-]$. The more Cl^- ion in solution leads to the larger active areathat can be generated and the lower the anodic overvoltage for a particular charging rate. The Figure also indicates that after a period of continuous dissolution prior to about 1.0 volt, all anodic polarization curves are characterized by a limiting current. The limiting current value increases with the increase of the concentration of Cl^- ions due to the ability of Cl^- to interfere with oxide formation and thereby decrease the repair efficiency. This causes further metal dissolution and even more Cl^- migration into the breakdown area [21]. The limiting current values could be considered as rate constant, k_0, in the absence of $[Cl^-]$ and as k_{obs} in the presence of different concentrations of Cl^- ion.

Figure 7 represents a plot of $(k_{obs} - k_o)$ vs $[Cl^-]$ concentrations, the Figure confirms the following rate expression [17,23].

$$(k_{obs} - k_o) = \frac{a[Cl^-]^2}{b + c[Cl^-]} \quad (9)$$

Table 2. Binding constant, number of active sites and change of free energy obtained from kinetic-thermodynamic model for 2,3,2-tet and cyclam.

Inhibitor	K	1/y	ΔG^{o}_{ads} (kJ/mol)
2,3,2-*tet*	931	0.97	–26.4
Cyclam	1442	1.00	–27.5

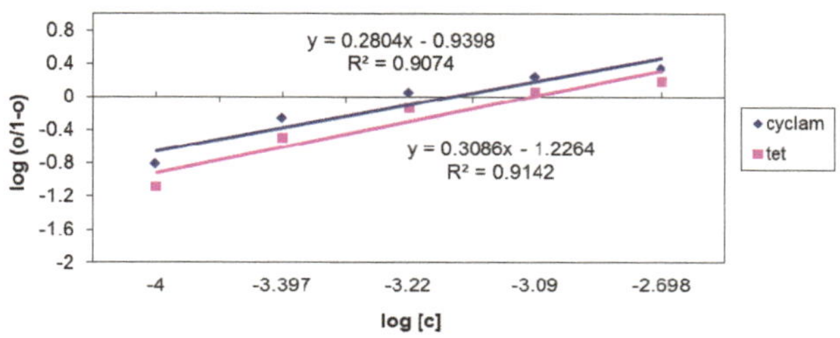

Figure 5. Test of kinetic-thermodynamic model.

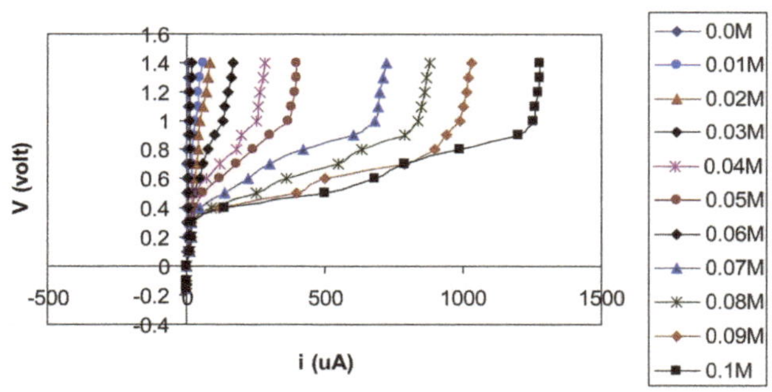

Figure 6. Anodic potentiodynamic polarization curves of nickel in 1.0 M Na_2SO_4 containing different concentrations of KCl.

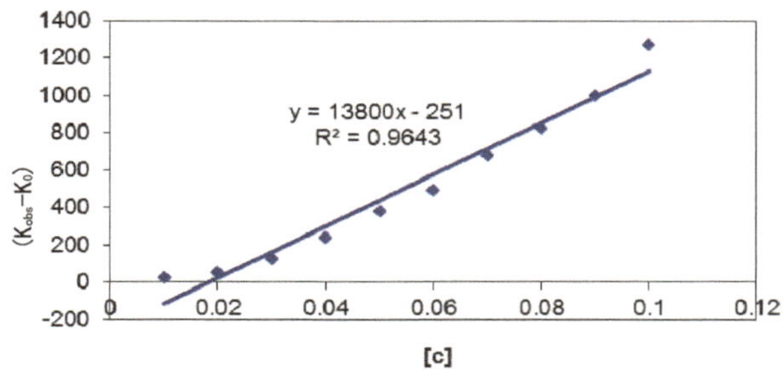

Figure 7. Plot of $(k_{obs} - k_o)$ vs $[Cl^-]$ for corrosion of nickel in 1.0 M Na_2SO_4.

Equation (9) refers to a number of possible mechanisms. These mechanisms could involve competitive anion adsorption, anion migration through the passive film, competition between metal dissolution and film repair or localized attack at preexisting flaws in the passive film. In other words, Cl^- ion is expected to enhance nickel dissolution probably through preferential adsorption of Cl^- on the Ni surface or it may delay passive film formation, probably via an interaction with soluble intermediate compound. Cl^- ion also could attack the precipitated oxide leading to damage its protective nature. In the pH range of the electrolyte considered in this study and according to the theoretical equilibrium diagram the main stable corrosion product in the passive film is expected to be $Ni_2O_3 \cdot H_2O$. Employing differential reflectometry in conjugation with ESCA, it has been proved that $Ni(OH)_2$ film is mainly observed on pure Ni in 0.15 N·Na_2SO_4 [22].

The proposed a mechanism in which Cl^- ions are adsorbed on the metal surface in competition with $(H_2O)_{ads}$, followed by a rate determined dissolution step (R.D.S) that is catalyzed by Cl^- ions.

$$Ni + H_2O \overset{K_1}{\Leftrightarrow} Ni(H_2O)_{ads}$$

$$Ni + Cl^- \overset{K_2}{\Leftrightarrow} Ni(Cl^-)_{ads}$$

$$Ni(Cl^-)_{ads} \Leftrightarrow NiCl + e^-$$

$$NiCl + Cl^- \underset{R.D.S}{\overset{k}{\Leftrightarrow}} NiCl_2\,(aq.) + e^-$$

or

$$Ni(Cl^-)_{ads} + Cl^- \underset{R.D.S}{\overset{k}{\Leftrightarrow}} NiCl_2 + 2e^-$$

This mechanism is similar to the mechanism of dissolution of Al in HCl [23] and leads to an observed rate constant dependence on Cl^- of the form:

$$k_{obs} = \frac{kK_2[Cl^-]^2}{1 + K_1(H_2O) + K_2[Cl^-]} \tag{10}$$

Under the conditions $K_1(H_2O); K_2[Cl^-] \gg 1$, Equation (10) reduces to:

$$k_{obs} = \frac{kK_2[Cl^-]^2}{1 + K_1(H_2O) + K_2[Cl^-]} \tag{11}$$

This condition is equivalent to saying that the surface of the metal is totally covered by either adsorbed H_2O or Cl^-. The inverse form of Equation (11) gives:

$$\frac{[Cl^-]}{k_{obs}} = \frac{1}{k} + \frac{K_1(H_2O)}{kK_2[Cl^-]} \tag{12}$$

A plot of the left-hand side of Equation (12) versus $1/[Cl^-]$ gave a good straight line (**Figure 8**). Employing the liner least-square fit of the data, the equilibrium constant, *k*, becomes 0.0519.

Kinetics of the inhibition of the pitting corrosion of nickel

Anodic potentiodynamic polarization curves of nickel in 1.0 M Na_2SO_4 + 0.05 M KCl solutions containing different concentrations of 2,3,2-*tet* and *cyclam* are represented in **Figures 9** and **10** respectively. Inspections of these Figures show that all potential-current curves are characterized by limiting current phenomena. It is observed that the limiting current decrease with the increase of the concentrations of the two inhibitors in the medium and *cyclam* offer lowering limiting current than 2,3,2-*tet*. These results confirm the previous data obtained from the percentage inhibition versus the logarithm of the concentration of the inhibitors in **Figure 3**.

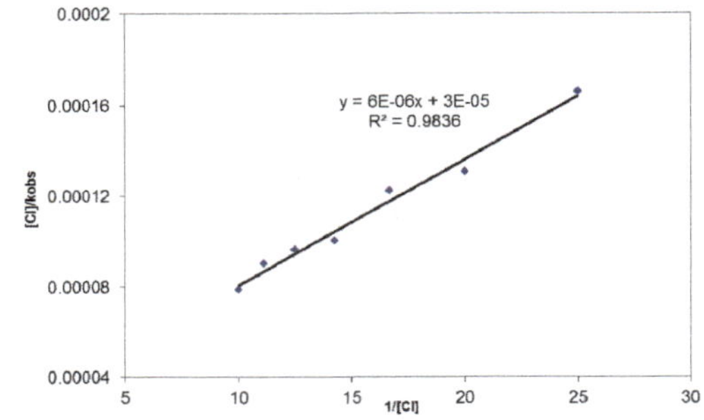

Figure 8. Plot of $[Cl^-]/k_{obs}$ vs $1/[Cl^-]$ for the Cl^- ion catalysis of nickel in 1.0 M Na_2SO_4.

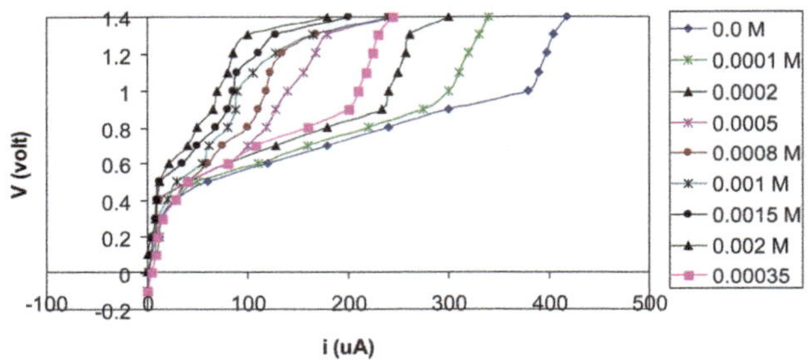

Figure 9. Anodic potentiodynamic polarization curves of nickel in 1.0 M Na_2SO_4 + 0.05 M KCl containing different concenrtrations of 2,3,2-*tet*.

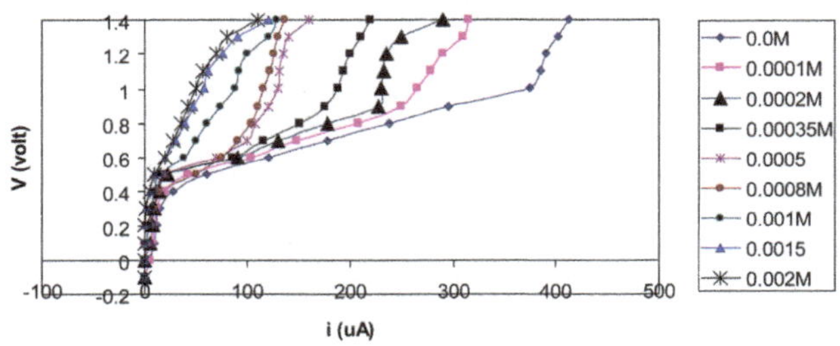

Figure 10. Anodic potentiodynamic polarization curves of nickel in 1.0 M Na_2SO_4 + 0.05 M KCl containing different concenrtrations of *cyclam*.

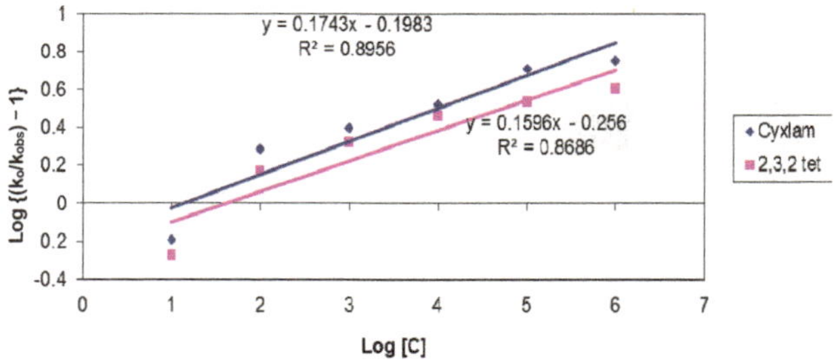

Figure 11. Plot of log[(k_o/k_{obs}) – 1] vs log [*I*] for corrosion of nickel in the presence of Cyclam and 2,3,2-*tet*.

The limiting dissolution current can be considered as k_o, k_{obs} in the absence and in the presence of inhibitors. The contribution of the inhibitor molecule, I, in the mechanism of dissolution of nickel in chloride media can be proposed as in the case of dissolution of aluminum in inhibited HCl solution [23], and can be given as:

$$Ni + H_2O \overset{K_1}{\Leftrightarrow} Ni(H_2O)_{ads}$$

$$Ni + Cl^- \overset{K_2}{\Leftrightarrow} Ni(Cl^-)_{ads}$$

$$Ni + yI \overset{K_3}{\Leftrightarrow} NiI_{y.ads}$$

$$Ni\left(Cl^-_{ads}\right) + Cl^- \underset{R.D.S}{\overset{k}{\Leftrightarrow}} NiCl_2 + 2e^-$$

This mechanism leads to the following rate equation:

$$\log\left(\frac{k_o}{k_{obs}} - 1\right) = \log\frac{K_3}{K_1[H_2O] + K_2\left[Cl^-\right]} + y\log[I] \quad (13)$$

Figure 11 represents a plot of $\log(k_o/k_{obs} - 1)$ vs. log [I] for *cyclam* and 2,3,2-*tet*. The plot gave a straight line with slope of *y* and an intercept of log ($K_3/K_2[Cl^-]$). The calculated values of $1/y$ (active site occupancy parameter or number of surface adsorbed water molecules replaced by an inhibitor molecule) obtained from a least-square fit are 1.05 and 1.18 for *cyclam* and 2,3,2-*tet* respectively which could be considered almost the same as in 0.1 M KCl solution. Of particular importance, however, is the ratio of the equilibrium constants for *cyclam* to that of 2,3,2-*tet* [K_3(*cyclam*)/K_3(2,3,2 *tet*)] which could be considered as a measure of the inhibition effi- ciency of the two ligands. The ratio is 2.3 indicating that *cyclam* is more efficient than 2,3,2-*tet* due to the macro- cyclic effect.

4. Conclusion

The results appear to demonstrate that adsorption-type organic inhibitors might be of interest in reducing the pitting corrosion of nickel. The corrosion inhibition of the qundridentate macrocyclic ligand (*cyclam*) is more efficient than its open chain analogous (2,3,2-*tet*) on nickel dissolution in neutral media.

REFERENCES

[1] E. Khamis, F. Bellucci, R. M. Latanision and E. S. H. El-Ashry, "Acid Corrosion Inhibition of Nickel by 2-(Triphenosphoranylidene) Succinic Anhydride," *Corrosion*, Vol. 47, No. 9, 1991, pp. 677-686. doi:10.5006/1.3585307

[2] J. Horvath, T. M. Salem, B. A. Abd-EL-Nabey, N. E. Khalil, L. Hackl and A. Rauscher, "Inhibition of the Pitting Corrosion of Stainless Steel in Presence of Cholride and Hydrogen Sulphide," *The 41th European Symposium on Corrosion Inhibitors*, Ferrara, 15-19 September 1975, p. 743.

[3] L. Jha, G. R. Parida, G. Kaur, and G. Singh, "Corrosion Inhibition of Nickel by Dicyandiamide in Neutral Medium," *Journal of the Electrochemical Society of India*, Vol. 39, No. 2, 1990, pp. 67-70.

[4] G. Singh, L. Jha, G. R. Parida and G. Kaur, "Study of Corrosion Inhibition Characteristics of Biguanide for Nickel in Neutral Medium," *Bulletin of Electrochemistry*, Vol. 6, No. 9, 1990, pp. 754-756.

[5] G. Singh, L. Jha, G. Kaur and G. R. Parida, "Corrosion Inhibition of Nickel by Guanylurea in Neutral Medium," *Transactions of the Society for Advancement of Electrochemical Science and Technology*, Vol. 25, No. 1-2, 1990, pp. 43-49.

[6] L. J. Jha, *Indian Journal of Science and Technology*, Vol. 28, 1990, p. 636.

[7] V. P. Grigoriev, G. N. Ekilik and E. N. Balakshina, *The 6th European Symposium on Corrosion Inhibitors*, Ferrara, 16-20 September 1985, p. 335.

[8] Y. I Kuznetsov and O. A. Luk'yanchikov, *Zashchita Metallov*, Vol. 24, 1988, p. 241.

[9] B. Bavarian, A. Moccari and D. D. Macdonald, "Inhibition of Stress Corrosion Cracking of Type 403 Stainless Steel in Sodium Sulfate Solutions," *Corrosion*, Vol. 38, No. 22, 1982, pp. 104-116.

[10] D. W. Deberry, G. R. Peyton and W. S. Clark, "Evaluation of Corrosion Inhibitors in SO_2 Scrubber Solutions," *Corrosion*, Vol. 40, No. 5, 1984, pp. 250-256. doi:10.5006/1.3581950

[11] A. Viehbeck and D. W. Deberry, "Inhibition of Localized Corrosion as Determined by a Galvanodynamic Method," *Journal of the Electrochemical Society*, Vol. 131, No. 8, 1984, pp. 1844-1846. doi:10.1149/1.2115973

[12] D. W. Deberry and A. Viehbeck, "Inhibition of Pitting of Type 304L Stainless Steel by N-Lauroylsarcosine," *Journal of The Electrochemical Society*, Vol. 133, No. 1, 1986, pp. 30-37. doi:10.1149/1.2108536

[13] D. W. Deberry and A. Viehbeck, "Inhibition of Pitting Corrosion of AISI 304L Stainless Steel by Surface Active Compounds," *Corrosion*, Vol. 44, No. 5, 1988, pp. 299-305. doi:10.5006/1.3583941

[14] B. A. Abd-El-Nabey, M. A. Khalifa , E. Khamis, A. F. Hefny and A. R. Michael, "Macrocyclic Effects in Corrosion Inhibition," *Corrosion Prevention and Control*, Vol. 35, 1988, p. 154.

[15] A. A. El-Awady, B. A. Abd-El-Nabey, S. G. Aziz, M. Khalifa, H. A. Al-Ghamidy and A. R. Michael, "Kinetics and

Thermodynamics of the inhibition of the Acid Corrosion of Steel by Some Macrocyclic Ligands," *International Journal of Chemical Kinetics*, Vol. 35, 1990, p. 169.

[16] B. A. Abd-El-Nabey and A. A. Samarkandy, "Inhibition of the Pitting Corrosion of Nickel by Macrocyclic Ligands," *International Journal of Chemistry*, Vol. 3, No. 2, 1991, p. 111.

[17] A. El-Awady, B. A. Abd-El Naby and S. G. Aziz, "Kinetic—Thermodynamic and Adsorption Isotherms Analyses for the Inhibition of the Acid Corrosion of Steel by Cyclic and Open—Chain Amines," *Journal of The Electrochemical Society*, Vol. 139, No. 8, 1992, pp. 2149-2154. doi:10.1149/1.2221193

[18] F. A. Cotton and G. Wilkinson, "Advanced Inorganic Chemistry," 4th Edition, Interscience Publication, New York, 1980.

[19] F. P. Hipz and D. W. Marerum, "Effect of Ligand Solvation on the Stability of Metal Complexes in Solution. Explanation of the Macrocyclic Effect," *Journal of the American Chemical Society*, Vol. 96, No. 15, 1974, pp. 4993-4994. doi:10.1021/ja00822a050

[20] G. Pallos and G. Wallwork, "Inhibition of Pitting Corrosion of Mild Steel in Neutral Solutions," *Corrosion*, Vol. 38, No. 6, 1982, pp. 305-313. doi:10.5006/1.3621690

[21] B. MacDougall and M. J. Graham, "Pitting of Nickel during Anodic Galvanostatic Charging in Na_2SO_4 Solutions," *Electrochemical Acta*, Vol. 27, No. 8, 1982, pp. 1093-1096. doi:10.1016/0013-4686(82)80115-1

[22] R. E. Hummel, R. J. Smith and E. D. Verink Jr, "The Passivation of Nickel in Aqueous Solutions I—The Identification of Insoluble Corrosion Products on Nickel Electrodes Using Optical and ESCA Techniques," *Corrosion Science*, Vol. 27, No. 8, 1987, pp. 803-813. doi:10.1016/0010-938X(87)90038-2

[23] A. A. El-Awady, B. A. Abd-El-Nabey and S. C Aziz, "Thermodynamic and Kinetic Factors in Ion Pitting and Nitrogen Donor Ligand Inhibition of Aluminium Metal Corrosion in Aggressive Acid Media," *Journal of the Chemical Society*, Vol. 89, 1993, pp. 795-802.

30

Surface Tension and Foam Stability Prediction of Polydimethylsiloxane-Polyol Systems

Oscar Arciniega Saavedra, Jesús Gracia Fadrique
Departamento de Fisicoquímica, Facultad de Química, Universidad Nacional Autónoma de México, México City, México
Email: jgraciaf@unam.mx

ABSTRACT

The Gibbs elasticity modulus represents an important tool to predict the foamability for transient and permanent foams like polyurethane flexible systems. Elasticity is related to foamability and so is used as a synonymous for the purpose of this paper. In this article we propose a method and a thermodynamic model to analyze the espumability of silicone surfactants in polyol binary mixtures using surface tension data. The present work describes foamability through the Gibbs elasticity modulus expressed in terms of first and second derivatives of surface pressure vs bulk composition. Furthermore, the Gibbs adsorption equation and the corresponding novel surface equation of state based on a modification of the Langmuir isotherm resulted in an elasticity equation with analytical solution. It is shown that according to foam model systems of surfactant solution in polyol used at commercial processes, optimum concentration level of surfactant obtained at this article by Gibbs adsorption equation and maximus on elasticity modulus finally match.

Keywords: Gibbs Elasticity Modulus; Foamability; Langmuir Isotherm; Gibbs Adsorption Equation

1. Introduction

Polyurethane foams are made by mixing an isocyanate and a polyol component [1]. Normally the polyol stream contains additives to achieve the end properties, among these additives are blowing agents, catalysts and surfactants. Silicone surfactant plays an important role in flexible polyurethane foam through decrement of the surface tension. Copolymers of a polydimethylsiloxane are well known and used as surfactants in polyurethane industry. Foam behaviour or foamability is a complex phenomena related with the composition and chemical structure of the amphiphile and depends on many factors such as type of the foam films, surface forces, disjoining pressure, temperature, kinetics of thinning and film rupture, surface and bulk rheology [2,3]. The mechanical properties of the cured polyurethane foam as air permeability and cell size are affected significantly by the structure of the silicone surfactant used in the formulation. Surfactants based Polydimethylsiloxane (PDMS) help with nucleation and in stabilizing foaming mixtures. *G. Besmans*, *L. Colman* and *R. Vandensande* used the maximum bubble pressure method to measure dynamic behavior of these silicone surfactants (PDMS) in the viscous polyol medium, a method which allowed fast screening of novel surfactants [4], that has been proved for measuring the dynamic response of surfactants in aqueous media. According to *G. Besmans*, *L. Colman* and *R. Vandensande*, dynamic surface tension response was measured in methyl capped polyethylene glycol PEG (MW 500 from ICI) using a surfactant loading of 0.1% by weight. Surfactants were obtained from Goldschmidt AG, Osi, and Air Products.

Commercially polydimethylsiloxane based surfactants are of a polymeric nature with a natural spread of molecular weight. The exact structure is part of the proprietary knowledge of the surfactant suppliers (**Figure 1**), but can be described with the formula where D is dimethylsiloxane and D is methylsiloxane with polyoxyalkylene, susbstitute R is composed of polyethylene oxide and polypropylene oxide [4].

2. Materials and Methods

Analytical grade nonyl phenol ethoxylates were purchased from Sigma and used with no further purification.

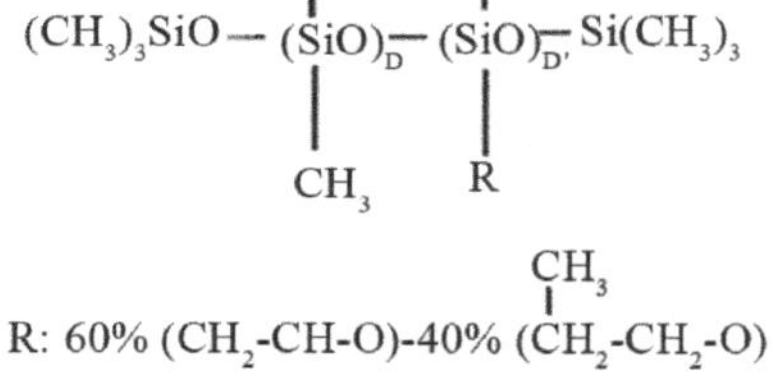

Figure 1. The structure of a typical surfactant used in polyurethane foam systems.

Silicone Tegostab BF-2370 provided by Goldschmidt AG (**Figure 2**), was used as surfactant with a molecular mass of 8800 g/mol [5]. The polyol polypropylene glycol was supplied by Sigma Aldrich, molecular mass is 2700 g/mol. Pluronic L61 (BASF) is a block copolymer, the average molecular mass is 2000 g/mol, consisting of 10% ethylene oxide and 90% propylene oxide.

Surfactant-polyol mixtures were prepared at controlled temperature of 30°C. The samples were placed in a thermostated vessel during the surface tension measurements and the temperature was regulated (within ± 0.1°C).

Surface tension methods

A dynamic method for surface tension was chosen, the maximum bubble pressure method displayes surface tension based on instantaneous bubble formation [6-8]. This is a dynamic method selected for surface tension measurements due to polyurethane foam formulation taking seconds during its production.

Surface tension vs composition was measured using QC3000 SensaDyne Surface Tensiometer (within ±0.1 dyne/cm) fitted with a bath (Haake K20/DC30) allowing to control the temperature (within ±0.1°C). An inert process gas (nitrogen or dry air) is bubbled slowly through two probes of different radio that are immersed in a test fluid. The bubbling of the nitrogen through the probes produces a differential pressure signal (ΔP) which is directly related to the fluid surface tension (σ).

The Young-Laplace equation rules the relation among curvature, surface energy and pressure difference between two phases; it has been used to describe spherical and non spherical shapes either in the absence or under the influenece of an external field. The general expression for the mechanical equilibrium between phases separated by spherical surface is

$$\Delta P = \frac{2\sigma}{r} \qquad (1)$$

where r is the sphere radius and ΔP is the difference in the pressure between phases. The difference in the pressure P_2 at the large probe from the pressure P_1 at the small probe results in a differential pressure equation keeping the two probes at the approximate same immersion depth cancelling the effects of liquid level (**Figure 3**), and the additional variation in the hydrostatic pressure caused by the gravitational field effect.

$$\Delta P = P_1 - P_2 = \left(\rho g h + \frac{2\sigma}{r_1}\right) - \left(\rho g h + \frac{2\sigma}{r_2}\right) = \frac{2\sigma}{r_1} - \frac{2\sigma}{r_2} \qquad (2)$$

Foam formation test

Foam was produced using a glass column of 1000 ml fitted at the bottom with a porous glass disk. Samples of the polydimethylsiloxane dispersions were carefully pured into the column and foam was produced at a constant temperature of 30°C by passing gas trough the porous glass (pore diameter 0.2 μm) at a controlled flux between 20 and 60 ml/min during ten minutes (**Figure 4**). Surfactant concentration ranged between 0.05 and 1.2 wt%. Results were expressed as foamability or the volume produced by a weight fraction of surfactant in polyol.

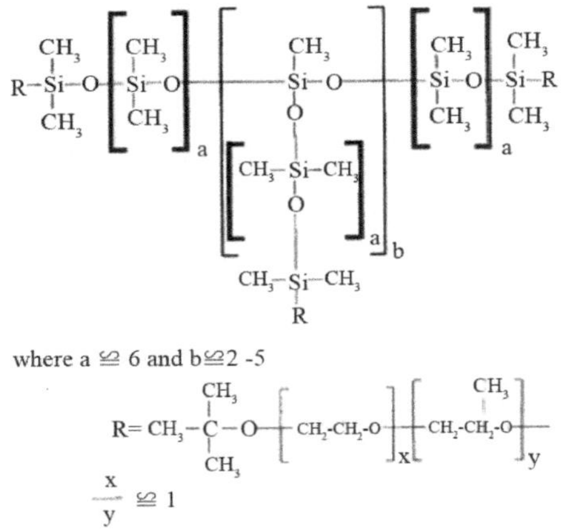

Figure 2. Structure of tegostab BF 2370.

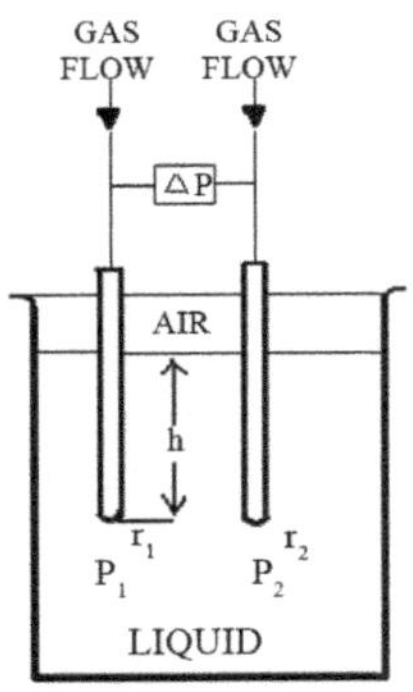

Figure 3. Differential pressure schematic.

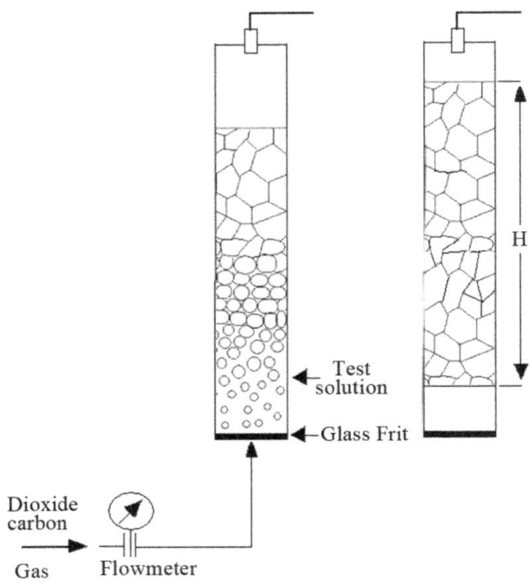

Figure 4. Schematics of the foam column test.

3. Results and discussion

Although surface properties for both kinds of polyols were similar (**Figure 5**), foaming behaviour presented differences (**Figure 6**). The foaming behaviour increased with concentration, with a maximum around the critical concentration and then decreased; an expected behaviour for transient foams.

Some works have been published describing methods to calculate equilibrium constants of molecular complexes in aqueous solution or to predict activity coefficients at infinite dilution from surface tension data [9, 10]. In this line it has been recently developed a model [11], able to describe the dependence of the surface tension on the concentration for a wide variety of solutions with different behavior (**Figures 7** and **8**), the model has been named Langmuir modified [12], that can be expressed as

$$\pi = \Gamma_s RT \frac{\beta x}{1+\beta x} \tag{3}$$

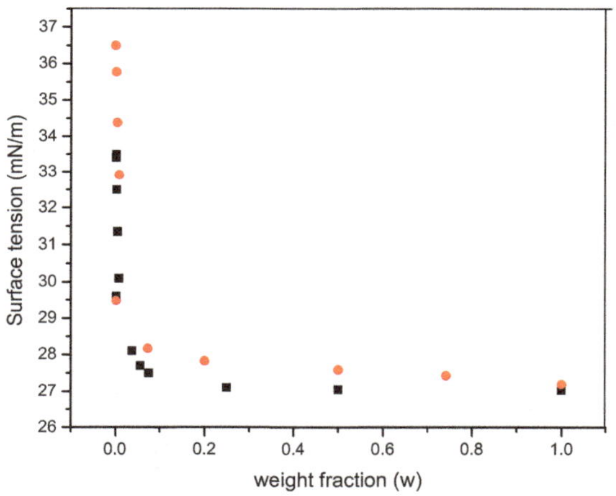

Figure 5. Surface tension at 30°C of PDMS at (•) Polypropylene glycol and (■) Pluronic L61.

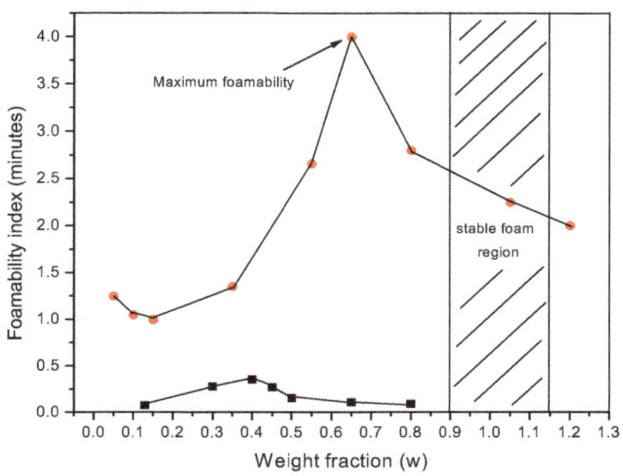

Figure 6. Foamability at 30°C of PDMS at (•) Polypropylene glycol and (■) Pluronic L61.

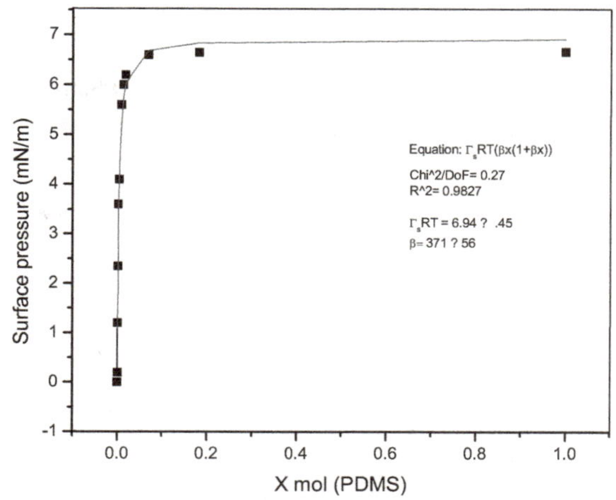

Figure 7. Langmuir modified. Pluronic L61-PDMS system.

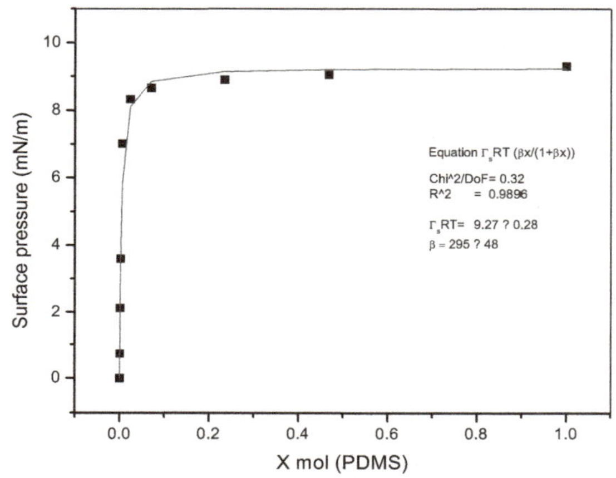

Figure 8. Langmuir modified. Polypropylene glycol-PDMS system.

where π is surface pressure, Γ is the excess surface concentration and is equall to the reciprocal area, Γ_s is the maximum surface concentration, R is the universal gas constant, T is the absolute temperature and β is a measure of the lyophobicity, so, based on the Langmuir isotherm, where θ is the surface coverage, $\theta = \Gamma / \Gamma_s$;

$$\theta = \frac{\beta x}{1+\beta x} \tag{4}$$

The Gibbs adsorption isotherm allows us to transform an isotherm to a surface equation of state [13]. The Langmuir isotherm in fluid-fluid interfaces, coupled with the Gibbs equation, leads to;

$$\theta = \frac{x}{\Gamma_s RT}\left(\frac{d\pi}{dx}\right) \tag{5}$$

The combination of Equations (4) and (5) and integral form leads to the corresponding surface equation of state SEOS

$$\pi = \Gamma_s RT \ln(1+\beta x) \tag{6}$$

from the Lagmuir-Frumkin equation;

$$\pi = -\Gamma_s RT \ln(1-\theta) \quad (7)$$

Equation (7) does not include atractive effects, so is proposed a general function $f(\theta)$;

$$\pi = -\Gamma_s RT \ln(1-\theta) - f(\theta) \quad (8)$$

That can be expressed as series expansion;

$$\pi = \Gamma_s RT\left(\theta + \frac{\theta^2}{2} + \frac{\theta^3}{3} + \frac{\theta^4}{4} + \cdots\right) - \Gamma_s RT\left(\frac{\theta^2}{2} + \frac{\theta^3}{3} + \frac{\theta^4}{4} + \cdots\right) \quad (9)$$

Obtaining the final Equation (3) to which the experimental data are fitted;

$$\pi = \Gamma_s RT \frac{\beta x}{1+\beta x}$$

Polidimethylsiloxanes form transient foams in polyols, meaning that the foam is present as long as the gas flux is present and depended on concentration. Pluronic L61 showed lower foamability with bubbles breaking up as soon as the gas flux was interrupted. On the other hand Polypropylene glycol showed better foamability with a maximum that coincides with the critical concentration just as predicted for transient foams.

Gibbs elasticity modulus is closely related to foamability and is expressed as;

$$\varepsilon = 2\frac{d\sigma}{d\ln A} \quad (10)$$

Γ is the surface concentration and is equal to the reciprocal of the molecular area;

$$A = \frac{1}{\Gamma} \quad (11)$$

leads to the equation;

$$\varepsilon = -2\Gamma\frac{d\sigma}{d\Gamma} \quad (12)$$

Surface tension expressed as surface pressure;

$$d\sigma = -d\pi \quad (13)$$

Expressions (12) and (13) can be substituted in Equation (10) providing;

$$\varepsilon = 2\Gamma\frac{d\pi}{d\Gamma} \quad (14)$$

Surface pressure can be used instead of surface tension and the associated area changed for specific area, thus

$$\varepsilon = 2\theta\frac{d\pi}{d\theta} \quad (15)$$

That can be expressed as;

$$\varepsilon = 2\theta\frac{(d\pi/dx)}{(d\theta/dx)} \quad (16)$$

From Gibbs adsorption equation;

$$\Gamma = \frac{x}{RT}\left(\frac{d\pi}{dx}\right)_T \quad (17)$$

Surface coverage can be used instead of surface tension, so

$$\theta = \frac{\Gamma}{\Gamma_s} = \frac{x}{\Gamma_s RT}\left(\frac{d\pi}{dx}\right)_T \quad (18)$$

The first derivative with respect to mol fraction is;

$$\frac{d\theta}{dx} = \frac{x}{\Gamma_s RT}\left(\frac{d^2\pi}{dx^2}\right)_T + \frac{1}{\Gamma_s RT}\left(\frac{d\pi}{dx}\right)_T \quad (19)$$

By substituing relationships (18) and (19) into Equation (16) to express Gibbs elasticity modulus as;

$$\varepsilon = 2x\frac{\left(\frac{d\pi}{dx}\right)^2}{x\left(\frac{d^2\pi}{dx^2}\right) + \left(\frac{d\pi}{dx}\right)} \quad (20)$$

The combination of Equation (3) and (20) leads to

$$\varepsilon = 2\Gamma_s RT\frac{\beta x}{(1+\beta x)(1-\beta x)} \quad (21)$$

Equation (21) can be expressed as follows;

$$\varepsilon = 2\Gamma_s RT\frac{\beta x}{1-\beta^2 x^2} \quad (22)$$

The bulk concentration x is related to the surface concentration θ and the foam goes to maximum (**Figure 9** and **Figure 10**), it happens when $x = 1/\beta$; beyond the critical concentration, the surface coverage is completed and the surface pressure remains constant thus Gibbs elasticity modulus goes to zero and the foam decays [14-17]. The concentration level where the maximum foam volume was obtained agrees with the concentration level on polyurethane foam formulations.

4. Conclusion

It has been shown that it is possible to characterize the surface tension of polydimethylsiloxane based surfactants in viscous model systems of a polyurethane foam forming medium. Even though both polyols don't share same characteristics, the surface tension equilibrium was very similar when PDMS is involved as surfactant. However Pluronic L61 presented poor foaming behaviour explained by the rectification process ocurring along the foam column. PDMS showed transient foam behaviour with maximum foamability around concentration level

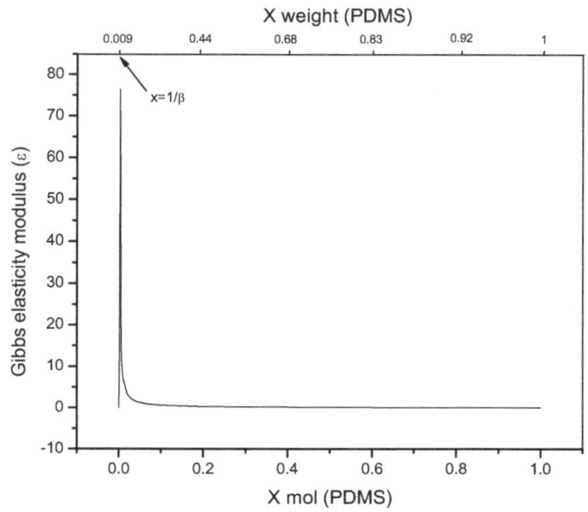

Figure 9. Gibbs elasticity modulus. Polypropylene glycol-PDMS system (Equation (22)).

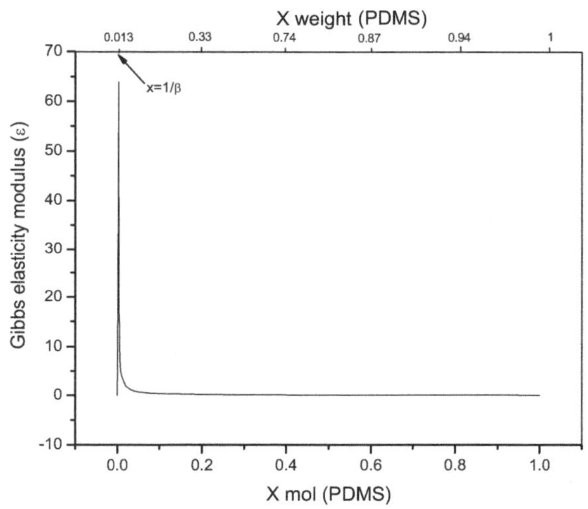

Figure 10. Gibbs elasticity modulus. Pluronic L61-PDMS system (Equation (22)).

used in polyurethane foam formulations. Polypropylene glycol presented better foaming properties compared to Pluronic L61 which had poor foaming behaviour explained by the rectification process ocurring along the foam causing Pluronic a lower capability to re-establish the surface concentration limiting the adsorption to the newly created surfaces. Even though foam is a dispersed system and hence out of equilibrium, the hypothesis of local chemical equilibrium between lamellar solution and foam's surface shows a proximity towards equilibrium conditions and allows for the thermodynamic description of elasticity; which for the purpose of this work was referred to foamability. The equation presented allows a direct relation to the Gibbs adsorption equation and is applicable for transient foams to predict concentration stability zone for polyurethane flexible systems that involve polidymethylsiloxane-poliol mixtures, where foamability decreases beyond a critical concentration. Foamability, or the foam volume produced by an amphiphile in a viscous medium like a poliol at a given concentration, is satisfactorily described by means of the Gibbs elasticity modulus (ε) presented in terms of the first and second derivatives of surface pressure and the bulk concentration.

5. Acknowledgements

This work was supported by Dirección General de Administración del Personal Académico (DGAPA) under the project PAPIIT-IT118711.

REFERENCES

[1] R. Herrington and K. Hock, "Flexible Polyurethane Foams," 2nd Edition, Dow Chemical Co., Midland, 1997.

[2] A. W. Adamson, "Physical Chemistry at Interfaces," 6th Edition, John Wiley & Sons Inc., New York, 1997.

[3] M. J. Rosen, "Surfactants and Interfacial Phenomena," 2nd Edition, Wiley Interscience, New York, 1989.

[4] G. Biesmans, L. Colman and R. Vandensande, "The Use of Principal Component. Analysis to Classify PDMS Surfactants Used to Make Rigid Polyurethane Foams Based on Their Dynamic Surface Tension Characteristics," *Journal of Colloid and Interface Science*, Vol. 199, No. 2, 1998, pp. 140-150. doi:10.1006/jcis.1997.5337

[5] X. D. Zhang, C. W. Macosko, H. T. Davis, A. D. Nikolov and D. T. Wasan, "Role of Silicone in Flexible Polyurethane Foam," *Journal of Colloid and Interface Science*, Vol. 215, No. 2, 1999, pp. 270-279. doi:10.1006/jcis.1999.6233

[6] B. D. Kaushiva, S. R. McCartney, G. R. Rossmy and G. L. Wilkes, "Surfactant Level Influences on Structure and Properties of Flexible Slabstock Polyurethane Foams," *Polymer*, Vol. 41, No. 1, 2000, pp. 285-310. doi:10.1016/S0032-3861(99)00135-4

[7] J. Eastoe, J. Dalton, G. Rogueda, E. Crooks, A. Pitt and E. Simister, "Dynamic Surface Tension of Nonionic Surfactant Solutions," *Journal of Colloid and Interface Science*, Vol. 188, No. 2, 1997, pp. 423-430. doi:10.1006/jcis.1997.4778

[8] P. Joos, J. P. Fang and G. Serrien, "Comments on Some Dynamic Tension Measurements by the Dynamic Bubble Pressure Method," *Journal of Colloid and Interface Science*, Vol. 151, No. 1, 1992, pp. 144-149. doi:10.1016/0021-9797(92)90245-H

[9] J. Eastoe and J. S. Dalton, "Dynamic Surface Tension and Adsorption Mechanisms of Surfactants at the Air-Water Interface," *Advances in Colloid and Interface Science*, Vol. 85, No. 2-3, 2000, pp. 103-144. doi:10.1016/S0001-8686(99)00017-2

[10] J. Viades-Trejo and J. Gracia, "A New Surface Equation of State Hydrophobic-Hydrophilic Contributions to the Activity Coefficient," *Fluid Phase Equilibria*, Vol. 264, No. 1-2, 2008, pp. 12-17.

[11] J. Viades-Trejo, A. Amigo and J. Gracia, "Activity Coef-

ficients at Infinite Dilution for Surfactants," *Fluid Phase Equilibria*, Vol. 250, No. 1-2, 2006, pp. 158-164. doi:10.1016/j.fluid.2006.10.015

[12] A. Piñeiro, P. Brocos, J. Gracia and A. Amigo, "Application of the Extended Langmuir Model to Surface Tension Data of Binary Liquid Mixtures," *Fluid Phase Equilibria*, Vol. 237, No. 1-2, 2005, pp. 140-151. doi:10.1016/j.fluid.2005.08.020

[13] E. O. Magaña, "Identificación de Fuerzas Intermoleculares en la Interfase Mediante Ecuaciones de Langmuir," Tesis de Licenciatura, U.N.A.M., 2008.

[14] J. Gracia, "Langmuir-BET Surface Equation of State in Fluid-Fluid Interfaces," *Langmuir*, Vol. 15, No. 9, 1999, pp. 3279-3282. doi:10.1021/la981244o

[15] J. G. Fadrique, "Cinética de Adsorción Superficial," Tesis de Maestría, U.N.A.M., 1979.

[16] G. Bleys and P. Joos, "Adsorption Kinetics of Bolaform Surfactants at the Air/Water Interface," *Journal of Physical Chemistry*, Vol. 89, No. 6, 1985, pp. 1027-1032. doi:10.1021/j100252a028

[17] F. Elizalde, J. Gracia and M. Costas, "Effect of Aggregates in Bulk and Surface Properties. Surface Tension, Foam Stability and Heat Capacities for 2 Butoxyethanol + Water," *Journal of Physical Chemistry*, Vol. 92, No. 12, 1988, pp. 3565-3568. doi:10.1021/j100323a048

Theoretical Simulation of the Infrared Absorption Spectrum of the Strong Hydrogen and Deuterium Bond in 2-Pyridone Dimer

Noureddine Issaoui*, **Houcine Ghalla**, **Brahim Oujia**
Quantum Physics Laboratory, Sciences Faculty of Monastir, Monastir, Tunisia
Email: *issaoui_noureddine@yahoo.fr

ABSTRACT

This work presents a theoretical simulation of the infrared spectra of strong hydrogen bond in alpha-phase 2-pyridone dimers, as well as in their deuterium derivatives at room temperature. The theory takes into account an adiabatic anharmonic coupling between the high-frequency N-H(D) stretching and the low-frequency intermolecular N...O stretching modes by considering that the effective angular frequency of the fast mode N-H(D) is assumed to be strongly dependent on the slow mode stretching coordinate N...O, the intrinsic anharmonicity of the low-frequency N...O mode through a Morse potential, Davydov coupling triggered by resonance exchange between the excited states of the fast modes of the two hydrogen bonds involved in the cyclic dimer, multiple Fermi resonances between the N-H(D) stretching and the overtone of the N-H(D) bending vibrations and the direct and indirect damping of the fast stretching modes of the hydrogen bonds and of the bending modes. The IR spectral density is computed within the linear response theory by Fourier transform of the autocorrelation function of the transition dipole moment operator of the N-H(D) bond. The theoretical line shapes of the $\upsilon_{N\text{-}H(D)}$ band of alpha-phase 2-pyridone dimers are compared to the experimental ones. The effect of deuteration is successfully reproduced.

Keywords: Strong Hydrogen Bound; Alpha-Phase 2-Pyridone Dimer; Infrared Spectral Density; Davydov Coupling; Morse Potential; Fermi Resonances

1. Introduction

The hydrogen bonding is responsible for the appearance of spectacular changes in IR spectra of associated molecules. This remark particularly concerns the $\upsilon_{X\text{-}H}$ proton stretching vibration bands, which are the attribute of the X-H atomic groups in the X-H…Y bridges. The main effects depend on a considerable low-frequency $\upsilon_{X\text{-}H}$ band shift, on the band integral intensity growth by even two orders of magnitude and on a noticeable increase of the band-width [1-3]. Numerous theoretical studies on hydrogen-bonded systems such as carboxylic acid have to explain the changes in the infrared spectrum induced by the formation of the H-bond Bridge.

In this work we present a theoretical approach for vibrational couplings in moderately strong hydrogen-bonded systems and use it for simulating experimental infrared spectra for strong hydrogen-bonded of alpha solid-state phase 2-pyridone crystal dimers. It is shown alpha phase of 2-pyridone forms approximately a centrosymmetric dimer. In fact, that from the results given in the original crystallographic work, it results that the dimers are not ideally centrosymmetric since one hydrogen bridge in a dimer is longer than the other. Without making significant errors, in the discussion of spectroscopic effects, one may assume that the dimers are approximately centrosymmetric. The alpha-phase crystals belong to the monoclinic system with Z = 8 [4]. However, we think that for the IR spectroscopy reasons you may safely consider these dimers as centrosymmetric ones. Several studies show that 2-pyridone form centrosymmetric dimers [5,6].

The studies have proved that in the case of cyclic dimers in this phase of 2-pyridone a single hydrogen bond is considerably stronger than a single hydrogen bond in a chain of molecules in the beta-phase. This fact suggests that the N^{+}-H…O^{-} bonds exist only in the alpha-phase and that these bonds are obviously stronger than the N-H…O hydrogen bonds in the beta-phase, since ionic hydrogen bonds are generally known to be stronger than the neutral ones [7,8]. 2-pyridone is an important precursor of antibiotics, which are used as inhibitors of the DNA gyrase. Thus 2-pyridone [9] and its

*Corresponding author.

derivatives [10,11] are widely investigated and exploited. Wuest and coworkers have proved that dipyridone can be formed by linking the two 2-pyridone using the functional groups like the acetylene or amine [12,13]. Asymmetric dipyridones are ready to form a dimer, whereas symmetric dipyridones are easy to self-assembly to form the linear and planar polymers in aprotic solvent [12]. The formation of the polymers via hydrogen bonding is an important approach of preparing liquid crystal and other functional polymers. There have been only a few reports on the hydrogen-bonding complexes formed by dipyridones. In addition, 2-pyridone dimer is an important reference system because it represents the rare case of a closed complex with two linear H-bonds. Recently, Wójcik presented a theoretical simulation of the bandshape and fine structure of $\upsilon_{N\text{-}H}$ stretching band for 2-pyridone-H and 2-pyridone-D [14], taking into account an adiabatic coupling between high-frequency N-H(D) stretching and low-frequency intermolecular N…O stretching modes, Davydov interactions and linear and quadratic distortions of potential energies for the low-frequency vibrations in the excited state of the N-H(D) stretching vibrations. The slow vibration modes were assumed as harmonic potentials. In other work, Wójcik [15] has presented a theoretical model for vibrational couplings in weak and moderately strong hydrogen-bonded systems and use it for modeling experimental infrared spectra for hydrogen-bonded crystals and hexagonal ice. The model is based on vibronic-type couplings between high and low frequency modes in hydrogen bridges, Davydov interactions [16] and Fermi resonance [17,18]. It allows calculation of energy and intensity distributions in the infrared spectra of hydrogen-bonded systems. The present theory is based on strong anharmonic coupling between the high-frequency hydrogen stretching vibration $\upsilon_{N\text{-}H}$ and low-frequency phonons, Davydov interactions and multiple Fermi resonances interaction between a fundamental vibration of a $\upsilon_{N\text{-}H}$ and one overtone or a combination tone of intermolecular vibration. Besides, this theory incorporated the intrinsic anharmonicity of the slow frequency mode through a Morse potential [19,20] whereas the fast mode was considered as harmonic. Note that, the Morse potential is undoubtedly more realistic than the harmonic one to describe this slow frequency mode. The adiabatic approximation [21] have been performed for each separate part of the dimer together with a strong non-adiabatic correction via the resonant exchange between the excited states of the two fast mode moieties. Both quantum direct (relaxation of the high-frequency modes) and indirect (relaxation of the H-bond bridges) dampings of the systems [22,23] were taken into account. This theory allows calculation of intensity distributions in the infrared spectra of hydrogen-bonded systems. The main purpose is to reproduce the experimental $\upsilon_{N\text{-}H}$ IR line shapes of hydrogen and deuterium bond in alpha-phase 2-pyridone crystal dimer at room temperature. We shall use infrared spectra of 2-pyridone in the alpha solid-state phase, measured by Flakus at the room temperature (**Figures 1** and **2** from [24]).

The numerical results show that this theoretical approach allows fitting the experimental $\upsilon_{N\text{-}H}$ infrared line shapes of cyclic alpha-phase 2-pyridone crystal dimer and its deuterium derivative by using a minimum number of parameters. With such tools, experimentalist should be able to compare experimental and theoretical data in an easy-to-use way.

2. Experimental Spectra

It is of interested to note that, 2-pyridone compound which has been used in this investigation was a commercial substance (Sigma-Aldrich). The single crystals of the alpha phase formed rectangular plates. For the purpose of the experiment they were exposed by using on a tin diaphragm with a 1.5 mm hole diameter. It was used without further purification. Crystals of 2-pyridone suitable for spectral studies were obtained by crystallization from

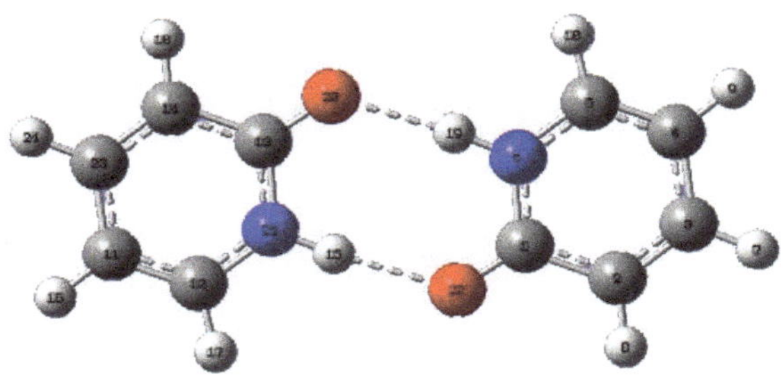

Figure 1. Geometry of the 2-pyridone dimer optimized at the HF/6-311^{++}G(d,p) level [26].

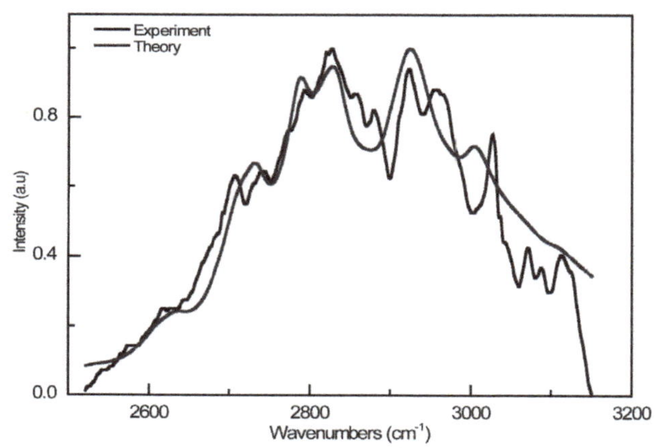

Figure 2. Comparison between the experimental and theoretical spectra of hydrogen-bonded in alpha-phase 2-pyridone at room temperature in the presence of three Fermi resonances.

melt between two closely spaced CaF2 windows. The solid-state spectra were measured by a transmission method at room temperature with the help of the FT-IR Nicolet Magna 560 spectrometer using a non polarized beam. The spectra were measured at 2 cm^{-1} resolution. Measurements of spectra were completed in a similar way for the crystals of deuterium derivative of 2-pyridone, which was synthesized by evaporation of the solution in D_2O under reduced pressure.

3. Theory

The present theory was developed in the framework of the adiabatic approximation [25]. The adiabatic approximation of the N-H stretching vibrations leads to the description of each moiety by effective Hamiltonians of the H-bond Bridge: for a single H-bond bridge, this effective Hamiltonian is either that of a harmonic oscillator, if the fast mode is in its ground state, or that of a driven harmonic oscillator if the fast mode is excited. When one of the two identical fast modes is excited, then, because of the symmetry of the cyclic dimer and because of coupling V_0 between the two degenerate fast-mode excited states, an interaction occurs (Davydov coupling) leading to an exchange between the two identical excited parts of the dimer, as considered by Maréchal and Witkowski in their pioneering work [21]. Of course, this interaction between degenerate excited states is of non-adiabatic nature, although the adiabatic approximation was performed to separate the high- and low-frequency motions. **Figure 1** present the geometry of the 2-pyridone dimer optimized at the HF/6-311^{++}G(d,p) level [26], which is formed by two hydrogen bonds (lengths of N6-H19… O20 and N21-H15…O22 are 2.89559 Å and 2.89558 Å, respectively).

Let q_1 and q_2, be the coordinates of the high-frequency N-H stretching vibrations, in the first and second hydrogen bond, and Q_1 and Q_2 coordinates of the two low-frequency intermolecular N…O stretching modes. The two moieties of the system are exchanged by the symmetry C2 parity operator.

For the theory dealing with this system, the basic physical parameters are 1) vibration angular frequency ω° of the two degenerates fast modes moieties when the H-bond bridge is at the equilibrium, 2) vibration angular frequency Ω of the two degenerates H-bond bridge moieties, 3) dimensionless anharmonic coupling parameter α° between the high frequency mode of one moiety and the H-bond bridge coordinate of the same moiety, 4) Davydov coupling V parameter between the degenerate first excited state of the two moieties high frequency mode, 5) direct and indirect damping parameters γ and γ, 6) coupling parameters fi involved in the Fermi resonances coupling between the first harmonics of some bending modes of angular frequency and the first excited state of the g symmetrized high frequency mode, 7) relaxations parameter of the first harmonic bending modes, and 8) absolute temperature T.

3.1. Full Hamiltonian of the System

The evaluation of the spectral density of the hydrogen bond system requires the knowledge of the full Hamiltonian of the hydrogen bond system. For this purpose, it is important to describe the basic various vibration modes in the dimer. In our present work, for each part of the dimer, we have described the slow-frequency mode by a Morse potential, which can be written as follows:

$$V_{\text{Morse}} = D_e \left[1 - \mathrm{e}^{\left(-\beta_e Q \sqrt{\frac{M\Omega}{\hbar}} \right)} \right]^2 \tag{1}$$

D_e is the dissociation energy of the Hydrogen Bond Bridge and β_e is given by:

$$\beta_e = \Omega \sqrt{(M/2D_e)} \sqrt{(\hbar/M\Omega)} \tag{2}$$

The fast frequency-mode is considered to be harmonic. It is important to note that in the majority of recent works, the slow-frequency mode was assumed as harmonic [25].

Within the strong anharmonic coupling theory and the anharmonic approximation for the H-bond bridge, the corresponding Hamiltonians of the slow and high-frequency modes of the two moieties of the dimer are, respectively given, using dimensionless operators, by:

$$[H_{\text{Slow}}]_i = (1/2) P_i^2 \hbar\Omega + D_e \left[1 - \mathrm{e}^{\left(-\beta_e Q_i \sqrt{M\Omega/\hbar} \right)} \right]^2 \tag{3}$$

$$[H_{\text{Fast}}]_i = (1/2)\left(p_i^2 + q_i^2\right) \hbar\omega(Q_i) \tag{4}$$

In these equations, P_i are the dimensionless conjugate momenta of the H-bond bridges coordinates Q_i of the two moieties, whereas p_i and q_i are the dimensionless coordinates and the conjugate momenta of the two degenerate high frequency modes of the two moieties. Ω is the angular frequency of the H-bond Bridge, whereas $\omega(Q_i)$ is that of the high frequency mode which is supposed to depend on the coordinate of the H bond bridge.

Using a Taylor development of the Morse potential, the Hamiltonian of the slow frequency modes given in Equation (3) can be rewritten as the sum of the Hamiltonian of a quantum harmonic oscillator and an anharmonic potential V:

$$[H_{\text{Slow}}]_i = [H_{\text{Slow}}]_i + V \tag{5}$$

where $[H_{\text{Slow}}]_i$ and V are respectively given by:

$$[H_{\text{Slow}}]_i = (1/2)\left(P_i^2 + Q_i^2\right) \hbar\Omega \tag{6}$$

$$V^{\circ\circ} = D_e \sum_{n=3}^{\infty} \left[\left((-1)^n \left(2^n - 2 \right) \right) / n! \right] \left(\beta_e Q_i \sqrt{M\Omega/\hbar} \right)^n \quad (7)$$

Expansion to first-order of the angular frequency of the fast mode with respect to the coordinate of the H-bond bridge leads to write:

$$\omega\left(Q_i\right) = \omega^\circ + \alpha^\circ \Omega Q_i \quad (8)$$

where ω° is the angular frequency of the two degenerate fast modes when the corresponding H-bond Bridge coordinates are at equilibrium, whereas α° is a dimensionless parameter which will appear to be an anharmonic coupling parameter.

In presence of damping, the thermal bath may be figured, by an infinite set of harmonic oscillators, and its coupling with the H-bond bridge are described by terms which are linear in the position coordinates of the bridge and of the bath oscillators:

$$H_{\text{Damp}} = \sum (1/2)\left[\tilde{p}_r^2 + \tilde{q}_r^2 \right] \hbar\omega_r + \sum \tilde{p}_r \left[Q_1 + Q_2 \right] \hbar g_r \quad (9)$$

Here, $\tilde{q}_r$ are the dimensionless position coordinate operators of the oscillators of the bath, $\tilde{p}_r$ are the corresponding conjugate moments, obeying the usual quantum commutation rules, ω_r are the corresponding angular frequencies and g_r are the coupling between the H-bond bridges and the oscillators of the bath. Within the adiabatic approximation, the Hamiltonian of each moiety of the dimer takes the form of sum of effective Hamiltonians which are depending on the degree of excitation of the fast mode according to:

$$\left[H_{\text{Adiab}} \right]_i = \sum \left[H^{\circ\{k\}} \right]_i \left| \{k\}_i \right\rangle \left\langle \{k\}_i \right|; \; i = 1,2; k = 0,1 \quad (10)$$

$$\left[H^{\circ\{k\}} \right]_i = (1/2) P_i^2 \hbar\Omega + D_e \left[1 - \exp\left(-\beta_e Q_i \sqrt{M\Omega/\hbar} \right) \right]^2 + k\alpha^\circ Q_i \hbar\Omega + k\hbar\omega^\circ \quad (11)$$

An excitation, of the fast mode of one moiety of the dimer is resonant with the excitation of the other moiety. Thus, a strong non-adiabatic correction [21] *i.e.* a Davydov coupling V_{12}, occurs between the two resonant states after excitation of one of the two fast modes, so that the full Hamiltonian of the two moieties is given by the equations:

$$H_{\text{Tot}} = \left[H_{\text{Adiab}} \right]_1 + \left[H_{\text{Adiab}} \right]_2 + V_{12} + H_{\text{Damp}} \quad (12)$$

with,

$$V_{12} = \hbar V^\circ \left[\left| \{1\}_1 \right\rangle \left\langle \{0\}_2 \right| + \left| \{0\}_1 \right\rangle \left\langle \{1\}_2 \right| \right] \quad (13)$$

3.2. Autocorrelation Functions and Spectral Density

The spectral density $I(\omega)$ of the $\upsilon_{\text{N-H}}$ mode is given, within the linear response theory [27,28], by the Fourier transform of the autocorrelation function $G(t)$ of transition moment operator of the fast mode :

$$I(\omega) = 2\,\text{Re} \int_0^{+\infty} G(t) \mathrm{e}^{-\mathrm{i}wt} \mathrm{d}t \quad (14)$$

Using the symmetry of the system, the ACF may be split into symmetric parts (g) and antisymmetric parts (u).

In the presence of Davydov coupling, the autocorrelation function (ACF) of the dipole moment operator of the fast mode may be written [25]:

$$G(t) \propto \left[G(t) \right]_g \left[\left[G^+(t) \right]_u + \left[G^-(t) \right]_u \right] \mathrm{e}^{-\gamma^\circ t} \quad (15)$$

Here, γ is the natural width of the excited state of the high frequency mode, the expression of which has been calculated by Rösh and Ratner [22]. $[G\,(t)]_g$ is the reduced ACF of the g part of the system related to Hamiltonian described the indirect damping. It's given by:

$$\left[G(t) \right]_g \propto \mathrm{e}^{\mathrm{i}\omega^\circ t} \mathrm{e}^{-\mathrm{i}\frac{\alpha^\circ}{2}\Omega t} \mathrm{e}^{-\mathrm{i}\left[\frac{\beta}{\sqrt{2}}\right]^2 \Omega t} \mathrm{e}^{-\mathrm{i}\left[\frac{\beta}{\sqrt{2}}\right]^2 \left[\langle n \rangle + 1/2\right]\left(2\mathrm{e}^{-\gamma t/2}\cos\Omega t - \mathrm{e}^{-\gamma t}\right)} \times \mathrm{e}^{\mathrm{i}|\beta|^2 \mathrm{e}^{(-\gamma t/2)} \sin\Omega t} \quad (16)$$

In these equation, <*n*> and β are respectively the thermal average of the occupation number of the quantum harmonic oscillator describing the H-bond bridge and the effective dimensionless anharmonic coupling parameter related to α°. These terms are given by:

$$\beta = \alpha^\circ \frac{\left[4\Omega^2 + \gamma^2\Omega^2 \right]^{1/2}}{2\sqrt{2}\left[\Omega^2 + \left(\gamma^2/4 \right) \right]} \quad (17)$$

$$\langle n \rangle = 1 / \left(\mathrm{e}^{(\hbar\Omega/kT)} - 1 \right) \quad (18)$$

Note that by "reduced", we mean that a partial trace has to be performed over the thermal bath because of the coupling between the symmetric Hamiltonian and the surrounding. In the last equation, $[G^+(t)]_u$ and $[G^-(t)]_u$ are the two (u) ACFs which are affected only by the Davydov coupling. They are given by [29]:

$$\left[G^\pm(t) \right]_u \propto \sum_n \sum_{n_\mu} \mathrm{e}^{-(n\hbar\Omega/kT)} \left[1 \pm (-1)^{n_\mu + 1} \right]^2 \times \left| C^\pm_{n_\mu,\mu} \right|^2 \mathrm{e}^{\mathrm{i}\omega^\pm_\mu t} \mathrm{e}^{-\mathrm{i}n_\mu \Omega t} \quad (19)$$

where, $C^\pm_{n_\mu,\mu}$ are the expansion coefficients of the eigenvectors on the basis of the eigenstates of the Hamiltonian of the quantum harmonic oscillator. One may observe that the angular frequency ω° of the high frequency mode must be decreased by a factor $1/\sqrt{2}$ on D isotopic substitution of the proton involved in the H-bonds. Besides, according to Maréchal and Witkowski theory

[21], the anharmonic coupling parameter α° must also be reduced by this factor upon this same substitution, whereas the frequency Ω of the H-bond bridge has no reason to be modified.

Following Equations (15) and (19), the spectral density of the dimer involving Davydov effect can be written as:

$$I(\omega) \propto 2\,\mathrm{Re}\int_0^{+\infty} \left[G(t)\right]_g \left[\left[G^+(t)\right]_u + \left[G^-(t)\right]_u\right] \times e^{-\gamma^\circ t} e^{-iwt}\,dt \quad (20)$$

That may be written formally:

$$I(\omega) \propto I^+(\omega) + I^-(\omega) \quad (21)$$

with respectively:

$$I^\pm(\omega) \propto \sum_m \sum_n \left\{P_{m,n}\right\} \sum_{n_\mu} \sum_n \left\{A_{m,n}^{n_\mu,\mu}\right\} \quad (22)$$

In view of the above equations, the components appearing here are given by the following equations [29]:

$$\left\{A_{m,n}^{n_\mu,\mu}\right\} = e^{\lambda \bar{n}_\mu} \left[1 \pm (-1)^{n_\mu+1}\right]^2 \times \left|C^\pm_{n_\mu,\mu}\right|^2 \left\{I^\pm_{mnn_\mu,\mu}(\omega)\right\} \quad (23)$$

$$\left\{I^\pm_{mnn_\mu,\mu}(\omega)\right\} \propto \left(\gamma_{m,n}\right) \Big/ \left\{\left(\omega - \Omega^\pm_{mnn_\mu,\mu}\right)^2 + \left(\gamma_{m,n}\right)^2\right\}, \quad (24)$$

$$\Omega^\pm_{mnn_\mu,\mu} = \omega^\circ - \left[(m - n + n_u)\Omega - \omega_u^\pm\right] - \alpha^\circ \sqrt{2}\Omega\ , \quad (25)$$

$$\gamma_{m,n} = (m+n)\gamma + \gamma^\circ, \quad (26)$$

$$P_{m,n} = \left\{\left[1 + \langle n\rangle\right]^m \langle n\rangle^n \left|\beta/\sqrt{2}\right|^{2(m+n)}\right\} \Big/ m!n! \quad (27)$$

Recall that $\langle n\rangle$ is the thermal average of the number occupation operator of the H-bond bridge vibration given by Equation (18), which is a function of the angular frequency Ω of the H-bond bridge and of the absolute temperature T. Besides, β is the effective anharmonic coupling parameter given by Equation (17), which is a function of the anharmonic coupling parameter α° between the slow and fast mode, and of the angular frequency Ω and the damping parameter γ of the H-bond bridge.

3.3. Situation with Fermi Resonances

The previous treatment was developed with the neglect of Fermi resonances. Now, suppose the situation where this later effect is taken into account. It is resulting from the interactions occurring between the first excited state of the high frequency mode and the first harmonic of some bending modes. As it was stated by Maréchal and Witkowski [21], if Fermi resonances are taken into account, one has to consider one fast mode, one slow mode and one bending mode for each hydrogen bond of the cyclic dimer. If we take into account Fermi resonances [29], they affect only the g state of the system. As a consequence, the autocorrelation functions $[G^-(t)]_u$ are not modified. In the presence of Davydov coupling and Fermi resonances, the ACF can be written following [29]:

$$\left[G(t)\right]_g \propto \sum_n \sum_\mu e^{-\lambda m} \left|a_{\{\mu,0,m\}g}\right|^2 \left[e^{i\{\omega_\mu\}_g t/\hbar}\right] e^{-im\Omega t} \quad (28)$$

Here, $\{\omega_\mu\}_g$ are the eigenvalues appearing in Equation (29) whereas, $a_{\{\mu,0,m\}g}$ are the expansion coefficients defined by Equation (30).

$$\left[H_D^F\right]_g \left|\varphi_\mu\right\rangle_g = \hbar \left|\varphi_\mu\right\rangle_g \left\{\omega_\mu\right\}_g \quad (29)$$

With

$$\left|\varphi_\mu\right\rangle_g = \sum_m \sum_g \left\{a_{\mu,l,m}\right\} \left|\psi_{l,m}\right\rangle_g \quad (30)$$

The g states involved in the above expansions are defined by:

$$\left|\left\{\Psi_{0,m}\right\}_g\right\rangle = \left|\{1\}_g\right\rangle \left|\{0_1\}_g\right\rangle \left|[0_2]_g\right\rangle \left|(m)_g\right\rangle \quad (31)$$

$$\left|\left\{\Psi_{1,m}\right\}_g\right\rangle = \left|\{0\}_g\right\rangle \left|\{2_1\}_g\right\rangle \left|[0_2]_g\right\rangle \left|(m)_g\right\rangle \quad (32)$$

$$\left|\left\{\Psi_{2,m}\right\}_g\right\rangle = \left|\{0\}_g\right\rangle \left|\{0_1\}_g\right\rangle \left|[2_2]_g\right\rangle \left|(m)_g\right\rangle \quad (33)$$

Here, the kets

$$\left|\{k\}_g\right\rangle,\ \left|(m)_g\right\rangle \text{ and } \left|[1]_g\right\rangle$$

are the g eigenstates of respectively the symmetrized high frequency quantum harmonic oscillator, the slow frequency quantum harmonic oscillator and the bending frequency quantum harmonic oscillator. The Fermi resonance mechanism, characterized by the coupling parameter f_i, is described by the following coupling operators $\hbar f$ which express the non-resonant exchanges between the state $\left|\{1\}_j\right\rangle$ of the jth fast mode and second damped excited state $\left|[2]_j\right\rangle$ of the jth bending mode. The introduction of the Fermi resonance coupling effects in the lineshape is presented by the complexes angular frequency gap Δ_i:

$$\Delta_i = \Delta_i^\circ - i\gamma_i^\delta \quad (34)$$

with

$$\Delta_i^\circ = -\omega^\circ - 2\omega_i^\delta \quad (35)$$

where ω_i^δ is the frequency of the bending modes $(i = 1, 2, \cdots, n_F)$.. The imaginary part in this gap is related to the lifetime of the corresponding excited states. Recall that the Hamiltonian of the dimer involving Davydov coupling, Fermi resonances between the g excited state of the fast mode and the g first harmonics of the bending mode, together with the damping of theses excited states is:

$$\left[H_{\text{Fermi+Dav}}\right]_g = \begin{pmatrix} \left[H^{\{1\}}\right] & \hbar f_1 & \hbar f_2 & \hbar f_3 & \dots & \hbar f_{nF} \\ \hbar f_1 & H^{\{0\}}+\hbar\Delta_1 & & & & \\ \hbar f_2 & & H^{\{0\}}+\hbar\Delta_2 & & & \\ \hbar f_3 & & & H^{\{0\}}+\hbar\Delta_3 & & \\ \dots & & & & \dots & \\ \hbar f_{nF} & & & & & H^{\{0\}}+\hbar\Delta_{nF} \end{pmatrix} \tag{36}$$

with

$$H^{\{0\}} = \left[a^+a+\frac{1}{2}\right]\hbar\Omega \tag{37}$$

and

$$\left[H^{\{1\}}\right]_g = \left[\left(a^+a+\frac{1}{2}\right)+\alpha\left(a^++a\right)\right]\hbar\left(\Omega-i\frac{\gamma}{2}\right) + \hbar\omega^\circ-\frac{\alpha}{2}\hbar\Omega \tag{38}$$

In these two last equations, a^+ and a are the boson operators obeying $[a^+, a] = 1$ and $i^2 = -1$.

4. Results and Discussions

In our theoretical approach, we have assumed that the 2-pyridone dimer crystal in alpha-phase is of approximately centrosymmetric type. In the case of the interpretation of the spectra of alpha-phase 2-pyridone crystal, this interpretation cannot be relied on since the N-H…O bonds are much shorter than the sulfur atoms containing hydrogen bonds. Most probably, the purely vibrational approach to solving the problem of the exciton interacttions in the hydrogen bond dimers should be abandoned. However, the contribution of electronic interactions to the vibrational exciton interactions between the hydrogen bonds in the dimers, expressed by the electronic coordinates, should be taken account.

From these considerations some consequences result for the electronic structure of 2-pyridone dimers in alpha-phase crystals. This is most probably the structure of a larger contribution of the zwitterion electronic structure. Such a continuous structure of the π-electron cloud in the pyridine rings, in cyclic N^+-H…O^- hydrogen bond dimers allows for an effective ("head-to-tail") coupling of the hydrogen bonds in the cyclic dimers. The experimental and theoretical results allow us to state that, for alpha-phase 2-pyridone studied the $\upsilon_{N\text{-}H}$ absorption bands of their dimers observed in the IR spectra are similar in structure and are very broad. This means that the mechanism of formation of these bands should, first of all, involve the participation of the cyclic structure, responsible for the formation of the hydrogen bond. Insignificant changes in the shape of the dimers and their deuterated derivatives bands observed in alpha-phase 2-pyridone. We believe that the interaction of the two intermolecular bonds N-H…O in the cyclic complex plays a determining role in the formation of the broad absorption band.

The main object in this work is to reproduce theoretically the IR spectra of alpha-phase 2-pyridone at 300 K. In a first place we have reproduced the $\upsilon_{N\text{-}H(D)}$ IR spectra of alpha-phase 2-pyridone dimer crystal by comparing theoretical line shapes. The spectral densities are computed by Equation (19) after construct and diagonalize the full Hamiltonian in a truncated basis. The stability of the computed spectra with respect to the size of this basis set was carefully checked. The stability of the spectra was also checked versus the order of the Taylor expansion of the Morse potential. Thus, in the first place, we have considered the general physical situation in which multiple Fermi resonances are taken into account and for which the spectral density is given by Equations (19), (20) and (28). The procedure we have used is the fitting of the experimental line shapes by optimizing the values of the basic parameters.

Figures 2 and **3** present the theoretical spectra of alpha-phase 2-pyridone dimer crystal and its deuterated derivative where multiple Fermi resonances have been introduced in our theoretical approach. We have performed numerical experimentation by increasing progressively the number of Fermi resonances. The spectral densities, presented in this Figure, are computed in the presence of three Fermi resonances. **Tables 1** and **2** show the parameters involved in the calculations.

The examination of the spectra obtained using our theoretical model shows that there is a good agreement between theory and experiment. It is important to emphasize that if Fermi resonances are acting, one observes the classical behavior of the Fermi resonance, which is well described in the literature: the displacement of the bands and the redistribution of the intensities in the bands. However, only the intensities at high frequencies are redistributed if one increases the number of Fermi resonances. In fact, a recent study on acetic acid dimers in the liquid phase [30] did not show any wavelength dependence of their dynamics.

Figures 4 and **5** show the theoretical line shapes of alpha-phase 2-pyridone dimer crystal and its deuterated derivative calculated by Equations (16), (19) and (20) when Fermi resonances are ignored. In both situations, the theoretical line shapes appear as red lines whereas the experimental ones are black lines.

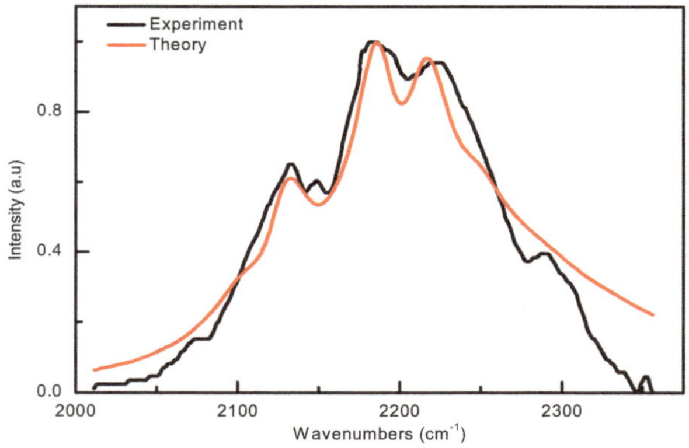

Figure 3. Comparison between the experimental and theoretical spectra of deuterium-bonded in alpha-phase 2-pyridone at room temperature in the presence of three Fermi resonances.

Table 1. Parameters used for fitting the experimental spectra of alpha-phase 2-pyridone at room temperature in the presence of three Fermi resonances.

Species	ω° (cm^{-1})	Ω (cm^{-1})	α°	V°	γ° (cm^{-1})	γ (cm^{-1})	η
O-H	2920	100	0.69	–0.48	0.16	0.1. Ω	0.9
O-D	2151	80	0.20	–0.44	0.11	0.1. Ω	0.5

Table 2. Fermi coupling parameters (in cm^{-1}) used for fitting experimental spectra of phase 2-pyridone at room temperature.

Species	f_1	f_2	f_3	Δ_1	Δ_2	Δ_3	$\gamma_{\delta 1}$	$\gamma_{\delta 2}$	$\gamma_{\delta 3}$
H	–80	90	100	10	10	10	0.1	0.1	0.1
D	–80	90	125	15	10	15	0.2	0.2	0.2

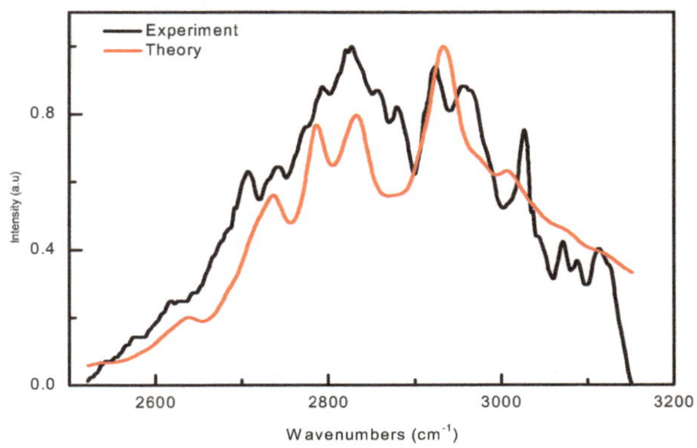

Figure 4. Comparison between the experimental and theoretical spectra of hydrogen-bonded in alpha-phase 2-pyridone at room temperature when the Fermi resonances are ignored.

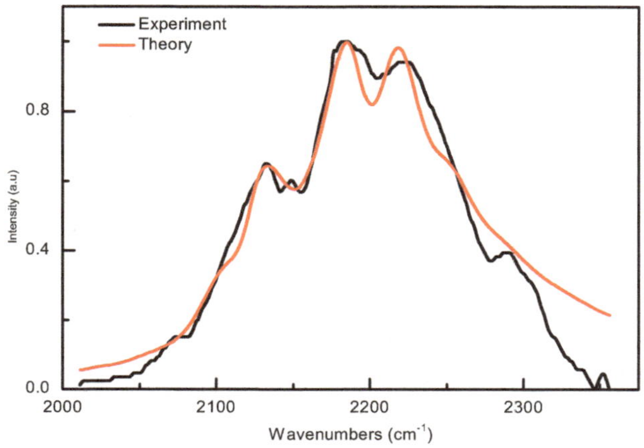

Figure 5. Comparison between the experimental and theoretical spectra of deuterium-bonded in alpha-phase 2-pyridone at room temperature in the absence of Fermi resonances.

In order to improve the validity of the present model, let us comment the magnitude of the parameters that we have used. Both the anharmonic coupling parameter and the angular frequencies of the high frequency mode for alpha-phase 2-pyridone in all treated situations, their magnitudes is decreased by a factor which is different from $\sqrt{2}$. This result is in agreement with theory when passing from the H to the D species and when the slow mode is assumed to be of Morse type (anharmonic potential). Indeed, according to [30,31], the isotope ratio

$$\chi = \frac{\langle \omega \rangle_H}{\langle \omega \rangle_D}$$

for hydrogen bonds O-H…O depends on the bond strength, it is close to the harmonic value $\sqrt{2}$ for weak bonds and decreases with the bond strength reaching minimum near $\chi = 0.9$ for low-barrier bonds. In our cases, to obtain good agreement between the theoretical and experimental one, we have used:

$$\frac{\alpha^\circ{}_H}{\alpha^\circ{}_D} = 3.415, \frac{\langle \omega_H \rangle}{\langle \omega_D \rangle} = 1.306 \text{ and } \frac{V_H}{V_D} = 1.118 \quad (39)$$

It is important to note that the introduction modulation of the equilibrium positions of the fast modes (q_e) and the quadratic dependence of their frequencies on Q_i which can represented by two expansions to the second order:

$$\omega(Q_i) = \omega^\circ + \alpha Q_i + \beta Q_i^2 ; q_e(Q_i) = q_e{}^\circ + f^\circ Q_i + g^\circ Q_i^2 \quad (40)$$

Recall that in present work and following Maréchal and Witkowski [21], we have only used the first order dependence of the angular frequency of the fast modes on the coordinate of the slow modes (Q_i) (see Equation (40) and we have neglected the modulation of the equilibrium positions of the fast modes and the quadratic dependence of their frequencies on (Q_i) [32] as conformed by several experimental correlations [33,34]. We have shown in recent [32] work that the fine structure of the IR υ_{X-H} stretching band are connected with the new anharmonic coupling parameters (*i.e.* β, f° and g°), whereas these parameters do not manifest markedly in effects of temperature and deuteration on the IR spectra. Besides, the account for all these parameters does not affect the similarity of the spectra in gas and condensed phases [35].

The isotope ratio χ of centers of gravity of band of light and deuterated alpha-phase 2-pyridone dimer is 1.306. The values obtained here are also in satisfactory agreement with the experimental data reported by Odinokov, *et al.* [36] for complexes of carboxylic acids with various bases and are comparable to those found in our recent work dealing with the H/D isotopic effects in H-bond spectra [33]. However, these ratios are different from those used by Blaise, *et al.* [37] since in their approaches dealing with theoretical interpretation of the IR line shapes of liquid and gaseous acetic acid [38] and gaseous propynoic and acrylic acid dimers [29], low and high-frequency hydrogen stretching vibrations in individual hydrogen bonds are assumed to be harmonic whereas in the present work we use a Morse potential in order to describe the anharmonicity of the H-bond Bridge. Recall that the removal of the harmonic approximation for the slow modes by introducing Morse potential in place of harmonic one has been done by Leviel and Maréchal [39] in a model similar to the present one, involving cyclic dimer, however without damping. They have shown that the value of the angular frequency of the slow mode which must be used to the experimental lineshape is more close to the experimental value when the anharmonicity of the bridge is introduced.

Now let us look to the strongest hydrogen bonds observed in alpha-phase 2-pyridone dimers. Generally this type of hydrogen bonds is observed in ion molecular complexes of the $(AHA)^-$ and $(BHB)^+$ types [40,41]; however, it is especially difficult to study the spectral properties of charged complexes under conditions of weak interactions with the medium. In the present approach, we take into account the natural width of the excited states of the fast mode due to the medium (direct relaxation) and the damping of the H-bond Bridge (indirect relaxation). The direct relaxation is included following quantum treatment of Rösch and Ratner [22] whereas the indirect one was taken into account via the approach of Louisell and Walker [42] dealing with the relaxation of driven damped quantum oscillators studied initially by Feynman and Vernon [43] and later by Louisell [44]. The values of the direct and indirect relaxation parameters reflecting the effect of the medium used presently for alpha-phase 2-pyridone are of the same magnitude as those used by Blaise, *et al.* [29,37,38] in their study dealing with acetic acid in the gas phase, whereas the indirect damping at 298 K for acetic acid in crystalline state is $\gamma = 1\ cm^{-1}$. One may ask whether the indirect damping used for crystalline state is weaker than that used for the gaseous phase since the indirect relaxation is ought to be larger in the solid state. The reason is that in the gas phase the indirect damping is an effective one which is the result of the combination of the indirect damping and of the rotational structure [45]. In neutral systems, the strongest hydrogen bonds of the N-H...O type are formed in self-associates of alpha-phase 2-pyridone and, seemingly, phosphinic acids [46], and it is only due to the high thermal stability of these associates that one is able to observe their IR spectra in the gas phase, at which cyclic dimers are in equilibrium with monomeric molecules [47,48]. This made it possible to analyze the dependence of the bandshape of the υ_{N-H} band, which serves as an important criterion in the choice of theoretical models. The parameters of this broad band and its shape turned out to be almost independent of the bonding frequency.

In order to obtain a good agreement with the experimental line shapes, we have take into account some breaking of the IR selection rule for the centrosymmetric cyclic dimer, via a large amount ($\eta = 0.5;\ 0.9$) of forbidden Ag transition. Recall that in general way the quality of the fitting is weakly improved by taking small values for η which lying between 0 and 1. This assumption was initially introduced by Flakus [49]. This is a general trend which has been observed recently in the cases of the centrosymmetric cyclic dimers of gaseous acetic acid [37]. Note that, according to Flakus hypothesis, the lack of forbidden transition ought to be stronger in the solid state (value near $\eta = 1$) than in the gaseous one and we must keep in mind that the Flakus assumption has been seen by this author to be unavoidable in diverse crystalline H-bonded carboxylic acids such as for instance glutaric [50] and cinnamic [51] acids and particularly in centrosymmetric H-bonded dimers.

Figure 6 presents the theoretical spectra of alpha-phase 2-pyridone dimer where multiple Fermi resonances have been introduced. The number of Fermi resonances n_F is increasing from zero (without Fermi resonances) to 3 ($n_F = 0$, 1, 2 and 3). The numerical results show that Fermi resonances appear to play an important role especially for the hydrogenated species whereas the Fermi resonances effect does not affect markedly the N-D derivative species. It may be explain by the fact that Fermi resonances mechanism may involve the N-H bond bending-in-plane vibrations in their first overtone states. So the coupling between the electronic systems with the electrons of the associated carboxyl groups implies a stabilization of the dimers. On the other hand, each deformation of the dimers provides a destruction of this

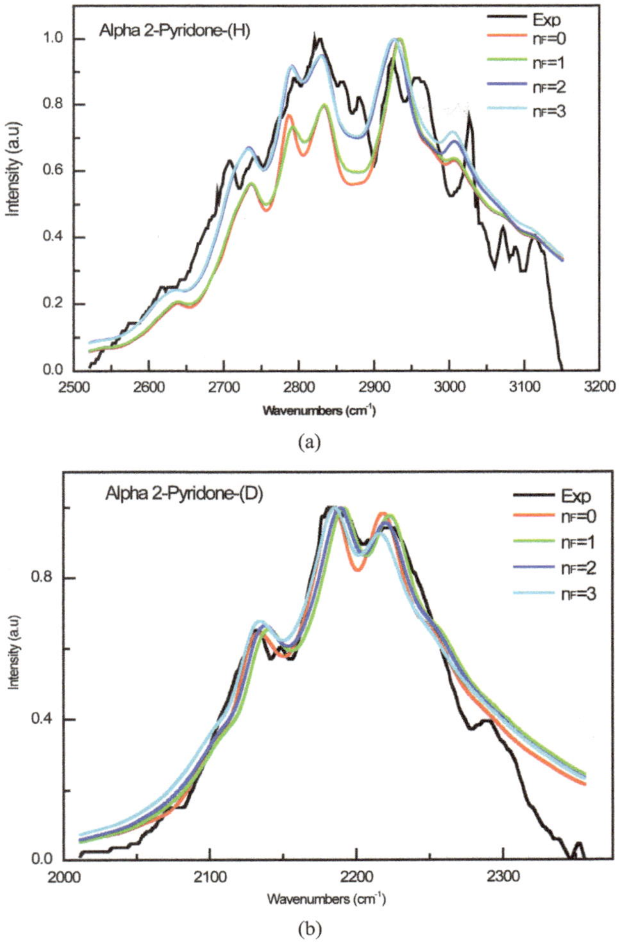

Figure 6. Multiple Fermi resonance effect on the IR spectral densities of hydrogen (a) and deuterium (b) bonded in alpha-phase 2-pyridone. The number of Fermi resonances nF is increasing from zero (without Fermi resonances) to 3.

stabilization mechanism. This is the most probable source of the anharmonic coupling involving the proton stretching and the proton bending-in-plane vibrations in the dimer of carboxylic acids. We think that the sensitive improvement of the theoretical line shape of strongly bound dimers of alpha-phase 2-pyridone by introduction of Fermi resonances might be due to some whole effect in which the Fermi resonances assisted by the strong anharmonic coupling would be augmented by the combination of the Davydov coupling and of the quantum direct and indirect damping. It appears from the present theoretical study that the simple model of Davydov coupling taking into account especially Fermi Resonances and quantum direct and indirect damping, is able to reproduce the experimental line shapes. When one of the two identical fast modes is excited, then, because of the symmetry of the cyclic dimer, and of Davydov coupling V_D between the two degenerate fast mode excited states, an interaction (Davydov coupling) occurs leading to an exchange between the two identical excited parts of the dimer of alpha-phase 2-pyridone. It is interesting to note that the Davydov coupling parameters used in the present work are similar to those used by Maréchal and Witkowski in their pioneering work dealing with the adipic acid [21]. However, we shall introduce in the Davydov coupling some further flexibility by assuming it to be dependent on the H-bond Bridge coordinate.

The complexity of the task to the experimental line shape of the cyclic dimer of alpha-phase 2-pyridone and to reproduce the corresponding isotope effect is determined by several factors, acting simultaneously. To take them into account, one needs to know a great number of parameters. For this idea, we think that the small residual discrepancies between theory and experiment are related to the neglect of some of these parameters: 1) assumption of a linear dependence of the angular frequency of the fast mode on the coordinate of the H-bond bridge [32], 2) assumption of the independence of the equilibrium position of the fast mode on the coordinate of the H-bond bridge [32], 3) assumption of a constant term for the Davydov coupling parameter [52], 4) neglect of electrical anharmonicity, 5) neglect of eventual weak tunneling through the potential barrier separating the two H-bond bridge minima, and 6) neglect of eventual relaxation mechanisms of non-adiabatic nature.

5. Conclusions

The presented theoretical approach is dealing within the strong anharmonic coupling theory according to which the high frequency mode and the H-bond bridge are anharmonically coupled through a linear dependence of the frequency of the fast mode on the elongation of the H-bond bridge and takes into account Davydov coupling, Fermi resonances, anharmonicity of the H-bond Bridge and direct and indirect quantum relaxations. The present approach contains as special instances the majority precedent theoretical approach [53] dealing with the subject.

This approach, which is applied to reproduce the $\upsilon_{N\text{-}H(D)}$ IR spectra of centrosymmetric strongly bound dimers of alpha-phase 2-pyridone and their deuterated derivatives in the solid-state, is comprised in the linear response theory calculates the line shape by aid of the Fourier transform of the autocorrelation function of the dipole moment operator. The model has been applied to alpha-phase 2-pyridone dimer crystal. It has been found that it is possible to correctly the experimental line shape of the hydrogenated compound and to predict satisfactorily the deuterium effect by using a set of spectral parameters. It appears from the present theoretical study that the simple model of Davydov coupling taking into account quantum indirect damping and the anharmonicity of the H-bond Bridge, is able to reproduce the experimental line shapes especially when the Fermi resonances were not ignored.

In the end, we can say that the Fermi resonances appear play an important role to reproduce the experimental line shapes of alpha-phase 2-pyridone dimer crystal.

6. Acknowledgements

The authors thank Professor Henryk T Flakus from Institute of Chemistry, University of Silesia, 9 Szkolna Street, Pl-40 006 Katowice, Poland for measuring the experimental spectra.

REFERENCES

[1] G. C. Pimentel and A. L. McClellan, "The Hydrogen Bond," W. H. Freeman, San Francisco, 1960.

[2] P. Schuster, G. Zundel and C. Sandorfy, "The Hydrogen Bond: Recent Developments in the Theory and Experiment, Parts I, II and III," Elsevier, Amsterdam, 1976.

[3] Y. Marechal, "The Hydrogen Bond and the Water Molecule," Elsevier, Amsterdam, Oxford, 2006.

[4] H. D. Arman, P. Poplaukhin and E. R. T. Tiekinkc, "2-Pyridone: Monoclinic Polymorph," *Acta Crystallographica Section E: Structure Reports Online*, Vol. 65, No. 12, 2009, p. 3187. doi:10.1107/S1600536809049496

[5] M. T. Boisdon, S. Castillo, J. F. Brazier, J. Favrot and C. J. Marsden, "2(1H)-Pyridinone(2-Pyridone): Self-Association and Association with Water Spectral and Structural Characteristics: Infrared Study and Ab Initio Calculations," *Spectrochimica Acta Part A*, Vol. 59, No. 14, 2003, pp. 3363-3377. doi:10.1016/S1386-1425(03)00162-8

[6] M. T. Boisdon, S. Castillo, J. F. Brazier and J. Favrot, "Vibrational Study of 2(1H)-Pyridinone(2-pyridone) in H_2O and 1-D-2(1H)-Pyridinone(2-pyridone ND) in D_2O," *Spectrochimica Acta Part A*, Vol. 55, No. 7-8, 1999, pp. 1379-1388. doi:10.1016/S1386-1425(98)00301-1

[7] H. Ratajczak and W. J. Orville-Thomas, "Molecular Interactions," John Wiley & Sons Inc., New York, 1980.

[8] G. R. Desiraju and T. Steiner, "The Weak Hydrogen Bond in Structural Chemistry and Biology," Oxford University Press, Oxford, 1999.

[9] M. Gallant, M. T. P. Viet and J. D. Wuest, "Hydrogen-Bonded Dimers. Direct Study of the Interconversion of Pyridone Dimers and Hydroxypyridine Monomers by Low-Temperature Nuclear Magnetic Resonance Spectroscopy," *Journal of the American Chemical Society*, Vol. 113, No. 2, 1991, pp. 721-723. doi:10.1021/ja00002a077

[10] C. Fuke, T. Arao, Y. Morinaga, H. Takaesu, K. Ameno and T. Miyazaki, "Analysis of Paraquat, Diquat and Two Diquat Metabolites in Biological Materials by High-Performance Liquid Chromatography," *Legal Medicine*, Vol. 4, No. 3, 2002, pp. 156-163. doi:10.1016/S1344-6223(02)00011-1

[11] M. El-Kemary, J. A. Organero and A. Douhal, "Assessment of Solvent Effect on the Relaxation Dynamics of Milrinone, " *Journal of Photochemistry and Photobiology A: Chemistry*, Vol. 187, No. 2-3, 2007, pp. 339-347. doi:10.1016/j.jphotochem.2006.10.032

[12] Y. Ducharme and J. D. Wuest, "Use of Hydrogen Bonds to Control Molecular Aggregation. Extensive, Self-Complementary Arrays of Donors and Acceptors," *Journal of Organic Chemistry*, Vol. 53, No. 24, 1988, pp. 5787-5789. doi:10.1021/jo00259a037

[13] M. Gallant, M. T. P. Viet and J. D. Wuest, "Use of Hydrogen Bonds to Control Molecular Aggregation. Association of Dipyridones Joined by Flexible Spacers," *Journal of Organic Chemistry*, Vol. 56, No. 7, 1991 pp. 2284-2286. doi:10.1021/jo00007a007

[14] M. J. Wójcik, W. Tatara, M. Boczar, A. Apola and S. Ikeda, "Spectroscopic and Theoretical Study of Vibrational Spectra of Hydrogen-Bonded 2-Pyridone," *Journal of Molecular Structure*, Vol. 596, No. 1-3, 2001, pp. 207-214. doi:10.1016/S0022-2860(01)00719-0

[15] M. J. Wójcik, "Theoretical Modeling of Vibrational Spectra and Multidimensional Proton Tunneling in Hydrogen-Bonded Systems," *Journal of Molecular Liquids*, Vol. 141, No. 1-2, 2008, pp. 39-46. doi:10.1016/j.molliq.2008.03.002

[16] M. J. Wójcik, "Theory of the Infrared Spectra of the Hydrogen Bond in Molecular Crystals," *International Journal of Quantum Chemistry*, Vol. 10, No. 5, 1976, pp. 747-760. doi:10.1002/qua.560100506

[17] A. Witkowski and M. J. Wójcik, "Infrared Spectra of Hydrogen Bond a General Theoretical Model," *Chemical Physics*, Vol. 1, No. 1, 1973, pp. 9-16. doi:10.1016/0301-0104(73)87017-X

[18] M. J. Wójcik, "Fermi Resonance in Dimers: A Model Study," *Molecular Physics*, Vol. 36, No. 6, 1978, pp. 1757-1767. doi:10.1080/00268977800102741

[19] P. M. Morse, "Diatomic Molecules According to the Wave Mechanics. II. Vibrational Levels," *Physics Reviews*, Vol. 34, No. 1, 1929, pp. 57-64. doi:10.1103/PhysRev.34.57

[20] N. Issaoui, N. Rekik, B. Oujia and M. J. Wójcik, "Theoretical Infrared Line Shapes of H-Bonds within the Strong Anharmonic Coupling Theory and Fermi Resonances Effects," *International Journal of Quantum Chemistry*, Vol. 110, No. 14, 2010, pp. 2583-2602. doi:10.1002/qua.22395

[21] Y. Maréchal and A. Witkowski, "Infrared Spectra of H-Bonded Systems," *Journal of Chemical Physics*, Vol. 48, No. 8, 1968, pp. 3697-3704. doi:10.1063/1.1669673

[22] N. Rösh and M. Ratner, "Model for the Effects of a Condensed Phase on the Infrared Spectra of Hydrogen-Bonded Systems," *Journal of Chemical Physics*, Vol. 61, No. 8, 1974, pp. 3344-3351. doi:10.1063/1.1682497

[23] B. Boulil, O. H. Rousseau and P. Blaise, "Infrared Spectra of Hydrogen Bonded Species in Solution," *Chemical Physics*, Vol. 126, No. 2-3, 1988, pp. 263-290. doi:10.1016/0301-0104(88)85038-9

[24] H. T. Flakus, A. Tyl and A. Maślankiewicz, "Electron-Induced Phase Transition in Hydrogen-Bonded Solid-State 2-Pyridone," *Journal of Physical Chemistry A*, Vol. 115, No. 6, 2011, pp. 1027-1039. doi:10.1021/jp108717v

[25] P. Blaise, M. J. Wójcik and O. H. Rousseau, "Theoretical Interpretation of the Line shape of the Gaseous Acetic Acid Dimer," *Journal of Chemical Physics*, Vol. 122, No.

6, 2005, pp. 64306-64317. doi:10.1063/1.1847491

[26] M. J. Frisch, "Gaussian 03, Revision B.03," Gaussian, Inc., Pittsburgh, 2003.

[27] R. Kubo, "Statistical-Mechanical Theory of Irreversible Processes. I. General Theory and Simple Applications to Magnetic and Conduction Problems," *Journal of Physics Society Japan*, Vol. 12, No. 6, 1957, pp. 570-586. doi:10.1143/JPSJ.12.570

[28] R. Kubo, "Lectures in Theoretical Physics I," In: W. E. Brittin and L. G. DUnham, Eds., *Interscience*, *1st Annual Summer Institute for Theoretical Physics*, Colorado, 16 June-22 August 1958, pp. 120-203.

[29] M. El-A. Benmalti, D. Chamma, P. Blaise and O. H. Rousseau, "Theoretical Interpretation of the Infrared Lineshape of Gaseous Propynoic and Acrylic Acid Dimers," *Journal of Molecular Structure*, Vol. 785, No. 1-3, 2006, pp. 27-31. doi:10.1016/j.molstruc.2005.09.036

[30] K. Heyne, N. Huse, J. Dreyer, E. T. J. Nibbering, T. Elsaesser and S. Mukamel, "Coherent Low-Frequency Motions of Hydrogen Bonded Acetic Acid Dimers in the Liquid Phase," *Journal of Chemical Physics*, Vol. 121, No. 2, 2004, pp. 902-913. doi:10.1063/1.1762873

[31] Z. Mielke and L. Sobczyk, "Vibrational Isotope Effects in Hydrogen Bonds," In: A. Kohen and H.-H. Limbach, Eds., *Isotope Effects in Chemistry and Biology*, CRC Press, Taylor and Francis Group, West Palm Beach, London, 2006, pp. 281-304.

[32] N. Issaoui, N. Rekik, B. Oujia and M. J. Wójcik, "Anharmonic Effects on Theoretical IR Line Shapes of Medium Strong H(D)Bonds," *International Journal of Quantum Chemistry*, Vol. 109, No. 3, 2009, pp. 483-499. doi:10.1002/qua.21839

[33] A. Novak, "Hydrogen Bonding in Solids Correlation of Spectroscopic and Crystallographic Data," *Structure and bonding*, Vol. 18, 1974, pp. 177-216.

[34] I. Olovsson and P. G. Jönsson, "Hydrogen Bond," In: P. Scheuster, G. Zundel and C. Sandorfy, Eds., Elsevier, Amsterdam, 1976, pp. 393-456.

[35] N. Rekik, B. Oujia and M. J. Wójcik, "Theoretical Infrared Spectral Density of H-bonds in Liquid and Gas Phases: Anharmonicities and Dampings Effect," *Chemical Physics*, Vol. 352, No. 1-3, 2008, pp. 65-76. doi:10.1016/j.chemphys.2008.05.009

[36] S. E. Odinokov and A. V. Iogansen, "Torsional γ(OH) Vibrations, Fermi Resonance [2γ(OH) ν(OH)] and Isotopic Effects in i.r. Spectra of H-Complexes of Carboxylic Acids with Strong Bases," *Spectrochimica Acta Part A: Molecular Spectroscopy*, Vol. 28, No. 12, 1972, pp. 2343-2350. doi:10.1016/0584-8539(72)80214-9

[37] P. Blaise, M. J. Wójcik and O. H. Rousseau, "Theoretical Interpretation of the Line Shape of the Gaseous Acetic Acid Cyclic Dimer," *Journal of Chemical Physics*, Vol. 122, No. 6, 2005, p. 064306. doi:10.1063/1.1847491

[38] M. El-A. Benmalti, P. Blaise, H. T. Flakus and O. H. Rousseau, "Theoretical Interpretation of the Infrared Line Shape of Liquid and Gaseous Acetic Acid," *Chemical Physics*, Vol. 320, No. 2-3, 2006, pp. 267-274. doi:10.1016/j.chemphys.2005.07.032

[39] J. L. Leviel and Y. Marechal, "Infrared Spectra of H-Bonded Systems: Anharmonicity oh the H-Bond Vibrations in Cyclic Dimers," *The Journal of Chemical Physics*, Vol. 54 No. 3, 1971, pp. 1104-1107. doi:10.1063/1.1674943

[40] D. Hadzi, "Infrared Spectra of Strongly Hydrogen-Bonded Systems," *Pure and Applied Chemistry*, Vol. 11, No. 3, 1965, pp. 435-454. doi:10.1351/pac196511030435

[41] J. Emsley, "Very Strong Hydrogen Bonding," *Chemical Society Reviews*, Vol. 9, No. 1, 1980, pp. 91-124. doi:10.1039/cs9800900091

[42] W. H. Louisell and L. R. Walker, "Density-Operator Theory of Harmonic Oscillator Relaxation," *Physical Review*, Vol. 137, No. 1B, 1965, pp. B204-B211. doi:10.1103/PhysRev.137.B204

[43] R. Feynmann and F. Vernon, "The Theory of a General Quantum System Interacting with a Linear Dissipative System," *Annals of Physics*, Vol. 24, 1963, pp.118-173. doi:10.1016/0003-4916(63)90068-X

[44] W. H. Louisell, "Quantum Statistical Properties of Radiations," Wiley, New York, 1973.

[45] K. Belhayara, D. Chamma, A. Velcescu and O. H. Rousseau, "On the Similarity of the IR Lineshapes of Weak H-Bonds in the Gas and Liquid Phases: Quantum Combined Effects of Strong an Harmonic Coupling, Multiple Fermi Resonances, Weak Dampings and Rotational Structure," *Journal of Molecular Structure*, Vol. 833, No. 1-3, 2007, pp. 65-73. doi:10.1016/j.molstruc.2006.09.001

[46] L. C. Thomas, R. A. Chittenden and H. E. Hartley, "Hydrogen Bonding in Alkyl Phosphonic and Alkyl Phosphonothionic Acids," *Nature*, Vol. 192, 1961, pp. 1283-1284. doi:10.1038/1921283a0

[47] K. G. Tokhadze, G. S. Denisov, M. Wierzejewska and M. Drozd, "First Example of the ABC ν(OH) Absorption Structure for Both Gaseous and Crystalline Phase: Infrared Studies of Dimethylphosphinic Acid," *Journal of Molecular Structure*, Vol. 404, No. 1-2, 1997, pp. 55-62. doi:10.1016/S0022-2860(96)09360-X

[48] R. E Asfin, G. S. Denisov and K. G. Tokhadze, "The Infrared Spectra and Enthalpies of Strongly Bound Dimers of Phosphinic Acids in the Gas Phase. $(CH_2Cl)_2$ POOH and $(C_6H_5)_2$ POOH," *Journal of Molecular Structure*, Vol. 608, No. 2-3, 2002, pp. 161-168. doi:10.1016/S0022-2860(01)00925-5

[49] H. T. Flakus, "On the Vibrational Transition Selection Rules for the Centrosymmetric Hydrogen-Bonded Dimeric Systems," *Journal of Molecular Structure: Theochem*, Vol. 187, 1989, pp. 35-53.

[50] H. T. Flakus and A. Miros, "Infrared Spectra of the Hydrogen Bonded Glutaric Acid Crystals: Polarization and Temperature Effects," *Journal of Molecular Structure*, Vol. 484, No. 1-3, 1999, pp. 103-115.

[51] H. T. Flakus and M. Jabonska, "Study of Hydrogen Bond Polarized IR Spectra of Cinnamic Acid Crystals," *Journal of Molecular Structure*, Vol. 707, No. 1-3, 2004, pp. 97-108.

[52] D. Chamma and O. H. Rousseau, "Infrared Spectra of

weak H-Bonds: Beyond an Adiabatic Description of Fermi Resonances," *Chemical Physics*, Vol. 248, No. 1, 1999, pp. 91-104.

[53] N. Issaoui, H. Ghalla and B. Oujia, "Theoretical Study of Hydrogen and Deuterium Bond in Glutaric Acid Crystal Dimer," *International Journal of Quantum Chemistry*, Vol. 112, No. 4, 2012, pp. 1006-1015.

32

The Physics of the Phenomena of Electrical Induced Elements Separation in Solution of Salts in Liquid Polar Dielectric

Mishik A. Kazaryan[1], Igor V. Shamanin[2*]
[1]National Academy of Sciences of Republic Armenia, Yerevan, Armenia
[2]Tomsk Polytechnic University, Tomsk, Russia
Email: *shiva@tpu.ru

ABSTRACT

In the work [1] is shown the possibility of the use the phenomena of the electrical-induced selective drift of cationic aquacomplexes in the water solution of the salts for creating of the new, non-electrode, power-saving technology of the solution enrichment by desired element. The development of the naturally-scientific basis of this technology has showed that inertial properties of formed in the solution over-molecular structural units—clusters depends from that's dimensions. Dimensions of these clusters are in limits from scores of angstrom to fractions of micron, at the same time liquid mediums are the base of functioning both biological and technological systems. Thus, water solution of salt contains nanoparticles and biological and many technological systems are functioning with the assistance of nanoparticles. For example, the ordinary blood circulatory system or many of chemical processing. The casual or intentional action of aperiodic electric field on systems containing salts solutions produces some positive or negative effects. This work shows new approach for theoretical describing of salts solution in dielectric liquids and contains some data on the realized experiments.

Keywords: Electromagnetic Field; Solution; Polar Liquid Dielectric; Solvation; Aqua Complex; Solvated Ion; Hydration Shell; Over-Molecular Cluster; Polarization; Polarized Charge; Directed Drift; Separation; Oscillations; Own Frequencies

1. Condition of the Problem

At dissolution occurs complex formation therefore is formed supermolecular structure—in water solutions of hydrated aquacomplexes. Aquacomplex may be considered as nanoparticle, formed from central ion and molecules of water.

At action of the external periodic electric field the solvated ions-clusters are drifting [1]. The oriented drift of the polarized solvated ions allows organizing the technological process of element enrichment of water solutions of salts [2]. Experimental researches have shown, that it not the sole appendix of the found out phenomenon (effect). In particular, the solutions of salts of metals in the polar, liquid dielectrics (water, spirits, ketones) show interesting properties at action of the magnetic fields.

As show theoretical and experimental researches, the size of cluster is the quantity determining the majority of physical and technological properties of solutions. Nowadays, two approaches were developed for definition of properties of solutions. The opportunity of use of the plasma concept of a condition of solutions of electrolytes is noted in the works of M. M. Baldanov [3]. The second approach assumes the modeling of processes of solvation on the basis of the incorporated quantum-classical theory of the functional of density [4].

Both in that, and in other case, for the obvious reasons, properties of solutions are defined mainly by energy of multiparty interactions of ions in the dielectric medium. And it uniquely assumes attraction of such individual characteristics, experimental definition of which is impossible (a condition of an electroneutrality), are probable only theoretical estimations, or semiempirical processing of an experimental material on the basis of some plausible, but arbitrary assumptions.

The validity of the assumptions can be certain experimentally, for example, at comparison of observable and theoretical values of any parameter which value directly depends on the solution characteristic of the defined within the limits of this or that approach. Such works

*Corresponding author.

have been begun in 30th years of the last century [5].

In the works [3,4] were received the radiuses values of the solvated ions, which compound units of angstroms. Use of such values at a theoretical explanation of the found out phenomenon (effect) of the selective drift of solvated cations under action of an asymmetric electric field [1] leads to estimated values of frequencies of the field (about 100 megahertz) at which it is necessary to expect display of the effect. Nevertheless, the phenomenon takes place at the frequencies which are not exceeding units of kilohertz, and at the certain concentration of salt in water—tens of hertz. Besides there are the experimental data specifying that clusters represent submicron formations [6].

In the work the solution is shown as quasi-neutral system of polarized charges which interacting with external electric field and among themselves. Consideration is spent with the purpose of definition of the geometrical sizes solvated ions—clusters and general characteristics of process of influence on the solution of an external electromagnetic field. Results will allow defining opportunities of various appendices of the found out phenomenon (effect).

2. The Self-Consistent Field in the Solution of Salt in the Liquid Polar Dielectric

Let' consider, that each cation (anion) causes the polarization of surrounded solvent (water): the "atmosphere" is forming around each cation (anion) with abundance of polarized molecules of the water. This "atmosphere" is shielding cation's (anion's) field. The generalized electronic shell of each polarized molecules of water is deformed concerning unperturbed configuration, when full spin of the molecule is a zero. The perturbation is caused by the action of the cation's (anion's) electric field. The deformation of the shell is cause that the part of nucleus' charge or electrons in the water's molecule will be uncompensated. This uncompensated part of a charge represents a polarizing charge of a molecule of water in a non-uniform electric field of the charged particle—cation (anion). According to representation about the self-coordinated field [7] there is such distribution of an electric field in system of the cooperating charged particles which creates the distribution of particles exciting a field. Salt solution is possible to consider as system of cooperating cations, anions, it is positive and negatively polarized molecules of water. For a finding of the self-coordinated field we shall use Poisson equation:

$$\Delta\varphi = -4\pi q \tag{1}$$

and Boltzman distribution:

$$n_k = \tilde{n}_k \exp\left(-\frac{Z_k e\varphi}{kT}\right) \tag{2}$$

where n_k—concentration of particles with charging number Z_k in a point with potential φ. For electrons, for example, $Z = -1$. But in a solution are not present free electrons. For cations $Z = m$, and for anions $Z = -m$, where m-valency of metal which salt is dissolved. $\tilde{n}_k$ in distribution (2) is a concentration of particles with charging number Z_k in a point with the zero potential, equal to average concentration of these particles on all volume of a solution.

Further an index "i" we shall designate cations, an index "a" – anions, "p^+"—polarized and being around of anions molecules of water, "p^-"—molecules of water around cations.

Average concentration $\tilde{n}_k$ satisfies to a condition of quasi-neutrality:

$$\sum_i \tilde{n}_i Z_i + \sum_a \tilde{n}_a Z_a + \sum_{p^+} \tilde{n}_{p^+} Z_{p^+} + \sum_{p^-} \tilde{n}_{p^-} Z_{p^-} = 0 \tag{3}$$

which show the fact, that the solution as a whole (from outside) is neutral.

Also following conditions take place:

$$\sum_i \tilde{n}_i Z_i = -\sum_a \tilde{n}_a Z_a, \quad \sum_a \tilde{n}_a Z_a = \sum_{p^+} \tilde{n}_{p^+} Z_{p^+}, \quad \sum_i \tilde{n}_i Z_i = \sum_{p^-} \tilde{n}_{p^-} Z_{p^-} \tag{4}$$

In which it is considered, that charging numbers of cation and anion are equal on absolute value, but opposite, and charging numbers of the polarized molecules of water are determined by around of what they associated: around anions $Z_{p^+} > 0$, and around cations $Z_{p^-} < 0$.

Thus,

$$\sum_k \tilde{n}_k Z_k = 0, \quad k = i, a, p^+, p^- \tag{5}$$

Also following connections take place:

$$Z_a = -Z_i = Z;\ Z_{p^-} = \frac{1}{N_a} Z_a;\ Z_{p^+} = \frac{1}{N_i} Z_i; \quad \tilde{n}_{p^-} = N_a \tilde{n}_a;\ \tilde{n}_{p^+} = N_i \tilde{n}_i;\ \tilde{n}_a = \tilde{n}_i = n_m; \tag{6}$$

where N_a—number of molecules of water, associated around of one anion; N_i—number of molecules of water, associated around of one cation; Z—valency of metal, which salt is dissolved; n_m—concentration of molecules of salt in a solution, if to consider they are not dissociated.

Volume charge thus,

$$q = e\sum_k Z_k n_k \tag{7}$$

where e—the module of a charge of electron (in system ***SI*** $1.6 \cdot 10^{-19}$ C). The correlation (7) is very similar to expression (5) with that essential difference, which in it there is a value of concentration of particles in that point in which the potential of the self-coordinated field is determined.

Equation of Poisson will copy in the form of:

$$\Delta\varphi = -4\pi e \sum_k Z_k \tilde{n}_k \exp\left(\frac{Z_k e\varphi}{kT}\right) \quad (8)$$

Using series expansion and considering connections (4) and (6), it is possible to receive function:

$$\sum_k Z_k \tilde{n}_k \exp\left(\frac{Z_k e\varphi}{kT}\right) \approx \frac{Z^2 e n_m \varphi}{kT}\left(2 + \frac{1}{N_i} + \frac{1}{N_a}\right) \quad (9)$$

Even if takes place no more, than secondary hydration, in hydrate shell it will be concentrated more than 10 molecules of water. Therefore it is possible to consider, that $2 \gg 1/N_i$ and $2 \gg 1/N_a$. In view of this, equation of Poisson will become:

$$\Delta\varphi = -\frac{8\pi Z^2 e}{kT}\cdot\varphi . \quad (10)$$

The decision (10) for spherical symmetric distribution of potential (cation or anion) looks like:

$$\varphi = \frac{C}{r}\exp(-\chi r), \quad (11)$$

where constant of insulation is:

$$\chi = \left(\frac{8\pi e^2}{kT} Z^2 n_m\right)^{0.5} \quad (12)$$

The inverse value $l = 1/\chi$ refers to as length of insulation and it is possible to consider, that its value defines radius of solvation sphere within the limits of which the polarized molecules of water are built. They also insulate a field of cation (anion). Thus, value of cluster radius (solvated cation or anion) can be estimated, using:

$$r_{cl} \approx \left(\frac{8\pi e^2}{kT} Z^2 n_m\right)^{-0.5} \quad (13)$$

3. The Characteristic Values of Frequencies of Excitation of Effect of Oriented Drift of the Solvated Ions

The own frequency of solvated ion (cluster) can be expressed of following equation: $\nu = \nu_e + \nu_k + \nu_r$, where items correspond of frequencies of excitation of electron shells (*e*), oscillation an ion and solvated groups (*k*) and rotations solvated groups concerning an ion (*r*). It is supposed, that excitation of electronic shells or oscillation of ion and solvated groups can be initiated only after deformation solvated shell (screen), which inhibits influence of an electric field. Deformations (rotation solvated groups concerning an ion) correspond to item ν_r.

Frequencies of transition of oscillatory motions in rotary are defined by equation:

$$\nu_0 \cong h/(8\pi^2 I) \quad (14)$$

where *h*—Planck's constant, *I*—the moment of inertia of system of the connected particles. The given equation is valid if cluster it is possible to consider as linear system of two rigidly connected particles (an ion and solvated shell), and oscillatory movement of any of particles causes rotation of cluster as the whole. These frequencies in conditions of the achievement of relative mobility of clusters, which is required for activation of process of the selective drift under influence of field, is possible to considered as the minimum frequencies of own free movements clusters. Broader range of frequencies of own motions of clusters, which is connected with free rotary-onward motions, can be presented in the form of set of the frequencies of the allowed quantum levels of free rotation by analogy to a spherical top (all of its three principal of the moment of inertia are equal):

$$\nu = J(J+1)h/(8\pi^2 I) \quad (15)$$

and frequencies of onward moving:

$$\nu = (n_x^2 + n_y^2 + n_z^2)h/(8\mu d^2) \quad (16)$$

where *J*—quantum number, $n_{x,y,z}$—the projections of the vector of the rotation to coordinate axes, *d*—the size of cluster, μ—the brought weight of cluster. The frequency of onward moving is meant frequency of rotation of all clusters (an ion together with the solvated shell) as the whole when sets of all three main directions of the fluctuations will be transformed in each other

If in Formulae (15)-(16) the form of cluster take as spherical—"zero" approach, it is possible to consider, that J = 1, $(n_x^2 + n_y^2 + n_z^2) = 1$. The number of the molecules of water (the solvated groups) in the cluster, having radius r_{cl}, is g, $d = 2r_{cl}$, $r_{cl} = r_g^{1/3}$, $r = (3V/(4))^{1/3}$, where *r*, *V*—radius and volume of the solvated groups, respectively. Thus $I = K_r \mu r_{cl}^2$, where K_r—the factor equal to the relation of the moment of inertia, corresponding rotation solvated shell around of an ion, to average of value of the projection of the oscillatory moment on the axis of rotation. This factor characterizes intensity of the Carioles interactions (the analogy to Carioles forces in the classical mechanics). It have value 0.4 for the rotation the solvated shell around cation (anion) and 1.4—for the rotation around axis, passing regarding to surface cluster. As the weight of the solvated shell considerably surpasses weight of the ion, positions of the centre of gravity (the center of inertia) and the ion practically coincide. The values of the frequencies received in the Formulae (14)-(16) for J = 1 (the minimal values), correspond to a case when only one aquacomplex is formed, and allow to estimate only values of the frequencies which should be expected in the experiments.

As follows from the ratio (13), the value of dimension

of the cluster is inversely to a square root from value of concentration of salt in water. Values of the frequencies are inversely to the value of the moment of inertia of the cluster. The moment of inertia is proportional to the equivalent mass of the cluster, that is the quantity of molecules of water in the solvated shell $g=(r_{cl}/r)^3$, and the square of radius of cluster r_{cl}^2. It turns out, that the moment of inertia $I \sim r_{cl}^5$ and values of the frequencies of excitation of various components of motions $\nu \sim r_{cl}^{-5}$. Thus, values of the frequencies $\nu \sim n_m^{2.5}$.

At present time there are several used approaches but only two forms that are checked in experiments. One of them was investigated for concentration of the salts in water solution, providing performing the condition $n_i \ll n_{H_2O}$, where n_{H_2O} —an amount of the molecules of water in unit of the volume, n_i —for amount ion ditto, it is possible to expect that: *potential becomes too small on distances, smaller distances between most close located ions* [1]. Values of frequencies, corresponding to different components of the rotary-onward motion of cationic aquacomplexes, and values of the frequencies, corresponding to the turning of the oscillatory motion in rotary, are in limits from several hundreds Hertz to several kiloHertz. If the solution is considered as quasi-neutral system of polarized charges which interacting with external electric field and among them, frequencies of oscillations are much less (no more than 100 Hertz).

4. The Fluctuations of the Polarized Charge in the Solution

In external electric field polarized due to deformation of the aquation shell of aquacomplexes are displaced. As a result there is a division of the aquacomplexes, displaced to a greater or lesser extent. Polarized charges of the divided aquacomplexes differ, at least, on absolute value. Thus, there is a division of charges and electrostatic forces in volume of a solution. Electrostatic forces should lead to excitation of own electrostatic fluctuations in volume on the average neutral (in an enough great volumes or for enough greater time intervals) a solution.

Let as a result of division of the polarized aquacomplexes in a solution there is a polarized charge q. Thus

$$q=-\boldsymbol{divP} \tag{17}$$

where $\boldsymbol{P}=\chi\cdot\varepsilon_0\cdot\boldsymbol{E}$ —a vector of polarization; χ —a dielectric susceptibility of a solution; ε_0 —an electric constant of vacuum; $\boldsymbol{E}$—a vector of intensity of the electric field acting on a solution. We shall remind that it is a question of solutions of salts of metals in polar dielectric liquids. Thus

$$\chi=\bar{n}\cdot\alpha \tag{18}$$

where α —factor of polarizability of the solvated ion—cluster; $\bar{n}$ —average number of clusters in unit of volume of a solution, and $\bar{n}=2n_m$, where n_m—number of dissociating molecules of salt in unit of volume of a solution.

The polarized charge of one cluster $(p=q/\bar{n})$ will be defined, thus, by expression:

$$p=-\frac{\chi\cdot\varepsilon_0}{2\cdot n_m}\boldsymbol{divE}=-\alpha\cdot\varepsilon_0\cdot\boldsymbol{divE} \tag{19}$$

The According to the law of preservation of a charge

$$\frac{\partial q}{\partial t}=-\boldsymbol{divj} \tag{20}$$

where $\boldsymbol{j}$—density of a current. Thus the current is carried by the polarized aquacomplexes. Then

$$\boldsymbol{j}=2\cdot n_m\cdot p\cdot\mathbf{v} \tag{21}$$

where $\boldsymbol{v}$—speed of the aquacomplexes, carrying a current. The equation of movement polarized aquacomplex, having weight m, will enter the name in the form of

$$m\frac{d\mathbf{v}}{dt}=p\boldsymbol{E} \tag{22}$$

Differentiating the Equations (20) and (21) and substituting (21) in (20), we shall receive

$$\frac{\partial^2 q}{\partial t^2}=-2n_m p\boldsymbol{div}\frac{\partial\mathbf{v}}{\partial t} \tag{23}$$

Substituting in (23) expression for $\partial\mathbf{v}/\partial t$, received of (22), and considering, that $\boldsymbol{divE}=4\pi q$, we shall receive

$$\frac{\partial^2 q}{\partial t^2}=-\frac{8\pi n_m p^2}{m}q \tag{24}$$

The received equation describes simple harmonious fluctuation with circular frequency

$$\omega_0=\sqrt{\frac{8\pi n_m p^2}{m}} \tag{25}$$

If the solution is placed between two flat electrodes, the distance between which is not great, consideration of process can be spent in one-dimensional geometry. We shall designate through E_x absolute value of intensity of the electric field acting on a solution. Then

$$\omega_0=\alpha\cdot\varepsilon_0\cdot\boldsymbol{divE}\cdot\sqrt{\frac{8\pi n_m}{m}} \tag{26}$$

or

$$\omega_0=\alpha\cdot\varepsilon_0\cdot\frac{E_x}{\Delta}\cdot\sqrt{\frac{8\pi n_m}{m}} \tag{27}$$

where Δ—thickness of a layer of a solution between the potential and earthed flat electrodes.

Thus, the density of a volumetric polarized charge in a solution will change with circular frequency ω_0 $(\nu_0=\omega_0/2\pi)$ after action on a solution of an external electric field. Thus frequency of fluctuations is defined not by amplitude of intensity of this field, and divergence of intensity of a field in volume of a solution.

5. Electromagnetic Wave Action on the Solution

As well as in plasma, any separation of charges in a solution of salt in liquid polar dielectric leads to fluctuations of density of a charge. On the average for many periods of fluctuations the solution behaves as quasi-neutral medium. Separation of polarizing charges is possible only on temporary intervals, smaller in comparison with temporary scale of separation of charges $t_0 \sim (\omega_0)^{-1}$. For spatial scale of separation of the charges *d* it is possible to accept distance which solvated the ion at the thermal movement migrate in time t_0, that is $d \sim \langle \upsilon \rangle / \omega_0$, where $\langle \upsilon \rangle$—the average speed of thermal movement of solvated ions (clusters), having weight *m*. In spatial scales, greater in comparison with *d*, it is observed quasi-neutral the solution.

By definition the dielectric constant of the continuous environment ε is the relation of intensity of external electric field E to intensity of the weakened field inside of this medium $E^{'}$ (dielectric). Thus ε always is more than 1.

In a solution $E/E' > 1$. Hence for a solution $\varepsilon < 1$. Thus the smaller frequency of an external electric field ω, $\nu = \omega/2\pi$, the more "range" of fluctuations polarized solvated ions, that is amplitude of their displacement. It turns out, that with diminution of ω the dielectric constant of a solution ε decreases also.

Also behave also free electrons in plasma that is used in techniques of sounding of plasma by radiowaves. In case of plasma is critical value of frequency ω_k of radiowaves, at which dielectric constant of plasma $\varepsilon = 0$. Value of this frequency coincides with value the Longmuir frequencies of plasma.

In our case the analogue of Longmuir frequencies is frequency of fluctuations of the polarized charge in volume of a solution ω_0.

The dielectric constant of plasma is defined by the ratio:

$$\varepsilon = 1 - (\omega_k/\omega) \tag{28}$$

where ω—frequency of an external radio emission. If this frequency $\omega < \omega_k$, then $\varepsilon < 0$.

In Maxwell's works it is established, that factor of refraction of an electromagnetic wave in substance $\gamma = \sqrt{\varepsilon}$. At $\varepsilon < 0$ electromagnetic waves cannot extend in substance and should be reflected completely from it. Therefore in relation to waves with frequency $\omega < \omega_k$ plasma is an ideal reflector. By analogy it is possible to assume, that a solution salt in liquid, polar dielectric will be a reflector for waves with frequency

$$\omega_0 < \alpha \cdot \varepsilon_0 \cdot \boldsymbol{divE} \cdot \sqrt{\frac{8\pi n_m}{m}} \tag{29}$$

where ***E***—intensity of an electric field in a wave.

6. Conclusions

The solvation process of ions in solutions of salts in liquid, polar dielectrics provides an opportunity of development of essentially new technologies and techniques. The conditions providing use of process mass transfer for the decision of various applied problems, first of all are defined by the sizes of solvated ions (clusters).

In particular, action of periodic electric fields with various combinations of frequency, intensity and the attitude of amplitudes of intensity in half-cycles on solutions of salts in polar dielectrics causes excitation of rotary-forward motion solvated ions (clusters) and, hence, mass transfer.

On the **Figure 1** is shown the scheme of experimental cell which was constructed for separation of solvated ions of cerium and lanthanum.

At submission on the input of the separating element on the basis of a technological cell of the base mixture of two component (a water solution of the mix of salts $Ce(NO_3)_3$ and $Y(NO_3)_3$ with concentration 3.5 and 3 g/l, accordingly) the separation factor in an element without circulation of the solution after 4 hours of influence of the field with intensity $E^+ = 14.3$ V/cm, factor of asymmetry $A^-/A^+ = 0,66$, with frequency 1.6 kHz have made: in the selection 1.00499; in the dump 1.00179; full 1.00679. In a solution there was an enrichment on the cations Ce^{3+}.

Checking of the condition of water in the solutions of sugars (α-methyl-D-glucopyranoside, α-methyl-D-mannopyranoside, α-methyl-D-glucopyranoside, Lactose and

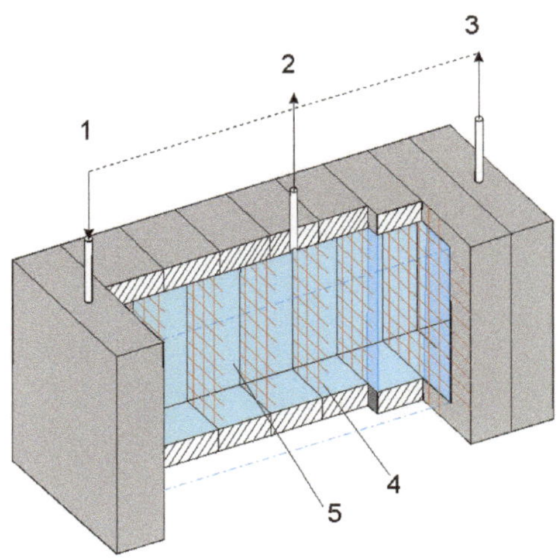

Figure 1. Technological cell: 1—nipple for the solution inlet; 2—nipple for the solution selection; 3—nipple for the solution selection and circulation; 4—insulated potential high-transparent nets; 5—solution.

saccharine) by means Spectrophotometer Perkin-Elmer Model 180 have showed, that a marked difference occurs between spectrums of light absorption by solution before and after asymmetrical electric field action. The received facts confirm that action of field on the solution causes either changes of certain conditions of water in the solution or changes of biological active compounds concentrations in the gap between potential and neutral electrodes.

Action of an electromagnetic wave on a solution salt in liquid, polar dielectric causes division of charges and leads to fluctuations of density of a charge. At the certain combinations of parameters of an electromagnetic wave and a solution it is possible to expect partial or its full reflection from the medium formed by a solution that allows considering the problem opportunities of use of electromagnetic waves for definition of parameters of solutions.

REFERENCES

[1] M. A. Kazarjan, I. V. Lomov and I. V. Shamanin, "The Physical Models and Applications of the Solvation Process in Salts Solution in Polar Dielectric," *International Scientific Journal for Alternative Energy and Ecology*, Vol. 11, No. 55, 2007, pp. 145-154.

[2] A. L. Gusev, M. A. Kazarjan, Yu. A. Trutnev, I. V. Shamanin and I. V. Lomov, "The Non-Electrode Technology of the Water Solution Enrichment by Desired Element in the Complex Conversion of Nucleus Raw Material," *International Scientific Journal for Alternative Energy and Ecology*, Vol. 3, No. 47, 2007, pp. 49-52.

[3] M. M. Baldanov, B. B. Tanganov and M. V. Mohoseev, "Conduction of Water Solution of the Weak Acids," *Doklady Akademii Nauk SSSR*, Vol. 299, No. 4, 1988, pp. 899-904.

[4] M. V. Fedorov and A. A. Kornishev, "Unravelling Solvent Response to Neutral Charged Solutes," *Molecular Physics*, Vol. 105, No. 1, 2007, pp. 1-7. doi:10.1080/00268970601110316

[5] A. N. Terenin, "Spectrums of the Absorption of Electrolytic Solutions," *Uspekhi Phyzicheskikh Nauk*, Vol. 17, No. 1, 1937, pp. 2-54.

[6] Yu. K. Stishkov and A. V. Steblianko, "The Breach of Homogeneity of the Weakly Conducting Liquids in Strong Electric Fields," *Zhurnal Tekhnicheskoi Fiziki*, Vol. 67, No. 10, 1997, pp.105-111.

[7] D. A. Frank-Kamenetskiy, "Lekcii po Phyzike Plazmi," Atomizdat, Moskow, 1968.

Electron—Proton Twins, Orderly Arranged in the Inside of Bioatoms

John C. Stavridis[1], George Baltatzis[1], Constantine N. Pattas[2]
[1]Institute of Biomedical Research & Biotechnology, Athens, Greece
[2]Department of Mechanical Engineering, Laboratory of Applied Thermodynamics, Polytechnic School of Thessaloniki, Thessaloniki, Greece
Email: sengir@ath.forthnet.gr

ABSTRACT

In the present study the first 20 microatoms of the periodic table are named as bioatoms, since they are needed for biochemical functions and services to life itself. The rationale behind this project is the detailed analysis of the ionization energy of the electrons in the inside of the bioatoms and their orderly arrangement at specific positions. Such position of the electrons is harmonized with the succession of their energy values in their logarithmic curves. The compelling arrangement of each electron at a particular place would not be feasible in the absence of an energy offset, which is a proton. The latter would hold electrons on their proper position. These fundamental aspects of our project receive such strong support from the quantum theory, according to which it is the electromagnetic interaction between electrons and protons by the exchange of photons, which hold them together in the atoms. According to our proposed model, the protons of the aforementioned proton—electron twins, are distributed on distinct positions which are the junctions of a primary network, coupled with their electrons, in a secondary network. The geometry and distance of the protons, in the plexus, is constant and is calculated at 8, 42 Å. This was estimated by a mathematical analysis of the proposed model, as discussed further. Our study has shown that electrons in the specific positions are moving in a symmetrical oscillation in the length of a channel, in vacuum, under the Coulomb forces. Moreover, the chemical evidence suggests that electrons, in an atom, have their own separate orbit, and that these orbits are closely interrelated.

Keywords: Bioatoms; Proton Network; Ionisation Energy; Helium

1. Introduction

The term "bioatom" is, for the first time, being used in the present study. Bioatoms are the original atoms (microatoms) of the periodic table, essential for biological activities. This new term lends itself to a distinction between the microatoms and the heavier atoms of the periodic table; In accordance with the existing knowledge concerning the origin of the atoms, the chemical elements are formed by a primary stellar nuclosynthesis, while the heavier components are born by the explosion of supernovae. If, according to Rutherford's model [1,2], electrons are placed at specific locations around the nucleus of the atom, the stability of the atom's structure presupposes an energy influence on each individual electron by an equal but opposite energy unit, which in the case in point is a proton. According to quantum theory, it is the electromagnetic interaction (by the exchange of photons) which holds electrons and nuclei (protons) together in the atoms [3]. But up till now quantum theory has never provided a practical model of how electrons and nuclei (protons) can absorb and emit photons [3]. This is because the completely successful application of the quantum theory to the atom structure has been limited to the hydrogen atom which disposes only one electron. There is, therefore, no certainty as to how this theory could be formulated for the case of a more complicated atom such as bioatoms. The interchange of photos, between electrons and protons, strengthens our point of view that all electrons, in their fixed position inside the electromagnetic field in the interior of the bioatoms are influenced equally by the same quantity of energy, due to protons. The magnetic properties of these subatomic elements furnish excellent proof of the electron coupling with protons.

Our proposal gives a concrete answer to this leading issue concerning the interrelationship between electrons and protons, by proving that protons and electrons participate mutually—by the exchange of photons—to the formation of twins, in a steady position and constant distance between them. The accuracy of this claim is affirmed by the quantity of the ionization energy necessary for the successive detachment of the electrons. Looking carefully at the absolute and logarithmic values of the successive ionization energies of electrons in different published papers [4-29] we can ascertain that in all atoms there are two central electrons that have ionization energy manifold greater as compared to the ionization energy of the rest of the peripheral electrons (**Figure 1**).

We suggest that these two electrons are not placed on a separate orbit but together with two protons and two neutrons form a helium atom placed at the centre of the atom (**Figure 2(c)**). For these two electrons we have estimated a progressive increase, in the values of the ionization energy for carbon to the calcium atom (**Table 1**). In order to study in detail these two interesting electrons we have scrutinized a possible attachment of a structure unit to a helium atom for its transition to another atom of different structure and different properties. A structure unit in this study is a new term, relevant to three subatomic particles: proton, electron and neutron, as a structural and functional entity (**Figures 2(a)** and **(b)**).

It appears that the structure units, during their attachment to helium atoms are placed in a mirror-image position, guided by Coulomb forces and form two networks:

The primary network, which consists of protons and neutrons and the secondary network which consists of electrons (**Figure 3**).

In our mathematical analysis we will show that the distance among protons and between protons and electrons are strictly equal, calculated at 8, 42 Å.

The exceptional upgrading accuracy of the values of the electron ionization energy in the literature as compared to the results of our new atomic structure for all electrons, especially for the two central electrons, shown in **Table 2**, is an extremely important finding as it substantiates the logic of the proposed new structure of bioatoms.

The attachment, for example, of a structure unit to a helium atom should be followed by the addition of certain quantity of energy which has been estimated in our study at 2.633 $MJ \cdot mol^{-1}$ for each of the two central electrons (**Table 1**). For the more peripheral eight electrons (octet of Landé) [30,31], there is also a need for the addition of a quantity of energy equal to 0.66 $MJ \cdot mol^{-1}$ for each electron (**Figure 4**).

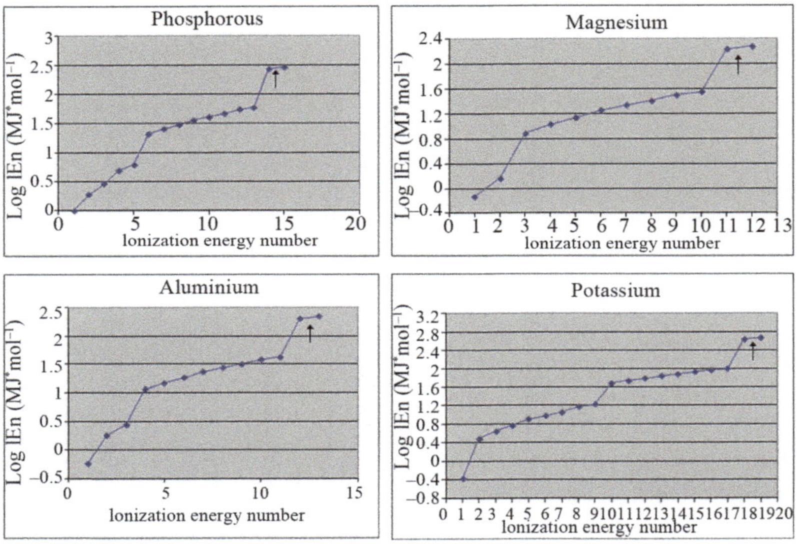

Figure 1. Two electrons (indicated by arrows) have ionization energy manifold greater than the ionization energy of other electrons.

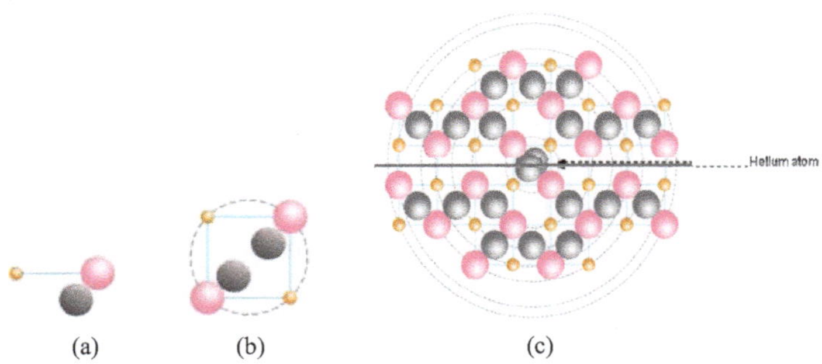

Figure 2. (a) Structural unit; (b) Helium atom; (c) Calcium atom showing protons (pink), electrons (orange), neutrons (grey) and a helium atom at the center of the atom.

Table 1. Upgrading of the ionization energies of the two central (α_1, α_2) electrons from the carbon atom to the calcium atom (I.E.: Ionization Energy).

Elements	I.E. MJ·mol^{-1}	Upgrading valus of the I.E. MJ·mol^{-1}	Rate of the upgrading values of the I.E. MJ·mol^{-1}
Beryllium	32.983	14.294	2.789
Carbon	47.277	17.083	2.735
Nitrogen	64.360	19.718	2.356
Oxygen	84.078	22.074	2.934
Fluoride	106.434	25.008	2.635
Neon	131.432	27.643	2.650
Sodium	159.075	30.293	2.654
Magnesium	189.368	32.947	2.658
Aluminum	222.315	35.605	2.665
Silicone	257.920	38.270	2.677
Phosphorous	296.19.	40.947	2.674
Sulful	337.137	43.621	2.686
Chlorine	380.758	46.307	2.689
Argon	427.065	48.996	2.703
Potassium	476.061	51.699	
Calcium	527.760		

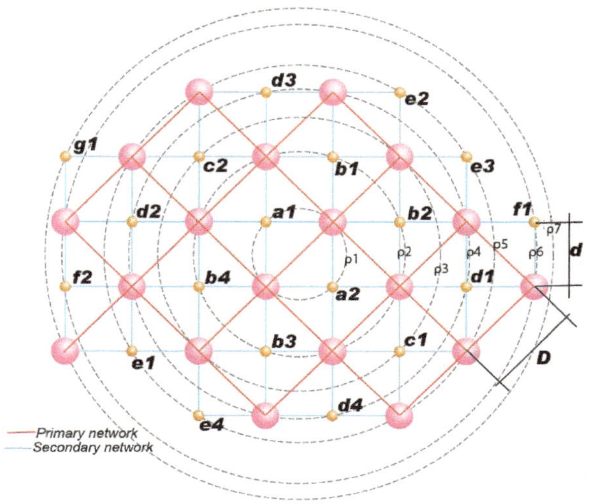

Figure 3. Primary and secondary network in a potassium atom.

For the most peripheral octet of electrons the quantity of energy for each electron is 0.33 MJ·mol^{-1}.This energy arrangement shows that the quantity of energy for each specific electron is strictly graduated in proportion to the degree of influence they receive according to their position inside the energy field in which they reside. Quantum theory maintains that among electrons and protons there is an electromagnetic reaction (interchange of photons) which holds electrons and protons in a fixed position [3]. This couldn't have happened by a central solitary energy nuclear "mass", if electrons are placed at different distances from the nucleus. In order for each proton to influence equally each peripheral electron it should be placed at equal distance from it. Thus, the progressive structure of bioatoms presupposes an arrangement in fixed positions. This kind of structure in bioatoms is confirmed by the mode of upgrading the electron ionization energy from one atom to the next (**Table 1**).

Table 2. Central electrons: exceptional accuracy of the values of their ionization energy in the literature as compared to the results of our study.

1	2	3	4	5	6	7
No.	Element	λ	μ	η	$E_{\alpha 1}$ Theory GJ/kmol	$E_{\alpha 1}$ Experiment GJ/kmol
1	Nitrogen	7	7	2	64.48	64.36
2	Oxygen	8	8	2	84.22	84.08
3	Fluoride	9	9	2	106.60	106.43
4	Sodium	11	11	2	159.24	159.07
5	Magnesium	12	12	2	189.50	189.37
6	Aluminum	13	13	2	222.40	222.31
7	Silicone	14	14	2	257.94	257.92
8	Phosphorous	15	15	2	296.10	296.19
9	Sulfur	16	16	2	336.90	337.14
10	Potassium	19	19	2	475.08	476.06
11	Calcium	20	20	2	526.40	527.76

λ= number of electrons in the atom, μ = number of neutrons in the atom, η = number of atoms in the molecule, $E_{\alpha 1}$ = detachment energy

Such an arrangement of electrons and protons in space could be ensured only with the formation of twins, which in all, form the networks and contribute to the formation of the electromagnetic field.

The accuracy of the proposed model is affirmed by the quantity of the ionization energy, necessary for the successive detachment of the electrons. The harmonization of the electron detachment with the arrangement of the structure unit into the specially shaped energy field, both accommodate the equal interaction between electrons and protons and create the energy decay of the energy field from the centre of the atom to the peripheral orbit.

In our proposal, electrons move in symmetrical rectilinear oscillation in the length of the electromagnetic channel, in vacuum, under the influence of Coulomb forces (**Figure 5**).

The quantity of ionization energy needed for the detachment of each electron was estimated even on the basis of its oscillation amplitude (**Table 3**).

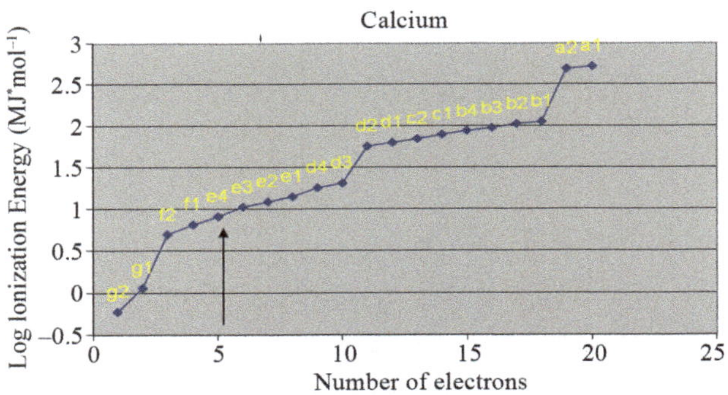

Figure 4. Grouping of electrons in octets (arrows).

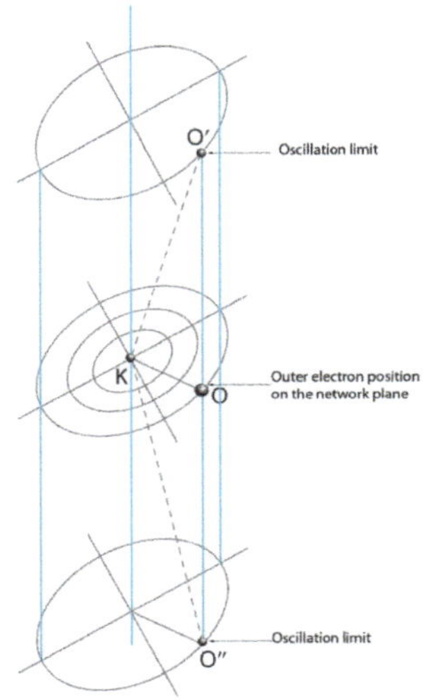

Figure 5. Electrons are in symmetrical oscillation in the length of a magnetic channel under the influence of Coulomb electric forces. The oscillation is not linear, therefore movement is not harmonic.

Table 3. Energies for each electron of the calcium atom.

i electron ascension number & energy level	r_i electron detachment distance & Thory—Experiment	ρ_i electron position distance Å Theory	α_i *electron oscillation amplitude Å Theory	ε'_i electron oscillation energy MJ/Kmol Theory	ε_i electron position energy MJ/Kmol Theory	E'_i electron detachment energy MJ/Kmol Theory	E_i electron detachment energy MJ/Kmol Experiment
1	4.21	4.21	≈ 0	526.40	526.40	526.40	527.76
2	4.25	4.21	0.58	495.37	500.08	495.37	494.85
3	17.86	9.41	15.18	111.67	211.95	111.67	111.71
4	17.96	9.41	15.30	104.88	200.18	104.88	104.90
5	18.87	9.41	16.36	93.95	188.40	93.95	94.00
6	19.27	9.41	16.82	86.25	176.63	86.25	86.31
7	19.67	12.63	15.08	78.87	122.83	78.87	78.89
8	20.56	12.63	16.22	70.06	114.05	70.06	70.11
9	20.98	15.18	14.48	63.37	87.59	63.37	63.41
10	21.35	15.18	15.01	57.09	80.29	57.09	57.11
11	54.38	15.18	52.22	20.37	72.99	20.37	20.38
12	54.85	15.18	52.71	17.35	62.69	17.35	18.19
13	62.43	17.36	59.97	14.20	51.06	14.20	14.21
14	63.25	17.36	60.82	12.26	44.68	12.26	12.27
15	63.37	17.36	60.95	10.49	38.29	10.49	10.49
16	67.99	17.36	65.74	8.15	31.91	8.15	8.15
17	68.32	21.13	64.97	6.48	20.97	6.48	6.49
18	67.70	21.13	64.32	4.909	15.73	4.909	4.912
19	193.57	22.62	192.24	1.144	9.79	1.144	1.145
20	187.96	22.62	186.59	0.5938	4.89	0.5938	0.5898

The important role of the quantum theory in supporting our project is clear: Quantum theory attempts to determine the way through which electrons and protons form twins and hold each other by the exchange of photons.

Our contribution is the determination of the arrangement of the proton/electron twins in space, the distance between electrons and protons and the necessary energy to achieve the union between the twins. Furthermore, it should be pointed out that up till now little is known on the role of neutrons in the completion of the structure and function of bioatoms.

Neutrons being in the interior of the energy field of the atoms, lend distinctive characteristics to them. For example, the addition of a structure unit to the carbon atom (one proton, one electron and one neutron) creates a nitrogen atom. However the addition of a hydrogen atom (one proton, one electron) to the carbon atom creates a simple chemical bond (**Figure 6**).

This finding shows that neutrons play an extremely important role in the chemical and physical properties of bioatoms.

The mathematical process of the proposed model substantiates our claims concerning the space-arrangement of protons and electrons in the form of two networks, estimates the distance between protons and electrons and offers a stable structure to the atoms.

2. Experimental Approach of the Proposed Structure of The Atom

The geometrical sizes of protons, neutrons and electrons are assumed as of spherical shape. The sizes of their mass and electrical charge are known. The electrical charge of protons is distributed on their maximum diameter D_p. Their distance D, as we shall prove, remains constant and depends only on the angular momentum (spin) of protons. The production of angular momentum from angular velocity ω and vice versa is obvious from **Figure 7(a)**.

It is: Polar inertia torque θ (kgm^2)

Proton angular momentum $L = \theta\omega$ (Js)

Current due to spin $I = e\omega$ (A)

Resulting magnetic field $H = I/D_p$

The conditions in the space of the atom are considered as conditions in vacuum, with a magnetic constant μ_o= 1.257 × 10^{-6} (Vs/Am) and dielectric constant ε_o= 8.854 × 10^{-12} (F/m).

Magnetic density is: $B = \mu_o H$

Magnetic flow is: $\Phi = Bq$

Where q is the surface run through by flow Φ.

The magnetic circle of a nuclear dipole is:

$$\sum_i H_i l_i = I$$

The attraction force of the dipoles' protons due to flow Φ is:

$$F_\mu = \frac{\Phi^2}{2\mu_o q} = \frac{qB^2}{2\mu_o} \qquad \textbf{(Figure 7(b))}$$

The principle of momentum conservation on the proton dipole in the closed integral in the whole circle Φ = constant.

Therefore:

$$\sum_i H_i l_i = I$$

or

$$\oint_{-D_p}^{D} H_D \mathrm{d}D = \oint_{-D_p}^{D} H_o \frac{1}{D^2} \mathrm{d}D = -H_o\left[\frac{1}{D} + \frac{1}{D_p}\right]$$

with $H_o = 2e\omega$ and $q = \frac{\pi}{4} D_p^2$,

the attraction force at a distance D between them is:

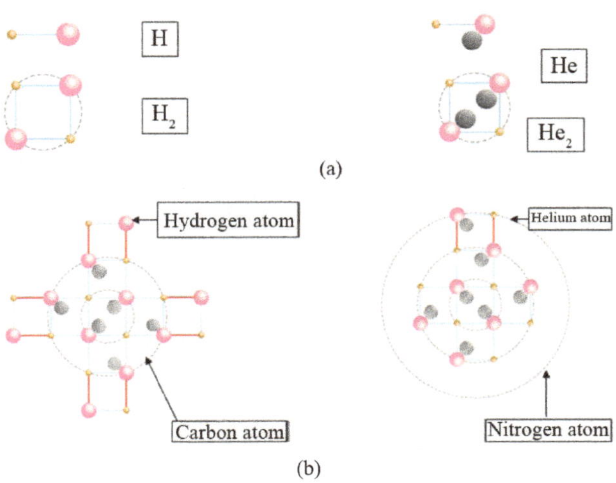

Figure 6. The addition of a structure unit to the carbon atom creates a nitrogen atom. The addition of a hydrogen atom to a carbon atom creates a simple chemical bond. H: Hydrogen atom, H_2: Hydrogen molecule, He: structure unit, He_2: helium atom.

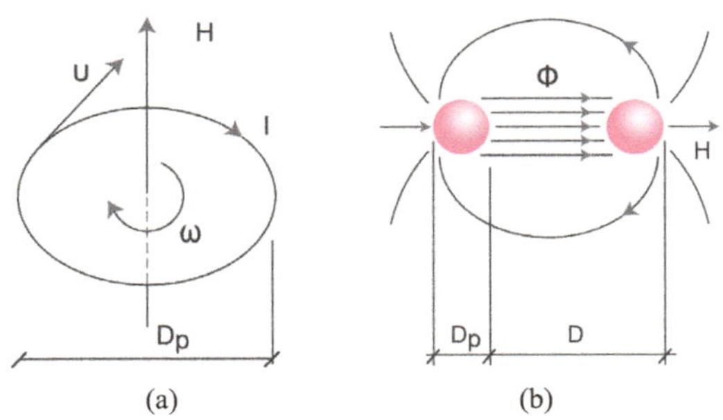

Figure 7. (a): Electromagnetic field due to angular momentum (*u* = angular velocity, *ω* = angular momentum, D_p = proton diameter, *I* = current, *H* = magnetic field); (b): Proton magnetic dipole (Φ = magnetic flow, *H* = magnetic field, D_p = proton diameter, *D* = proton distance).

$$F_\mu = \frac{\Phi_D^2}{2\mu_o q} = 2\mu_o q^2 H_o^2 \left[\frac{1}{D} + \frac{1}{D_p}\right]^2 \text{(N)} \quad (1)$$

They are repulsed with a force F_e due to their positive charge based on Coulomb's law.

$$F_e = -\frac{1}{D^2} 23\times10^{-29}\ \text{(N)} \quad (2)$$

The balance between electrical and magnetic forces is achieved through

$$F_\mu = -F_e \quad (3)$$

From (1), (2) and (3) the eliminant is derived

$$\frac{D}{D_p} = 1.28\times10^{-4}\frac{c}{u} - 1 \quad (4)$$

where: $c = 3\times10^8$ m/s the speed of light and u (m/s) the tangent velocity of the proton. $D_p = 1.3\times10^{-14}$ (m) is the proton's diameter.

From (4) the following observations are derived:

- At an angular momentum with $u = 3.84\times10^4$ m/s the two protons come together since $D/D_p = 0 \to D = 0$
- At an angular momentum with $u = 1.92\times10^4\ m/s$ a rupture of Coulomb's barrier occurs. Now the protons balance between them at a contact position.
- At an angular momentum with velocities of the positive charges smaller than $1.92\times10^4\ m/s$ protons balance at a distance D which can be calculated by Equation (4).
- The magnetic dipole of two or more protons does not depend on the measure of their positive charges, because Equation (4) involves only measures of velocity and dimensions. Therefore for any Q^+ is D = constant. This means that the positive charge can also associate with other protons or electrons, without change in the distance between protons.

Following the above observations we conclude that the space arrangement and geometry of the atom/molecule according to **Figure 7(c)** is stable.

The denoted electrons of diameter D_e are attracted by the protons with diameter D_p. These forces give a resultant at the drawing level vertical to the flow of the magnetic field which also lies or is parallel to the drawing level. Therefore the exercised forces on the electrons are, as the external vectorial product of the latter, vertical to the drawing level. In this way the magnetic field creates a channel vertical to the drawing level for each electron. Hence the electrons are in a symmetric linear oscillation in the length of this channel under the influence of Coulomb electric forces.

The oscillation amplitude is defined by the electric potential for every specific system and for each electron. This is derived by the analysis of ionized atoms and is considered by the following, where the proton distance D is also defined.

Based on the value which derives from Equation (10), $D = 8.42\times10^{-10}\ m$ and from (4) we can define the protons' tangent velocity of angular momentum:

$$u = 0.59\ \text{(m/s)}$$

Angular momentum (spin) is calculated by $L = \theta\omega$ with the protons' polar torque

$$\theta = 0.56\times10^{-55}\ (\text{Kg}\cdot\text{m}^2)$$

and angular velocity

$$\omega = \sqrt{2u/D_p} = 2.95\times10^{18}\ (\text{s}^{-1})$$

$$L = 0.53\times10^{-48}\ \text{(JS/proton)}$$

The above applies for all protons, at least for the elements that are incorporated in the context of this study.

2.1. Energy Analysis for The Bioatomic Structure

The analytical elaboration of the proposed model from an energy point of view, has taken into consideration the following:

- Protons together with neutrons are distributed on a network. The same applies for electrons which are on a secondary network.
- The distance of the junctions of the network does not change.
- Dynamic balances due to developing magnetic or electrical forces were not taken into account since the networks were considered firm.
- The attempted, in the following, energy analysis aims to define the geometric sizes through calculations of the electric potentials and their comparison with measurements established in current bibliography.
- It was assumed that in the distributed nucleus, the positive charges are concentrated at the centre of symmetry of the network which is occupied by a helium (He) molecule in all the elements.
- The appearing homocentric circles are part of the characteristics of the network. All the electrons are present at their peripheries and are affected by the positive charges of the centre of positive charges *K*.

The proposed bioatomic structure of the atom's nucleus in all elements, for example the atom of calcium, can be seen in **Figure 1(c)**. Protons and neutrons are at the junctions of the rectangular network at distances *D*. This is the primary network.

Electrons are located at the junctions of the diagonals of the primary network by dividing

$$d = \frac{D}{\sqrt{2}}$$

which we characterize as secondary network.

The ratio of the radials of the circles as a geometric location of the junctions of the networks ρ_i/D has the following values:

0.500/1.118/1.500/1.803/2.062/2.510/2.686/...

The spherical electric potential which resists the detachment of electrons is due to Coulomb forces, which in vacuum give:

$$F_i = -10^{-29} 23\nu/\rho_i^2 \text{ (N/atom)} \quad (5)$$

ρ_i (m) distance of electrons from the centre of positive charges

ν (–) number of positive charges that correspond to the detached electron.

The detachment energy of the i-electron can be calculated by the integral:

$$E_i' = \int_{\rho_i}^{\infty} F_i d\rho = -F_i\rho_i \text{ (J/atom)} \quad (6)$$

The energy E_i' is independent of the escape path from ρ_i to ∞ of the electron.

It is obvious for a substance atom with λ electrons the following applies:

$$I = \lambda + 1 - \nu$$

with $I = \lambda$ for the first valency electron of the outer orbital and $I = 1$ for the last electron of the inner orbital.

The above electron detachment energies from a substance atom are compared with values for detachment of electrons by ionization from the respective bibliography. Those for 11 elements are shown in **Table 3**, column 7 and refer to kmol of substance.

For the comparison (6) is transformed to:

$$E_i = E_i'(\lambda + \mu)\eta N \times 10^{-9} \text{ (GJ/kmol)} \quad (7)$$

We observe that the values of column 7 in **Table 1** are reflected with extreme accuracy by the relationship:

$$E = 0.329\lambda(\lambda + \mu)\eta \text{ (GJ/kmol)} \quad (8)$$

where: μ number of neutrons in the atom, η number of atoms in the molecule (single atoms are counted as dual atoms)

N Avogadro's constant 6.022×10^{26} number of atoms/kmol

Relationship (3) for the electron of the inner orbital with $I = 1$ can be written:

$$E_1 = -\left[10^{29} 23\lambda/\rho_1\right](\lambda + \mu)\eta N \times 10^{-9} \text{ (GJ/kmol)} \quad (9)$$

Because $E = -E_I$ from (8) and (9) we can derive for all elements:

$$D = 8.42 \times 10^{-10} \text{ (m)} = 8.42 \text{ (Å)} \quad (10)$$

This means that the distance between the junctions of the primary network of protons of the arranged nucleus according to the proposed bioatomic structure is constant, at least for all the elements that are under consideration in this analysis and are included in **Table 1**.

The values for the detachment of an electron of the inner orbital that are calculated by (9) are included in column 6 of **Table 3**.

The extreme accuracy between theory and experiment (columns 6 and 7) is important for the required detachment energy of the last electron of all the elements that are referred in **Table 3**.

With $\mu = \lambda$ and $\eta = 2$, something that occurs in all the substances in question, we derive through (8):

$$E_{a1} = 1.316\lambda^2 \text{ (GJ/kmol)} \quad (11)$$

Relationships (8) and (9) constitute an analytical statement of the experimental measurements.

The detachment energy of the first valency electron of the outer orbital of the nucleus demands significantly less energy.

As it is derived by (11), the ratio between the maximum and minimum detachment energy of an electron is approximately equal to the square of the number of electrons of the atom of the element; in other words λ^2.

The detailed calculation of detachment energy of all the electrons of an atom and its comparison with that measured by ionization, takes into account the oscillation of electrons and therefore the oscillation amplitude α_i, apart from the distance ρ_i from the oscillation channel of electrons. Hence the distance r_i that defines the outermost position of electrons from the centre of positive charges in the field of the spherical Coulomb potential is:

$$r_i = \sqrt{\rho_i^2 + \alpha_i^2} \text{ (m)} \quad (12)$$

With the distances ρ_i and the oscillation amplitude α_i we have calculated in **Table 5** the required detachment energies for each electron of oxygen, as an example, and they were compared to the energies that have been derived experimentally. They came out:

$$\alpha_1/\rho_1 = 0, \quad \alpha_2/\rho_2 = 1.86 \text{ and } \alpha_3/\rho_3 = 2.48.$$

3. Conclusions

Our proposal is a timely reminder of the importance of the fundamental aspects of subatomic structure. Advances in the field in the past century have been considerable, and have resulted in multidisciplinary research [32-40]. The results of the present study enforce the association of the initial revolutional scientific thought of Rutherford [1,2] to the later stages of the quantum mechanics attempt to explain the stability of matter.

In the present study we went off in quest of the concrete space arrangement of the electrons and protons in bioatoms. We have also, formulated the quantitative and qualitative stability requirements for the bioatomic structure, namely the space arrangement of protons, electrons and neutrons, in a manner that: 1) can ensure a steady equilibrium and 2) confirm the relationship between energy and matter. These two parameters are extremely important.

As each electron and proton is charged with the same quantity of negative or positive energy—respectively—their interaction must equally be affected. In this case distances between them are to be equal and influence their interaction with the same intensity. This is not feasible if electrons are disorderly dispersed around a positively charged massive nucleus and not amenable to ordinary rules. Besides, it is not understandable how the structuring mechanism of the atoms could work properly with the electrons in disarray around the nucleus. This could show an unruly structure of the atoms since in the same time their energy condition shows an absolutely concrete and definite energy upgrading in the course of our structuring model, as shown in **Table 1**. Moreover, in our model a trinity of proton, electron and neutron is proposed as a structure unit, which justifies the existing order inside the atomic structure. Furthermore, the orderly structure in bioatoms, distinguishes for the first time the master role of the neutrons. The strictly definite energy upgrading of electrons during succession of one atom to the next, substantiates the existence of a structural order, which serves the stability of the proposed model and strengthens its rational.

In conformity with the ionization energy, the distance between protons was estimated as $D = 8.42 \times 10^{-10}$ m. This derives in large part from the realization that the angular momentum of protons must be to the extent of satisfying the estimation of D by measuring the ionization energy.

The magnetic bipolar of two or more protons does not depend on their positive charge, because the relation $D/D_p = 1.28 \times 10^{-4} \times c/v^{-1}$ (where $c = 3 \times 10^8$ m/s = speed of light) contains only velocity and dimension. Thus, no matter what Q^+ is, D = constant and depends exclusively on the angular velocity of protons. This means that the positive charge is able to interact with other protons or electrons without changing the distance between protons. The above observations show that the space arrangement and the geometry of bioatoms (**Figure 3**) is constant. Electrons of diameter D_e are attracted by protons of diameter D_p. These forces give a resultant at the level of planning vertical to the constant flow of the magnetic field or parallel to the planning level. Thus, the forces exercised on the electrons are the external vectorial product of the previous ones, vertical to the planning level. In this way the magnetic field creates a channel vertical to the planning level for each electron. Electrons are now in a symmetrical linear oscillation along this channel due to the influence of the electric Coulomb forces.

The results specify and define the geometric and energy characteristics of the subatomic particles with the use of generally accepted laws of physics, in agreement with established experimental observations.

REFERENCES

[1] E. Rutherford, "The Scattering of α and β Particles by Matter and the Structure of the Atom," *Philosophical Magazine Series* 6, Vol. 21, No. 125, 1911, pp. 669-688. doi:10.1080/14786440508637080

[2] E. Rutherford, "The Stability of Atoms," *Proceedings of the Physical Society of London*, Vol. 33, 1920, pp. 389-394. doi:10.1088/1478-7814/33/1/337

[3] H. Torres-Silva, "New Interpretation of the Atomic Spectra of the Hydrogen Atom: Amixed Mechanism of Classical LC Circuits and Quantum Wave-Particle Duality. Ingeniare," *Revista Chilena Ingenieria*, Vol. 16, No. 1, 2008, pp. 669-688.

[4] V. Trimble, "The Origin and Abundances of the Chemical Elements," *Reviews of Modern Physics*, Vol. 47, No. 877, 1975, pp. 877-976. doi:10.1103/RevModPhys.47.877

[5] D. R. Lide, "Atomic, Molecular, and Optical Physics; Ionization Potentials of Atoms and Atomic Ions," 84th Edition, CRC Handbook of Chemistry and Physics, CRC Press, Boca Raton, 2003.

[6] G. H. Wannier, "The Threshold Law for Single Ionization of Atoms or Ions by Electrons," *Physical Review*, Vol. 190, No. 5, 1953, pp. 817-825. doi:10.1103/PhysRev.90.817

[7] E. P. Wigner, "On the Behavior of Cross Sections Near Thresholds," *Physical Review*, Vol. 73, No. 9, 1948, pp. 1002-1009. doi:10.1103/PhysRev.73.1002

[8] L. Johansson, *Ark. Fys*, Vol. 31, 1966, p. 489.

[9] K. B. S. Eriksson and J. E. Pettersson, "New Measurements in the Spectrum of the Neutral Nitrogen Atom," *Physica Scripta*, Vol. 3, No. 5, 1971, p. 211. doi:10.1088/0031-8949/3/5/003

[10] K. B. S. Eriksson and H. B. S. Isberg, *Ark. Fys*, Vol. 37, No. 17, 1968, p. 221.

[11] B. Edlen, "On the Identification of Ar *x* and Ar *xiv* in the Solar Corona and the Origin of the Unidentified Coronal Lines," *Solar Physics*, Vol. 9, No. 2, 1969, pp. 439-445. doi:10.1007/BF02391668

[12] K. Liden, *Ark. Fys*, Vol. 1, 1949, pp. 229-267.

[13] V. Kaufman and L. Minnhagen, "Accurate Ground-Term Combinations in Ne I," *Journal of the Optical Society of America*, Vol. 62, No. 1, 1972, pp. 92-95. doi:10.1364/JOSA.62.000092

[14] J. F. Baugh, C. E. Burkhardt, J. J. Leventhal and T. Bergeman, "Precision Stark Spectroscopy of Sodium ^2P and ^2D States," *Physical Review A*, Vol. 58, No. 2, 1998, pp. 1585-1588. doi:10.1103/PhysRevA.58.1585

[15] M. Ciocca, C. E. Burkhardt, J. J. Leventhal, and T. Bergeman, "Precision Stark Spectroscopy of Sodium: Improved Values for the Ionization Limit and Bound States," *Physical Review A*, Vol. 45, No. 7, 1992, pp. 4720-4730. doi:10.1103/PhysRevA.45.4720

[16] V. Kaufman and W. C. Martin, "Wavelengths and Energy Level Classifications of Magnesium Spectra for All Stages of Ionization (Mg I through Mg XII)," *Journal of Physical and Chemical Reference Data*, Vol. 20, No. 1, 1991, pp. 83-153. doi:10.1063/1.555879

[17] E. S. Chang, "Solar Emission Lines Revisited: Extended Study of Magnesium," *Physica Scripta*, Vol. 35, No. 6, 1987, pp. 792-797. doi:10.1088/0031-8949/35/6/006

[18] V. Kaufman and W. C. Martin, "Wavelengths and Energy Level Classifications for the Spectra of Aluminum (Al_i through Al_{xiii})," *Journal of Physical and Chemical Reference Data*, Vol. 20, No. 5, 1991, pp. 775-858. doi:10.1063/1.555895

[19] E. S. Chang, "Energy Levels of Atomic Aluminum with Hyperfine Structure," *Journal of Physical and Chemical Reference Data*, Vol. 19, No. 1, 1990, pp. 119-125. doi:10.1063/1.555870

[20] W. C. Martin and R. Zalubas, "Energy Levels of Silicon, Si I through Si XIV," *Journal of Physical and Chemical Reference Data*, Vol. 12, No. 2, 1983, pp. 323-379. doi:10.1063/1.555685

[21] N. Svendenius, "The Spectrum and Term Analysis of Neutral Phosphorus, Part I," *Physica Scripta*, Vol. 22, No. 3, 1980, pp. 240-287. doi:10.1088/0031-8949/22/3/010

[22] W. C. Martin, R. Zalubas and A. Musgrove, "Energy Levels of Sulfur, S I Through S XVI," *Journal of Physical and Chemical Reference Data*, Vol. 19, No. 4, 1990, pp. 821-881. doi:10.1063/1.555862

[23] L. J. Radziemski and V. Kaufman, "Wavelengths, Energy Levels, and Analysis of Neutral Atomic Chlorine (Cl I)," *Journal of the Optical Society of America*, Vol. 59, No. 4, 1969, pp. 424-443. doi:10.1364/JOSA.59.000424

[24] I. Velchev, W. Hogervorst and W. Ubachs, "Precision VUV spectroscopy of Ar I at 105 nm," *Journal of Physics B: Atomic, Molecular and Optical Physics*, Vol. 32, No. 17, 1999, p. L511. doi:10.1088/0953-4075/32/17/105

[25] J. Sugar and C. Corliss, "Thermodynamic Properties of Key Organic Oxygen Compounds in the Carbon Range C_1 to C_4. Part 1. Properties of Condensed Phases," *Journal of Physical and Chemical Reference Data*, Vol. 14, No. 1, 1985, pp. 1-664. doi:10.1063/1.555747

[26] C.J. Lorenzen and K. Niemax, "Quantum Defects of the $n^2P_{1/2,3/2}$ Levels in ^{39}K I and ^{85}Rb I," *Physica Scripta*, Vol. 27, No. 4, 1983, pp. 300-305. doi:10.1088/0031-8949/27/4/012

[27] V. Appendix, "Dubbel Taschenbuch fur den Masckinenbau," Auflage Springer-Verlag, Berlin, 1990.

[28] W. N. Cottingham and D. A. Greenwood, "Introduction to Nuclear Physics," 2nd Edition, Cambridge University Press, Cambridge, 2001.

[29] D. W. Oxtoby, H. P. Gilis and A. Campion, "Principles of Modern Chemistry," 6th Edition, Thomson, Madrid, 2008.

[30] A. Landé, *Phys. Ge.*, Vol. 21, No. 2, 1919, pp. 644-653.

[31] L. Zeit, *Physik*, Vol. 83, No. 2, 1920, p. 380.

[32] L. Parson, *Smithsonian Institute Publication Miscel, Collections*, Vol. 65, No. 11, 1915.

[33] J. Thomson, "XXIV. On the Structure of the Atom: An Investigation of the Stability and Periods of Oscillation of a Number of Corpuscles Arranged at Equal Intervals around the Circumference of a Circle; with Application of the Results to the Theory of Atomic Structure," *Philosophical Magazine Series* 6, Vol. 7, No. 39, 1904, pp. 237-265. doi:10.1080/14786440409463107

[34] P. Lenard, "Uber Die Absorption der Kathodenstrahlen Verschiedener Geschwindigkeit," *Annalen der Physik*, Vol. 317, No. 12, 1903, pp. 714-744. doi:10.1002/andp.19033171203

[35] H. Nagaoka, "Kinetics of a System of Particles Illustrating the Line and the Band Spectrum and the Phenomena of Radioactivity," *Philosophical Magazine Series* 6, Vol. 7, No. 41, 1904, pp. 445-455. doi:10.1080/14786440409463141

[36] L. Rayleigh, "On Electrical Vibration and the Constitution of the Atom," *Philosophical Magazine Series* 6, Vol. 11, No. 61, 1906, pp. 117-123. doi:10.1080/14786440609463428

[37] J. H. Jeans, "On the Constitution of the Atom," *Philosophical Magazine Series* 6, Vol. 11, 1906, pp. 604-607.

[38] G. A. Schott, "On the Electron Theory of Matter and Explanation of the Fine Spectrum Lines of Gravitation," *Philosophical Magazine Series* 6, Vol. 12, No. 67, 1906, pp. 21-29. doi:10.1080/14786440609463507

[39] J. Stark, "Prinzipien der Atomdynami: Die Electrishen Quanten," Verlag Von S. Hirsel, Leipzig, 1910.

[40] R. Hofstadter, "Structure of Nuclei and Nucleons," *Science*, Vol. 136, No. 3521, 1962, pp. 1013-1022. doi:10.1126/science.136.3521.1013

34

Investigation of Chiral Molecular Micelles by NMR Spectroscopy and Molecular Dynamics Simulation

Kevin F. Morris[1], Eugene J. Billiot[2], Fereshteh H. Billiot[2], Kenny B. Lipkowitz[3], William M. Southerland[4], Yayin Fang[4*]
[1]Department of Chemistry, Carthage College, Kenosha, USA
[2]Department of Physical and Environmental Sciences, Texas A & M University, Corpus Christi, USA
[3]Office of Naval Research, Arlington, USA
[4]Department of Biochemistry and Molecular Biology, Howard University College of Medicine, Howard University, Washington DC, USA
Email: *yfang@howard.edu

ABSTRACT

NMR spectroscopy and Molecular Dynamics (MD) simulation analyses of the chiral molecular micelles poly-(Sodium Undecyl-(L,L)-Leucine-Valine) (poly-SULV) and poly-(Sodium Undecyl-(L,L)-Valine-Leucine) (poly-(SUVL)) are reported. Both molecular micelles are used as chiral selectors in electrokinetic chromatography and each consists of covalently linked surfactant chains with chiral dipeptide headgroups. To provide experimental support for the structures from MD simulations, NOESY spectra were used to identify protons in close spatial proximity. Results from the NOESY analyses were then compared to radial distribution functions from MD simulations. In addition, the hydrodynamic radii of both molecular micelles were calculated from NMR-derived diffusion coefficients. Corresponding radii from the MD simulations were found to be in agreement with these experimental results. NMR diffusion experiments were also used to measure association constants for polar and non-polar binaphthyl analytes binding to both molecular micelles. Poly (SUVL) was found to bind the non-polar analyte enantiomers more strongly, while the more polar analyte enantiomers interacted more strongly with poly(SULV). MD simulations in turn showed that poly(SULV) had a more open structure that gave greater access for water molecules to the dipeptide headgroup region.

Keywords: NOESY; Chiral Recognition; Polymeric Surfactant; Molecular Modeling

1. Introduction

The specificity and efficacy of many biologically important reactions are based on chiral interactions, a fundamental phenomenon that is observed in all biological systems. More often than not, the pharmaceutical properties and toxicities of enantiomers are quite different. Therefore, an understanding of the factors governing chiral recognition is extremely important. While significant progress has been made, from both a fundamental and application-oriented perspective, a complete understanding of chiral recognition has yet to materialize.

Since the Federal Food and Drug Administration mandated in 1992 that producers of pharmaceuticals evaluate the effects of individual enantiomers separately and verify the enantiomeric purity of chiral drugs, the number of publications focused on chiral separations has skyrocketed. This research has lead to the development of many chiral separation media for the enantiomeric resolution of chiral compounds including chiral cyclodextrins, surfactants, and polymers [1]. The focus of this investigation is two chiral polymers formed by covalently linking amino acid based surfactant monomers. These materials are known as Molecular Micelles (MM) because they contain a hydrophobic hydrocarbon core and hydrophillic amino acid headgroups. However, unlike conventional micelles covalent linkages attach the surfactant chains into a single macromolecular structure. Here MD simulations and NMR spectroscopy were used to investigate the structures of MM with headgroups containing the amino acids leucine and valine.

The use of amino acid based MM as pseudostationary phases for the enantiomeric separation of chiral compounds in capillary electrokinetic chromatography (EKC) was initially introduced by Wang and Warner in 1994 [2]. In EKC, analyte enantiomers are separated based upon their differential interactions with the chiral MM amino acid headgroups as they are both pulled through a capillary under the influence of an electric field. Since this

*Corresponding author.

initial report of chiral EKC separations with amino acid based MM, subsequent studies have shown that MM have several advantages over conventional micelles as chiral separation media [3-15]. First, the covalent linkages between the monomer units of the MM provides a rigid structure which improves the mass-transfer rate between the MM and the solute, thereby reducing peak broadening in EKC. In addition, the covalent linkages eliminate the dynamic equilibrium between micelles and monomers and since MM have no critical micelle concentration, they can be used as pseudostationary phases over a wider range of concentrations than unpolymerized surfactants. Finally, MM are more stable than conventional micelles in separations utilizing high concentrations of organic modifiers [16].

In order to characterize more fully the structures of these chiral selectors in aqueous solution, we have studied the properties of two MM with dipeptide headgroups containing the same amino acids, but differing in order in which the amino acids are connected. The rationale for this choice is as follows. It has been observed experimentally that dipeptide based MM are, in general, better chiral selectors than comparable polymers with a single amino acid headgroup [4]. Also, experimental results have shown that differences in amino acid order and thus the position of the chiral centers in the dipeptide headgroup have a dramatic effect on chiral selectivity and the physical properties of the MM [6]. It was hypothesized that these differences were due in part to the steric and hydrophobic forces governing the orientation of the R-groups of the amino acids. From these results, a structural model of the dipeptide surfactants was proposed [9]. Part of the purpose of this investigation was to examine the validity of this model and to develop a more detailed picture of the effect that amino acid order has on the dipeptide MM structure.

The specific MM examined in this study are poly-(sodium undecyl-(L,L)-leucine-valine) (poly-SULV) and poly-(sodium undecyl-(L,L)-valine-leucine) (poly-SUVL)). **Figure 1** shows the chemical structures of these MM along with the proton labels used in the NMR analyses. These two compounds were chosen because previous experimental work has shown that poly(SULV) is a relatively versatile chiral selector as compared to poly (SUVL). For example, in a study done by Shamsi, *et al.* to examine the versatility of poly(SULV) as a chiral selector, successful enantioseparation of 58 out of 75 different racemic compounds was reported using poly(SULV) as the chiral selector [17]. In contrast, chiral selectivity with poly(SUVL) is generally relatively poor [9]. Thus comparison of the structures of these two MM would be expected to highlight MM structural features that correlate with chiral selectivity. This project is also part of a

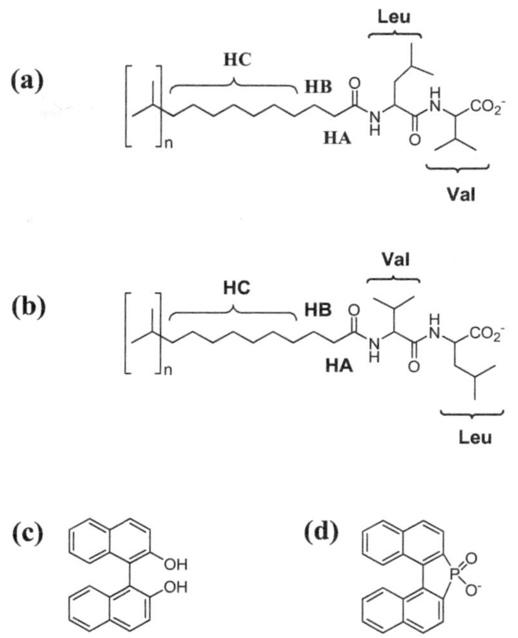

Figure 1. Molecular structures of (a) poly(SULV); (b) poly(SUVL); (c) BOH; and (d) BNP. The proton labels in the molecular micelle structures are used in the NMR analyses.

broader effort to use MD simulations to investigate chiral recognition by MM. Here we present the first stage of this effort. The goal of this study was to establish MD simulations as a viable method for investigating MM structures by comparing poly(SULV) and poly(SUVL) MD simulation results to experiment. Subsequent MD simulation analyses will then be used to probe MM structures in greater detail and to investigate the structural and dynamic properties of chiral analyte: MM intermolecular complexes.

While, many studies have employed MD and Monte Carlo simulations to examine the structural and dynamic properties of micelles [18-22], micelle formation [23-28], and the behavior of water in micellar solutions [29], few simulation studies have investigated MM. The MM investigated here were synthesized by gamma-irradiating amino acid based surfactant solutions at a concentrations ~5 - 20 times the critical micelle concentration. The gamma radiation causes free radical polymerization of the vinyl group at the end of the hydrophobic tail. Subsequent fluorescence quenching studies of the MM showed that the average aggregation number (number of repeat units for the polymers) was ~18 - 23 at room temperature [30]. While the exact nature of the connectivity of the surfactant monomers is not known, here we assumed that the hydrophobic tails are covalently linked one to another in the core of the micelle.

The computational results obtained for poly(SULV) and poly(SUVL) were compared to experimental results from NMR spectroscopy. NMR was first applied to the study of amino acid based surfactants by Ruggart, *et al.* [31]. This study showed that the dipeptide headgroup of poly(SULV) adopted a folded conformation containing a chiral groove or pocket into which chiral analytes inserted. NMR spectroscopy has also been used to measure the free energies of binding of chiral analytes to MM and to identify the primary chiral recognition site on the dipeptide headgroup [32-34]. Here two-dimensional NOESY spectra were used to identify MM headgroup atoms that were in close spatial proximity. The separation of corresponding atoms predicted by the MD simulations were then compared to these experimental results. In addition, the MM radii from the MD simulations were compared to hydrodynamic radii calculated from NMR-derived diffusion coefficients. Finally, association constants for the binding of both polar and non-polar chiral compounds to the MM were measured with NMR. These results were then used to rationalize the number of water molecules occupying the MM hydrocarbon core and dipeptide headgroup regions in the MD simulations.

Both NMR and MD simulation methods were utilized to obtain complimentary information about the molecular micelle systems. NMR techniques, especially the NOESY methods utilized here, are able to probe interactions between nuclei that are within approximately 5 Å [32]. However, in the macromolecular poly(SULV) and poly (SUVL) spectra some of the NMR resonances are broad and overlapping and NMR analyses alone cannot be used to extract definitive MM structures. MD simulation analyses in contrast provide unambiguous atomic scale information about the structures of the MM. Combining both experimental and computational methods, therefore, provided a clearer picture of MM structure than could be obtained by either method alone. In addition, since to our knowledge this is the first application of MD simulation methods to the study of dipeptide terminated MM, the results from the NOESY spectra also provided important experimental validation of the MD simulation structures.

2. Experimental Details

2.1. Materials

Poly-(sodium undecyl-(L, L)-leucine-valine) and (poly-SULV) and poly-(sodium undecyl-(L, L)-valine-leucine) (poly-SUVL)] were synthesized by the method of Wang and Warner [2]. Deuterium oxide (99.9 at % D), boric acid (ACS reagent grade), and (R) and (S)-1,1'bi-2-naphthol (BOH) were purchased from the Sigma-Aldrich, Inc. and used as received.

The MM solutions for the NMR analyses were prepared gravimetrically in a 0.100 M boric acid buffer solution at pH 10.0. The solvent was either D_2O or a 90% H_2O-10% D_2O mixture. The pH of the solutions was adjusted by adding small aliquots of NaOD. Solutions were sonicated and allowed to equilibrate for at least an hour before NMR experiments were performed. The MM equivalent monomer concentration of each sample was 25.0 mM. In the NMR diffusion experiments with (R) or (S)-BOH, the analyte concentration was 2.00 mM.

2.2. NMR Experiments

NMR experiments were performed on a Bruker DPX 300 MHz spectrometer with a 5 mm quad (^1H, ^{13}C, ^{19}F, ^{31}P) probe and a shielded z-gradient coil. All spectra were collected at 298 K. The NOESY spectra of poly(SULV) and poly(SUVL) were collected in phase sensitive mode. The mixing time was 400 ms and the WATERGATE method was used to remove the solvent signal but retain the resonances from the NH protons in the 90% H_2O-10% D_2O solutions [35]. Each two-dimensional spectrum contained 256 increments and 2 k data points in the f_1 and f_2 dimensions, respectively. After acquisition, the data sets were zero filled to a 1 K × 2 K matrix, apodized with a $\pi/2$-shifted sine-bell-squared function and then Fourier transformed and baseline corrected in both dimensions. The NH region of the poly(SULV) NOESY spectrum is shown in **Figure 2**.

NMR diffusion measurements were performed with mixtures containing either (R) or (S)-BOH and poly (SUVL). The other diffusion coefficients and association constants reported are from the literature. The spectrometer's probe was found to produce a gradient of 50.3 $G \cdot cm^{-1}$ at 100% gradient strength by carrying out a pulsed gradient diffusion experiment with a β-cyclodextrin sample in D_2O. The known D value for β-cyclodextrin was used to calculate the coil constant [36]. The Bipolar Pulse Pair Longitude Encode-Decode (BPPLED) pulse sequence was used in the diffusion coefficient experiments [37]. Each measurement was done by collecting

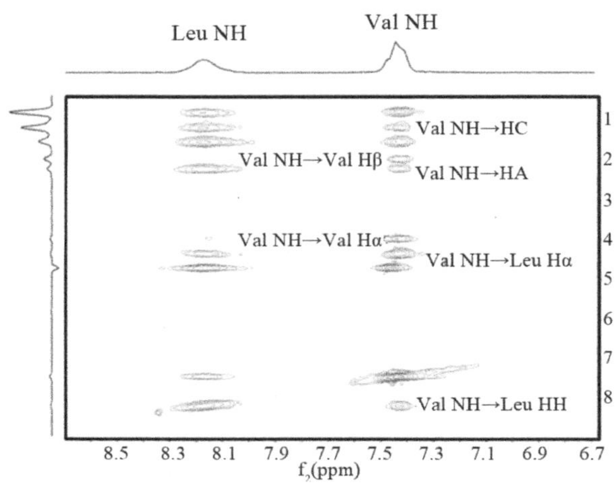

Figure 2. NH region of the poly(SULV) NOESY spectrum.

18 spectra with gradient amplitudes ranging from 5.2 to 35.2 $G\cdot cm^{-1}$. Three replicate trials were performed in each measurement. FID's contained 16 k data points and had a spectral width of 6172.8 Hz.

In the BPPLED experiment the intensity, I, of each resonance in the spectrum changes with increasing gradient amplitude, G, according to Equation (1) [37].

$$I = I_o \cdot \exp\left[-(\gamma \cdot G \cdot \delta)^2 \cdot (\Delta - \delta/3 - \tau/2) \cdot D\right] \quad (1)$$

I_o is the resonance intensity at zero gradient amplitude, γ is the magnetogyric ratio, δ and Δ are respectively the duration of the gradient pulse and the diffusion time, τ is a delay between the bipolar gradient pulses, and D is the diffusion coefficient. The values of δ, Δ, and τ were 4.0, 250, and 0.2 ms, respectively. The echo storage delay at the end of the pulse sequence was 10.0 ms.

Each free induction decay collected in the diffusion experiments was apodized with 1.0 Hz line broadening, Fourier transformed, phased, and baseline corrected. The BOH aromatic and polymer hydrocarbon chain resonances were integrated and the natural log of the integrals was plotted versus the quantity

$$(\gamma \cdot G \cdot \delta)^2 (\Delta - \delta/3 - \tau/2)$$

resulting in a line with slope $-D$. Linear regression analyses were performed to determine the diffusion coefficient. R^2 values for the regression analyses exceeded 0.99. A representative NMR diffusion plot for a mixture containing (R)-BOH and poly(SUVL) is shown in **Figure 3**.

In the BOH:MM mixtures, it was assumed that the BOH molecules experience fast exchange between the bound (D_b) and free (D_{free}) states and that when bound to the polymer, the BOH and MM have the same diffusion coefficient. Under this fast exchange approximation, the BOH diffusion coefficient in the presence of the MM

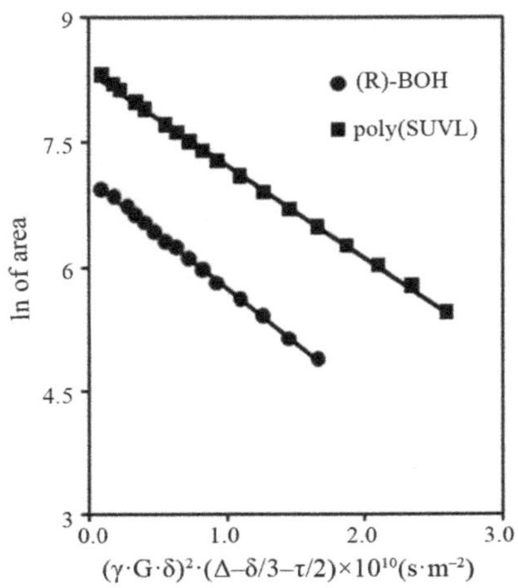

Figure 3. ln of peak area versus $(\gamma \cdot G \cdot \delta) \cdot (\Delta - \delta/3 - \tau/2)$ diffusion plot for (R)-BOH: poly(SUVL) mixture.

(D_{obs}) is given by Equation (2):

$$D_{obs} = f_b \cdot D_b + (1 - f_b) \cdot D_{free} \quad (2)$$

where f_b is the mole fraction of bound BOH molecules [38]. D_{free} was taken as the free solution analyte diffusion coefficient ($(5.40 \pm 0.05) \times 10^{-10}$ $m^2 \cdot s^{-1}$) and D_b was the diffusion coefficient of the MM [32]. Therefore, measurements of D_{obs} and D_b in experiments with the BOH: poly(SUVL) mixture along with the literature D_{free} value allowed f_b to be calculated via Equation (2). Once f_b was established, the BOH: MM association constant, K, was calculated with Equation (3):

$$K = \frac{f_b}{(1 - f_b) \cdot [MM]} \quad (3)$$

[MM] is the equivalent monomer concentration of the molecular micelles [32].

2.3. Computational Methods

The MM systems poly(SULV) and poly(SUVL) were selected for the MD simulations. Since the connectivity of the monomers in the MM has not been fully characterized, in the MD simulations we limited ourselves to the "simplest" way to make the covalent linkages. In this model, the monomers of both SULV and SUVL were connected by covalent bonds at the end of each monomer's hydrocarbon chain in an end-to-end fashion, making each initial structure resemble an extended, linear polymer. The average aggregation number for these dipeptide surfactants has been reported as ~18 - 23 at room temperature, so in the MD simulation experiments, both poly(SULV) and poly(SUVL) were built with 20 monomer units per MM [30]. Next 20 sodium conterions were added near each carboxylate group. The TIP3P water model was used in MD simulations with both molecular micelles. The poly(SULV) and poly(SUVL) MD simulations contained 8360 and 8689 water residues, respectively providing a 10 Å buffer region between the edge of micelle and the edge of the periodic box.

MD simulations were performed using AMBER 10 [39] and the parm 99 [40] force field. In the calculations, a truncated octahedral periodic boundary condition was used. Non-bonded cutoffs were fixed at 12.0 Å and the long-range electrostatic interactions were calculated by the Particle Mesh Ewarld technique [41]. During the MD simulations, the SHAKE algorithm was utilized to constrain the bonds lengths containing the hydrogen atoms. Both systems were optimized before MD simulations were carried out. Then 20 ps MD simulations were done to heat the systems from 0 K to 300 K with weak restraints on the position of the micelle. Later 1 ns MD simulations were performed to allow the system to reach equilibration ($P = 1$ atm, $T = 300$ K). Finally, a 10.0 ns production run was used to collect the statistical data, resulting in a total simulation time of 11.0 ns. The time step was 2 fs and the structures were saved every 0.2 ps. MD simulation results were analyzed using the PTRAJ utility in AMBER 10 [39].

3. Results and Discussion

Two representative structures from the MD simulations of poly(SULV) and poly(SUVL) are shown in **Figures 4(a)** and **(b)**, respectively. These were chosen by first calculating the average structure over all the structures collected during the MD simulations. The Root Mean Squared Deviation (RMSD) of each MD simulation structure with respect to the average was then calculated. Representative structures were selected as those with the lowest RMSD with respect to the average.

The two representative structures in **Figures 4(a)** and **(b)** suggest that both MM adopt a slightly elongated or oval shape. **Figures 4(c)** and **(d)** show surfactant chains extracted from the center of the respective poly(SULV) and poly(SUVL) micelles. These representative chains illustrate differences between the two MM structures. For example, as shown in **Figure 4(d)**, most of the dipeptide headgroups in poly(SUVL) are rotated toward the core of the micelle, leading to a relatively closed, compact micelle structure. However in poly(SULV), significantly more of the dipeptide headgroups are rotated away from the core giving the MM an overall more open shape (see **Figure 4(c)**). As mentioned above, a model of the poly (SULV) headgroup structure has been developed based upon EKC and NMR results [9,31]. In this model, the dipeptide headgroups were hypothesized to adopt a conformation with the two R-groups rotated in opposite directions. The conformation shown in **Figure 4(c)** has the R-groups oriented in this fashion and is thus consistent with the proposed model. In order to further assess the validity of these structures, MD simulation results were compared to experimental NOESY and diffusion coefficient analyses from NMR spectroscopy.

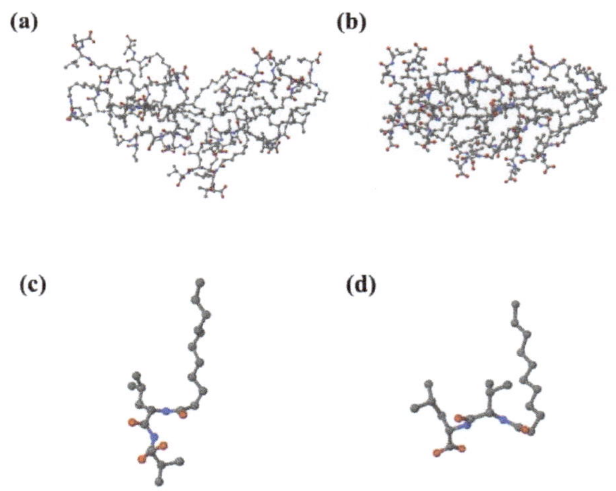

Figure 4. Representative MD simulation structures for (a) poly(SULV) and (b) poly(SUVL). Representative chains from the center of the (c) poly(SULV) and (d) poly(SUVL) molecular micelles.

3.1. NOESY Spectra and Radial Distribution Functions

Figures 1(a) and **(b)** show the structures of poly(SULV) and poly(SUVL) and denote proton labels for the hydrocarbon chain protons (HA, HB, and HC) used in the NOESY analyses. In NOESY spectra, cross peaks are detected between resonances corresponding to protons that are within approximately 5 Å [42]. **Table 1** presents the NOESY cross peaks observed in spectra of both molecular micelles. The NH region of the poly(SULV) NOESY spectrum is shown in **Figure 2**. All NOESY cross peaks were negative, as expected for the macromolecules investigated here. Negative NOESY cross peaks were also previously reported in an NMR study of analyte binding to poly(SULV) [32]. Only MM intramolecular NOESY interactions are presented in **Table 1** because the goal of the analyses was to validate the MD simulations. NOESY investigations of intermolecular MM: chiral analyte association have been reported previously [32-34].

The NOESY spectrum in **Figure 2** shows Val NH and Leu NH diagonal resonances with chemical shifts of, respectively, 7.43 ppm and 8.16 ppm in both dimensions. The cross peaks to the Val NH resonance are assigned on the spectrum. For example, resonances at (7.43 ppm, 8.16 ppm), (7.43 ppm, 4.36 ppm), and (7.43 ppm, 3.98 ppm) are assigned as NOE interactions between the ValNH and the Leu NH, Leu Hα, and Val Hα protons, respectively. The remaining peaks in the NOESY spec-

Table 1. NOE interactions observed for poly(SULV) and poly(SUVL). A (+) denotes the presence of and NOE cross peak and a (–) denotes the absence of an NOE interaction.

Poly(SULV) NOESY Interactions					
	Val NH	Val H	Val H	HA	HC
Leu NH	+	+	+	+	+
Leu H	+	+	–	+	+
HA	+	–	–	n/a	+
HC	+	–	–	+	n/a

Poly(SUVL) NOESY Interactions				
	Val NH	Val H	HA	HC
Leu NH	+	+	+	+
HA	+	+	n/a	+
HC	+	+	+	n/a

trum were assigned in this manner and used to generate the results reported in **Table 1**.

In **Table 1**, a (+) denotes the presence of an NOESY cross peak and a (–) indicates the absence of an NOE interaction. Therefore, based upon the **Table 1** data it can be concluded that the leucine NH and valine NH protons in poly(SULV) are within 5 Å of one another because a cross peak is detected between these proton resonances in the NOESY spectrum (note the (+) designation in **Table 1**). No NOESY cross peak, though is detected between the poly(SULV) valine Hα and hydrocarbon chain HA protons, therefore the separation of these protons would be expected to be greater than the approximately 5 Å cut off for NOESY interactions.

Table 1 does not show results for all possible headgroup and hydrocarbon chain interactions because a number of the MM proton resonances overlap. For example, in NOESY spectra of both MM, the leucine Hα and Hα and hydrocarbon chain HB proton resonances overlap at ~1.5 ppm. Therefore, NOE cross peaks to this resonance in the spectrum could not be unambiguously assigned and are not reported. Likewise, the leucine Hα and valine Hα resonances overlap at ~0.8 ppm, so cross peaks to this chemical shift could not be definitively assigned to the either the leucine or valine methyl groups. Finally, in the poly(SULV) spectrum, the leucine and valine Hα protons are well resolved with chemical shifts of 4.36 and 3.98 ppm, respectively. In the poly(SUVL) spectrum though, the valine and leucine Hα resonances overlap in the 4.1 ppm region of the spectrum. Therefore, NOE interactions with the headgroup Hα protons are not reported because they could not be unambiguously assigned to either the valine or leucine amino acid.

Radial distribution functions (RFD's) from the MD simulations for each of the proton-proton interactions in **Table 1** were generated to make comparisons between the experimental NMR and computational results. Each RDF was generated by monitoring distances throughout the MD simulations between specified protons on the same and four adjacent surfactant monomer chains. **Figure 5** shows four representative RDF's for poly(SULV). The appearances of the remaining RDF's generated from the MD simulations along with distances corresponding to the RDF maxima are summarized in **Table 2**. Overall, the RDF results show good agreement between the MD simulation results and the NOESY spectrum. For example, an NOE interaction was detected between the leucine and valine Hα atoms. The RDF for these two protons shown in **Figure 5(a)** is relatively sharp and shows an internuclear separation in 4 and 5 Å range. An NOE interaction was also observed between the leucine Hα and the hydrocarbon chain HA protons. The RDF in **Figure 5(b)** is consistent with the NOESY results, with the internuclear separation between these atoms again falling

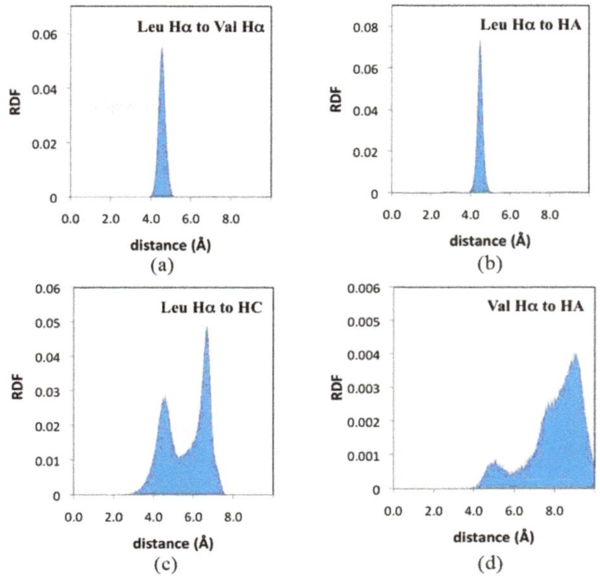

Figure 5. Poly(SULV) radial distribution functions from MD simulations for the (a) leucine Hα and valine Hα, (b) leucine Hα and HA, (c) leucine Hα and HC, and (d) valine Hα and HA atoms. See Figure 1 for hydrocarbon chain proton designations.

Table 2. Summary of radial distribution functions for poly (SULV) and poly(SUVL).

Poly(SULV) Radial Distribution Functions		
Interaction	Description	Rdf Maxima (Å)
Leu NH to Val NH	Single, sharp peak	2.1
Leu NH to Val H	Single sharp peak	5.1
Leu NH to Val H	Bimodal	4.8, 5.6
Leu NH to HA	Single sharp peak	2.2
Leu NH to HC	Broad, bimodal distribution	2.9, 4.6
Leu Hα to Val NH	Single, sharp peak	3.3
Leu Hα to Val H	Three sharp peaks	5.0, 5.8, 6.4
Val NH to HA	Single sharp peak	3.7
Val NH to HC	Broad bimodal distribution	2.4, 6.2
Val Hα to HC	Broad bimodal distribution	5.0, 9.1
Val Hα to HA	Broad bimodal distribution	5.9, 7.1
Val Hα to HC	Broad, single peak	8.5
Poly(SUVL) Radial Distribution Functions		
Interaction	Description	Rdf Maxima (Å)
Val NH to Leu NH	Single, sharp peak	1.9
Val NH to HA	Single, sharp peak	2.1
Val Hα to Leu NH	Single, sharp peak	4.1
Val Hα to HC	Broad distribution, multiple maxima	2.9, 4.8, 6.3, 7.6
Leu NH to HA	Single, sharp peak	3.7
Leu NH to HC	Single peak, wide distribution	6.3

between 4 Å and 5 Å. Finally, the leucine Hα resonance of poly(SULV) showed a NOESY cross peak to the hydrocarbon chain protons HC. The RDF for these protons (**Figure 5(c)**) has a bimodal shape with peaks centered around 4.5 Å and 6.5 Å. The 4.5 Å peak suggests that during the MD simulation the leucine Hα and HC protons spend sufficient time close to one another to exhibit NOE activity. The bimodal shape of this RDF is discussed in more detail below.

Furthermore, in cases where no poly(SULV) NOESY cross peak was observed, the RDF's show relatively large internuclear separations. For example, no NOESY interaction was observed between the valine Hα proton and either the HA or HC atoms of the hydrocarbon chain. The RDF's in **Figure 5(d)** (HA) and **Table 2** (HC) show that during the MD simulation the distances between the valine Hα and hydrocarbon chain protons exceeded the 5 Å separation required for NOESY interaction. Overall, the MD simulation results predicted relatively short internuclear separations when NOE activity was observed and much longer separations for cases where no NOE interaction was detected.

Table 1 also reports the NOESY cross peaks observed in the poly(SUVL) spectrum. Because of the spectral overlap discussed above, fewer interactions are reported for this MM. **Figure 6(a)** shows the RDF between the valine NH proton and the hydrocarbon chain proton HC. As in the poly(SULV) leucine Hα:HC RDF, a bimodal distribution is observed. In **Figure 6(a)**, both peaks in the RDF lie within the distance range expected for NOE activity. **Figure 6(b)** shows the RDF between the VHα and HA protons. An NOE interaction was observed between these two atoms and again the larger peak in the RDF is in the 4 - 5 Å distance range. The relatively short distance between the valine Hα and hydrocarbon chain protons suggests that a significant number of the poly(SUVL) chains adopt a conformation with the valine R group pointed toward the hydrocarbon chain. The representative structure in **Figure 4(d)** shows a poly(SUVL) chain in this orientation.

The RDF's for the leucine Hα and HC atoms in poly (SULV) and valine Hα and HC atoms in poly(SUVL) are shown in **Figures 5(c)** and **6(c)**, respectively. Note that each RDF has a bimodal shape. It was initially thought that the two maxima centered at approximately four and eight Angstroms represented intrachain and interchain interactions. However, the same bimodal pattern was observed when the RDF analyses are performed for the corresponding atoms on the same hydrocarbon chain (data not shown). Therefore, it seems more likely that the two peaks in the Hα to HC RDF's represent different headgroup conformations that place the Hα atom close to or farther from the hydrocarbon chain. The poly(SULV) structure in **Figure 4(c)** with the more open wing-like headgroup conformation places the leucine Hα proton in relatively close proximity to the hydrocarbon chain. However, the folded headgroup conformation that predominates for poly(SUVL) (see **Figure 4(d)**) has the valine R-group pointing toward the hydrocarbon chain and the Hα proton pointing away. Therefore, the RDF plots for the Hα:HC distances likely report the relative number of open wing-like and folded surfactant headgroup conformations. Also note that the 4 Å peak in the bimodal distribution, which is attributed to the wing-like headgroup conformation, is larger in the poly(SULV) RDF than in the poly(SUVL) plot. This result suggests that in poly(SULV) a larger number of surfactant chains that adopt this conformation.

This RDF analysis also provides possible clues as to why poly(SULV) is generally a better chiral selector than poly(SUVL). Interactions of chiral analytes with MM likely involve a combination of non-stereoselective and chiral interactions. The later requires that the solvent and chiral compounds have access to the MM chiral centers. The folded surfactant chain conformations detected in poly(SUVL), however, limit analyte access to the chiral centers on the dipeptide headgroup. Thus solvent and chiral compounds in EKC have less opportunity to interact with the region of the MM containing the chiral centers, thereby reducing the number of possible chiral interactions. In contrast, the larger number of more open headgroup conformations observed for the poly(SULV) chains allows solvent and chiral analytes better access to the chiral centers on the dipeptide headgroup [9]. Water shell results presented below in fact show that poly

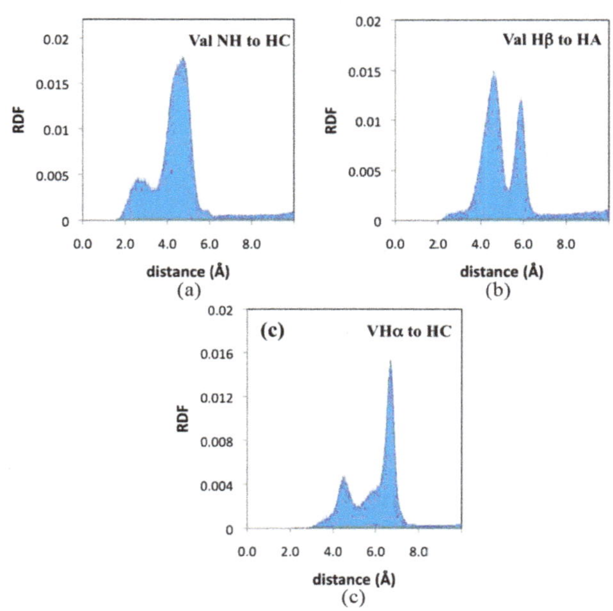

Figure 6. Poly(SUVL) radial distribution functions from MD simulations for the (a) valine NH and HC; (b) valine Hα and HA; and (c) valine Hα and HC atoms.

(SULV) has more water molecules than poly(SUVL) in the MM headgroup region. Characterization of the specific intermolecular interactions that lead to chiral recognition and chiral selectivity in EKC, though, will require that MD simulations be performed on chiral analyte: MM complexes. This work is currently underway and will be presented in a subsequent report.

3.2. Molecular Micelle Size

Along with comparisons of RDF's and NOESY spectra, the overall size or radius of the MM predicted by the MD simulations was compared to hydrodynamic radii from NMR diffusion experiments. Diffusion coefficients of $(1.08 \pm 0.01) \times 10^{-10}$ and $(1.14 \pm 0.01) \times 10^{-10}$ $m^2 \cdot s^{-1}$ have been previously reported for poly(SULV) and poly (SUVL), respectively [34]. Substitution of these diffusion coefficients into the Stokes-Einstein equation yielded a hydrodynamic radius, R_h of 18.4 Å for poly(SULV) and an R_h value of 17.2 Å for poly(SUVL). These hydrodynamic radii correspond to the radius of a sphere with the same diffusion coefficient as the MM [43]. A D_2O viscosity of 1.100 mPa·s at 25.0°C was used in these calculations [44]. The relatively small difference in R_h values for poly(SULV) and poly(SUVL) may be within the experimental error of the method employed, given that D values depend on solution viscosity and obstruction effects [45]. The radii of ~17 - 18 Å calculated for poly (SULV) and poly(SUVL), though, do provide an experimental estimate of micelle size that can be compared to the values predicted by the MD simulations.

The following procedure was used to extract average MM radii from the MD simulations. Here the goal was to calculate a single effective spherical radius from the MD simulations that would represent the dimensions of the more ellipsoid-shaped MM and that could in turn be compared to the hydrodynamic radii from NMR. The method utilized is depicted in **Figure 7**. The dotted line represents the shape of the MM and the solid line represents the effective spherical radius. The first step of the analysis was to use representative structures for both MM to locate the atom nearest the center of mass of each structure. These atoms were found to be the terminal hydrocarbon chain carbons on chains 10 and 13 for poly (SULV) and poly(SUVL), respectively. The average distance from this carbon atom to each of the twenty carboxylate oxygens was then measured over all of the structures collected during the MD simulations. These distances are depicted by the arrows in **Figure 7**. The resulting distances for the twenty surfactant chains of each MM were then averaged. This analysis resulted in an average radius of 18.0 Å for poly(SULV) and 17.3 Å for poly(SUVL), which agree well with the hydrodynamic radii from NMR diffusion measurements.

3.3. Water Distribution and Analyte Binding

NMR diffusion experiments have also been used to investigate the binding of the chiral binaphthyl compounds BOH and 1,1'-binaphthyl-2,2'-diyl hydrogen phosphate (BNP) to poly(SULV) and poly(SUVL) [31-34]. The chemical structures of these compounds are shown in **Figures 1(c)** and **(d)**, respectively. In the NMR binding measurements, the diffusion coefficient of the chiral analyte in the presence of the MM, D_{obs}, is assumed to be a weighted average of the micelle bound, D_b, and free solution values (D_{free}). NMR diffusion measurements of D_{obs} and D_b in an analyte:MM mixture and of D_{free} in the absence of the MM then allow the mole fraction of bound analyte, f_b, and the analyte:MM association constant, K, to be calculated using Equations (2) and (3) [32]. **Table 3** presents D_{obs}, D_b, f_b, and K values for BOH and BNP enantiomers interacting with both MM. The BOH: poly(SUVL) results are from this study and the remaining values are from the literature [32,34]. A BOH D_{free} literature value of $(5.40 \pm 0.05) \times 10^{-10}$ $m^2 \cdot s^{-1}$ was used to calculate the poly(SUVL) f_b and K values from the diffusion coefficients listed in **Table 3** [32].

Examination of the K values in **Table 3** shows that overall BOH interacts more strongly than BNP with both MM. This result is not surprising because under the conditions of the measurements, BNP was an anion and would be expected to experience some electrostatic repulsion from the anionic MM headgroups. The association constants also show that both enantiomers of the less polar analyte BOH bind more strongly to poly(SUVL). For example, the (R)-BOH association constant is 590 ± 19 for poly(SUVL) and 425 ± 11 for poly(SULV). In contrast, association constants of 125 ± 3 and 73.8 ± 3.5 were obtained for (S)-BNP binding to poly(SULV) and poly(SUVL), respectively, showing that the more polar analyte interacts more strongly with poly(SULV). It should also be noted that NOESY studies of BOH and BNP association with both MM have shown cross peaks between resonances corresponding to the analyte and the MM hydrocarbon chain protons (HC). Interactions were

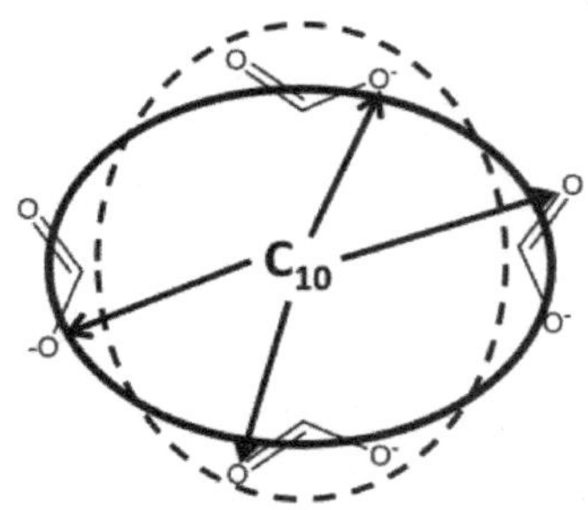

Figure 7. Calculation of the molecular radius from the MD simulations. The calculation for poly(SULV) is illustrated with the terminal hydrocarbon chain carbon of chain number 10 at the center of the MM.

Table 3. D_{obs}, D_b, f_b, **and** K **values for BOH and BNP binding to poly(SULV) and poly(SUVL). The BOH: poly(SULV) and BNP data are from the literature [32,34].**

Analyte	$D_{obs}\times 10^{10}$ ($m^2\cdot s^{-1}$)	$D_b\times 10^{10}$ ($m^2\cdot s^{-1}$)	f_b	K (M^{-1})
Poly(SUVL) Association Constants				
(R)-BOH	1.28 ± 0.07	1.14 ± 0.01	0.967 ± 0.015	590 ± 19
(S)-BOH	1.30 ± 0.06	1.14 ± 0.01	0.963 ± 0.017	514 ± 15
(R)-BNP [34]	2.00 ± 0.08	1.16 ± 0.02	0.789 ± 0.022	73.8 ± 3.5
(S)-BNP [34]	2.01 ± 0.05	1.14 ± 0.01	0.783 ± 0.014	72.1 ± 2.3
Poly(SULV) Association Constants				
Analyte	$D_{obs}\times 10^{10}$ ($m^2\cdot s^{-1}$)	$D_b\times 10^{10}$ ($m^2\cdot s^{-1}$)	f_b	K (M^{-1})
(R)-BOH [32]	1.12 ± 0.01	1.00 ± 0.01	0.955 ± 0.020	425 ± 11
(S)-BOH [32]	1.27 ± 0.01	1.06 ± 0.01	0.952 ± 0.020	393 ± 11
(R)-BNP [32]	1.77 ± 0.01	1.11 ± 0.01	0.837 ± 0.019	103 ± 5
(S)-BNP [32]	1.64 ± 0.01	1.08 ± 0.03	0.862 ± 0.016	125 ± 3

also observed between the analyte resonances and resonances from protons on the MM dipeptide headgroups [32-34]. These results suggest that both analytes interact in a stereoselective fashion with the MM headgroups and experience non-stereoselective hydrophobic interactions with the micelle core. In other words, when both BNP and BOH bind to the MM, they sample or spend time not only in the core but also in the headgroup regions.

Results from the MD simulations were compared to these analyte binding results by examining the number of water molecules in each MM core and dipeptide headgroup region. The cores of both MM were assigned as the regions extending from the terminal carbon of each hydrocarbon chain to the carbonyl carbon separating the hydrocarbon chain from the dipeptide headgroup. RDF's for these carbon atoms were generated and the average distance separating the two atoms averaged over the 20 monomer chains was found to be 9.5 Å. The number of water molecules in each hydrocarbon core was then quantified by calculating the number of waters in the region extending from each terminal hydrocarbon chain carbon atom outward 9.5 Å. A plot of the results for both MM is shown in **Figure 8(a)**. Examination of the water distributions in the cores of the two MM shows that there are on average fewer water molecules found in the poly (SUVL) core and more water molecules found in the core of poly(SULV). This result is consistent with the analyte binding data presented above in that the poly(SUVL) core excludes water to a larger extent than poly(SULV) and at the same time has a greater binding affinity for the non-polar BOH molecule.

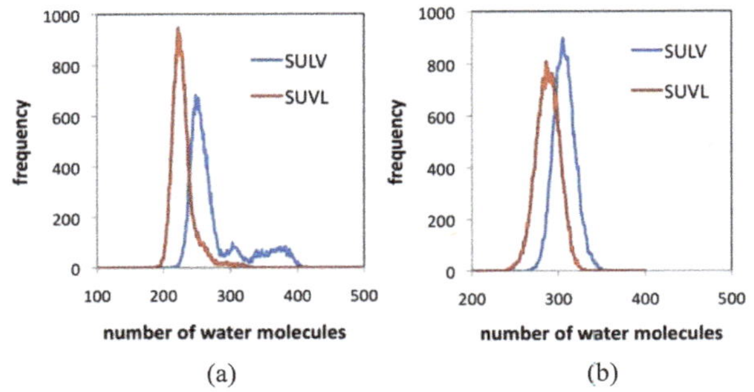

Figure 8. Distribution of water molecules in the hydrocarbon core and headgoup regions of (a) poly(SULV) and (b) poly(SUVL).

A similar comparison can be made of the number of water molecules in each MM headgroup region. This region for both MM was assumed to extend from the carbonyl carbon separating the headgroup and hydrocarbon chain to the C-terminal carboxylate of the dipeptide headgroup. RDF's for the distance between these atoms were generated as above and the headgroup was estimated to extend 6.5 Å beyond the reference carbonyl atom. The number of water molecules in this 6.5 Å region was then calculated and the results are plotted in **Figure 8(b)**. This analysis showed that there are more water molecules in the headgroup region of poly(SULV). The NMR binding data in turn show that the polar analyte BNP binds more strongly to poly(SULV) than to poly(SUVL). Therefore, the MD simulation results suggest that the structure and headgroup conformation of poly(SULV) give both polar water molecules and the polar analyte BNP greater access to the hydrophillic headgroup regions of the MM.

Taken together, the distribution of water molecules around both MM also gives clues as to why poly(SULV) is the better chiral selector. The same steric factors that cause poly(SUVL) to adopt a more compact structure that prevents H_2O penetration into both the core and headgroup regions may also limit the access chiral compounds have to these regions. This would in turn result in fewer stereoselective interactions between analyte enantiomers and the MM chiral centers, thus reducing chiral selectivity. Poly(SULV) in contrast has a more open structure that gives greater access for both water molecules and chiral analytes to the dipeptide headgroup.

4. Conclusion

NMR spectroscopy and MD simulations were used to investigate the structures of two molecular micelles in aqueous solution. Radial distribution functions from the MD simulations were found be in good agreement with NMR NOESY spectra. The RDF's in general showed short internuclear separations for protons connected by NOESY cross peaks and longer separations between protons for which no NOESY interaction was observed. Hydrodynamic radii of the two MM calculated from NMR-derived diffusion coefficients also corresponded well to radii measured in the MD simulations. Finally, NMR studies of the binding of BNP and BOH enantiomers to poly(SULV) and poly(SUVL) showed that BOH bound more strongly than BNP to both MM. However, the association constant for the non-polar BOH analyte with poly(SUVL) was larger than the corresponding poly(SULV) value. The more polar BNP molecule was found to associate more strongly with poly(SULV). The MD simulations in turn showed that poly(SULV) also had more water molecules in both the MM hydrocarbon core and dipeptide headgroup regions. Overall, given the relatively close correspondence between MD simulation and experimental NMR results, it can be concluded that meaningful MM properties can be extracted from the MD simulation.

5. Acknowledgements

This work was supported by grant # 8G12 MD007597 from NIMHD, NIH to the RCMI program at Howard University, an NSF CAREER grant to Dr. Eugene Billiot (No. 0449742), Howard University College of Medicine Bridge Funds and Pilot Study Awards program (BFPSAP) to Dr. Yayin Fang (No. U400040) and a Robert A. Welch Chemistry Departmental Grant to the Chemistry Program at Texas A & M University-Corpus Christi. We also acknowledge the Donors of the American Chemical Society Petroleum Research Fund (46707-B4) and the generosity of the Ralph E. Klingenmeyer family.

REFERENCES

[1] T. J. Ward and K. D. Ward, "Chiral Separations: Fundamental Review," *Analytical Chemistry*, Vol. 82, No. 12, 2010, pp. 4712-4722.

[2] J. Wang and I. M. Warner, "Chiral Separations Using Micellar Electrokinetic Capillary Chromatography and a Polymerized Chiral Micelle," *Analytical Chemistry*, Vol. 66, No. 21, 1994, pp. 3773-3776.

[3] J. Wang and I. M. Warner, "Combined Polymerized Chiral Micelle and γ-Cyclodextrin for Chiral Separation in Capillary Electrophoresis," *Journal of Chromatography: A*, Vol. 711, No. 2, 1995, pp. 297-304.

[4] S. A. Shamsi and I. M. Warner, "Improved Chiral Separations Using a Polymerized Dipeptide Anionic Chiral Surfactant in Electrokinetic Chromatography: Separations of Basic, Acidic, and Neutral Racemates," *Analytical Chemistry*, Vol. 69, No. 15, 1997, pp. 2980-2987.

[5] E. J. Billiot, K. Macossay, S. A. Shamsi and I. M. Warner, "Chiral Separations Using Dipeptide Polymerized Surfactants: Effect of Amino Acid Order," *Analytical Chemistry*, Vol. 70, No. 7, 1998, pp. 1375-1381.

[6] E. J. Billiot, R. A. Agbaria, S. A. Shamsi and I. M. Warner, "Amino Acid Order in Polymeric Dipeptide Surfactants: Effect on Physical Properties and Enantioselectivity," *Analytical Chemistry*, Vol. 71, No. 7, 1999, pp. 1252-1256.

[7] E. J. Billiot, S. Thibodeaux, S. A. Shamsi and I. M. Warner, "Evaluating Chiral Separation Interactions by Use of Diastereomeric Polymeric Dipeptide Surfactants," *Analytical Chemistry*, Vol. 71, No. 18, 1999, pp. 4044-4049.

[8] H. H. Yarabe, S. A. Shamsi and I. M. Warner, "Characterization and Thermodynamic Studies of the Interactions of Two Chiral Polymeric Surfactants with Model Substances: Phenylthiohydantoin Amino Acids," *Analytical Chemistry*, Vol. 71, No. 18, 1999, pp. 3992-3999.

[9] E. J. Billiot and I. M. Warner, "Examination of Structural Changes of Polymeric Amino Acid-Based Surfactants on Enantioselectivity: Effect of Amino Acid Order, Steric Factors, and Number and Position of Chiral Centers," *Analytical Chemistry*, Vol. 72, No. 8, 2000, pp. 1740-1748.

[10] H. H. Yarabe, E. J. Billiot and I. M. Warner, "Enantiomeric Separations by use of Polymeric Surfactant Electrokinetic Chromatography," *Journal of Chromatography A*, Vol. 875, No. 1, 2000, pp. 179-206.

[11] C. Akbay, N. L. Gill, R. A. Agbaria and I. M. Warner, "Copolymerized Polymeric Surfactants: Characterization and Application in Micellar Electrokinetic Chromatography," *Electrophoresis*, Vol. 24, No. 24, 2003, pp. 4209-4220.

[12] C. Akbay, R. A. Agbaria and I. M. Warner, "Monomeric and Polymeric Anionic Gemini Surfactants and Mixed Surfactant Systems in Micellar Electrokinetic Chromatography Part II: Characterization of Chemical Selectivity Using Two Linear Solvation Energy Relationship Models," *Electrophoresis*, Vol. 26, No. 3, 2005, pp. 426-445.

[13] S. A. Rizvi and S. A. Shamsi, "Polymeric Alkenoxy Amino Acid Surfactants: IV: Effects of Hydrophobic Chain Length and Degree of Polymerization of Molecular Micelles on Chiral Separation of β-Blockers," *Electrophoresis*, Vol. 26, No. 21, 2005, pp. 4172-4186.

[14] S. A. Rizvi and S. A. Shamsi, "Polymeric Alkenoxy Amino Acid Surfactants: V. Comparison of Carboxylate and Sulfate Head Group Polymeric Surfactants for Enantioseparation in MEKC," *Electrophoresis*, Vol. 28, No. 11, 2007, pp. 1762-78.

[15] K. Otsuka and S. Terabe, "Enantiomer Separation of Drugs by Micellar Electrokinetic Chromatography using Chiral Surfactants," *Journal of Chromatography A*, Vol. 875, No. 1-2, 2000, pp. 163-178.

[16] C. P. Palmer and S. Terabe, "Micelle Polymers as Pseudostationary Phases in MEKC: Chromatographic Performance and Chemical Selectivity," *Analytical Chemistry*, Vol. 69, No. 10, 1997, pp. 1852-1860.

[17] S. A. Shamsi, B. C. Valle, F. H. Billiot and I. M. Warner, "Polysodium *N*-Undecanoyl-l-Leucylvalinate: A Versatile Chiral Selector for Micellar Electrokinetic Chromatography," *Analytical Chemistry*, Vol. 75, No. 3, 2003, pp. 379-387.

[18] A. D. MacKerell Jr., "Molecular Dynamics Simulation Analysis of a Sodium Dodecyl Sulfate Micelle in Aqueous

Solution: Decreased Fluidity of the Micelle Hydrocarbon Interior," *Journal of Physical Chemistry*, Vol. 99, No. 7, 1995, pp. 1846-1855.

[19] C. D. Bruce, M. L. Berkowitz, L. Perera and M. D. E. Forbes, "Molecular Dynamics Simulation of Sodium Dodecyl Sulfate Micelle in Water: Micellar Structural Characteristics and Counterion Distribution," *Journal of Physical Chemistry B*, Vol. 106, No. 15, 2002, pp. 3788-3793.

[20] S. Bogusz, R. M. Venable and R.W. Pastor, "Molecular Dynamics Simulations of Octyl Glucoside Micelles: Structural Properties," *Journal of Physical Chemistry B*, Vol. 104, No. 23, 2000, pp. 5462-5470.

[21] D. P. Tieleman, D. van der Spoel and H. J. C. Berendsen, "Molecular Dynamics Simulations of Dodecylphosphocholine Micelles at Three Different Aggregate Sizes: Micellar Structure and Chain Relaxation," *Journal of Physical Chemistry B*, Vol. 104, No. 27, 2000, pp. 6380-6388.

[22] S. Bogusz, R. M. Venable and R. W. Pastor, "Molecular Dynamics Simulations of Octyl Glucoside Micelles: Dynamic Properties," *Journal of Physical Chemistry B*, Vol. 105, No. 35, 2001, pp. 8312-8321.

[23] S. J. Marrink, D. P. Tieleman and A. E. Mark, "Molecular Dynamics Simulation of the Kinetics of Spontaneous Micelle Formation," *Journal of Physical Chemistry B*, Vol. 104, No. 51, 2000, pp. 12165-12173.

[24] J. Gao, W. Ge, G. Hu and J. Li, "From Homogeneous Dispersion to Micelles A Molecular Dynamics Simulation on the Compromise of the Hydrophilic and Hydrophobic Effects of Sodium Dodecyl Sulfate in Aqueous Solution," *Langmuir*, Vol. 21, No. 11, 2005, pp. 5223-5229.

[25] T. Lazaridis, B. Mallik and Y. Chen, "Implicit Solvent Simulations of DPC Micelle Formation," *Journal of Physical Chemistry B*, Vol. 109, No. 31, 2005, pp. 15098-15106.

[26] B. C. Stephenson, K. Beers and D. Blankschtein, "Complementary Use of Simulations and Molecular-Thermodynamic Theory to Model Micellization," *Langmuir*, Vol. 22, No. 4, 2006, pp. 1500-1513.

[27] A. Cavallo, M. Müller and K. Binder, "Formation of Micelles in Homopolymer-Copolymer Mixtures: Quantitative Comparison between Simulations of Long Chains and Self-Consistent Field Calculations," *Macromolecules*, Vol. 39, No. 26, 2006, pp. 9539-9550. doi:10.1021/ma061493g

[28] B. C. Stephenson, A. Goldsipe, K. J. Beers and D. Blankschtein, "Quantifying the Hydrophobic Effect. 2. A Computer Simulation-Molecular-Thermodynamic Model for the Micellization of Nonionic Surfactants in Aqueous Solution," *Journal of Physical Chemistry B*, Vol. 111, No. 5, 2007, pp. 1045-1062.

[29] C. D. Bruce, S. Senapati, M. L. Berkowitz, L. Perera and M. D. E. Forbes, "Molecular Dynamics Simulations of Sodium Dodecyl Sulfate Micelle in Water: The Behavior of Water," *Journal of Physical Chemistry B*, Vol. 106, No. 42, 2002, pp. 10902-10907.

[30] E. J. Billiot, "Chiral Recognition with Polymerized Dipeptide Surfactants in Capillary Electrophoresis," Ph.D. Dissertation, Louisiana State University, Baton Rouge, 1998.

[31] J. K. Rugutt, E. J. Billiot and I. M. Warner, "NMR Study of the Interaction of Monomeric and Polymeric Chiral Surfactants with (*R*)- and (*S*)-1,1'-Binaphthyl-2,2'-Diyl Hydrogen Phosphate," *Langmuir*, Vol. 16, No. 7, 2000, pp. 3022-3029.

[32] K. F. Morris, B. A. Becker, B. C. Valle, I. M. Warner and C. K. Larive, "Use of NMR Binding Interaction Mapping Techniques to Examine Interactions of Chiral Molecules with Molecular Micelles," *Journal of Physical Chemistry B*, Vol. 110, No. 35, 2006, pp. 17359-17369. doi:10.1021/jp0627224

[33] B. C. Valle, K. F. Morris, K. A. Fletcher, V. Fernand, D. M. Sword, S. Eldridge, C. K. Larive and I. M. Warner, "Understanding Chiral Molecular Micellar Separations Using Steady-State Fluorescence Anisotropy, Capillary Electrophoresis, and NMR," *Langmuir*, Vol. 23, No. 2, 2007, pp. 425-435. doi:10.1021/la0612623

[34] S. A. Kingsbury, C. J. Ducommun, B. M. Zahakaylo, E. H. Dickinson and K. F. Morris, "NMR Characterization of 1,1'-Binaphthyl-2,2'-Diyl Hydrogen Phosphate Binding to Chiral Molecular Micelles," *Magnetic Resonance in Chemistry*, Vol. 48, No. 3, 2010, pp. 184-191.

[35] M. Piotto, V. Saudek and V. Skienar, "Gradient-Tailored Excitation for Single Quantum NMR Spectroscopy of Aqueous Solutions," *Journal of Biomolecular NMR*, Vol. 2, No. 6, 1992, pp. 661-665.

[36] M. Lin, D. A. Jayawickrama, R. A. Rose, J. A. DelViscio and C.K. Larive, "Nuclear Magnetic Resonance Spectroscopic Analysis of the Selective Complexation of the *Cis* and *Trans* Isomers of Phenylalanylproline by β-Cyclodextrin," *Analytica Chimica Acta*, Vol. 307, No. 3, 1995, pp. 449-457.

[37] D. H. Wu, A. D. Chen and C. S. Johnson. Jr. "An Improved Diffusion Ordered Spectroscopy Experiment Incorporating Bipolar Gradient Pulses," *Journal of Magnetic Resonance*, Vol. A115, No. 2, 1995, pp. 260-264.

[38] P. Stilbs, "Fourier Transform Pulsed-Gradient Spin-Echo Studies of Molecular Diffusion," *Progress in NMR Spectrosopy*, Vol. 19, No. 1, 1987, pp. 1-45.

[39] D. A. Case, T. A. Darden, T. E. Cheatham, C. L. Simmerling, J. Wang, R. E. Duke, R. Luo, M. Crowley, R. C. Walker, W. Zhang, K. M. Merz, B. Wang, S. Hayik, A. Roitberg, G. Seabra, I. Kolossváry, K. F.Wong, F. Paesani, J. Vanicek, X. Wu, S. R. Brozell, T. Steinbrecher, H. Gohlke, L. Yang, C. Tan, J. Mongan, V. Hornak, G. Cui, D. H. Mathews, M. G. Seetin, C. Sagui, V. Babin and P. A. Kollman, "AMBER 10," University of California, San Francisco, 2008.

[40] J. Wang, P. Cieplak and P. A. Kollman, "How Well Does a Restrained Electrostatic Potential (RESP) Model Perform in Calculating Conformational Energies of Organic and Biological Molecules," *Journal of Computational Chemistry*, Vol. 21, No. 12, 2000, pp. 1049-1074.

[41] U. Essmann, L. Perera, M. L. Berkowitz, T. A. Daden and L. H. Pedersen, "A Smooth Particle Mesh Ewald Method," *Journal of Chemical Physics*, Vol. 103, No. 19, 1995, pp. 8577-8593.

[42] K. Wüthrich, "NMR of Proteins and Nucleic Acids," John Wiley & Sons, Inc., New York, 1986.

[43] D. K. Wilkins, S. B. Grimshaw, V. Receveur, C. M. Dobson, J. A. Jones and L. J. Smith, "Hydrodynamic Radii of Native and Denatured Proteins Measured by Pulse Field Gradient NMR Techniques," *Biochemistry*, Vol. 38, No. 50, 1999, pp. 16424-16431.

[44] C. H. Cho, J. Urquidi, S. Singh and G. W. Robinson, "Thermal Offset Viscosities of H_2O, D_2O, and T_2O," *Journal of Physical Chemistry B*, Vol. 102, No. 11, pp. 1991-1994.

[45] S. A. Rogers-Sanders, D. V. Velde and C.K. Larive, "Evaluation of NMR Diffusion Measurements for the Conformational Analysis of Flexible Peptides," *Journal of Analytical Chemistry*, Vol. 369, No. 3, 2001, pp. 308-331.

Permissions

All chapters in this book were first published in OJPC, by Scientific Research Publishing; hereby published with permission under the Creative Commons Attribution License or equivalent. Every chapter published in this book has been scrutinized by our experts. Their significance has been extensively debated. The topics covered herein carry significant findings which will fuel the growth of the discipline. They may even be implemented as practical applications or may be referred to as a beginning point for another development.

The contributors of this book come from diverse backgrounds, making this book a truly international effort. This book will bring forth new frontiers with its revolutionizing research information and detailed analysis of the nascent developments around the world.

We would like to thank all the contributing authors for lending their expertise to make the book truly unique. They have played a crucial role in the development of this book. Without their invaluable contributions this book wouldn't have been possible. They have made vital efforts to compile up to date information on the varied aspects of this subject to make this book a valuable addition to the collection of many professionals and students.

This book was conceptualized with the vision of imparting up-to-date information and advanced data in this field. To ensure the same, a matchless editorial board was set up. Every individual on the board went through rigorous rounds of assessment to prove their worth. After which they invested a large part of their time researching and compiling the most relevant data for our readers.

The editorial board has been involved in producing this book since its inception. They have spent rigorous hours researching and exploring the diverse topics which have resulted in the successful publishing of this book. They have passed on their knowledge of decades through this book. To expedite this challenging task, the publisher supported the team at every step. A small team of assistant editors was also appointed to further simplify the editing procedure and attain best results for the readers.

Apart from the editorial board, the designing team has also invested a significant amount of their time in understanding the subject and creating the most relevant covers. They scrutinized every image to scout for the most suitable representation of the subject and create an appropriate cover for the book.

The publishing team has been an ardent support to the editorial, designing and production team. Their endless efforts to recruit the best for this project, has resulted in the accomplishment of this book. They are a veteran in the field of academics and their pool of knowledge is as vast as their experience in printing. Their expertise and guidance has proved useful at every step. Their uncompromising quality standards have made this book an exceptional effort. Their encouragement from time to time has been an inspiration for everyone.

The publisher and the editorial board hope that this book will prove to be a valuable piece of knowledge for researchers, students, practitioners and scholars across the globe.

List of Contributors

Feridoun Samavat, Mohammad Hossein Tavakoli, Safdar Habibi, Babak Jaleh and Parisa Taravati Ahmad
Department of Physics, Bu-Ali Sina University, Hamedan, Iran

Sa'ib J. Khouri
Department of Chemistry, American University of Madaba, Madaba, Jordan

Volker Buss
Fachbereich Chemie, Universität Duisburg-Essen, Essen, Germany

Rabindra N. Roy, Lakshmi N. Roy, Blake M. Bodendorfer, Zachary M. Downs, Stephen D. Rocchio, Joshua T. Wollen, Jessica M. Stegner, Isaac B. Henson, Nicholas W. Grove and Lauren A. Dieterman
Hoffman Department of Chemistry, Drury University, Springfield, USA

Diego E. Pissinis, Leónides E. Sereno and Juan M. Marioli
Departamento de Química, Universidad Nacional de Río Cuarto, Río Cuarto, Argentina

Mohd Zul Helmi Rozaini
School of Environmental Sciences, University of East Anglia, Norwich, Norfolk, UK
Department of Chemical Sciences, University Malaysia Terengganu, Kuala Terengganu, Terengganu, Malaysia

Oussama Kamal, Abdelaziz Benlyamani, Farid Serdaoui, Mohammed Riri and Abdelmjid Cherif
Laboratoire d'Interface Matériaux d'Electrochimie et Chimie de l'Environnement (LIME), Université Hassan II Faculté des Sciences Aïn Chock, Casablanca, Morocco

Miloudi Hlaïbi
Laboratoire d'Interface Matériaux d'Electrochimie et Chimie de l'Environnement (LIME), Université Hassan II Faculté des Sciences Aïn Chock, Casablanca, Morocco
Laboratoire des Polymères, Biopolymères, Surfaces, UMR 6270 du CNRS, Faculté des Sciences, Université de Rouen, Mont-Saint-Aignan, France

Jacek Lubczak
Faculty of Chemistry, Rzeszów University of Technology, Rzeszów, Poland

Sherine N. Khattab and Ezzat A. Hamed
Department of Chemistry, Faculty of Science, Alexandria University, Ibrahimia, Egypt

Mohamed A. H. Kharaba and Amal El-Hawary
Department of Chemistry and Physics, Faculty of Education, Alexandria University, Alexandria, Egypt

Ayman El-Faham
Department of Chemistry, Faculty of Science, Alexandria University, Ibrahimia, Egypt
Institute for Research in Biomedicine, Barcelona Science Park, Baldiri Reixac, Spain
Department of Chemistry, College of Science, King Saud University, Riyadh, KSA

Jalal Dweik
Institut Européen des Membranes, CNRS UMR 5635, Université Montpellier II, ENSCM, Montpellier, France
Département de Physique et Chimie, Faculté des Sciences, Université Libanaise, Nabatieh, Liban

Mahmoud Srour, Khaled Karaky, Marwan Kobeissi and Khalil Abou-Saleh
Département de Physique et Chimie, Faculté des Sciences, Université Libanaise, Nabatieh, Liban

Wissam Joumaa
Laboratoire de Physiotoxcicité Environnementale EDST017, Faculté des Sciences, Université Libanaise, Nabatieh, Liban

Amaria Ould-Abbas, Mama Bouchaour and Nasr-Eddine Chabane Sari
Research Unit Materials and Renewable Energy (RUMRE), University of Abou Bekr Belkaid of Tlemcen, Tlemcen, Algeria

Jorge Galvez
Department of Physical Chemistry, Faculty of Pharmacy, University of Valencia, Valencia, Spain

Anil B. Naik and L. R. Naik
Department of Studies in Physics, Karnatak University, Dharwad, India

V. Jayathirtha Rao
Organic Division II, Indian Institute of Chemical Technology, Hyderabad, India

Haridas Pal
Radiation and Photochemistry Division, Bhabha Atomic Research Centre (BARC), Mumbai, India

Konstantin A. Yakimovich
Joint Institute for High Temperatures Russian Academy of Sciences, Moscow, Russia

Tatiana Biryukova
The Institute of Informatics Problems, Russian Academy of Sciences, Moscow, Russia

Thorsten Wilke, Daniela Schricker, Josefin Rolf and Karl Kleinermanns
Institute of Physical Chemistry, Heinrich-Heine-University, Duesseldorf, Germany

S. Natarajan Balasubrahmanyam
Department of Organic Chemistry, Indian Institute of Science, Bangalore, India

Jawaharlal Nehru Centre for Advanced Scientific Research, Bangalore, India

Irishi Narayanan Namboothiri
Department of Chemistry, Indian Institute of Technology, Mumbai, India

Kuruvilla Pius
Department of Chemistry, Mahatma Gandhi University, Kottayam, India

Kalliat T. Valsaraj
Department of Chemical Engineering, Louisiana State University, Baton Rouge, USA

Daniel Ogermann, Thorsten Wilke and Karl Kleinermanns
Institute of Physical Chemistry, Heinrich Heine University, Duesseldorf, Germany

Min-Hsien Liu and Ming-Yung Wu
Department of Chemical and Materials Engineering, Chung Cheng Institute of Technology National Defense University, Taiwan, China

Adel Saadi, Kahina Lanasri, Djamila Halliche and Chérifa Rabia
Laboratoire de Chimie du Gaz Naturel, Faculty of Chemistry, University of Sciences and Technology Houari Boumediene, Algiers, Algeria

Khaldoun Bachari
Laboratoire de Chimie du Gaz Naturel, Faculty of Chemistry, University of Sciences and Technology Houari Boumediene, Algiers, Algeria
Centre de Recherche Scientifique et Technique en Analyses Physico-Chimiques, Algiers, Algeria

Feridoun Samavat, Somayeh Solgi and P. Taravati Ahmad
Physics Department, Bu-Ali Sina University, Hamedan, Iran

Ebrahim Haji Ali
Physics Department, Imam Hossein University, Tehran, Iran

Gholam Hossein Shafiee
Department of Chemistry, Islamic Azad University of Abadeh, Abadeh, Iran

Thomas Eckert
Institut für Biochemie und Endokrinologie, Veterinärmedizinische Fakultät, Justus-Liebig-Universität, Gießen, Germany
Klinik für Geburtshilfe, Gynäkologie und Andrologie, Veterinärmedizinische Fakultät, Justus-Liebig-Universität, Gießen, Germany

Sabine Stötzel, Monika Burg-Roderfeld and Thomas Lütteke
Institut für Biochemie und Endokrinologie, Veterinärmedizinische Fakultät, Justus-Liebig-Universität, Gießen, Germany

Judith Sewing
Coastal Research & Management, Kiel-Holtenau, Germany
Institut für Virologie und Zellbiologie, Lübeck Universität, Lübeck, Germany

Nikolay E. Nifantiev
N. D. Zelinsky Institute of Organic Chemistry, Russian Academy of Sciences, Moscow, Russian Federation

Johannes F. G. Vliegenthart
Department of Bio-Organic Chemistry, Bijvoet Center for Biomolecular Research, Utrecht University, Utrecht, The Netherlands

Hans-Christian Siebert
Institut für Biochemie und Endokrinologie, Veterinärmedizinische Fakultät, Justus-Liebig-Universität, Gießen, Germany
RI-B-NT - Research Institute of Bioinformatics and Nanotechnology, Kiel, Germany

Arianna Gladney, Changyong Qin and Helene Tamboue
Department of Biology, Chemistry and Environmental Health Science Benedict College, Columbia, USA

Flavia Cristina Sertori Paschoalino and Elizabete Jordão
Department of Chemical Systems Engineering, Laboratory of Catalytic Processes Development (LDPC), Faculty of Chemical Engeneering, UNICAMP-University of Campinas, Campinas, Brazil

Matheus Paes Paschoalino and Wilson de Figueiredo Jardim
Department of Analytical Chemistry Institute, UNICAMP-University of Campinas, Campinas, Brazil

Shiraz M. Mammadov, Adil A. Garibov, Sehrana A. Rzayeva, Akif K. Salehov and Sahile S. Ahmadova
Institute of Radiation Problems of ANAS, Baku, Azerbaijan

Oktay A. Akparov
Baku State University, Baku, Azerbaijan

Feridoun Samavat, Farzane Mahmoodi, Parisa Taravati Ahmad, Mohammad Faraz Samavat, Mohammad Hossein Tavakoli and Sanaz Hadidchi
Department of Physics, Bu-Ali Sina University, Hamedan, Iran

Ahmad Alshammari and Abdulaziz Bagabas
National Nanotechnology Research Center, King Abdulaziz City for Science and Technology, Riyadh, Saudi Arabia

Angela Köckritz, Venkata Narayana Kalevaru and Andreas Martin
Leibniz-Institute for Catalysis at University of Rostock, Rostock, Germany

Fatma Mohamed Mahgoub
Department of Chemistry, Faculty of Science, King Abdaziz University, Jeddah, Saudi Arabia
Department of Materials Science, Institute of Graduate Studies and Research, Alexandria University, Alexandria, Egypt

Ahmed Mohamed Hefnawy
Department of Materials Science, Institute of Graduate Studies and Research, Alexandria University, Alexandria, Egypt

Oscar Arciniega Saavedra and Jesús Gracia Fadrique
Departamento de Fisicoquímica, Facultad de Química, Universidad Nacional Autónoma de México, México City, México

Noureddine Issaoui, Houcine Ghalla and Brahim Oujia
Quantum Physics Laboratory, Sciences Faculty of Monastir, Monastir, Tunisia

Mishik A. Kazaryan
National Academy of Sciences of Republic Armenia, Yerevan, Armenia

Igor V. Shamanin
Tomsk Polytechnic University, Tomsk, Russia

John C. Stavridis and George Baltatzis
Institute of Biomedical Research & Biotechnology, Athens, Greece

Constantine N. Pattas
Department of Mechanical Engineering, Laboratory of Applied Thermodynamics, Polytechnic School of Thessaloniki, Thessaloniki, Greece

Kevin F. Morris
Department of Chemistry, Carthage College, Kenosha, USA

Eugene J. Billiot and Fereshteh H. Billiot
Department of Physical and Environmental Sciences, Texas A & M University, Corpus Christi, USA

Kenny B. Lipkowitz
Office of Naval Research, Arlington, USA

William M. Southerland and Yayin Fang
Department of Biochemistry and Molecular Biology, Howard University College of Medicine, Howard University, Washington DC, USA